D1202158

SPECIAL FUNCTIONS

The subject of special functions is often presented as a collection of disparate results, which are rarely organized in a coherent way. This book answers the need for a different approach to the subject. The authors' main goals are to emphasize general unifying principles and to provide clear motivation, efficient proofs, and original references for all of the principal results.

The book covers standard material, but also much more, including chapters on discrete orthogonal polynomials and elliptic functions. The authors show how a very large part of the subject traces back to two equations – the hypergeometric equation and the confluent hypergeometric equation – and describe the various ways in which these equations are canonical and special.

Each chapter closes with a summary that provides both a convenient guide to the logical development and a useful compilation of the formulas. This book serves as an ideal graduate-level textbook as well as a convenient reference.

Richard Beals is Professor Emeritus of Mathematics at Yale University.

Roderick Wong is Professor of Mathematics and Vice President for Research at City University of Hong Kong.

Special Functions
A Graduate Text

RICHARD BEALS
Yale University

RODERICK WONG
City University of Hong Kong

CAMBRIDGE
UNIVERSITY PRESS

CAMBRIDGE UNIVERSITY PRESS

Cambridge, New York, Melbourne, Madrid, Cape Town, Singapore,
São Paulo, Delhi, Dubai, Tokyo, Mexico City

Cambridge University Press
The Edinburgh Building, Cambridge CB2 8RU, UK

Published in the United States of America by Cambridge University Press, New York

www.cambridge.org
Information on this title: www.cambridge.org/9780521197977

First published 2010

Printed in the United Kingdom at the University Press, Cambridge

A catalogue record for this publication is available from the British Library

Library of Congress Cataloging-in-Publication Data
Beals, Richard, 1938–
Special functions : a graduate text / Richard Beals, Roderick Wong.
p.   cm. – (Cambridge studies in advanced mathematics ; 126)
Includes bibliographical references and indexes.
ISBN 978-0-521-19797-7
1. Functions, Special–Textbooks.   I. Wong, R. (Roderick), 1944–   II. Title.
QA351.B34 2010
515′.5–dc22
2010017815

ISBN 978-0-521-19797-7 Hardback

m 8

Contents

Preface

The subject of special functions is one that has no precise delineation. This book includes most of the standard topics and a few that are less standard. The subject does have a long and distinguished history, which we have tried to highlight through remarks and numerous references. The unifying ideas are easily lost in a forest of formulas. We have tried to emphasize these ideas, especially in the early chapters.

To make the book useful for self-study we have included introductory remarks for each chapter, as well as proofs, or outlines of proofs, for almost all the results. To make it a convenient reference, we have concluded each chapter with a concise summary, followed by brief remarks on the history, and references for additional reading.

We have tried to keep the prerequisites to a minimum: a reasonable familiarity with power series and integrals, convergence, and the like. Some proofs rely on the basics of complex function theory, which are reviewed in Appendix A. The necessary background from differential equations is covered in Chapter 3. Some familiarity with Hilbert space ideas, in the L^2 framework, is useful but not indispensable. Chapter 11 on elliptic functions relies more heavily than the rest of the book on concepts from complex analysis. Appendix B contains a quick development of basic results from Fourier analysis.

The first-named author acknowledges the efforts of some of his research collaborators, especially Peter Greiner, Bernard Gaveau, Yakar Kannai, and Jacek Szmigielski, who managed over a period of years to convince him that special functions are not only useful but beautiful. The authors are grateful to Jacek Szmigielski, Mourad Ismail, Richard Askey, and an anonymous reviewer for helpful comments on the manuscript.

1

Orientation

The concept of "special function" is one that has no precise definition. From a practical point of view, a special function is a function of one variable that is (a) not one of the "elementary functions" – algebraic functions, trigonometric functions, the exponential, the logarithm, and functions constructed algebraically from these functions – and is (b) a function about which one can find information in many of the books about special functions. A large amount of such information has been accumulated over a period of three centuries. Like such elementary functions as the exponential and trigonometric functions, special functions arise in numerous contexts. These contexts include both pure mathematics and applications, ranging from number theory and combinatorics to probability and physical science.

The majority of the special functions that are treated in many of the general books on the subject are solutions of certain second-order linear differential equations. Indeed, these functions were discovered through the study of physical problems: vibrations, heat flow, equilibrium, and so on. The associated equations are partial differential equations of second order. In some coordinate systems, these equations can be solved by separation of variables, leading to the second-order ordinary differential equations in question. (Solutions of the analogous *first-order* linear differential equations are elementary functions.)

Despite the long list of adjectives and proper names attached to this class of special functions (hypergeometric, confluent hypergeometric, cylinder, parabolic cylinder, spherical, Airy, Bessel, Hankel, Hermite, Kelvin, Kummer, Laguerre, Legendre, Macdonald, Neumann, Weber, Whittaker, . . .), each of them is closely related to one of two families of equations: the confluent hypergeometric equation(s)

$$x\,u''(x) + (c - x)\,u'(x) - a\,u(x) = 0 \qquad (1.0.1)$$

1

and the hypergeometric equation(s)

$$x(1 - x) u''(x) + [c - (a + b + 1)x] u'(x) - ab\, u(x) = 0. \qquad (1.0.2)$$

The parameters a, b, c are real or complex constants.

Some solutions of these equations are polynomials: up to a linear change of variables, they are the "classical orthogonal polynomials." Again, there are many names attached: Chebyshev, Gegenbauer, Hermite, Jacobi, Laguerre, Legendre, ultraspherical. In this chapter we discuss one context in which these equations, and (up to normalization) no others, arise. We also shall see how two equations can, in principle, give rise to such a menagerie of functions.

Some special functions are *not* closely connected to linear second-order differential equations. These exceptions include the gamma function, the beta function, and the elliptic functions. The gamma and beta functions evaluate certain integrals. They are indispensable in many calculations, especially in connection with the class of functions mentioned earlier, as we illustrate below.

Elliptic functions arise as solutions of a simple *nonlinear* second-order differential equation, and also in connection with integrating certain algebraic functions. They have a wide range of applications, from number theory to integrable systems.

1.1 Power series solutions

The general homogeneous linear second-order equation is

$$p(x) u''(x) + q(x) u'(x) + r(x) u(x) = 0, \qquad (1.1.1)$$

with p not identically zero. We assume here that the coefficient functions $p, q,$ and r are holomorphic (analytic) in a neighborhood of the origin.

If a function u is holomorphic in a neighborhood of the origin, then the function on the left-hand side of (1.1.1) is also holomorphic in a neighborhood of the origin. The coefficients of the power series expansion of this function can be computed from the coefficients of the expansions of the functions $p, q, r,$ and u. Under these assumptions, equation (1.1.1) is equivalent to the sequence of equations obtained by setting the coefficients of the expansion of the left-hand side equal to zero. Specifically, suppose that the coefficient functions p, q, r have series expansions

$$p(x) = \sum_{k=0}^{\infty} p_k x^k, \quad q(x) = \sum_{k=0}^{\infty} q_k x^k, \quad r(x) = \sum_{k=0}^{\infty} r_k x^k,$$

and u has the expansion

$$u(x) = \sum_{k=0}^{\infty} u_k \, x^k.$$

Then the constant term and the coefficients of x and x^2 on the left-hand side of (1.1.1) are

$$2p_0 u_2 + q_0 u_1 + r_0 u_0, \tag{1.1.2}$$

$$6p_0 u_3 + 2p_1 u_2 + 2q_0 u_2 + q_1 u_1 + r_1 u_0 + r_0 u_1,$$

$$12p_0 u_4 + 6p_1 u_3 + 2p_2 u_2 + 3q_0 u_3 + 2q_1 u_2 + q_2 u_1 + r_0 u_2 + r_1 u_1 + r_2 u_0,$$

respectively. The sequence of equations equivalent to (1.1.1) is the sequence

$$\sum_{j+k=n,\, k\geq 0} (k+2)(k+1)p_j u_{k+2} + \sum_{j+k=n,\, k\geq 0} (k+1)q_j u_{k+1}$$

$$+ \sum_{j+k=n,\, k\geq 0} r_j u_k = 0, \quad n = 0, 1, 2, \ldots \tag{1.1.3}$$

We say that equation (1.1.1) is *recursive* if it has a nonzero solution u holomorphic in a neighborhood of the origin, and equations (1.1.3) determine the coefficients $\{u_n\}$ by a simple recursion: the nth equation determines u_n in terms of u_{n-1} alone. Suppose that (1.1.1) is recursive. Then the first of equations (1.1.2) should involve u_1 but not u_2, so $p_0 = 0$, $q_0 \neq 0$. The second equation should not involve u_3 or u_0, so $r_1 = 0$. Similarly, the third equation shows that $q_2 = r_2 = 0$. Continuing, we obtain

$$p_0 = 0, \quad p_j = 0, \ j \geq 3; \quad q_j = 0, \ j \geq 2; \quad r_j = 0, \ j \geq 1.$$

After collecting terms, the nth equation is then

$$\left[(n+1)np_1 + (n+1)q_0\right] u_{n+1} + \left[n(n-1)p_2 + nq_1 + r_0\right] u_n = 0.$$

For special values of the parameters p_1, p_2, q_0, q_1, r_0 one of these coefficients may vanish for some value of n. In such a case either the recursion breaks down or the solution u is a polynomial, so we assume that this does not happen. Thus

$$u_{n+1} = -\frac{n(n-1)p_2 + nq_1 + r_0}{(n+1)np_1 + (n+1)q_0} u_n. \tag{1.1.4}$$

Assume $u_0 \neq 0$. If $p_1 = 0$ but $p_2 \neq 0$, the series $\sum_{n=0}^{\infty} u_n x^n$ diverges for all $x \neq 0$ (ratio test). Therefore, up to normalization – a linear change of coordinates and a multiplicative constant – we may assume that $p(x)$ has one of the two forms $p(x) = x(1-x)$ or $p(x) = x$.

If $p(x) = x(1 - x)$ then equation (1.1.1) has the form

$$x(1 - x) u''(x) + (q_0 + q_1 x) u'(x) + r_0 u(x) = 0.$$

Constants a and b can be chosen so that this is (1.0.2).

If $p(x) = x$ and $q_1 \neq 0$ we may replace x by a multiple of x and take $q_1 = -1$. Then (1.1.1) has the form (1.0.1).

Finally, suppose $p(x) = x$ and $q_1 = 0$. If also $r_0 = 0$, then (1.1.1) is a first-order equation for u'. Otherwise we may replace x by a multiple of x and take $r_0 = 1$. Then (1.1.1) has the form

$$x u''(x) + c u'(x) + u(x) = 0. \tag{1.1.5}$$

This equation is not obviously related to either of (1.0.1) or (1.0.2). However, it can be shown that it becomes a special case of (1.0.1) after a change of variable and a "gauge transformation" (see exercises).

Summarizing: up to certain normalizations, an equation (1.1.1) is recursive if and only if it has one of the three forms (1.0.1), (1.0.2), or (1.1.5). Moreover, (1.1.5) can be transformed to a case of (1.0.1).

Let us note briefly the answer to the analogous question for a homogeneous linear *first-order* equation

$$q(x) u'(x) + r(x) u(x) = 0 \tag{1.1.6}$$

with q not identically zero. This amounts to taking $p = 0$ in the argument above. The conclusion is again that q is a polynomial of degree at most one, with $q_0 \neq 0$, while $r = r_0$ is constant. Up to normalization, $q(x)$ has one of the two forms $q(x) = 1$ or $q(x) = x - 1$. Thus the equation has one of the two forms

$$u'(x) - a u(x) = 0; \quad (x - 1) u'(x) - a u(x) = 0,$$

with solutions

$$u(x) = c e^{ax}; \quad u(x) = c (x - 1)^a,$$

respectively.

Let us return to the confluent hypergeometric equation (1.0.1). The power series solution with $u_0 = 1$ is sometimes denoted $M(a, c; x)$. It can be calculated easily from the recursion (1.1.4). The result is

$$M(a, c; x) = \sum_{n=0}^{\infty} \frac{(a)_n}{(c)_n\, n\,!}\, x^n, \quad c \neq 0, -1, -2, \ldots \tag{1.1.7}$$

Here the "shifted factorial" or "Pochhammer symbol" $(a)_n$ is defined by

$$(a)_0 = 1, \quad (a)_n = a(a+1)(a+2)\cdots(a+n-1), \tag{1.1.8}$$

so that $(1)_n = n!$. The series (1.1.7) converges for all complex x (ratio test), so M is an entire function of x.

The special nature of equation (1.0.1) is reflected in the special nature of the coefficients of M. It leads to a number of relationships among these functions when the parameters (a, b) are varied. For example, a comparison of coefficients shows that the three "contiguous" functions $M(a, c; x)$, $M(a+1, c; x)$, and $M(a, c-1; x)$ are related by

$$(a - c + 1) M(a, c; x) - a M(a+1, c, x) + (c-1) M(a, c-1; x) = 0. \tag{1.1.9}$$

Similar relations hold whenever the respective parameters differ by integers. (In general, the coefficients are rational functions of (a, c, x) rather than simply linear functions of (a, b).)

1.2 The gamma and beta functions

The gamma function

$$\Gamma(a) = \int_0^\infty e^{-t} t^{a-1} \, dt, \quad \operatorname{Re} a > 0,$$

satisfies the functional equation $a \Gamma(a) = \Gamma(a+1)$. More generally, the shifted factorial (1.1.8) can be written

$$(a)_n = \frac{\Gamma(a+n)}{\Gamma(a)}.$$

It is sometimes convenient to use this form in series like (1.1.7).

A related function is the beta function, or beta integral,

$$B(a, b) = \int_0^1 s^{a-1} (1-s)^{b-1} \, ds, \quad \operatorname{Re} a > 0, \ \operatorname{Re} b > 0,$$

which can be evaluated in terms of the gamma function

$$B(a, b) = \frac{\Gamma(a) \, \Gamma(b)}{\Gamma(a+b)};$$

see the next chapter. These identities can be used to obtain a representation of
the function M in (1.1.7) as an integral, when $\operatorname{Re} c > \operatorname{Re} a > 0$. In fact

$$
\begin{aligned}
\frac{(a)_n}{(c)_n} &= \frac{\Gamma(a+n)}{\Gamma(a)} \cdot \frac{\Gamma(c)}{\Gamma(c+n)} \\
&= \frac{\Gamma(c)}{\Gamma(a)\,\Gamma(c-a)} \mathrm{B}(a+n, c-a) \\
&= \frac{\Gamma(c)}{\Gamma(a)\,\Gamma(c-a)} \int_0^1 s^{n+a-1}(1-s)^{c-a-1}\,ds.
\end{aligned}
$$

Therefore

$$
\begin{aligned}
M(a, c; x) &= \frac{\Gamma(c)}{\Gamma(a)\,\Gamma(c-a)} \int_0^1 \left\{ s^{a-1}(1-s)^{c-a-1} \sum_{n=0}^{\infty} \frac{(sx)^n}{n!} \right\} ds \\
&= \frac{\Gamma(c)}{\Gamma(a)\,\Gamma(c-a)} \int_0^1 s^{a-1}(1-s)^{c-a-1} e^{sx}\,ds. \qquad (1.2.1)
\end{aligned}
$$

This integral representation is useful in obtaining information that is not evi-
dent from the power series expansion (1.1.7).

1.3 Three questions

First question: *How can it be that so many of the functions mentioned in the
introduction to this chapter can be associated with just two equations* (1.0.1)
and (1.0.2)*?*

Part of the answer is that different solutions of the same equation may have
different names. An elementary example is the equation

$$
u''(x) - u(x) = 0. \qquad (1.3.1)
$$

One might wish to normalize a solution by imposing a condition at the origin
like

$$
u(0) = 0 \quad \text{or} \quad u'(0) = 0,
$$

leading to $u(x) = \sinh x$ or $u(x) = \cosh x$ respectively, or a condition at
infinity like

$$
\lim_{x \to -\infty} u(x) = 0 \quad \text{or} \quad \lim_{x \to +\infty} u(x) = 0,
$$

leading to $u(x) = e^x$ or $u(x) = e^{-x}$ respectively. Similarly, Bessel, Neumann, and both kinds of Hankel functions are four solutions of a single equation, distinguished by conditions at the origin or at infinity.

The rest of the answer to the question is that one can transform solutions of one second-order linear differential equation into solutions of another, in two simple ways. One such transformation is a change of variables. For example, starting with the equation

$$u''(x) - 2x\,u'(x) + \lambda\,u(x) = 0, \tag{1.3.2}$$

suppose $u(x) = v(x^2)$. It is not difficult to show that (1.3.2) is equivalent to the equation

$$y\,v''(y) + \left(\frac{1}{2} - y\right) v'(y) + \frac{1}{4}\lambda\,v(y) = 0,$$

which is the case $a = -\frac{1}{4}\lambda$, $c = \frac{1}{2}$ of (1.0.1). Therefore, even solutions of (1.3.2) can be identified with certain solutions of (1.0.1). The same is true of odd solutions: see the exercises. An even simpler example is the change $u(x) = v(ix)$ in (1.3.1), leading to $v'' + v = 0$, and the trigonometric and complex exponential solutions $\sin x$, $\cos x$, e^{ix}, e^{-ix}.

The second type of transformation is a "gauge transformation". For example, if the function u in (1.3.2) is written in the form

$$u(x) = e^{x^2/2}\,v(x),$$

then (1.3.2) is equivalent to an equation with no first-order term:

$$v''(x) + (1 + \lambda - x^2)\,v(x) = 0. \tag{1.3.3}$$

Each of the functions mentioned in the third paragraph of the introduction to this chapter is a solution of an equation that can be obtained from (1.0.1) or (1.0.2) by one or both of a change of variable and a gauge transformation.

Second question: *What does one want to know about these functions?*

As we noted above, solutions of an equation of the form (1.1.1) can be chosen uniquely through various normalizations, such as behavior as $x \to 0$ or as $x \to \infty$. The solution (1.1.7) of (1.0.1) is normalized by the condition $u(0) = 1$. Having explicit formulas, like (1.1.7) for the function M, can be very useful. On the other hand, understanding the behavior as $x \to +\infty$ is not always straightforward. The integral representation (1.2.1) allows one to compute this behavior for M (see exercises). This example illustrates why it can be useful to have an integral representation (with an integrand that is well understood).

Any three solutions of a second-order linear equation (1.1.1) satisfy a linear relationship, and one wants to compute the coefficients of such a relationship. An important tool in this and in other aspects of the theory is the computation of the Wronskian of two solutions u_1, u_2:

$$W(u_1, u_2)(x) \equiv u_1(x)\, u_2'(x) - u_2(x)\, u_1'(x).$$

In particular, these two solutions are linearly independent if and only if the Wronskian does not vanish.

Because of the special nature of equations (1.0.1) and (1.0.2) and the equations derived from them, solutions satisfy various linear relationships like (1.1.9). One wants to determine a set of relationships that generate all such relationships.

Finally, the coefficient of the zero-order term in equations like (1.0.1), (1.0.2), or (1.3.3) is an important parameter, and one often wants to know how a given normalized solution like $M(a, c; x)$ varies as the parameter approaches $\pm\infty$. In (1.3.3), denote by v_λ the even solution normalized by $v_\lambda(0) = 1$. As $1 + \lambda = \mu^2 \to +\infty$, over any bounded interval the equation looks like a small perturbation of the equation $v'' + \mu^2 v = 0$. Therefore it is plausible that

$$v_\lambda(x) \sim A_\lambda(x) \cos(\mu x + B_\lambda) \quad \text{as} \quad \lambda \to +\infty,$$

with $A_\lambda(x) > 0$. We want to compute the "amplitude function" $A_\lambda(x)$ and the "phase constant" B_λ. Some words about notation like that in the preceding equation: the meaning of the statement

$$f(x) \sim A\, g(x) \quad \text{as} \quad x \to \infty$$

is

$$\lim_{x \to \infty} \frac{f(x)}{g(x)} = A.$$

This is in slight conflict with the notation for an *asymptotic series expansion*:

$$f(x) \sim g(x) \sum_{n=0}^{\infty} a_n x^{-n} \quad \text{as} \quad x \to \infty$$

means that for every positive integer N, truncating the series at $n = N$ gives an approximation to order x^{-N-1}:

$$\frac{f(x)}{g(x)} - \sum_{n=0}^{N} a_n x^{-n} = O\left(x^{-N-1}\right) \quad \text{as} \quad x \to \infty.$$

As usual, the "big O" notation

$$h(x) = O\big(k(x)\big) \quad \text{as} \quad x \to \infty$$

means that there are constants A, B such that

$$\left| \frac{h(x)}{k(x)} \right| \le A \quad \text{if} \quad x \ge B.$$

The similar "small o" notation

$$h(x) = o\big(k(x)\big)$$

means

$$\lim_{x \to \infty} \frac{h(x)}{k(x)} = 0.$$

Third question: *Is this list of functions or related equations exhaustive, in any sense?*

A partial answer has been given: the requirement that the equation be "recursive" leads to just three cases, (1.0.1), (1.0.2), and (1.1.5), and the third of these three equations reduces to a case of the first equation. Two other answers are given in Chapter 3.

The first of the two answers in Chapter 3 starts with a question of mathematics: given that a differential operator of the form that occurs in (1.1.1),

$$p(x) \frac{d^2}{dx^2} + q(x) \frac{d}{dx} + r(x),$$

is self-adjoint with respect to a weight function on a (bounded or infinite) interval, under what circumstances will the eigenfunctions be polynomials? An example is the operator in (1.3.2), which is self-adjoint with respect to the weight function $w(x) = e^{-x^2}$ on the line:

$$\int_{-\infty}^{\infty} \left[u''(x) - 2x\, u'(x) \right] v(x)\, e^{-x^2}\, dx = \int_{-\infty}^{\infty} u(x) \left[v''(x) - 2x\, v'(x) \right] e^{-x^2}\, dx.$$

The eigenvalues are $\lambda = 2, 4, 6, \ldots$ in (1.3.2) and the Hermite polynomials are eigenfunctions. Up to normalization, the equation associated with such an operator is one of the three equations (1.0.1), (1.0.2) (after a simple change of variables), or (1.3.2). Moreover, as suggested above, (1.3.2) can be converted to two cases of (1.0.1).

The second of the two answers in Chapter 3 starts with a question of mathematical physics: given the Laplace equation

$$\Delta u(\mathbf{x}) = 0$$

or the Helmholtz equation

$$\Delta u(\mathbf{x}) + \lambda u(\mathbf{x}) = 0$$

say in three variables, $\mathbf{x} = (x_1, x_2, x_3)$, what equations arise by separating variables in various coordinate systems (cartesian, cylindrical, spherical, parabolic–cylindrical)? Each of the equations so obtained can be related to one of (1.0.1) and (1.0.2) by a gauge transformation and/or a change of variables.

1.4 Elliptic functions

The remaining special functions to be discussed in this book are also associated with a second-order differential equation, but not a linear equation. One of the simplest nonlinear second-order differential equations of mathematical physics is the equation that describes the motion of an ideal pendulum, which can be normalized to

$$2\theta''(t) = -\sin\theta(t). \tag{1.4.1}$$

Multiplying equation (1.4.1) by $\theta'(t)$ and integrating gives

$$[\theta'(t)]^2 = a + \cos\theta(t) \tag{1.4.2}$$

for some constant a. Let $u = \sin\frac{1}{2}\theta$. Then (1.4.2) takes the form

$$[u'(t)]^2 = A[1 - u(t)^2][1 - k^2 u(t)^2]. \tag{1.4.3}$$

By rescaling time t, we may take the constant A to be 1. Solving for t as a function of u leads to the integral form

$$t = \int_{u_0}^{u} \frac{dx}{\sqrt{(1 - x^2)(1 - k^2 x^2)}}. \tag{1.4.4}$$

This is an instance of an elliptic integral: an integral of the form

$$\int_{u_0}^{u} R\left(x, \sqrt{P(x)}\right) dx, \tag{1.4.5}$$

where P is a polynomial of degree 3 or 4 with no repeated roots and R is a rational function (quotient of polynomials) in two variables. If P had degree 2, then (1.4.5) could be integrated by a trigonometric substitution. For example

$$\int_0^u \frac{dx}{\sqrt{1 - x^2}} = \sin^{-1} u;$$

equivalently,

$$t = \int_0^{\sin t} \frac{dx}{\sqrt{1 - x^2}}.$$

Elliptic functions are the analogues, for the case when P has degree 3 or 4, of the trigonometric functions in the case of degree 2.

1.5 Exercises

1.1 Suppose that u is a solution of (1.0.1) with parameters (a, c). Show that the derivative u' is a solution of (1.0.1) with parameters $(a + 1, c + 1)$.

1.2 Suppose that u is a solution of (1.0.2) with parameters (a, b, c). Show that the derivative u' is a solution of (1.0.1) with parameters $(a + 1, b + 1, c + 1)$.

1.3 Show that the power series solution to (1.1.5) with $u_0 = 1$ is

$$u(x) = \sum_{n=0}^{\infty} \frac{1}{(c)_n \, n!} (-x)^n, \quad c \neq 0, -1, -2, \ldots$$

1.4 In Exercise 1.3, suppose $c = \frac{1}{2}$. Show that

$$u(x) = \cosh \left[2(-x)^{\frac{1}{2}} \right].$$

1.5 Compare the series solution in Exercise 1.3 with the series expansion (7.1.1). This suggests that if u is a solution of (1.1.5), then

$$u(x) = x^{-\frac{1}{2}\nu} v\left(2\sqrt{x}\right), \quad \nu = c - 1,$$

where v is a solution of Bessel's equation (3.6.12). Verify this fact directly. Together with results in Section 3.7, this confirms that (1.1.5) can be transformed to (1.0.1).

1.6 Show that the power series solution to (1.0.1) with $u_0 = 1$ is given by (1.1.7).

1.7 Show that the power series solution to (1.0.2) with $u_0 = 1$ is the *hypergeometric function*

$$u(x) = F(a, b, c; x) = \sum_{n=0}^{\infty} \frac{(a)_n (b)_n}{(c)_n \, n!} x^n.$$

1.8 In the preceding exercise, suppose that $a = c$. Find a closed-form expression for the sum of series. Relate this to the differential equation (1.0.2) when $a = c$.

1.9 Consider the first-order equation (1.1.6) under the assumption that q and r are polynomials and neither is identically zero. Show that all solutions have the form

$$u(x) = P(x) \exp R(x),$$

where P is a polynomial and R is a rational function (quotient of polynomials).

1.10 Suppose that an equation of the form (1.1.1) with holomorphic coefficients has the property that the nth equation (1.1.3) determines u_{n+2} from u_n alone. Show that up to normalization, the equation can be put into one of the following two forms:

$$u''(x) - 2x\,u'(x) + 2\lambda\,u(x) = 0;$$

$$(1 - x^2)\,u''(x) + ax\,u'(x) + b\,u(x) = 0.$$

1.11 Determine the power series expansion of the even solution $(u(-x) = u(x))$ of the first equation in Exercise 1.10, with $u(0) = 1$. Write $u(x) = v(x^2)$ and show that $v(y) = M(a, c; y)$ for suitable choices of the parameters a and c.

1.12 Determine the power series expansion of the odd solution $(u(-x) = -u(x))$ of the first equation in Exercise 1.10, with $u'(0) = 1$. Write $u(x) = x\,v(x^2)$ and show that $v(y) = M(a, c; y)$ for suitable choices of the parameters a and c.

1.13 Let $x = 1 - 2y$ in the second equation of Exercise 1.10 and show that in terms of y, the equation takes the form (1.0.2) for some choice of the parameters a, b, c.

1.14 When does (1.0.1) have a (nonzero) polynomial solution? What about (1.0.2)?

1.15 Show that $\Gamma(n) = (n - 1)!$, $n = 1, 2, 3, \ldots$

1.16 Show that one can use the functional equation for the gamma function to extend it as a meromorphic function on the half-plane $\{\mathrm{Re}\,a > -1\}$ with a simple pole at $a = 0$.

1.17 Show that the gamma function can be extended to a meromorphic function on the whole complex plane with simple poles at the non-positive integers.

1.18 Use the change of variables $t = (1 - s)x$ in the integral representation (1.2.1) to prove the asymptotic result

$$M(a, c; x) \sim \frac{\Gamma(c)}{\Gamma(a)}\,x^{a-c}\,e^x \quad \text{as} \quad x \to +\infty.$$

1.19 Derive an integral representation for the series solution to (1.0.2) with $u(0) = 1$ (Exercise 1.7), assuming $\operatorname{Re} c > \operatorname{Re} a > 0$.

1.20 Show that the change of variables $u(x) = v(2\sqrt{x})$ converts (1.0.1) to an equation with leading coefficient 1 and zero-order coefficient of the form $r(y) - a$.

1.21 Suppose the coefficient p in (1.1.1) is positive. Suppose that $y = y(x)$ satisfies the equation $y'(x) = p(x)^{\frac{1}{2}}$. Show that the change of variables $u(x) = v(y(x))$ converts (1.1.1) to an equation with leading coefficient 1 and zero-order coefficient of the form $r_1(y) + r_0(x(y))$.

1.22 Suppose the coefficient p in (1.1.1) is positive. Show that there is a gauge transformation $u(x) = \varphi(x)v(x)$ so that the equation takes the form

$$p(x) v''(x) + r_1(x) v(x) = 0,$$

with no first-order term. (Assume that a first-order equation $p(x) f'(x) = g(x) f(x)$ has a nonvanishing solution f for any given function g.)

1.23 Eliminate the first-order term in (1.0.1) and in (1.0.2).

1.24 Show that the Wronskian of two solutions of (1.1.1) satisfies a homogeneous first-order linear differential equation. What are the possible solutions of this equation for (1.0.1)? For (1.0.2)?

1.25 Find the asymptotic series $\sum_{n=0}^{\infty} a_n x^{-n}$ for the function e^{-x}, $x \to \infty$. What does the result say about whether a function is determined by its asymptotic series expansion?

1.26 Use the method of Exercise 1.18 to determine the full asymptotic series expansion of the function $M(a, c; x)$ as $x \to \infty$, assuming that $\operatorname{Re} c > \operatorname{Re} a > 0$.

1.27 Verify that (1.4.2) follows from (1.4.1).

1.28 Verify that (1.4.3) follows from (1.4.2).

1.29 Show that the change of variables in the integral (1.4.4) given by the linear fractional transformation $w(z) = (kz + 1)/(z + k)$ converts (1.4.4) to the form (1.4.5) with a polynomial $P(w)$ of degree 3. What are the images $w(z)$ of the four roots $z = \pm 1$ and $z = \pm k$ of the original polynomial?

1.30 Suppose that P in (1.4.5) has degree 3, with no repeated roots. Show that there is a linear fractional transformation that converts (1.4.5) to the same form but with a polynomial of degree 4 instead.

1.6 Summary

Many special functions are solutions of one of two families of equations: confluent hypergeometric

$$x u''(x) + (c - x) u'(x) - a u(x) = 0$$

or hypergeometric

$$x(1 - x) u''(x) + [c - (a + b + 1)x] u'(x) - ab u(x) = 0.$$

In particular, up to a linear change of variables, this is true of the "classical orthogonal polynomials." Other special functions include the gamma function, the beta function, and elliptic functions.

1.6.1 Power series solutions

We say that the homogeneous linear second-order equation

$$p(x) u''(x) + q(x) u'(x) + r(x) u(x) = 0$$

with holomorphic p, q, r is recursive if it has a solution with a power series expansion

$$u(x) = \sum_{n=0}^{\infty} u_n x^n, \quad u_0 = 1$$

whose coefficients can be computed by a simple recursion

$$u_{n+1} = f_n(u_n)$$

using the coefficients in the expansions of p, q, and r. Up to normalization, a recursive equation is either the confluent hypergeometric equation, the hypergeometric equation, or an equation that can be transformed to the former.

In the confluent hypergeometric case, the solution is the Kummer function

$$M(a, c; x) = \sum_{n=0}^{\infty} \frac{(a)_n}{(c)_n \, n!} x^n, \quad c \neq 0, -1, -2, \ldots,$$

where $(a)_n$ denotes the shifted factorial

$$(a)_0 = 1, \quad (a)_n = a(a + 1)(a + 2) \cdots (a + n - 1).$$

These functions satisfy numerous relationships, e.g.

$$(a - c + 1) M(a, c; x) - a M(a + 1, c, x) + (c - 1) M(a, c - 1; x) = 0.$$

1.6.2 The gamma and beta functions

The gamma function

$$\Gamma(a) = \int_0^\infty e^{-t} t^{a-1} \, dt, \quad \operatorname{Re} a > 0,$$

satisfies

$$(a)_n = \frac{\Gamma(a+n)}{\Gamma(a)}.$$

The beta function

$$\mathrm{B}(a,b) = \int_0^1 s^{a-1}(1-s)^{b-1} \, ds, \quad \operatorname{Re} a > 0, \ \operatorname{Re} b > 0,$$

satisfies

$$\mathrm{B}(a,b) = \frac{\Gamma(a)\,\Gamma(b)}{\Gamma(a+b)}.$$

It follows that the solution $M(a, c; x)$ of the confluent hypergeometric equation has an integral representation

$$M(a, c; x) = \frac{\Gamma(c)}{\Gamma(a)\,\Gamma(c-a)} \int_0^1 s^{a-1}(1-s)^{c-a-1} e^{sx} \, ds, \quad \operatorname{Re} c > \operatorname{Re} a > 0.$$

1.6.3 Three questions

Q. *How can it be that so many of the functions mentioned in the introduction to this chapter can be associated with just two equations* (1.0.1) *and* (1.0.2)?

A. One equation can be transformed to another by a change in the independent variable or by a gauge transformation $u(x) = \varphi(x)\, v(x)$ that changes the dependent variable. Examples: the conversion of

$$u''(x) - 2x\, u'(x) + \lambda\, u(x) = 0$$

to either of

$$y\, v''(y) + \left(\frac{1}{2} - y\right) v'(y) + \frac{1}{4}\lambda\, v(y) = 0 \quad \text{or}$$

$$v''(x) + (1 + \lambda - x^2)\, v(x) = 0.$$

Q. *What does one want to know about these functions?*

A. Explicit formulas such as series expansions and integral representations; behavior at finite or infinite endpoints of an interval; relations among solutions; asymptotic behavior with respect to a parameter.

Q. *Is this list of functions or related equations exhaustive, in any sense?*

A. The "recursive" requirement leads to just three cases and the third reduces to the first, leaving (1.0.1) and (1.0.2). Asking what second-order equations that are symmetric with respect to a weight function have polynomials as eigenvalues also leads to three cases, and again, ultimately, to (1.0.1) and (1.0.2). A number of equations that arise from writing basic equations of mathematical physics in special coordinates and separating variables reduce to these same two equations, after gauge transformations and/or changes of variable.

1.6.4 Elliptic functions

The pendulum equation

$$\theta''(t) = -2\sin\theta(t)$$

leads to the elliptic integral

$$t = \int_{u_0}^{u} \frac{dx}{\sqrt{(1 - x^2)(1 - k^2 x^2)}},$$

a particular case of

$$\int_{u_0}^{u} R\left(x, \sqrt{P(x)}\right) dx,$$

where P is a polynomial of degree 3 or 4 with no repeated roots and R is a rational function. Elliptic functions are the analogues of the trigonometric functions: the case of degree 2.

1.7 Remarks

A comprehensive classical reference for the theory and history of special functions is Whittaker and Watson [315]. The theory is also treated in Andrews, Askey, and Roy [7], Hochstadt [133], Lebedev [179], Luke [190], Nikiforov and Uvarov [219], Rainville [236], Sneddon [263], and Temme [284]. Many historical references are found in [315] and in [7]. The connection with differential equations of the type (1.0.1) and (1.0.2) is emphasized in [219]. There are a very large number of identities of the type (1.1.9) and (1.2.1). Extensive lists of such identities and other basic formulas are found in the Bateman Manuscript Project [82, 83], Jahnke and Emde [144], Magnus and Oberhettinger [195], Abramowitz and Stegun [3], Gradshteyn and Ryzhik [118], Magnus, Oberhettinger, and Soni [196], and Brychkov [36]. A revised and updated version of [3] is now available: Olver *et al.* [223, 224]. For representations as continued fractions, see Cuyt *et al.* [60]. For emphasis on applications, see Andrews [8] and Carlson [41]. For calculations, see van der Laan and Temme [294] and Zhang and Jin [321]. For a short history and critical review of the handbooks, see Askey [15].

Some other organizing principles for approaching special functions are: integral equations, Courant and Hilbert [59]; differential–difference equations, Truesdell [289]; Lie theory, Miller [204]; group representations, Dieudonné [70], Müller [209], Talman [281], Varchenko [298], Vilenkin [299], Vilenkin and Klimyk [300], Wawrzyńczyk [307]; generating functions, McBride [200]; Painlevé functions, Iwasaki *et al.* [138]; singularities, Slavyanov and Lay [262]; zeta-functions, Kanemitsu and Tsukada [151].

One of the principal uses of special functions is to develop in series the solutions of equations of mathematical physics. See Burkhardt [38] for an exhaustive historical account to the beginning of the 20th century, and Higgins [127] and Johnson and Johnson [149] for related questions.

Most of the theory of special functions was developed in the 18th and 19th centuries. For a general introduction to the history of mathematics in that period, see Dieudonné [69].

A comment on terminology: the fact that a mathematician's name is attached to a particular equation or function is often an indication that the equation or function in question was first considered by someone else, e.g. by one of the Bernoullis or by Euler. (There are exceptions to this.) Nevertheless, we generally adhere to standard terminology.

2

Gamma, beta, zeta

The first two of the functions discussed in this chapter are due to Euler. The third is usually associated with Riemann, though it was also studied by Euler. Collectively they are of great importance historically, theoretically, and for purposes of calculation.

Historically and theoretically, study of these functions and their properties provided a considerable impetus to the study and understanding of fundamental aspects of mathematical analysis, including limits, infinite products, and analytic continuation. They also motivated advances in complex function theory, such as the theorems of Weierstrass and of Mittag–Leffler on representations of entire and meromorphic functions. The zeta function and its generalizations are intimately connected with questions of number theory.

From the point of view of calculation, many of the explicit constants of mathematical analysis, especially those that come from definite integrals, can be evaluated by means of the gamma and beta functions.

There is much to be said for proceeding historically in discussing these and other special functions, but we shall not make it a point to do so. In mathematics it is often, even usually, the case that later developments cast a new light on earlier ones. One result is that later expositions can often be made both more efficient and, one hopes, more transparent than the original derivations.

After introducing the gamma and beta functions and their basic properties, we turn to a number of important identities and representations of the gamma function and its reciprocal. Two characterizations of the gamma function are established, one based on complex analytic properties, the other one based on a geometric property. Asymptotic properties of the gamma function are considered in detail. The psi function and the incomplete gamma function are introduced.

The identity that evaluates the beta integral in terms of gamma funtions has important modern generalizations due to Selberg and to Aomoto. Aomoto's proof is sketched.

The zeta function, its functional equation, and Euler's evaluation of $\zeta(n)$ for $n = 2, 4, 6, \ldots$ are the subject of the last section.

2.1 The gamma and beta functions

The gamma function was obtained by Euler in 1729 [84] in answer to the question of finding a function that takes the value $n!$ at each non-negative integer n. At that time, "function" was understood as a formula expressed in terms of the standard operations of algebra and calculus, so the problem was not trivial. Euler's first solution was in the form of the limit of certain quotients of products, which we discuss below. For many purposes the most useful version is a function that takes the value $(n - 1)!$ at the positive integer n and is represented as an integral (also due to Euler [84]):

$$\Gamma(z) = \int_0^\infty e^{-t} t^z \frac{dt}{t}, \quad \operatorname{Re} z > 0. \tag{2.1.1}$$

The integral is holomorphic as a function of z in the right half-plane. An integration by parts gives the *functional equation*

$$z \, \Gamma(z) = \Gamma(z + 1), \quad \operatorname{Re} z > 0. \tag{2.1.2}$$

This extends inductively to

$$(z)_n \, \Gamma(z) = \Gamma(z + n), \tag{2.1.3}$$

where, as in (1.1.8), the shifted factorial or Pochhammer symbol $(z)_n$ is

$$(z)_n = z(z + 1) \cdots (z + n - 1). \tag{2.1.4}$$

Since $\Gamma(1) = 1$, it follows that

$$\Gamma(n + 1) = (1)_n = n!, \quad n = 0, 1, 2, 3, \ldots \tag{2.1.5}$$

Theorem 2.1.1 *The gamma function extends to a meromorphic function on* **C**. *Its poles are simple poles at the non-positive integers. The residue at* $-n$ *is* $(-1)^n/n!$. *The extension continues to satisfy the functional equations* (2.1.2) *and* (2.1.3) *for* $z \neq 0, -1, -2, -3, \ldots$

Proof The extension, and the calculation of the residues, can be accomplished by using the extended functional equation (2.1.3), which can be used to define $\Gamma(z)$ for $\mathrm{Re}\, z > -n$. Another way is to write the integral in (2.1.1) as the sum of two terms:

$$\Gamma(z) = \int_0^1 e^{-t} t^z \frac{dt}{t} + \int_1^\infty e^{-t} t^z \frac{dt}{t}.$$

In the first term, the power series representation of e^{-t} converges uniformly, so the series can be integrated term by term. Thus

$$\Gamma(z) = \int_0^1 \sum_{n=0}^\infty \frac{(-1)^n}{n!} t^{z+n} \frac{dt}{t} + \int_1^\infty e^{-t} t^z \frac{dt}{t}$$

$$= \sum_{n=0}^\infty \frac{(-1)^n}{n!(z+n)} + \int_1^\infty e^{-t} t^z \frac{dt}{t}.$$

The series in the last line converges for $z \neq 0, -1, -2, \ldots$ and defines a meromorphic function which has simple poles and residues $(-1)^n/n!$. The integral in the last line extends as an entire function of z. The functional equation represents a relationship between $\Gamma(z)$ and $\Gamma(z+1)$ that necessarily persists under analytic continuation.   □

Euler's first definition of the gamma function started from the observation that for any positive integers k and n,

$$(n+k-1)! = (n-1)!\,(n)_k = (k-1)!\,(k)_n.$$

As $n \to \infty$ with k fixed, $(n)_k \sim n^k$, so

$$(k-1)! = \lim_{n\to\infty} \frac{(n-1)!\,(n)_k}{(k)_n} = \lim_{n\to\infty} \frac{(n-1)!\,n^k}{(k)_n}.$$

Thus, for k a positive integer,

$$\Gamma(k) = \lim_{n\to\infty} \frac{\Gamma(n)\,n^k}{(k)_n}. \tag{2.1.6}$$

As we shall see, the limit of this last expression exists for any complex k for which $(k)_n$ is never zero, i.e. $k \neq 0, -1, -2, \ldots$ Therefore there is another way to solve the original problem of extending the factorial function in a natural way: define the function for arbitrary k by (2.1.6). In the next section we show that this gives the same function Γ.

The beta function, or beta integral, occurs in many contexts. It is the function of two complex variables defined first for $\mathrm{Re}\, a > 0$ and $\mathrm{Re}\, b > 0$ by

$$B(a, b) = \int_0^1 s^{a-1} (1-s)^{b-1}\, ds. \tag{2.1.7}$$

Taking $t = 1 - s$ as the variable of integration shows that B is symmetric:

$$B(a, b) = B(b, a).$$

Taking $u = s/(1 - s)$ as the variable of integration gives the identity

$$B(a, b) = \int_0^\infty u^a \left(\frac{1}{1 + u} \right)^{a+b} \frac{du}{u}.$$

Both the beta integral and Euler's evaluation of it in terms of gamma functions [85] come about naturally when one seeks to evaluate the product $\Gamma(a) \Gamma(b)$:

$$\begin{aligned}
\Gamma(a) \Gamma(b) &= \int_0^\infty \int_0^\infty e^{-(s+t)} s^a t^b \frac{ds}{s} \frac{dt}{t} \\
&= \int_0^\infty \int_0^\infty e^{-t(1+u)} u^a t^{a+b} \frac{dt}{t} \frac{du}{u} \\
&= \int_0^\infty \int_0^\infty e^{-x} u^a \left(\frac{x}{1+u} \right)^{a+b} \frac{dx}{x} \frac{du}{u} \\
&= \Gamma(a + b) \int_0^\infty u^a \left(\frac{1}{1+u} \right)^{a+b} \frac{du}{u} \\
&= \Gamma(a + b) B(a, b).
\end{aligned}$$

Summarizing, we have shown that any of three different expressions may be used to define or evaluate the beta function.

Theorem 2.1.2 *The beta function satisfies the following identities for* $\mathrm{Re}\, a > 0$, $\mathrm{Re}\, b > 0$:

$$\begin{aligned}
B(a, b) &= \int_0^1 s^{a-1} (1 - s)^{b-1} \, ds \\
&= \int_0^\infty u^a \left(\frac{1}{1 + u} \right)^{a+b} \frac{du}{u} \\
&= \frac{\Gamma(a) \Gamma(b)}{\Gamma(a + b)}.
\end{aligned} \tag{2.1.8}$$

The beta function has an analytic continuation to all complex values of a *and* b *such that* $a, b \neq 0, -1, -2, \ldots$

Proof The identities were established above. The analytic continuation follows immediately from the continuation properties of the gamma function. $\square$

As noted above, for a fixed positive integer k,

$$\frac{(n+k-1)!}{(n-1)!} = n^k \left[1 + O\left(n^{-1}\right) \right]$$

as $n \to \infty$. The beta integral allows us to extend this asymptotic result to non-integer values.

Proposition 2.1.3 *For any complex a,*

$$\frac{\Gamma(x+a)}{\Gamma(x)} = x^a \left[1 + O\left(x^{-1}\right) \right] \tag{2.1.9}$$

as $x \to +\infty$.

Proof The extended functional equation (2.1.3) can be used to replace a by $a + n$, so we may assume that $\operatorname{Re} a > 0$. Then

$$\frac{\Gamma(x)}{\Gamma(x+a)} = \frac{B(x,a)}{\Gamma(a)} = \frac{1}{\Gamma(a)} \int_0^1 s^{a-1}(1-s)^{x-1}\, ds$$

$$= \frac{x^{-a}}{\Gamma(a)} \int_0^x t^{a-1} \left(1 - \frac{t}{x}\right)^x \left(1 - \frac{t}{x}\right)\, dt \sim \frac{x^{-a}}{\Gamma(a)} \int_0^\infty t^{a-1} e^{-t}\, dt$$

$$= x^{-a}.$$

This gives the principal part of (2.1.9). The error estimate is left as an exercise.

$\square$

The extended functional equation (2.1.3) allows us to extend this result to the shifted factorials.

Corollary 2.1.4 *If $a, b \neq 0, -1, -2, \ldots$ then*

$$\lim_{n \to \infty} \frac{(a)_n}{(b)_n} n^{b-a} = \frac{\Gamma(b)}{\Gamma(a)}. \tag{2.1.10}$$

2.2 Euler's product and reflection formulas

The first result here is Euler's product formula [84].

Theorem 2.2.1 *The gamma function satisfies*

$$\Gamma(z) = \lim_{n \to \infty} \frac{(n-1)! \, n^z}{(z)_n} \tag{2.2.1}$$

for $\mathrm{Re}\, z > 0$ *and*

$$\Gamma(z) = \frac{1}{z} \prod_{n=1}^{\infty} \left(1 + \frac{1}{n}\right)^z \left(1 + \frac{z}{n}\right)^{-1} \qquad (2.2.2)$$

for all complex $z \neq 0, -1, -2, \ldots$ *In particular, the gamma function has no zeros.*

Proof The identity (2.2.1) is essentially the case $a = 1$, $b = z$ of (2.1.10), since

$$(1)_n\, n^{z-1} = (n-1)!\, n^z.$$

To prove (2.2.2) we note that

$$\frac{(n-1)!\, n^z}{(z)_n} = \frac{n^z}{z} \prod_{j=1}^{n-1} \left(1 + \frac{z}{j}\right)^{-1}. \qquad (2.2.3)$$

Writing

$$n = \frac{n}{n-1} \frac{n-1}{n-2} \cdots \frac{2}{1} = \left(1 + \frac{1}{n-1}\right) \left(1 + \frac{1}{n-2}\right) \cdots (1+1),$$

we can rewrite (2.2.3) as

$$\frac{1}{z} \prod_{j=1}^{n-1} \left(1 + \frac{1}{j}\right)^z \left(1 + \frac{z}{j}\right)^{-1}.$$

The logarithm of the jth factor is $O(j^{-2})$. It follows that the product converges uniformly in any compact set that excludes $z = 0, -1, -2, \ldots$ Therefore (first for $\mathrm{Re}\, z > 0$ and then by analytic continuation), taking the limit gives (2.2.2) for all $z \neq 0, -1, -2, \ldots$ $\qquad\square$

The reciprocal of the gamma function is an entire function. Its product representation can be deduced from (2.2.1), since

$$\frac{(z)_n}{(n-1)!\, n^z} = z \exp\left(z \left[\sum_{k=1}^{n-1} \frac{1}{k} - \log n\right]\right) \prod_{k=1}^{n-1} \left(1 + \frac{z}{k}\right) e^{-z/k}. \qquad (2.2.4)$$

The logarithm of the kth factor in the product (2.2.4) is $O(k^{-2})$, so the product converges uniformly in bounded sets. The coefficient of z in the exponential is

$$\sum_{k=1}^{n-1} \frac{1}{k} - \log n = \sum_{k=1}^{n-1} \int_k^{k+1} \left[\frac{1}{k} - \frac{1}{t}\right] dt = \sum_{k=1}^{n-1} \int_k^{k+1} \frac{t-k}{tk}\, dt.$$

The kth summand in the last sum is $O(k^{-2})$, so the sum converges as $n \to \infty$. The limit is known as *Euler's constant*

$$\gamma = \lim_{n \to \infty} \left\{ \sum_{k=1}^{n-1} \frac{1}{k} - \log n \right\} = \lim_{n \to \infty} \sum_{k=1}^{n-1} \int_k^{k+1} \left[\frac{1}{k} - \frac{1}{t} \right] dt. \qquad (2.2.5)$$

We have shown that (2.2.1) has the following consequence.

Corollary 2.2.2 *The reciprocal of the gamma function has the product representation*

$$\frac{1}{\Gamma(z)} = z \, e^{\gamma z} \prod_{n=1}^{\infty} \left(1 + \frac{z}{n} \right) e^{-z/n}, \qquad (2.2.6)$$

where γ is Euler's constant (2.2.5).

This (or its reciprocal) is known as the "Weierstrass form" of the gamma function, although it was first proved by Schlömilch [251] and Newman [215] in 1848.

The next result is *Euler's reflection formula* [89].

Theorem 2.2.3 *For z not an integer,*

$$\Gamma(z) \, \Gamma(1 - z) = \frac{\pi}{\sin \pi z}. \qquad (2.2.7)$$

Proof For $0 < \operatorname{Re} z < 1$,

$$\Gamma(z) \, \Gamma(1 - z) = B(z, 1 - z) = \int_0^\infty \frac{t^{z-1} \, dt}{1 + t}.$$

The integrand is a holomorphic function of t for t in the complement of the positive real axis $[0, +\infty)$, choosing the argument of t in the interval $(0, 2\pi)$. Let C be the curve that comes from $+\infty$ to 0 along the "lower" side of the interval $[0, \infty)$ ($\arg t = 2\pi$) and returns to $+\infty$ along the "upper" side of the interval ($\arg t = 0$). Evaluating

$$\int_C \frac{t^{z-1} \, dt}{1 + t}$$

by the residue calculus gives

$$(1 - e^{2\pi i z}) \int_0^\infty \frac{t^{z-1} \, dt}{1 + t} = 2\pi i \, \operatorname{res}(t^{z-1}, -1) = -2\pi i \, e^{i\pi z};$$

see Appendix A. This proves (2.2.7) in the range $0 < \operatorname{Re} z < 1$. Once again, analytic continuation gives the result for all non-integer complex z. $\qquad \square$

Corollary 2.2.4 (Hankel's integral formula) *Let C be an oriented contour in the complex plane that begins at* $-\infty$, *continues on the real axis* (arg $t = -\pi$) *to a point* $-\delta$, *follows the circle* $\{|t| = \delta\}$ *in the positive (counterclockwise) direction, and returns to* $-\infty$ *along the real axis* (arg $t = \pi$). *Then*

$$\frac{1}{\Gamma(z)} = \frac{1}{2\pi i} \int_C e^t \, t^{-z} \, dt. \tag{2.2.8}$$

Here t^{-z} *takes its principal value where C crosses the positive real axis.*

Proof The function defined by the integral in (2.2.8) is entire, so it suffices to prove the result for $0 < z < 1$. In this case we may take $\delta \to 0$. Setting $s = -t$, the right-hand side is

$$\frac{1}{2\pi i} \int_0^\infty e^{-s} \left\{ \left(se^{-i\pi}\right)^{-z} - \left(se^{i\pi}\right)^{-z} \right\} ds = \frac{\sin \pi z}{\pi} \int_0^\infty e^{-s} s^{-z} \, ds$$

$$= \frac{1}{\Gamma(z)\,\Gamma(1-z)} \cdot \Gamma(1-z),$$

where we have used (2.2.7). $\qquad\square$

The contour C described in Corollary 2.2.4 is often referred to as a *Hankel loop*.

Corollary 2.2.5 (Euler's product for sine)

$$\frac{\sin \pi z}{\pi z} = \prod_{n=1}^{\infty} \left(1 - \frac{z^2}{n^2}\right). \tag{2.2.9}$$

Proof This follows from (2.2.7) together with (2.2.2). $\qquad\square$

Corollary 2.2.6 $\Gamma\left(\frac{1}{2}\right) = \sqrt{\pi}$.

Proof Take $z = \frac{1}{2}$ in (2.2.7). $\qquad\square$

This evaluation can be obtained in a number of other ways, for example

$$\Gamma\left(\frac{1}{2}\right)^2 = B\left(\frac{1}{2}, \frac{1}{2}\right) = \int_0^\infty \frac{t^{\frac{1}{2}}}{1+t} \frac{dt}{t}$$

$$= 2 \int_0^\infty \frac{du}{1+u^2} = \tan^{-1} u \Big|_{-\infty}^\infty = \pi;$$

$$\Gamma\left(\frac{1}{2}\right) = \int_0^\infty e^{-t} \frac{dt}{\sqrt{t}} = 2 \int_0^\infty e^{-u^2} \, du$$

$$= \left[\int_{-\infty}^\infty \int_{-\infty}^\infty e^{-(x^2+y^2)} \, dx \, dy\right]^{\frac{1}{2}} = \left[\int_0^{2\pi} \int_0^\infty e^{-r^2} r \, dr \, d\theta\right]^{\frac{1}{2}} = \sqrt{\pi}.$$

2.3 Formulas of Legendre and Gauss

Observe that

$$\Gamma(2n) = (2n-1)! = 2^{2n-1} \frac{1}{2} \cdot 1 \cdot \left(\frac{1}{2}+1\right) \cdot 2 \cdots (n-1)\left(\frac{1}{2}+n-1\right)$$

$$= 2^{2n-1}(n-1)! \left(\frac{1}{2}\right)_n = 2^{2n-1}\Gamma(n)\frac{\Gamma\left(n+\frac{1}{2}\right)}{\Gamma\left(\frac{1}{2}\right)} = \frac{2^{2n-1}}{\sqrt{\pi}}\,\Gamma(n)\Gamma\left(n+\frac{1}{2}\right).$$

This is the positive integer case of *Legendre's duplication formula* [182]:

$$\Gamma(2z) = \frac{2^{2z-1}}{\sqrt{\pi}}\,\Gamma(z)\,\Gamma\left(z+\frac{1}{2}\right), \quad z \neq 0, -1, -2, \ldots \qquad (2.3.1)$$

We give two proofs of (2.3.1). The second proof generalizes to give Gauss's formula for $\Gamma(mz)$. The first proof begins with $\mathrm{Re}\, z > 0$ and uses the change of variables $t = 4s(1-s)$ on the interval $0 \leq s \leq \frac{1}{2}$, as follows:

$$\frac{\Gamma(z)^2}{\Gamma(2z)} = \mathrm{B}(z,z) = \int_0^1 [s(1-s)]^z \frac{ds}{s(1-s)}$$

$$= 2\int_0^{\frac{1}{2}} [s(1-s)]^z \frac{ds}{s(1-s)} = 2\int_0^1 \left(\frac{t}{4}\right)^z \frac{dt}{t\sqrt{1-t}}$$

$$= 2^{1-2z}\,\mathrm{B}\left(z,\frac{1}{2}\right) = 2^{1-2z}\frac{\Gamma(z)\,\Gamma\left(\frac{1}{2}\right)}{\Gamma\left(z+\frac{1}{2}\right)}.$$

This gives (2.3.1) for $\mathrm{Re}\, z > 0$; analytic continuation gives the result for $z \neq 0, -1, -2, \ldots$

The second proof uses (2.2.1), which implies

$$\Gamma(2z) = \lim_{n\to\infty} \frac{(2n)!\,(2n)^{2z-1}}{(2z)_{2n}}. \qquad (2.3.2)$$

Now

$$(2n)! = 2^{2n}\left(\frac{1}{2}\right)_n (1)_n = 2^{2n}\frac{\Gamma\left(\frac{1}{2}+n\right)\Gamma(n+1)}{\Gamma\left(\frac{1}{2}\right)} \qquad (2.3.3)$$

and

$$(2z)_{2n} = 2^{2n}(z)_n \left(z+\frac{1}{2}\right)_n = 2^{2n}\frac{\Gamma(z+n)\,\Gamma\left(z+\frac{1}{2}+n\right)}{\Gamma(z)\,\Gamma\left(z+\frac{1}{2}\right)}$$

$$= 2^{2n}\frac{\Gamma(n)}{\mathrm{B}(z,n)} \cdot \frac{\Gamma(z)\,\Gamma\left(\frac{1}{2}+n\right)}{\Gamma\left(z+\frac{1}{2}\right)\mathrm{B}\left(z,\frac{1}{2}+n\right)}. \qquad (2.3.4)$$

According to the calculation at the beginning of Section 2.2,

$$B(z, n) \sim n^{-z}\Gamma(z)$$

as $n \to \infty$, so in taking the limit in (2.3.2) we may replace the expression in (2.3.4) with

$$2^{2n}n^{2z}\frac{\Gamma(n)\,\Gamma\left(\frac{1}{2}+n\right)}{\Gamma(z)\,\Gamma\left(z+\frac{1}{2}\right)}. \tag{2.3.5}$$

Multiplying the quotient of (2.3.3) and (2.3.5) by $(2n)^{2z-1}$ gives (2.3.1).

The previous proof can be adapted to prove the first version of the following result of Gauss [103]:

$$\Gamma(mz) = m^{mz-1}\frac{\Gamma(z)\,\Gamma\left(z+\frac{1}{m}\right)\Gamma\left(z+\frac{2}{m}\right)\cdots\Gamma\left(z+\frac{m-1}{m}\right)}{\Gamma\left(\frac{1}{m}\right)\Gamma\left(\frac{2}{m}\right)\cdots\Gamma\left(\frac{m-1}{m}\right)}$$

$$= m^{mz-\frac{1}{2}}\frac{\Gamma(z)\,\Gamma\left(z+\frac{1}{m}\right)\Gamma\left(z+\frac{2}{m}\right)\cdots\Gamma\left(z+\frac{m-1}{m}\right)}{(2\pi)^{\frac{1}{2}(m-1)}}. \tag{2.3.6}$$

To prove the second version we evaluate the denominator in the first version, using the reflection formula (2.2.7):

$$\left[\Gamma\left(\frac{1}{m}\right)\Gamma\left(\frac{2}{m}\right)\cdots\Gamma\left(\frac{m-1}{m}\right)\right]^2$$

$$= \prod_{k=1}^{m-1}\frac{\pi}{\sin\left(\frac{\pi k}{m}\right)} = \pi^{m-1}\prod_{k=1}^{m-1}\frac{2i}{e^{\pi ik/m}(1-e^{-2\pi ik/m})}.$$

Since $1 + 2 + \cdots + m - 1 = \frac{1}{2}m(m-1)$,

$$\prod_{k=1}^{m-1}\frac{2i}{e^{\pi ik/m}} = 2^{m-1},$$

and since $\omega = e^{2\pi i/m}$ is a primitive mth root of 1 it follows that

$$\prod_{k=1}^{m-1}\left(1-\omega^k\right) = \lim_{t\to 1}\prod_{k=1}^{m-1}\left(t-\omega^k\right) = \lim_{t\to 1}\frac{t^m - 1}{t-1} = m.$$

Therefore

$$\Gamma\left(\frac{1}{m}\right)\Gamma\left(\frac{2}{m}\right)\cdots\Gamma\left(\frac{m-1}{m}\right) = \frac{(2\pi)^{\frac{1}{2}(m-1)}}{m^{\frac{1}{2}}} \tag{2.3.7}$$

and we get the second version of (2.3.6).

2.4 Two characterizations of the gamma function

The gamma function is not uniquely determined by the functional equation
$\Gamma(z+1) = z\,\Gamma(z)$; in fact, if f is a function such as $\sin 2\pi x$ that is periodic
with period 1, then the product $f\,\Gamma$ also satisfies the functional equation. Here
we give two characterizations, one as a function holomorphic on the half-plane
$\{\mathrm{Re}\, z > 0\}$ and one as a function on the positive reals.

Theorem 2.4.1 (Wielandt) *Suppose that G is holomorphic in the half-plane
$\{\mathrm{Re}\, z > 0\}$, bounded on the closed strip $\{1 \leq \mathrm{Re}\, z \leq 2\}$, and satisfies the
equations $G(1) = 1$ and $z\,G(z) = G(z+1)$ for $\mathrm{Re}\, z > 0$. Then $G(z) = \Gamma(z)$.*

Proof Let $F(z) = G(z) - \Gamma(z)$. Then F satisfies the functional equation and
vanishes at $z = 1$, so it vanishes at the positive integers. This implies that
F extends to an entire function. In fact, the functional equation allows us to
extend by defining

$$F(z) = \frac{F(z+n)}{(z)_n}, \quad -n < \mathrm{Re}\, z \leq 2 - n, \quad n = 1, 2, 3, \ldots$$

This is clearly holomorphic where $(z)_n \neq 0$, i.e. except for $z = 0, -1, \ldots,$
$1 - n$. These values of z are zeros of $F(z + n)$ and are simple zeros of $(z)_n$, so
they are removable singularities for F. Therefore, the extension of F is entire.

The functional equation and the fact that F is regular at 0 and bounded
on the strip $\{1 \leq \mathrm{Re}\, z \leq 2\}$ imply that F is bounded on the wider strip $S = \{0 \leq \mathrm{Re}\, z \leq 2\}$. Therefore, the function $f(z) = F(z)\,F(1 - z)$ is entire and
bounded on S. Moreover,

$$f(z + 1) = z\,F(z)\,F(-z) = -F(z)(-z)F(-z) = -f(z).$$

Thus $f(z + 2) = f(z)$. Since f is bounded on a vertical strip of width 2, this
implies that f is a bounded entire function, hence constant. But $f(1) = 0$, so
$f \equiv 0$. It follows that $F \equiv 0$. $\square$

The next characterization starts from the observation that $\log \Gamma$ is a convex
function on the interval $\{x > 0\}$. In fact, (2.2.6) implies that

$$\log \Gamma(x) = -\gamma x - \log x + \sum_{k=1}^{\infty}\left[\frac{x}{k} - \log\left(1 + \frac{x}{k}\right)\right], \tag{2.4.1}$$

from which it follows that

$$(\log \Gamma)''(x) = \sum_{k=0}^{\infty}\frac{1}{(x+k)^2}, \tag{2.4.2}$$

which is positive for $x > 0$. This implies convexity. The following theorem is the converse.

Theorem 2.4.2 (Bohr–Mollerup) *Suppose that $G(x)$ is defined and positive for $x > 0$, satisfies the functional equation $x\,G(x) = G(x + 1)$, and suppose that $\log G$ is convex and $G(1) = 1$. Then $G(x) = \Gamma(x)$ for all $x > 0$.*

Proof Let $f = \log G$. In view of the functional equation, it is enough to prove $G(x) = \Gamma(x)$ for $0 < x < 1$. The assumptions imply that $G(1) = G(2) = 1$ and $G(3) = 2$, so $f(1) = f(2) < f(3)$. Convexity of f implies that f is increasing on the interval $[2, \infty)$, and that for each integer $n \geq 2$,

$$f(n) - f(n-1) \leq \frac{f(n+x) - f(n)}{x} \leq f(n+1) - f(n).$$

By the functional equation and the definition of f, this is equivalent to

$$(n-1)^x \leq \frac{G(x+n)}{G(n)} \leq n^x.$$

The functional equation applied to the quotient leads to

$$\frac{(n-1)!\,(n-1)^x}{(x)_n} \leq \frac{G(x)}{G(1)} \leq \frac{(n-1)!\,n^x}{(x)_n}.$$

By (2.2.1), the expressions on the left and the right have limit $\Gamma(x)$. $\square$

   Wielandt's theorem first appeared in Knopp [157]; see Remmert [237]. The theorem of Bohr and Mollerup is in [34].

2.5 Asymptotics of the gamma function

Suppose x is real and $x \to +\infty$. The integrand $e^{-t}\,t^x$ for $\Gamma(x+1)$ has its maximum $(x/e)^x$ at $t = x$, which suggests a change of variables $t = xu$:

$$\Gamma(x) = \frac{1}{x}\,\Gamma(x+1) = x^x \int_0^\infty \left(u\,e^{-u}\right)^x du$$

$$= \left(\frac{x}{e}\right)^x \int_0^\infty \left(u\,e^{1-u}\right)^x du.$$

Now $u e^{1-u}$ attains its maximum 1 at $u = 1$, so the part of the last integral over a region $|u - 1| > \delta > 0$ decays exponentially in x as $x \to \infty$. Note that $u e^{1-u}$ agrees to second-order at $u = 1$ with $e^{-(u-1)^2}$; it follows that we can change variables $u \to s$ near $u = 1$ in such a way that

$$u\,e^{1-u} = e^{-\frac{1}{2}s^2}, \quad u'(s) = \sum_{n=0}^{\infty} a_n s^n,$$

with $a_0 = 1$. Combining these remarks, we obtain an asymptotic expansion

$$\int_0^{\infty} \left(u\,e^{1-u} \right)^x du \sim \sum_{n=0}^{\infty} a_n \int_{-\infty}^{\infty} e^{-\frac{1}{2}xs^2} s^n\, ds$$

$$= \sum_{m=0}^{\infty} a_{2m}\, x^{-m-\frac{1}{2}} \int_{-\infty}^{\infty} e^{-\frac{1}{2}s^2} s^{2m}\, ds.$$

(The odd terms vanish because their integrands are odd functions.) Since $a_0 = 1$, the first term in this expansion is $\sqrt{2\pi/x}$. The same considerations apply for complex z with $\mathrm{Re}\,z \to +\infty$. Thus we have *Stirling's formula* [270]:

$$\Gamma(z) = \frac{z^z}{e^z} \left\{ \left(\frac{2\pi}{z}\right)^{\frac{1}{2}} + O\left(z^{-\frac{3}{2}}\right) \right\} \quad \text{as} \quad \mathrm{Re}\,z \to +\infty. \qquad (2.5.1)$$

The previous can be made more explicit. Binet [31] proved the following.

Theorem 2.5.1 *For* $\mathrm{Re}\,z > 0$,

$$\Gamma(z) = \frac{z^z}{e^z} \left(\frac{2\pi}{z}\right)^{\frac{1}{2}} e^{\theta(z)}, \qquad (2.5.2)$$

where

$$\theta(z) = \int_0^{\infty} \left(\frac{1}{e^t - 1} - \frac{1}{t} + \frac{1}{2} \right) e^{-zt}\, \frac{dt}{t}. \qquad (2.5.3)$$

Proof We follow Sasvari [248]. By definition, (2.5.2) holds with $\theta(z)$ replaced by

$$\varphi(z) = \log \Gamma(z) + z(1 - \log z) + \frac{1}{2} \log z - \frac{1}{2} \log(2\pi).$$

The functional equation implies that

$$\varphi(z) - \varphi(z+1) = \left(z + \frac{1}{2} \right) \log\left(\frac{z+1}{z} \right) - 1,$$

and (2.5.1) implies that $\varphi(z) \to 0$ as $\operatorname{Re} z \to +\infty$. Then

$$\varphi'(z) - \varphi'(z+1) = -\frac{1}{2}\left(\frac{1}{z} + \frac{1}{z+1}\right) + \log\left(1 + \frac{1}{z}\right);$$

$$\varphi''(z) - \varphi''(z+1) = -\frac{1}{2}\left(\frac{1}{z} + \frac{1}{z+1}\right)' + \frac{1}{z+1} - \frac{1}{z}.$$

With θ given by (2.5.3), we have

$$\theta(z) - \theta(z+1) = \int_0^\infty \left[e^{-t} + \left(\frac{1}{2} - \frac{1}{t}\right)(1 - e^{-t})\right] e^{-zt}\, \frac{dt}{t};$$

$$\theta'(z) - \theta'(z+1) = -\int_0^\infty \left[e^{-t} + \left(\frac{1}{2} - \frac{1}{t}\right)(1 - e^{-t})\right] e^{-zt}\, dt$$

$$= -\frac{1}{2}\left(\frac{1}{z} + \frac{1}{z+1}\right) + \int_0^\infty \frac{1}{t}(1 - e^{-t}) e^{-zt}\, dt;$$

$$\theta''(z) - \theta''(z+1) = -\frac{1}{2}\left(\frac{1}{z} + \frac{1}{z+1}\right)' + \frac{1}{z+1} - \frac{1}{z}.$$

Since the various functions and derivatives have limit zero as $\operatorname{Re} z \to \infty$, it follows that $\theta = \varphi$.   □

The integrand in (2.5.3) includes the function

$$\frac{1}{e^t - 1} - \frac{1}{t} + \frac{1}{2} = \frac{1}{2}\frac{e^{t/2} + e^{-t/2}}{e^{t/2} - e^{-t/2}} - \frac{1}{t}.$$

This function is odd and is holomorphic for $|t| < 2\pi$, so it has an expansion

$$\frac{1}{e^t - 1} - \frac{1}{t} + \frac{1}{2} = \sum_{m=1}^\infty \frac{B_{2m}}{(2m)!} t^{2m-1}, \quad |t| < 2\pi. \tag{2.5.4}$$

The coefficients B_{2m} are known as the *Bernoulli numbers* [27]. They can be computed recursively; see the exercises. Putting partial sums of this expansion into (2.5.3) gives the asymptotic estimates

$$\theta(z) = \sum_{m=1}^N \frac{B_{2m}}{2m(2m-1)} z^{1-2m} + O\left(z^{-2N-1}\right), \quad \operatorname{Re} z \to +\infty. \tag{2.5.5}$$

Stirling's formula (2.5.1) for positive integer z can be used in conjuction with the product formula (2.2.6) to extend (2.5.1) to the complement of the negative real axis. The method here is due to Stieltjes.

Theorem 2.5.2 *For* $|\arg z| < \pi$,

$$\Gamma(z) = \frac{z^z}{e^z}\left\{\left(\frac{2\pi}{z}\right)^{\frac{1}{2}} + O\left(z^{-\frac{3}{2}}\right)\right\} \tag{2.5.6}$$

as $|z| \to \infty$, *uniformly for* $|\arg z| \le \pi - \delta < \pi$.

Proof Let $[s]$ denote the greatest integer function. Then

$$\int_k^{k+1} \frac{\frac{1}{2} - s + [s]}{s+z}\, ds = \int_k^{k+1}\left(\frac{\frac{1}{2}+k+z}{s+z} - 1\right) ds$$

$$= \left(k + \frac{1}{2} + z\right)\left[\log(k+1+z) - \log(k+z)\right] - 1$$

$$= \left[\left(k + \frac{1}{2} + z\right)\log(k+1+z)\right.$$

$$\left. - \left(k - \frac{1}{2} + z\right)\log(k+z)\right] - \log(k+z) - 1.$$

Summing,

$$\int_0^n \frac{\frac{1}{2} - s + [s]}{s+z}\, ds$$

$$= \left(n - \frac{1}{2} + z\right)\log(n+z) - \left(-\frac{1}{2} + z\right)\log z - \sum_{k=0}^{n-1}\log(k+z) - n.$$

Now

$$-\sum_{k=1}^{n-1}\log(k+z) = -\sum_{k=1}^{n-1}\log\left(1 + \frac{z}{k}\right) - \log\Gamma(n)$$

$$= \sum_{k=1}^{n-1}\left[\frac{z}{k} - \log\left(1 + \frac{z}{k}\right)\right] - z\sum_{k=1}^{n-1}\frac{1}{k} - \log\Gamma(n),$$

and

$$\log(n+z) = \log n + \frac{z}{n} + O(n^{-2}).$$

It follows from (2.2.5), (2.5.1), and (2.2.6), respectively, that

$$\sum_{k=1}^{n-1}\frac{1}{k} = \log n + \gamma + O\left(n^{-1}\right);$$

$$\log \Gamma(n) = \left(n - \frac{1}{2}\right) \log n - n + \frac{1}{2} \log 2\pi + O\left(n^{-1}\right);$$

$$\log \Gamma(z) = -z\gamma - \log z + \sum_{k=1}^{\infty} \left[\frac{z}{k} - \log\left(1 + \frac{z}{k}\right)\right].$$

Combining the previous formulas gives

$$\int_0^\infty \frac{\frac{1}{2} - s + [s]}{s + z} \, ds = \log \Gamma(z) - \left(z - \frac{1}{2}\right) \log z - \frac{1}{2} \log 2\pi + z. \quad (2.5.7)$$

We estimate the integral by integrating by parts: let

$$f(s) = \int_0^s \left(\frac{1}{2} - t + [t]\right) dt.$$

This function has period 1 and is therefore bounded. Then

$$\int_0^\infty \frac{\frac{1}{2} - s + [s]}{s + z} \, ds = \int_0^\infty \frac{f(s)}{(s + z)^2} \, ds = O\left(|z|^{-1}\right)$$

uniformly for $1 + \cos(\arg z) \geq \delta > 0$, since $z = re^{i\theta}$ implies

$$|s + z|^2 = s^2 + r^2 + 2sr \cos \theta \geq (s^2 + r^2) \min\{1, 1 + \cos \theta\}.$$

Therefore, exponentiating (2.5.7) gives (2.5.6). $\qquad\qquad\square$

Corollary 2.5.3 *For real x and y,*

$$\left|\Gamma(x + iy)\right| = \sqrt{2\pi} \, |y|^{x - \frac{1}{2}} \, e^{-\frac{1}{2}\pi|y|} \left[1 + O\left(|y|^{-1}\right)\right], \quad (2.5.8)$$

as $|y| \to \infty$.

2.6 The psi function and the incomplete gamma function

The logarithmic derivative of the gamma function is denoted by $\psi(z)$:

$$\psi(z) = \frac{d}{dz} \log \Gamma(z) = \frac{\Gamma'(z)}{\Gamma(z)}.$$

Most of the properties of this function can be obtained directly from the corresponding properties of the gamma function. For instance, the only singularities of $\psi(z)$ are simple poles with residue -1 at the points $z = 0, -1, -2, \ldots$

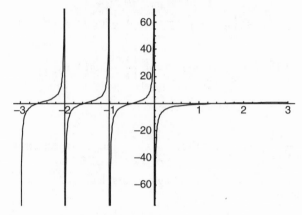

Figure 2.1 The ψ function.

The product formula (2.2.6) for $1/\Gamma(z)$ implies that

$$\psi(z) = -\gamma + \sum_{n=0}^{\infty}\left(\frac{1}{n+1} - \frac{1}{n+z}\right), \quad z \neq 0, -1, -2, \ldots, \qquad (2.6.1)$$

and therefore, as noted in Section 2.4,

$$\psi'(z) = \sum_{n=0}^{\infty} \frac{1}{(z+n)^2}. \qquad (2.6.2)$$

Thus ψ is a meromorphic function which has simple poles with residue -1 at the non-positive integers.

The graph of $\psi(x)$ for real x is shown in Figure 2.1.

Using (2.6.1), the recurrence relation

$$\psi(z+1) = \psi(z) + \frac{1}{z}$$

is readily verified and we have

$$\psi(1) = -\gamma, \quad \psi(k+1) = -\gamma + 1 + \frac{1}{2} + \frac{1}{3} + \cdots + \frac{1}{k}, \quad k = 1, 2, 3, \ldots$$

Taking the logarithmic derivative of (2.5.3) gives Binet's integral formula for ψ:

$$\psi(z) = \log z - \frac{1}{2z} - \int_0^{\infty}\left(\frac{1}{e^t-1} - \frac{1}{t} + \frac{1}{2}\right)e^{-zt}\,dt, \quad \operatorname{Re} z > 0.$$

The *incomplete gamma function* $\gamma(\alpha, z)$ is defined by

$$\gamma(\alpha, z) = \int_0^z t^{\alpha-1} e^{-t} \, dt, \quad \text{Re}\,\alpha > 0. \tag{2.6.3}$$

It is an analytic function of z in the right half-plane. We use the power series expansion of e^{-t} and integrate term by term to obtain

$$\gamma(\alpha, z) = z^\alpha \sum_{n=0}^\infty \frac{(-z)^n}{n!\,(n+\alpha)}, \quad \text{Re}\,\alpha > 0.$$

The series converges for all z, so we may use this formula to extend the function in z and α for $\alpha \neq 0, -1, -2, \ldots$ If α is fixed, then the branch of $\gamma(\alpha, z)$ obtained after z encircles the origin m times is given by

$$\gamma\left(\alpha, z e^{2m\pi i}\right) = e^{2m\alpha\pi i} \gamma(\alpha, z), \quad \alpha \neq 0, -1, -2, \ldots \tag{2.6.4}$$

The *complementary incomplete gamma function* is defined by

$$\Gamma(\alpha, z) = \int_z^\infty t^{\alpha-1} e^{-t} \, dt. \tag{2.6.5}$$

In (2.6.5) there is no restriction on α if $z \neq 0$, and the principal branch is defined in the same manner as for $\gamma(\alpha, z)$. Combining with (2.6.3) gives

$$\gamma(\alpha, z) + \Gamma(\alpha, z) = \Gamma(\alpha). \tag{2.6.6}$$

It follows from (2.6.4) and (2.6.6) that

$$\Gamma\left(\alpha, z e^{2m\pi i}\right) = e^{2m\alpha\pi i} \Gamma(\alpha, z) + \left(1 - e^{2m\alpha\pi i}\right) \Gamma(\alpha), \quad m = 0, \pm 1, \pm 2, \ldots \tag{2.6.7}$$

The *error function* and *complementary error function* are defined respectively by

$$\text{erf}\,z = \frac{2}{\sqrt{\pi}} \int_0^z e^{-t^2} \, dt, \quad \text{erfc}\,z = \frac{2}{\sqrt{\pi}} \int_z^\infty e^{-t^2} \, dt.$$

Both are entire functions. Clearly

$$\text{erf}\,z + \text{erfc}\,z = 1$$

and

$$\text{erf}\,z = \frac{2}{\sqrt{\pi}} \sum_{n=0}^\infty \frac{(-1)^n z^{2n+1}}{n!\,(2n+1)}.$$

By a simple change of variable, one can show that

$$\operatorname{erf} z = \frac{1}{\sqrt{\pi}} \gamma \left(\frac{1}{2}, z^2 \right), \quad \operatorname{erfc} z = \frac{1}{\sqrt{\pi}} \Gamma \left(\frac{1}{2}, z^2 \right).$$

2.7 The Selberg integral

The *Selberg integral* [257] is

$$S_n(a, b, c) = \int_0^1 \cdots \int_0^1 \prod_{i=1}^{n} x_i^{a-1} (1 - x_i)^{b-1} \prod_{1 \le j < k \le n} |x_j - x_k|^{2c} \, dx_1 \cdots dx_n,$$

(2.7.1)

where convergence of the integral is assured by the conditions

$$\operatorname{Re} a > 0, \quad \operatorname{Re} b > 0, \quad \operatorname{Re} c > \max \left\{ -\frac{1}{n}, -\frac{\operatorname{Re} a}{n-1}, -\frac{\operatorname{Re} b}{n-1} \right\}. \quad (2.7.2)$$

In particular, $S_n(a, b, 0) = \mathrm{B}(a, b)^n$. Thus Selberg's evaluation of S_n, stated in the following theorem, generalizes Euler's evaluation of the beta function in Theorem 2.1.2. Note that

$$S_n(a, b, c) = S_n(b, a, c).$$

Theorem 2.7.1 *Under the conditions* (2.7.2),

$$S_n(a, b, c) = \prod_{j=0}^{n-1} \frac{\Gamma(a + jc) \, \Gamma(b + jc) \, \Gamma(1 + [j+1]c)}{\Gamma(a + b + [n-1+j]c) \, \Gamma(1 + c)}. \quad (2.7.3)$$

For convenience we let $\mathbf{x} = (x_1, x_2, \ldots, x_n)$, $d\mathbf{x} = dx_1 \cdots dx_n$, and denote the integrand in (2.7.1) by

$$w(\mathbf{x}) = \prod_{i=1}^{n} x_i^{a-1} (1 - x_i)^{b-1} \prod_{1 \le j < k \le n} |x_j - x_k|^{2c}.$$

Note that $w(\mathbf{x})$ is symmetric in $x_1, \ldots, x_n$. Let C_n denote the n-dimensional cube $[0, 1] \times \cdots \times [0, 1]$. Then (2.7.1) is

$$S_n(a, b, c) = \int_{C_n} w(\mathbf{x}) \, d\mathbf{x}.$$

The proof of Theorem 2.7.1 outlined here is due to Aomoto [9]. It makes use of the related integrals

$$I_k = \int_{C_n} w(\mathbf{x}) \prod_{i=1}^{k} x_i \, d\mathbf{x}, \quad k = 0, 1, \ldots, n. \tag{2.7.4}$$

Note that $I_0 = S_n(a, b, c)$ and $I_n = S_n(a+1, b, c) = S_n(b, a+1, c)$. We may find a relation between I_k and I_{k-1} by integrating the identity

$$\frac{\partial}{\partial x_1} \left[(1-x_1) \, w(\mathbf{x}) \prod_{j=1}^{k} x_j \right] = a \, (1-x_1) \, w(\mathbf{x}) \prod_{j=2}^{k} x_j - b \, w(\mathbf{x}) \prod_{j=1}^{k} x_j$$

$$+ (1-x_1) \, 2c \, w(\mathbf{x}) \left[\sum_{j=2}^{n} \frac{1}{x_1 - x_j} \right] \prod_{j=1}^{k} x_j. \tag{2.7.5}$$

The left-hand side integrates to zero. The first two terms on the right integrate to $aI_{k-1} - (a+b)I_k$. The remaining terms can be integrated with the use of the following lemma, which is left as an exercise.

Lemma 2.7.2 *Let I_k be the integral (2.7.4). Then*

$$\int_{C_n} \frac{w(\mathbf{x})}{x_1 - x_j} \prod_{i=1}^{k} x_i \, d\mathbf{x} = \begin{cases} 0, & 2 \le j \le k; \\ \dfrac{1}{2}I_{k-1}, & k < j \le n, \end{cases} \tag{2.7.6}$$

and

$$\int_{C_n} \frac{x_1 \, w(\mathbf{x})}{x_1 - x_j} \prod_{i=1}^{k} x_i \, d\mathbf{x} = \begin{cases} \dfrac{1}{2}I_k, & 2 \le j \le k; \\ I_k. & k < j \le n. \end{cases} \tag{2.7.7}$$

Applying these identities to the integral of (2.7.5) gives the identity

$$I_k = \frac{a + (n-k)c}{a + b + (2n - k - 1)c} I_{k-1}. \tag{2.7.8}$$

It follows that

$$S_n(a+1, b, c) = \prod_{j=0}^{n-1} \frac{a + jc}{a + b + (n-1+j)c} S_n(a, b, c).$$

Iterating this k times with a and b interchanged gives

$$S_n(a, b, c) = \prod_{j=0}^{n-1} \frac{(a + b + [n-1+j]c)_k}{(b + jc)_k} S_n(a, b+k, c).$$

We let $k \to \infty$ and use (2.1.10) to conclude that

$$S_n(a, b, c) =$$

$$\prod_{j=0}^{n-1} \frac{\Gamma(b + jc)}{\Gamma(a + b + [n - 1 + j]c)} \cdot \lim_{k \to \infty} \left[k^{na+n(n-1)c} S_n(a, b + k, c) \right].$$

The expression in the limit can be rewritten as

$$k^{na+n(n-1)c} \int_0^k \cdots \int_0^k \prod_{i=1}^n \left(\frac{x_i}{k} \right)^{a-1} \left(1 - \frac{x_i}{k} \right)^{k+b-1} \prod_{i<j} \left| \frac{x_i}{k} - \frac{x_j}{k} \right|^{2c} \frac{d\mathbf{x}}{k^n}$$

$$= \int_0^k \cdots \int_0^k \prod_{i=1}^n x_i^{a-1} \left(1 - \frac{x_i}{k} \right)^{k+b-1} \prod_{i<j} |x_i - x_j|^{2c} \, d\mathbf{x}.$$

Taking the limit as $k \to \infty$, we obtain

$$S_n(a, b, c) = \prod_{j=0}^{n-1} \frac{\Gamma(b + jc)}{\Gamma(a + b + [n - 1 + j]c)}$$

$$\times \int_0^\infty \cdots \int_0^\infty \prod_{i=1}^n x_i^{a-1} e^{-x_i} \prod_{i<j} |x_i - x_j|^{2c} \, d\mathbf{x}. \qquad (2.7.9)$$

Denoting the last integral by $G_n(a, c)$ and using once again the symmetry in (a, b), we find that

$$\frac{G_n(a, c)}{\prod_{j=0}^{n-1} \Gamma(a + jc)} = \frac{G_n(b, c)}{\prod_{j=0}^{n-1} \Gamma(b + jc)}.$$

We denote the common value by $D_n(c)$ and note that $D_1(c) \equiv 1$. Returning to (2.7.9), we obtain

$$S_n(a, b, c) = \prod_{j=0}^{n-1} \frac{\Gamma(a + jc)\,\Gamma(b + jc)}{\Gamma(a + b + [n - 1 + j]c)} D_n(c). \qquad (2.7.10)$$

To complete the proof of Theorem 2.7.1, we need to evaluate $D_n(c)$. This will be done by using the following two lemmas, which are left as exercises.

Lemma 2.7.3 *For any function f continuous on the interval $[0, 1]$,*

$$\lim_{a \to 0+} \int_0^1 a\, t^{a-1} f(t) \, dt = f(0).$$

Lemma 2.7.4 *For any symmetric function* $f(\mathbf{x})$,

$$\int_{C_n} f(\mathbf{x})\, dx = n!\int_0^1 \int_{x_n}^1 \cdots \int_{x_2}^1 f(\mathbf{x})\, dx_1\, dx_2\cdots dx_n.$$

It follows that

$$\lim_{a\to 0+} \frac{a}{n!} S_n(a,b,c) = \lim_{a\to 0+} a\int_0^1 x_n^{a-1}(1-x_n)^{b-1}$$

$$\times \int_{x_n}^1 \cdots \int_{x_2}^1 \prod_{i=1}^{n-1} x_i^{a-1}(1-x_i)^{b-1} \prod_{i<j\leq n} |x_i - x_j|^{2c}\, d\mathbf{x}$$

$$= \int_0^1 \int_{x_{n-1}}^1 \cdots \int_{x_2}^1 \prod_{i=1}^{n-1} x_i^{2c-1}(1-x_i)^{b-1} \prod_{i<j<n} |x_i - x_j|^{2c}\, dx_1\cdots dx_{n-1}$$

$$= \frac{1}{(n-1)!}\, S_{n-1}(2c,b,c). \tag{2.7.11}$$

Since $\lim_{a\to 0} a\Gamma(a) = 1$, it follows from (2.7.11) and (2.7.10) that

$$D_n(c) = \frac{n\,\Gamma(nc)}{\Gamma(c)}\, D_{n-1}(c) = \frac{\Gamma(nc+1)}{\Gamma(c+1)}\, D_{n-1}(c).$$

But $D_1 = 1$, so

$$D_n(c) = \prod_{j=1}^{n} \frac{\Gamma(jc+1)}{\Gamma(c+1)}.$$

Combining this with (2.7.10) gives (2.7.3).

The relationship (2.7.8) together with (2.7.3) and the fact that $I_0 = S_n(a,b,c)$ gives Aomoto's generalization of (2.7.3):

$$\int_0^1 \cdots \int_0^1 \prod_{i=1}^{k} x_i \prod_{i=1}^{n} x_i^{a-1}(1-x_i)^{b-1} \prod_{1\leq i<j\leq n} |x_i - x_j|^{2c}\, dx_1\cdots dx_n$$

$$= \prod_{i=1}^{k} \frac{a+(n-i)c}{a+b+(2n-i-1)c}\, S_n(a,b,c)$$

$$= \prod_{i=1}^{k} \frac{a+(n-i)c}{a+b+(2n-i-1)c} \prod_{j=0}^{n-1} \frac{\Gamma(a+jc)\,\Gamma(b+jc)\,\Gamma(1+[j+1]c)}{\Gamma(a+b+[n-1+j]c)\,\Gamma(1+c)},$$

if

$$\operatorname{Re} a > 0, \quad \operatorname{Re} b > 0, \quad \operatorname{Re} c > \max\left\{ -\frac{1}{n}, -\frac{\operatorname{Re} a}{n-1}, -\frac{\operatorname{Re} b}{n-1} \right\}.$$

2.8 The zeta function

The zeta function is of particular importance in number theory. We mention it briefly here because its functional equation is closely connected with the gamma function.

For Re $z > 1$, the *Riemann zeta function* is defined by

$$\zeta(z) = \sum_{n=1}^{\infty} \frac{1}{n^z}. \qquad (2.8.1)$$

The uniqueness of prime factorization implies that each n^{-z} occurs exactly once in the product over all primes p of the series

$$1 + \frac{1}{p^z} + \frac{1}{p^{2z}} + \cdots + \frac{1}{p^{mz}} + \cdots = \left(1 - \frac{1}{p^z}\right)^{-1},$$

so we obtain Euler's product formula

$$\zeta(z) = \prod_{p \text{ prime}} \left(1 - \frac{1}{p^z}\right)^{-1}, \quad \text{Re } z > 1. \qquad (2.8.2)$$

Euler evaluated the zetà function at even positive integer z [91]:

$$\zeta(2m) = \frac{(-1)^{m-1}}{2} \frac{(2\pi)^{2m}}{(2m)!} B_{2m}, \qquad (2.8.3)$$

where the Bernoulli numbers are given by (2.5.4). In particular,

$$1 + \frac{1}{4} + \frac{1}{9} + \cdots + \frac{1}{n^2} + \cdots = \frac{\pi^2}{6}; \qquad (2.8.4)$$

$$1 + \frac{1}{16} + \frac{1}{81} + \cdots + \frac{1}{n^4} + \cdots = \frac{\pi^4}{90};$$

$$1 + \frac{1}{64} + \frac{1}{729} + \cdots + \frac{1}{n^6} + \cdots = \frac{\pi^6}{945}.$$

The evaluation (2.8.3) follows from the product formula (2.2.9) for sine: taking the logarithm gives

$$\log(\sin \pi x) = \log(\pi x) + \sum_{n=1}^{\infty} \log\left(1 - \frac{x^2}{n^2}\right).$$

Differentiating both sides gives

$$\frac{\pi \cos \pi x}{\sin \pi x} = \frac{1}{x} + \sum_{n=1}^{\infty} \frac{2x}{x^2 - n^2} = \frac{1}{x} + \sum_{n=1}^{\infty} \left[\frac{1}{x+n} + \frac{1}{x-n} \right]. \qquad (2.8.5)$$

The function

$$f(x) = \frac{\pi \cos \pi x}{\sin \pi x} - \frac{1}{x}$$

is holomorphic near the origin. It follows from (2.8.5) that the derivative

$$f^{(k)}(x) \Big|_{x=0} = \sum_{n=1}^{\infty} (-1)^k k! \left[\frac{1}{(x+n)^{k+1}} + \frac{1}{(x-n)^{k+1}} \right] \Big|_{x=0}$$

$$= \begin{cases} 0, & k = 2m; \\ -2 \displaystyle\sum_{m=1}^{\infty} \frac{(2m-1)!}{n^{2m}}, & k = 2m-1. \end{cases}$$

Therefore, the McLaurin expansion is

$$\frac{\pi \cos \pi x}{\sin \pi x} - \frac{1}{x} = -2 \sum_{m=1}^{\infty} \zeta(2m) x^{2m-1}. \qquad (2.8.6)$$

On the other hand, expressing $\cos \pi x$ and $\sin \pi x$ in terms of exponentials and making use of (2.5.4) gives

$$\frac{\pi \cos \pi x}{\sin \pi x} - \frac{1}{x} = i\pi \frac{e^{i\pi x} + e^{-i\pi x}}{e^{i\pi x} - e^{-i\pi x}} - \frac{1}{x}$$

$$= 2i\pi \left[\frac{1}{e^{2\pi i x} - 1} + \frac{1}{2} - \frac{1}{2\pi i x} \right]$$

$$= 2i\pi \sum_{m=1}^{\infty} \frac{B_{2m}}{(2m)!} (2\pi i x)^{2m-1}. \qquad (2.8.7)$$

Comparing coefficients of x^{2m-1} in the expansions (2.8.6) and (2.8.7) gives (2.8.3).

A change of variables in the integral defining $\Gamma(z)$ shows that

$$\frac{1}{n^z} = \frac{1}{\Gamma(z)} \int_0^{\infty} e^{-nt} t^z \frac{dt}{t}.$$

Therefore

$$\zeta(z) = \frac{1}{\Gamma(z)} \int_0^{\infty} \frac{e^{-t}}{1 - e^{-t}} t^z \frac{dt}{t} = \frac{1}{\Gamma(z)} \int_0^{\infty} \frac{t^{z-1} dt}{e^t - 1}.$$

The zeta function can be extended so as to be meromorphic in the plane. Set

$$f(z) = \int_C \frac{(-t)^{z-1}\, dt}{e^t - 1},$$

where C comes from $+\infty$ along the "lower" real axis ($\arg t = -2\pi$), circles zero in the negative (clockwise) direction at distance $\delta > 0$, and returns to $+\infty$ along the "upper" real axis ($\arg t = 0$); we choose the branch of $\log(-t)$ that is real for $t < 0$. The function f is entire. If $\operatorname{Re} z > 0$ we may let $\delta \to 0$ and evaluate the integral by the residue theorem to conclude that

$$f(z) = \left[e^{-i\pi z} - e^{i\pi z}\right] \Gamma(z)\, \zeta(z) = -2i \sin \pi z\, \Gamma(z)\, \zeta(z)$$

(see Appendix A). Thus

$$\zeta(z) = -\frac{f(z)}{2i\, \Gamma(z)\, \sin \pi z}. \tag{2.8.8}$$

This provides the analytic continuation to $\operatorname{Re} z \leq 1$, and shows that the only pole of ζ in that half-plane, hence in $\mathbf{C}$, is a simple pole at $z = 1$.

The *functional equation* for the zeta function can be obtained from (2.8.8). We evaluate the function f by expanding the circle in the path of integration. The integrand has simple poles at $t = \pm 2n\pi i$, $n \in \mathbf{N}$, and the residue of $1/(e^t - 1)$ at each pole is 1. The residues of the integrand of f at $2n\pi i$ and $-2n\pi i$ sum to

$$(2n\pi)^{z-1} i \left(e^{-\frac{1}{2}\pi z i} - e^{\frac{1}{2}\pi z i}\right) = 2^z (n\pi)^{z-1} \sin\left(\frac{1}{2}\pi z\right).$$

Suppose now that $\operatorname{Re} z < 0$, so $\operatorname{Re}(z - 1) < -1$. Letting the circle expand (between successive poles), we get

$$f(z) = (-2\pi i) \cdot 2^z \pi^{z-1} \sin\left(\frac{1}{2}\pi z\right) \sum_{n=1}^{\infty} \frac{1}{n^{1-z}}$$

$$= -i(2\pi)^z\, 2 \sin\left(\frac{1}{2}\pi z\right) \zeta(1 - z). \tag{2.8.9}$$

This holds for other values by analytic continuation. Combining (2.8.8) and (2.8.9) we obtain the functional equation

$$\zeta(1 - z) = \frac{2}{(2\pi)^z} \cos\left(\frac{1}{2}\pi z\right) \Gamma(z)\, \zeta(z). \tag{2.8.10}$$

2.9 Exercises

2.1 For $\operatorname{Re} a > 0$ and $\operatorname{Re} b > 0$, show that

$$\int_0^1 t^{a-1} \left(1 - t^2\right)^{b-1} dt = \frac{1}{2} \operatorname{B}\left(\frac{1}{2}a, b\right).$$

2.2 For $\operatorname{Re} a > 0$ and $\operatorname{Re} b > 0$, show that

$$\int_0^{\pi/2} \sin^{a-1}\theta \cos^{b-1}\theta \, d\theta = \frac{1}{2} \operatorname{B}\left(\frac{1}{2}a, \frac{1}{2}b\right).$$

2.3 Prove the functional relation

$$\operatorname{B}(a, b) = \frac{a+b}{b} \operatorname{B}(a, b+1).$$

2.4 Complete the proof of Proposition 2.1.3 by showing that the error in the appproximation is $O(x^{a-1})$.

2.5 Carry out the integration in the proof of Theorem 2.2.3.

2.6 Verify the contour integral representation of the beta function, for $\operatorname{Re} a > 0$ and any complex b:

$$\operatorname{B}(a, b) \frac{\sin \pi b}{\pi} = \frac{\Gamma(a)}{\Gamma(a+b)\,\Gamma(1-b)} = \frac{1}{2\pi i} \int_C s^{a-1}(s-1)^{b-1} \, ds.$$

Here the contour C is a counterclockwise loop that passes through the origin and encloses the point $s = 1$. We take the arguments of s and $s - 1$ to be zero for $s > 1$. Hint: assume first that $\operatorname{Re} b > 0$ and move the contour to run along the interval $[0, 1]$ and back.

2.7 Use (2.2.7) to verify that for $\operatorname{Re} a < 1$ and $\operatorname{Re} (a + b) > 0$,

$$\frac{\Gamma(a+b)}{\Gamma(a)\,\Gamma(b)} = (a+b-1) \frac{e^{-i\pi a}}{2\pi i} \int_C t^{-a}(1+t)^{-b} \, dt,$$

where the curve C runs from $+\infty$ to 0 along the upper edge of the cut on $[0, \infty)$ and returns to $+\infty$ along the lower edge, and the principal branch of t^{-a} is taken along the upper edge.

2.8 Use (2.2.2) and the evaluation of $\Gamma\left(\frac{1}{2}\right)$ to prove *Wallis's formula* [302]:

$$\frac{\pi}{4} = \frac{2}{3} \cdot \frac{4}{3} \cdot \frac{4}{5} \cdot \frac{6}{5} \cdots \frac{2n}{2n+1} \cdot \frac{2n+2}{2n+1} \cdots$$

2.9 One of the verifications of the evaluation of $\Gamma\left(\frac{1}{2}\right)$ obtained the identity

$$\int_{-\infty}^{\infty} e^{-x^2} dx = \sqrt{\pi},$$

by evaluating

$$\int_{-\infty}^{\infty} \int_{-\infty}^{\infty} e^{-(x^2+y^2)} \, dx \, dy$$

in polar coordinates. Turned around, this can be viewed as a way of evaluating the length of the unit circle. Use a "polar coordinate" computation of the n-dimensional version of this integral to show that the $(n-1)$-volume (area) of the unit $(n-1)$-sphere is $2\pi^{n/2}/\Gamma\left(\frac{1}{2}n\right)$.

2.10 Use the Wielandt Theorem or the Bohr–Mollerup Theorem to prove Gauss's multiplication theorem, the first line of (2.3.6). Hint: replace z by z/m on the right-hand side and define $G(z)$ as the result.

2.11 For real $y \neq 0$, show that

$$|\Gamma(iy)|^2 = \frac{\pi}{y \sinh \pi y}, \quad \left|\Gamma\left(\frac{1}{2}+iy\right)\right|^2 = \frac{\pi}{\cosh \pi y}.$$

2.12 If x and y are real, prove that

$$\left|\frac{\Gamma(x)}{\Gamma(x+iy)}\right|^2 = \prod_{k=0}^{\infty}\left\{1 + \frac{y^2}{(x+k)^2}\right\}, \quad x \neq 0, -1, -2, \ldots,$$

and hence $|\Gamma(x+iy)| \leq |\Gamma(x)|$.

2.13 If a and b are not negative integers, show that

$$\prod_{k=1}^{\infty} \frac{k(a+b+k)}{(a+k)(b+k)} = \frac{\Gamma(a+1)\Gamma(b+1)}{\Gamma(a+b+1)}.$$

2.14 Prove (2.5.5).

2.15 Find the value

$$\int_0^{\infty} t^{x-1} e^{-\lambda t \cos \theta} \cos(\lambda t \sin \theta) \, dt,$$

where $\lambda > 0$, $x > 0$, and $-\frac{1}{2}\pi < \theta < \frac{1}{2}\pi$.

2.16 Let z be positive, and integrate $\pi w^{-z} \operatorname{cosec} \pi z / \Gamma(1-z)$ around the rectangle with vertices $c \pm iR$, $-R \pm iR$, where $c > 0$. Show that

$$\frac{1}{2\pi i} \int_{c-i\infty}^{c+i\infty} w^{-z} \Gamma(z) \, dz = e^{-w}.$$

Prove that this formula holds for $|\arg w| \leq \frac{1}{2}\pi - \delta$, where $\delta > 0$.

2.17 (Saalshütz) Prove that for $\operatorname{Re} z < 0$,

$$\Gamma(z) = \int_0^{\infty} t^{z-1} \left\{ e^{-t} - 1 + t - \frac{t^2}{2!} + \cdots + (-1)^{k+1} \frac{t^k}{k!} \right\} \, dt,$$

where k is an integer between $\operatorname{Re}(-z)$ and $\operatorname{Re}(-z-1)$.

2.18 Show that for $s > 1$,

$$\log \zeta(s) = \sum_{p \text{ prime}} \sum_{m=1}^{\infty} \frac{1}{m \, p^{ms}}.$$

2.19 (a) The *Hurwitz zeta function* is defined by

$$\zeta(x, s) = \sum_{n=0}^{\infty} \frac{1}{(n+x)^s} \quad \text{for } x > 0.$$

Show that $\zeta(x+1, s) = \zeta(x, s) - x^{-s}$, and hence

$$\left(\frac{\partial \zeta(x+1, s)}{\partial s} \right)_{s=0} - \left(\frac{\partial \zeta(x, s)}{\partial s} \right)_{s=0} = \log x.$$

(b) Establish that for $\operatorname{Re} s > 1$,

$$\frac{\partial^2 \zeta(x, s)}{\partial x^2} = s(s+1) \sum_{n=0}^{\infty} \frac{1}{(n+x)^{s+2}}$$

and

$$\frac{d^2}{dx^2} \left(\frac{\partial \zeta(x, s)}{\partial s} \right)_{s=0} = \sum_{n=0}^{\infty} \frac{1}{(x+n)^2}.$$

(c) Show that the results (a) and (b), together with (2.4.2), imply that

$$\left(\frac{\partial \zeta(x, s)}{\partial s} \right)_{s=0} = C + \log \Gamma(x).$$

(d) Prove that

$$\zeta'(0) = -\frac{1}{2} \log 2\pi.$$

Use this and the result in (c) to prove *Lerch's theorem*:

$$\left(\frac{\partial \zeta(x, s)}{\partial s} \right)_{s=0} = \log \frac{\Gamma(x)}{\sqrt{2\pi}}.$$

2.20 First prove that

$$\zeta(x, s) = \frac{1}{\Gamma(s)} \int_0^{\infty} \frac{t^{s-1} e^{-xt}}{1 - e^{-t}} \, dt.$$

Then use the idea of Hankel's loop integral for the gamma function to derive the contour integral representation

$$\zeta(x, s) = \frac{e^{-i\pi s} \Gamma(1-s)}{2\pi i} \int_C \frac{t^{s-1} e^{-xt}}{1 - e^{-t}} \, dt,$$

where C starts at infinity on the positive real axis, encircles the origin once in the positive direction, excluding the points $\pm 2n\pi i$, $n \geq 1$, and returns to positive infinity.

2.21 Deduce from (2.6.2) that for $x > 0$, $\Gamma(x)$ has a single minimum, which lies between 1 and 2.

2.22 Show that for $z \neq 0, -1, -2, \ldots$,

$$\psi(1 - z) - \psi(z) = \pi \cot \pi z$$

and

$$\psi(z) + \psi\left(z + \frac{1}{2}\right) + 2\log 2 = 2\psi(2z).$$

2.23 Prove that

$$\psi\left(\frac{1}{2}\right) = -\gamma - 2\log 2, \quad \psi'\left(\frac{1}{2}\right) = \frac{1}{2}\pi^2.$$

2.24 Show that for $\operatorname{Re} z > 0$,

$$\psi(z) = \int_0^\infty \left(\frac{e^{-t}}{t} - \frac{e^{-zt}}{1 - e^{-t}}\right) dt.$$

This is known as Gauss's formula. Deduce that

$$\psi(z + 1) = \frac{1}{2z} + \log z - \int_0^\infty \left[\frac{1}{2}\coth\left(\frac{1}{2}t\right) - \frac{1}{t}\right] e^{-zt}\, dt.$$

2.25 Show that $\gamma(\alpha, z)/z^\alpha \Gamma(\alpha)$ is entire in both α and z. Furthermore, prove that

$$\frac{\gamma(\alpha, z)}{z^\alpha \Gamma(\alpha)} = e^{-z} \sum_{n=0}^\infty \frac{z^n}{\Gamma(\alpha + n + 1)}.$$

2.26 Prove that

$$\frac{2}{\pi} \int_0^\infty \frac{e^{-zt^2}}{1 + t^2}\, dt = e^z \left[1 - \operatorname{erf}\left(\sqrt{z}\right)\right].$$

2.27 The generalized exponential integral is defined by

$$E_n(z) = \int_1^\infty \frac{e^{-zt}}{t^n}\, dt, \quad n = 1, 2, \ldots$$

Show that

(a) $E_n(z) = z^{n-1}\Gamma(1-n, z)$;

(b) $E_n(z) = \int_z^\infty E_{n-1}(t)\,dt = \cdots = \int_z^\infty \int_{t_1}^\infty \cdots \int_{t_{n-1}}^\infty \frac{e^{-t_n}}{t_n}dt_n \cdots dt_2\,dt_1$;

(c) $E_n(z) = \frac{e^{-z}}{(n-1)!} \int_0^\infty \frac{e^{-t}t^{n-1}}{z+t}\,dt$.

2.28 Let $0 < \lambda < 1$ and $k = 1, 2, \ldots$ Put

$$f(k) = \frac{\Gamma(k+\lambda)}{\Gamma(k+1)}(k+\alpha)^{1-\lambda}$$

and

$$g(k) = \frac{f(k+1)}{f(k)}.$$

(a) Show that

$$\lim_{k\to\infty} g(k) = \lim_{k\to\infty} f(k) = 1$$

and

$$g(k) = \frac{k+\lambda}{k+1}\left(\frac{k+\alpha+1}{k+\alpha}\right)^{1-\lambda}.$$

Considering k as a continuous variable, show also that

$$g'(k) = \frac{A_k(\lambda; \alpha)}{(k+\alpha)^{2-\lambda}(k+1)^2(k+\alpha+1)^\lambda},$$

where

$$A_k(\lambda; \alpha) = (1-\lambda)\left(-\lambda k + 2\alpha k - \lambda + \alpha^2 + \alpha\right).$$

(b) When $\alpha = 0$, prove (i) $g'(k)$ is negative for $0 < \lambda < 1$ and $k = 1, 2, \ldots$; (ii) $g(k) > 1$ and $f(k) < 1$ for $k = 1, 2, \ldots$; and (iii)

$$\frac{\Gamma(k+\lambda)}{\Gamma(k+1)} < \frac{1}{k^{1-\lambda}}.$$

(c) When $\alpha = 1$, prove (i) $g'(k)$ is positive for $0 < \lambda < 1$ and $k = 1, 2, \ldots$; (ii) $g(k) < 1$ and $f(k) > 1$ for $k = 1, 2, \ldots$; and (iii)

$$\frac{\Gamma(k+\lambda)}{\Gamma(k+1)} > \frac{1}{(k+1)^{1-\lambda}}.$$

These inequalities were first given by Gautschi [107], but the argument outlined here is taken from Laforgia [169].

2.29 Use the symmetry of the function $w(\mathbf{x})$ to prove Lemma 2.7.2.

2.30 Prove Lemma 2.7.3.

2.31 Prove Lemma 2.7.4. Hint: partition the domain C_n and use symmetry.

2.32 Use the product formula (2.2.9) to prove the first of the equations (2.8.4).

2.33 Use the product formula (2.2.9) to prove the second of the equations (2.8.4).

2.34 Multiply both sides of (2.5.4) by $t(e^t - 1)$ and show that B_{2m} satisfies

$$B_{2m} = -\left[\binom{2m}{2}\frac{B_{2m-2}}{3} + \binom{2m}{4}\frac{B_{2m-4}}{5} + \cdots + \frac{B_0}{2m+1}\right] + \frac{1}{2},$$

$$m = 1, 2, \ldots,$$

with $B_0 = 1$.

2.35 Compute the Bernoulli numbers B_2, B_4, and B_6 and use (2.8.3) to verify each of the equations (2.8.4).

2.36 The *Bernoulli polynomials* $\{B_n(x)\}$ are defined by the identity

$$\frac{t\,e^{xt}}{e^t - 1} = \sum_{n=0}^{\infty} \frac{B_n(x)}{n!} t^n, \quad |t| < 2\pi.$$

Thus $B_{2m}(0) = B_{2m}$. Set $B_{-1}(x) = 0$. Prove the identities

(a) $B_n'(x) = n\,B_{n-1}(x)$;

(b) $\displaystyle\int_0^1 B_0(x)\,dx = 1$;

(c) $\displaystyle\int_0^1 B_n(x)\,dx = 0$;

(d) $B_n(x+1) = B_n(x) + n\,x^{n-1}$;

(e) $1 + 2^n + 3^n + \cdots + m^n = \dfrac{B_{n+1}(m+1) - B_{n+1}(0)}{n+1}$.

2.37 Verify that the first six Bernoulli polynomials are

$$B_0(x) = 1;$$

$$B_1(x) = \frac{2x - 1}{2};$$

$$B_2(x) = \frac{6x^2 - 6x + 1}{6};$$

$$B_3(x) = \frac{2x^3 - 3x^2 + x}{2};$$

$$B_4(x) = \frac{30x^4 - 60x^3 + 30x^2 - 1}{30};$$

$$B_5(x) = \frac{6x^5 - 15x^4 + 10x^3 - x}{6}.$$

2.38 The *Euler polynomials* $\{E_n(x)\}$ are defined by the identity

$$\frac{2e^{xt}}{e^t + 1} = \sum_{n=0}^{\infty} \frac{E_n(x)}{n!} t^n, \quad |t| < \pi.$$

Prove that

(a) $E_n'(x) = n\,E_{n-1}(x);$

(b) $E_n(x+1) + E_n(x) = 2\,x^n;$

(c) $1 - 2^n + 3^n - \cdots + (-1)^{m-1}m^n = \dfrac{E_n(1) + (-1)^{m-1}E_n(m+1)}{2}.$

2.39 Verify that the first six Euler polynomials are

$$E_0(x) = 1;$$

$$E_1(x) = \frac{2x-1}{2};$$

$$E_2(x) = x^2 - x;$$

$$E_3(x) = \frac{4x^3 - 6x^2 + 1}{4};$$

$$E_4(x) = x^4 - 2x^3 + x;$$

$$E_5(x) = \frac{2x^5 - 5x^4 + 5x^2 - 1}{2}.$$

2.40 The *Euler numbers* $\{E_n\}$ are defined by

$$E_n = 2^n\,E_n\left(\frac{1}{2}\right).$$

Prove that

$$\sum_{n=0}^{\infty} \frac{E_n}{n!} t^n = \frac{2}{e^t + e^{-t}}, \quad |t| < \frac{\pi}{2}.$$

2.41 Prove that $\zeta(-2m) = 0,\ m = 1, 2, 3 \ldots$

2.10 Summary

2.10.1 The gamma function

The gamma function can be defined by the integral

$$\Gamma(z) = \int_0^\infty e^{-t} t^z \frac{dt}{t}, \quad \operatorname{Re} z > 0.$$

It is holomorphic on the right half-plane, takes the values

$$\Gamma(n) = (n-1)!$$

on the positive integers, and extends to a meromorphic function on the complex plane, with simple poles at the non-positive integers and residue $(-1)^n/n!$ at $-n$. It satisfies the functional equations

$$z\Gamma(z) = \Gamma(z+1), \quad (z)_n \Gamma(z) = \Gamma(z+n), \quad z \neq 0, -1, -2, \ldots,$$

where $(z)_n$ is the shifted factorial

$$(z)_0 = 1, \quad (z)_n = z(z+1) \cdots (z+n-2)(z+n-1), \quad n = 1, 2, \ldots,$$

and has the representation

$$\Gamma(z) = \sum_{n=0}^\infty \frac{(-1)^n}{n!(z+n)} + \int_1^\infty e^{-t} t^z \frac{dt}{t}.$$

The beta function or beta integral is

$$B(a, b) = B(b, a) = \int_0^1 s^{a-1} (1-s)^{b-1} \, ds$$

$$= \int_0^\infty u^a \left(\frac{1}{1+u} \right)^{a+b} \frac{du}{u}, \quad \operatorname{Re} a > 0, \operatorname{Re} b > 0.$$

The identity

$$B(a, b) = \frac{\Gamma(a)\,\Gamma(b)}{\Gamma(a+b)}$$

allows analytic continuation to all $a, b \neq 0, -1, -2, \ldots$ The beta function can be used to show that for any fixed a,

$$\frac{\Gamma(z+a)}{\Gamma(z)} = z^a + O(z^{a-1}) \quad \text{as} \quad \operatorname{Re} z \to +\infty.$$

2.10.2 Euler's product and reflection formulas

Two more characterizations of the gamma function are

$$\Gamma(z) = \lim_{n \to \infty} \frac{(n-1)! \, n^z}{(z)_n}, \quad \operatorname{Re} z > 0;$$

$$= \frac{1}{z} \prod_{n=1}^{\infty} \left(1 + \frac{1}{n}\right)^z \left(1 + \frac{z}{n}\right)^{-1}, \quad z \neq 0, -1, -2, \ldots$$

The reciprocal of the gamma function is an entire function with product representation

$$\frac{1}{\Gamma(z)} = z \, e^{\gamma z} \prod_{n=1}^{\infty} \left(1 + \frac{z}{n}\right) e^{-z/n},$$

where γ is Euler's constant

$$\gamma = \lim_{n \to \infty} \left\{ \sum_{k=1}^{n-1} \frac{1}{k} - \log n \right\}.$$

Euler's reflection formula is

$$\Gamma(z) \, \Gamma(1-z) = \frac{\pi}{\sin \pi z}, \quad z \neq 0, \pm 1, \pm 2, \ldots$$

Hankel's integral formula is

$$\frac{1}{\Gamma(z)} = \frac{1}{2\pi i} \int_C e^t \, t^{-z} \, dt,$$

where C is an oriented contour in the complex plane that begins at $-\infty$, continues on the real axis, circles the origin in the counterclockwise direction, and returns to $-\infty$.

The reflection and product formulas give Euler's product for the sine function

$$\frac{\sin \pi z}{\pi z} = \prod_{n=1}^{\infty} \left(1 - \frac{z^2}{n^2}\right).$$

Taking $z = \frac{1}{2}$ in the reflection formula is one of several ways to obtain the evaluation

$$\Gamma\left(\frac{1}{2}\right) = \sqrt{\pi}.$$

2.10.3 Formulas of Legendre and Gauss

Legendre's duplication formula is

$$\Gamma(2z) = \frac{2^{2z-1}}{\sqrt{\pi}} \, \Gamma(z) \, \Gamma\left(z + \frac{1}{2}\right), \quad z \neq 0, -1, -2, \dots$$

A more general result is due to Gauss:

$$\Gamma(mz) = m^{mz-\frac{1}{2}} \frac{\Gamma(z) \, \Gamma\left(z + \frac{1}{m}\right) \Gamma\left(z + \frac{2}{m}\right) \cdots \Gamma\left(z + \frac{m-1}{m}\right)}{(2\pi)^{\frac{1}{2}(m-1)}}.$$

2.10.4 Two characterizations of the gamma function

Wielandt's characterization: Suppose that G is holomorphic in the half-plane $\{\operatorname{Re} z > 0\}$, bounded on the closed strip $\{1 \leq \operatorname{Re} z \leq 2\}$, $G(1) = 1$, and $z \, G(z) = G(z + 1)$ for $\operatorname{Re} z > 0$. Then $G(z) = \Gamma(z)$.

The gamma function is logarithmically convex on $\{x > 0\}$: (2.2.6) implies that

$$(\log \Gamma)''(x) = \sum_{k=0}^{\infty} \frac{1}{(x+k)^2} > 0 \quad \text{for } x > 0.$$

Conversely, Bohr and Mollerup proved that if a function G is such that $G(x)$ is defined and positive for $x > 0$, $G(1) = 1$, G satisfies the functional equation $x \, G(x) = G(x + 1)$, and $\log G$ is convex, then $G(x) \equiv \Gamma(x)$.

2.10.5 Asymptotics of the gamma function

Stirling's formula is

$$\Gamma(z) = \frac{z^z}{e^z} \left\{ \left(\frac{2\pi}{z}\right)^{\frac{1}{2}} + O\left(z^{-\frac{3}{2}}\right) \right\} \quad \text{as} \quad \operatorname{Re} z \to +\infty.$$

A more precise version is due to Binet:

$$\Gamma(z) = \frac{z^z}{e^z} \left(\frac{2\pi}{z}\right)^{\frac{1}{2}} e^{\theta(z)}, \quad \operatorname{Re} z > 0,$$

where

$$\theta(z) = \int_0^{\infty} \left(\frac{1}{e^t - 1} - \frac{1}{t} + \frac{1}{2}\right) e^{-zt} \, \frac{dt}{t}.$$

The integrand in (2.5.3) includes the function

$$\frac{1}{e^t - 1} - \frac{1}{t} + \frac{1}{2} = \sum_{m=1}^{\infty} \frac{B_{2m}}{(2m)!} t^{2m-1}, \quad |t| < 2\pi,$$

where the coefficients B_{2m} are the Bernoulli numbers. Putting partial sums of this expansion into Binet's formula gives estimates

$$\theta(z) = \sum_{m=1}^{N} \frac{B_{2m}}{2m(2m-1)} z^{1-2m} + O\left(z^{-2N-1}\right), \quad \text{as} \quad \mathrm{Re}\, z \to +\infty.$$

An asymptotic result valid on the complement of the negative real axis is the uniform estimate

$$\Gamma(z) = \frac{z^z}{e^z} \left\{ \left(\frac{2\pi}{z}\right)^{\frac{1}{2}} + O\left(z^{-\frac{3}{2}}\right) \right\}, \quad |\arg z| \le \pi - \delta < \pi, \quad \delta > 0.$$

In particular, for real x and y,

$$|\Gamma(x+iy)| = \sqrt{2\pi}\, |y|^{x-\frac{1}{2}} e^{-\frac{1}{2}\pi|y|} \left[1 + O\left(|y|^{-1}\right)\right], \quad \text{as} \quad |y| \to \infty.$$

2.10.6 The psi function and the incomplete gamma function

The psi function is the logarithmic derivative of the gamma function:

$$\psi(z) = \frac{d}{dz} \log \Gamma(z) = -\gamma + \sum_{n=0}^{\infty} \left(\frac{1}{n+1} - \frac{1}{n+z}\right), \quad z \ne 0, -1, -2, \ldots$$

It satisfies the recurrence relation

$$\psi(z+1) = \psi(z) + \frac{1}{z},$$

so

$$\psi(1) = -\gamma, \quad \psi(k+1) = -\gamma + 1 + \frac{1}{2} + \frac{1}{3} + \cdots + \frac{1}{k}, \quad k = 1, 2, 3, \ldots$$

Binet's integral formula is

$$\psi(z) = \log z - \frac{1}{2z} - \int_0^{\infty} \left(\frac{1}{e^t - 1} - \frac{1}{t} + \frac{1}{2}\right) e^{-zt}\, dt, \quad \mathrm{Re}\, z > 0.$$

The incomplete gamma function is

$$\gamma(\alpha, z) = \int_0^z t^{\alpha-1} e^{-t}\, dt, \quad \mathrm{Re}\, \alpha > 0.$$

It is analytic in the complement of the negative real axis and has the series expansion

$$\gamma(\alpha, z) = z^\alpha \sum_{n=0}^{\infty} \frac{(-z)^n}{n!\,(\alpha+n)}, \quad \alpha \neq 0, -1, -2, \dots$$

for all z.

The complementary incomplete gamma function is

$$\Gamma(\alpha, z) = \int_z^\infty t^{\alpha-1} e^{-t} \, dt$$

$$= \Gamma(\alpha) - \gamma(\alpha, z).$$

The error function and the complementary error function are

$$\operatorname{erf} z = \frac{2}{\sqrt{\pi}} \int_0^z e^{-t^2} \, dt, \quad \operatorname{erfc} z = \frac{2}{\sqrt{\pi}} \int_z^\infty e^{-t^2} \, dt.$$

They satisfy

$$\operatorname{erf} z = 1 - \operatorname{erfc} z = \frac{2}{\sqrt{\pi}} \sum_{n=0}^{\infty} \frac{(-1)^n z^{2n+1}}{n!(2n+1)} = \frac{1}{\sqrt{\pi}} \gamma\left(\frac{1}{2}, z^2\right);$$

$$\operatorname{erfc} z = \frac{1}{\sqrt{\pi}} \Gamma\left(\frac{1}{2}, z^2\right).$$

2.10.7 The Selberg integral

The Selberg integral is

$$S_n(a, b, c) = \int_0^1 \cdots \int_0^1 \prod_{i=1}^n x_i^{a-1} (1-x_i)^{b-1} \prod_{1 \leq j < k \leq n} |x_j - x_k|^{2c} \, dx_1 \cdots dx_n,$$

$$\operatorname{Re} a > 0, \quad \operatorname{Re} b > 0, \quad \operatorname{Re} c > \max\left\{-\frac{1}{n}, -\frac{\operatorname{Re} a}{n-1}, -\frac{\operatorname{Re} b}{n-1}\right\}.$$

The value is given by

$$S_n(a, b, c) = \prod_{j=0}^{n-1} \frac{\Gamma(a + jc)\,\Gamma(b + jc)\,\Gamma(1 + [j+1]c)}{\Gamma(a + b + [n-1+jc])\,\Gamma(1+c)}.$$

Aomoto's generalization, under the same conditions on the indices (a, b, c), is

$$\int_0^1 \cdots \int_0^1 \prod_{i=1}^k x_i \prod_{i=1}^n x_i^{a-1} (1-x_i)^{b-1} \prod_{1 \le i < j \le n} |x_i - x_j|^{2c} \, dx_1 \cdots dx_n$$

$$= \prod_{i=1}^k \frac{a + (n-i)c}{a + b + (2n-i-1)c} \, S_n(a, b, c)$$

$$= \prod_{i=1}^k \frac{a + (n-i)c}{a + b + (2n-i-1)c} \prod_{j=0}^{n-1} \frac{\Gamma(a+jc)\,\Gamma(b+jc)\,\Gamma(1+[j+1]c)}{\Gamma(a+b+[n-1+j]c)\,\Gamma(1+c)}.$$

2.10.8 The zeta function

The Riemann zeta function is

$$\zeta(z) = \sum_{n=1}^\infty \frac{1}{n^z}, \quad \text{Re}\, z > 1,$$

and is also given by Euler's product formula

$$\zeta(z) = \prod_{p \text{ prime}} \left(1 - \frac{1}{p^z}\right)^{-1}, \quad \text{Re}\, z > 1.$$

Euler's evaluation of the zeta function at even positive integers is

$$\zeta(2m) = \frac{(-1)^{m-1}}{2} \frac{(2\pi)^{2m}}{(2m)!} B_{2m},$$

where the B_{2m} are the Bernoulli numbers (2.5.4). In particular,

$$1 + \frac{1}{4} + \frac{1}{9} + \cdots + \frac{1}{n^2} + \cdots = \frac{\pi^2}{6};$$

$$1 + \frac{1}{16} + \frac{1}{81} + \cdots + \frac{1}{n^4} + \cdots = \frac{\pi^4}{90};$$

$$1 + \frac{1}{64} + \frac{1}{729} + \cdots + \frac{1}{n^6} + \cdots = \frac{\pi^6}{945}.$$

An integral representation is

$$\zeta(z) = \frac{1}{\Gamma(z)} \int_0^\infty \frac{t^{z-1}\, dt}{e^t - 1}, \quad \text{Re}\, z > 0.$$

The formula

$$\zeta(z) = -\frac{1}{2i\,\Gamma(z)\sin\pi z} \int_C \frac{(-t)^{z-1}\, dt}{e^t - 1},$$

where C comes from $+\infty$ along the real axis, circles zero in the negative (clockwise) direction, and returns to $+\infty$ along the real axis provides the analytic continuation to $\operatorname{Re} z \le 1$ and shows that the only pole of ζ in the plane is a simple pole at $z = 1$.

The functional equation for the zeta function is

$$\zeta(1 - z) = \frac{2}{(2\pi)^z} \cos\left(\frac{1}{2}\pi z\right) \Gamma(z) \zeta(z).$$

2.11 Remarks

The history and properties of the gamma function are discussed in detail in the book by Nielsen [217]; see also articles by Davis [64], Dutka [76], and Gautschi [109], and the books by Artin [13], Campbell [39], and Godefroy [115]. Copson's text [56] has an extensive list of exercises with identities involving the gamma function and Euler's constant.

It has been remarked that the beta function might better be termed the "beta integral." Selberg's generalization and its further elaborations have been utilized in a number of areas, including random matrix theory, statistical mechanics, combinatorics, and integrable systems. See the chapter on the Selberg integral in [7] and the extensive survey by Forrester and Warnaar [98]. For a probabilistic proof of Selberg's formula, see the book by Mehta [202].

The literature on the zeta function and its generalizations is copious. See in particular the books of Titchmarsh [285], Edwards [79], Ivić [137], Patterson [226], and Motohashi [207]. The celebrated *Riemann hypothesis* is that all the nontrivial zeros of $\zeta(s)$ lie on the line $\left\{\operatorname{Re} s = \frac{1}{2}\right\}$. (The "trivial zeros" are at $s = -2, -4, -6, \ldots$) A number of consequences concerning analytic number theory would follow from the truth of the Riemann hypothesis.

3

Second-order differential equations

As noted in Chapter 1, most of the functions commonly known as "special functions" are solutions of second-order linear differential equations. These equations occur naturally in certain physical and mathematical contexts. In a certain sense there are exactly two (families of) equations in question: the confluent hypergeometric equation (Kummer's equation)

$$xu''(x) + (c - x)u'(x) - a u(x) = 0, \qquad (3.0.1)$$

with indices (a, c), and the hypergeometric equation

$$x(1 - x)u''(x) + \left[c - (a + b + 1)x\right]u'(x) - ab u(x) = 0, \qquad (3.0.2)$$

with indices (a, b, c), where a, b, c are constants. The various other equations (Bessel, Whittaker, Hermite, Legendre, . . .) are obtained from these by specialization, by standard transformations, or by analytic continuation in the independent variable.

In this chapter we give a brief general treatment of some questions concerning second-order linear differential equations, starting with gauge transformations, L^2 symmetry with respect to a weight, and the Liouville transformation.

The basic existence and uniqueness theorems are proved, followed by a discussion of the Wronskian, independence of solutions, comparison theorems, and zeros of solutions.

A natural classification question is treated: classify symmetric problems whose eigenfunctions are polynomials. This question leads, up to certain normalizations, to the equations (3.0.1) and (3.0.2). General results on local maxima and minima of solutions of homogeneous equations are obtained and applied to some of these polynomials.

Another source of equations related to (3.0.1) and (3.0.2) is physics: problems involving the Laplace operator, when one seeks to find solutions by separating variables in special coordinate systems. We discuss the various equations that arise in this way and how they are related to (3.0.1) and (3.0.2).

3.1 Transformations, symmetry

Throughout this chapter, $p, q, r, f, u, v, \ldots$ will denote real functions defined on the finite or infinite open real interval $I = (a, b) = \{x : a < x < b\}$, and we assume that $p(x) > 0$, all $x \in I$. All functions will be assumed to be continuous and to have continuous first and second derivatives as needed.

The general linear second-order differential equation on the interval I is

$$p(x) u''(x) + q(x) u'(x) + r(x) u(x) = f(x). \qquad (3.1.1)$$

The corresponding *homogeneous equation* is the equation with right-hand side $f \equiv 0$:

$$p(x) u''(x) + q(x) u'(x) + r(x) u(x) = 0. \qquad (3.1.2)$$

The associated differential *operator* is

$$L = p(x) \frac{d^2}{dx^2} + q(x) \frac{d}{dx} + r(x). \qquad (3.1.3)$$

In (3.1.3) the functions p, q, and r are identified with the operations of multiplication by p, by q, and by r.

A *gauge transformation* of (3.1.1) is a transformation of the form

$$u(x) = \varphi(x) v(x), \quad \varphi(x) \neq 0. \qquad (3.1.4)$$

The function u satisfies (3.1.1) if and only if v satisfies

$$p(x) v''(x) + \left[2p(x) \frac{\varphi'(x)}{\varphi(x)} + q(x) \right] v'(x)$$

$$+ \left[p(x) \frac{\varphi''(x)}{\varphi(x)} + q(x) \frac{\varphi'(x)}{\varphi(x)} + r(x) \right] v(x) = \frac{f(x)}{\varphi(x)}. \qquad (3.1.5)$$

Note that the left-hand side of this equation is not changed if φ is replaced by $C\varphi$, C constant, $C \neq 0$. The corresponding transformed operator is

$$L_\varphi = p(x) \frac{d^2}{dx^2} + \left[2p(x) \frac{\varphi'(x)}{\varphi(x)} + q(x) \right] \frac{d}{dx}$$

$$+ \left[p(x) \frac{\varphi''(x)}{\varphi(x)} + q(x) \frac{\varphi'(x)}{\varphi(x)} + r(x) \right]. \qquad (3.1.6)$$

The usefulness of gauge transformations comes from the fact that the homogeneous linear first-order differential equation

$$\varphi'(x) = h(x)\,\varphi(x) \tag{3.1.7}$$

always has a solution

$$\varphi(x) = \exp\left\{ \int_{x_0}^x h(y)\,dy \right\}, \tag{3.1.8}$$

where x_0 is any point of the interval I. This solution has no zeros in the interval. Note that if ψ is a second solution of (3.1.7), then the quotient ψ/φ has derivative 0 and therefore is constant.

In particular, a gauge transformation can be used to eliminate the first-order term qu' of (3.1.1) by taking φ such that $\varphi'/\varphi = -q/2p$. A second use is to *symmetrize* the operator (3.1.3). Suppose that $w > 0$ on I. The associated weighted L^2 space L^2_w consists of all measurable real-valued functions f such that

$$\int_a^b f(x)^2\, w(x)\,dx < \infty.$$

The *inner product* (f, g) between two such functions is

$$(f, g) = (f, g)_w = \int_a^b f(x)\,g(x)\,w(x)\,dx.$$

The operator L of (3.1.3) is said to be *symmetric* with respect to the *weight function* w if

$$(Lu, v) = (u, Lv)$$

for every pair of twice continuously differentiable functions u, v that vanish outside some closed subinterval of I. The proofs of the following propositions are sketched in the exercises.

Proposition 3.1.1 *The operator L is symmetric with respect to the weight w if and only if it has the form*

$$L = p\,\frac{d^2}{dx^2} + \frac{(pw)'}{w}\,\frac{d}{dx} + r = \frac{1}{w}\,\frac{d}{dx}\left(pw\,\frac{d}{dx} \right) + r. \tag{3.1.9}$$

Proposition 3.1.2 *If L has the form (3.1.3) then there is a weight function w, unique up to a multiplicative constant, such that L is symmetric with respect to w.*

Proposition 3.1.3 *Given an operator (3.1.3) and a weight function w on the interval I, there is a gauge transformation (3.1.4) such that the corresponding operator L_φ is symmetric with respect to w.*

An invertible transformation T from an L^2 space with weight w_1 to an L^2 space with weight w_2 is said to be *unitary* if

$$(Tf, Tg)_{w_2} = (f, g)_{w_1}$$

for every pair f, g in $L^2_{w_1}$. Operators L_1 and L_2 in the respective spaces are said to be *unitarily equivalent* by T if

$$L_2 = T\, L_1\, T^{-1}.$$

Proposition 3.1.4 *An operator symmetric with respect to a weight w on an interval I is unitarily equivalent, by a gauge transformation, to an operator that is symmetric with respect to the weight 1 on the interval I.*

A second useful method for transforming a differential equation like (3.1.1) is to make a change of the independent variable. If $y = y(x)$ and $u(x) = v\big(y(x)\big)$, then

$$u'(x) = y'(x)\, v'\big(y(x)\big), \quad u''(x) = \big[y'(x)\big]^2 v''\big(y(x)\big) + y''(x)\, v'\big(y(x)\big).$$

In particular, we may eliminate the coefficient p by taking

$$y(x) = \int_{x_0}^x \frac{dt}{\sqrt{p(t)}}. \tag{3.1.10}$$

Then equation (3.1.1) becomes

$$v'' + \left[\frac{q}{\sqrt{p}} - \frac{p'}{2\sqrt{p}} \right] v' + r\, v = f.$$

This involves an abuse of notation: the primes on v refer to derivatives with respect to y, while the prime on p refers to the derivative with respect to x. To rectify this we consider p, q, r, and f as functions of $y = y(x)$ by taking $p(x) = p_1(y(x))$, etc. The previous equation becomes

$$v''(y) + \left[\frac{q_1(y)}{\sqrt{p_1(y)}} - \frac{p_1'(y)}{2p_1(y)} \right] v'(y) + r_1(y)\, v(y) = f_1(y). \tag{3.1.11}$$

If we then eliminate the first-order term by a gauge transformation, the resulting composite transformation is known as the *Liouville transformation*.

3.2 Existence and uniqueness

The following standard fact about equations of the form (3.1.2) is crucial.

Theorem 3.2.1 *The set of solutions of the homogeneous equation*

$$p(x)\,u''(x) + q(x)\,u'(x) + r(x)\,u(x) = 0 \tag{3.2.1}$$

is a vector space of dimension two.

Proof The set of solutions of (3.2.1) is a vector space, since a linear combination of solutions is again a solution (the "superposition principle").

To simplify notation, let us assume that the interval I contains the point $x = 0$. A gauge transformation is an invertible linear map, so it does not change the dimension. Therefore, we may assume that $q \equiv 0$ and write the equation in the form

$$u''(x) = s(x)\,u(x), \quad s(x) = -\frac{r(x)}{p(x)}. \tag{3.2.2}$$

We show first that there are two solutions u and v characterized by the conditions

$$u(0) = 1, \quad u'(0) = 0; \quad v(0) = 0, \quad v'(0) = 1. \tag{3.2.3}$$

Solutions of (3.2.2) that satisfy these conditions would be solutions of the integral equations

$$u(x) = 1 + \int_0^x \int_0^y s(z)\,u(z)\,dz\,dy, \tag{3.2.4}$$

$$v(x) = \int_0^x \left\{ 1 + \int_0^y s(z)\,v(z)\,dz \right\} dy, \tag{3.2.5}$$

respectively. Conversely, solutions of these integral equations would be solutions of (3.2.2).

The equations (3.2.4) and (3.2.5) can be solved by the *method of successive approximations*. Let $u_0 = 1$, $v_0 = 0$, and define inductively

$$u_{n+1}(x) = 1 + \int_0^x \int_0^y s(z)\,u_n(z)\,dz\,dy;$$

$$v_{n+1}(x) = \int_0^x \left\{ 1 + \int_0^y s(z)\,v_n(z)\,dz \right\} dy, \quad n \geq 0.$$

It is enough to show that each of the sequences $\{u_n\}$ and $\{v_n\}$ converges uniformly on each bounded closed subinterval $J \subset I$. We may assume $0 \in J$. Let

$$C = \sup_{x \in J} |s(x)|.$$

It is easily proved by induction that for $x \in J$,

$$|u_{n+1}(x) - u_n(x)| \leq \frac{C^n x^{2n}}{(2n)!}; \tag{3.2.6}$$

$$|v_{n+1}(x) - v_n(x)| \leq \frac{C^n |x|^{2n+1}}{(2n+1)!}.$$

It follows that the sequences $\{u_n\}$ and $\{v_n\}$ are Cauchy sequences. They converge uniformly on J to the desired solutions u and v. The conditions (3.2.3) imply that u and v are linearly independent, so the dimension of the space of solutions of (3.2.2) is at least two.

Suppose now that w is a solution of (3.2.2). Replacing w by

$$w(x) - w(0) u(x) - w'(0) v(x),$$

we may assume that $w(0) = w'(0) = 0$. The proof can be completed by showing that $w \equiv 0$ on each subinterval J as before. Let

$$M = \sup_{x \in J} |w(x)|.$$

Now

$$w(x) = \int_0^x \int_0^y s(z) w(z) \, dz \, dy.$$

It follows that for $x \in J$,

$$|w(x)| \leq \frac{C M x^2}{2}.$$

Inductively,

$$|w(x)| \leq \frac{C^n M x^{2n}}{(2n)!} \tag{3.2.7}$$

for all n. The right-hand side has limit 0 as $n \to \infty$, so $w(x) = 0$. $\qquad\square$

These arguments lead to a sharpened form of Theorem 3.2.2.

Theorem 3.2.2 *Given a point x_0 in the interval I and two constants c_0, c_1, there is a unique solution of the homogeneous equation*

$$p(x) u''(x) + q(x) u'(x) + r(x) u(x) = 0 \tag{3.2.8}$$

that satisfies the conditions

$$u(x_0) = c_0; \quad u'(x_0) = c_1.$$

In particular, if $u(x_0) = 0$, then either it is a simple zero, i.e. $u'(x_0) \neq 0$, or else $u \equiv 0$ on I. Moreover, if u is not identically zero, then wherever $r(x) \neq 0$, zeros of u' are simple: if $u'(x)$ and $u''(x)$ both vanish, then (3.2.8) implies that $u(x) = 0$. This proves the following.

Corollary 3.2.3 *If u is a solution of* (3.2.8) *that does not vanish identically, then any zero of u in I is a simple zero. Moreover, u' has only simple zeros wherever $r \neq 0$.*

3.3 Wronskians, Green's functions, comparison

Suppose that u_1 and u_2 are two differentiable functions on the interval I. The *Wronskian* $W(u_1, u_2)$ is the function

$$W(u_1, u_2)(x) = \begin{vmatrix} u_1(x) & u_2(x) \\ u_1'(x) & u_2'(x) \end{vmatrix} = u_1(x) u_2'(x) - u_1'(x) u_2(x).$$

Proposition 3.3.1 *Suppose that u_1 and u_2 are solutions of the homogeneous equation* (3.1.2). *The Wronskian $W(u_1, u_2)$ is identically zero if u_1 and u_2 are linearly dependent and nowhere zero if u_1 and u_2 are independent.*

Proof The assumption on u_1 and u_2 implies

$$p W' = p (u_1 u_2'' - u_1'' u_2) = -q W,$$

so W is the solution of the first-order homogeneous equation $W' = -qp^{-1}W$. It follows that W is either identically zero or never zero. Clearly, $W \equiv 0$ is implied by linear dependence. Conversely, if $W \equiv 0$, then in any subinterval where $u_1 \neq 0$ we have

$$\left[\frac{u_2}{u_1} \right]' = \frac{W(u_1, u_2)}{u_1^2} = 0.$$

Therefore, u_2/u_1 is constant. □

Let us look for a solution of equation (3.1.1) that has the form

$$u(x) = \int_{x_0}^{x} G(x, y) f(y) \, dy. \tag{3.3.1}$$

Then

$$u'(x) = G(x, x) f(x) + \int_{x_0}^x G_x(x, y) f(y) \, dy.$$

To get rid of $f'(x)$ in taking the second derivative, we need $G(x, x) = 0$. Then

$$u''(x) = G_x(x, x) f(x) + \int_{x_0}^x G_{xx}(x, y) f(y) \, dy.$$

Therefore, $Lu = f$ provided

$$LG(x, \cdot) = 0, \quad G(x, x) = 0; \quad p(x) G_x(x, x) = 1. \tag{3.3.2}$$

Suppose that u_1 and u_2 are linearly independent homogeneous solutions of $Lu = 0$ on the interval. The first equation in (3.3.2) implies that for each $y \in I$,

$$G(x, y) = v_1(y) u_1(x) + v_2(y) u_2(x).$$

The remaining two conditions in (3.3.2) give linear equations whose solution is

$$G(x, y) = \frac{u_1(y)u_2(x) - u_2(y)u_1(x)}{p(y) W(y)}, \tag{3.3.3}$$

where $W = W(u_1, u_2)$ is the Wronskian.

We may now generalize Theorem 3.2.2 to the inhomogeneous case.

Theorem 3.3.2 *Suppose that x_0 is a point of I. For any two real constants c_0, c_1, there is a unique solution u of (3.1.1) that satisfies the conditions*

$$u(x_0) = c_0, \quad u'(x_0) = c_1. \tag{3.3.4}$$

Proof The solution (3.3.1), (3.3.3) satisfies the conditions $u(x_0) = 0$, $u'(x_0) = 0$. We may add to it any linear combination of u_1 and u_2. The Wronskian is not zero, so there is a unique linear combination that yields the conditions (3.3.4). □

In order to satisfy more general boundary conditions, we look for a solution of the form

$$u(x) = \int_{y<x} G_+(x, y) f(y) \, dy + \int_{y>x} G_-(x, y) f(y) \, dy, \tag{3.3.5}$$

where $G_-(\cdot, y)$ satisfies a condition to the left and $G_+(\cdot, y)$ satisfies a condition to the right. If $u_\pm$ are linearly independent solutions that satisfy such conditions, then $G_\pm(x, y) = v_\pm(y)u_\mp(x)$, and the previous argument shows that

$$G_+(x, y) = \frac{u_+(y)\,u_-(x)}{p(y)\,W(y)}, \tag{3.3.6}$$

$$G_-(x, y) = \frac{u_-(y)\,u_+(x)}{p(y)\,W(y)}.$$

The Wronskian also plays a role in the proof of the following important result of Sturm [277].

Theorem 3.3.3 (Sturm comparison theorem) *Suppose that u_1 and u_2 are solutions of the equations*

$$p(x)\,u_j''(x) + q(x)\,u_j'(x) + r_j(x)\,u_j(x) = 0, \quad j = 1, 2, \tag{3.3.7}$$

on the interval I, neither u_1 nor u_2 is identically zero, and

$$r_1(x) < r_2(x), \quad all \quad x \in I.$$

Suppose that $u_1 = 0$ at points c, d in I, $c < d$. Then $u_2(x) = 0$ at some point x of the interval (c, d).

Proof The assumptions and the conclusion are unchanged under gauge transformations and under division of the equations by p, so we may assume for simplicity that $p \equiv 1$ and $q \equiv 0$. We may assume that u_1 has no zeros in (c, d); otherwise, replace d by the first zero. If u_2 has no zeros in (c, d), then up to a change of sign we may assume that u_1 and u_2 are positive in the interval. The Wronskian $W = W(u_1, u_2)$ satisfies

$$W' = (r_1 - r_2)\,u_1\,u_2,$$

so it is non-increasing on the interval. Our assumptions to this point imply that $u_1'(c) > 0$ and $u_1'(d) < 0$, so

$$W(c) = -u_1'(c)u_2(c) \le 0, \quad W(d) = -u_1'(d)u_2(d) \ge 0.$$

It follows that $W \equiv 0$ on (c, d), so u_2/u_1 is constant. But this implies that u_2 satisfies both of the equations (3.3.7), $j = 1, 2$, which is incompatible with the assumptions that $r_1 < r_2$ and $u_2 \ne 0$ on the interval. $\square$

This proof also serves to prove the following generalization:

Theorem 3.3.4 *Suppose that u_1 and u_2 are solutions of the equations*

$$p(x)\,u_j''(x) + q(x)\,u_j'(x) + r_j(x)\,u_j(x) = 0, \quad j = 1, 2,$$

on an open interval I, neither is identically zero, and

$$r_1(x) < r_2(x), \quad all \quad x \in I.$$

Suppose that $u_1(c) = 0$ at a point c in I and that the Wronskian $u_1(x)u_2'(x) - u_2(x)u_1'(x)$ has limit zero as x approaches one of the endpoints of I. Then $u_2(x) = 0$ at some point between c and that endpoint.

Corollary 3.3.5 *Suppose that $u(x, t)$, $0 \le t < T$, is a solution of the equation*

$$p(x)\, u''(x, t) + q(x)\, u'(x, t) + r(x, t)\, u(x, t) = 0,$$

on an open interval I, where the primes denote derivatives with respect to x. Suppose that at a point $a \in I$,

$$u(a, t) \equiv 0 \quad or \quad u'(a, t) \equiv 0.$$

Let $a < x_1(t) < x_2(t) < \ldots$ denote the zeros of $u(x, t)$ to the right of a. If $r(x, t)$ is continuous and increases with t, then $x_k(t)$ decreases as t increases.

Another useful result about zeros is the following.

Theorem 3.3.6 *Suppose that w is positive and r is negative on (c, d), and the real function u satisfies*

$$[w\, u']'(x) + r(x)\, u(x) = 0, \quad c < x < d,$$

and is not identically zero. Then u has at most one zero in (c, d).

If uu' is positive in a subinterval $(c, c + \varepsilon)$ or if $\lim_{x \to d} u(x) = 0$, then u has no zeros in (c, d).

Proof Suppose that $u(a) = 0$ for some a in the interval. Replacing u by its negative if necessary, we may assume that $u'(a) > 0$. Then $u(x)$ and $u'(x)$ are positive on some interval $a < x < b \le d$. The equation shows that wu' is increasing on the interval, so $u' > 0$ on (a, b). Taking b to be maximal with respect to these properties, it is clear that $b = d$. It follows that u has at most one zero in (c, d). This argument also shows that either of the additional conditions implies that there are no zeros. $\qquad\square$

3.4 Polynomials as eigenfunctions

It is of interest to extend the symmetry condition of Section 3.1 to a largest possible "allowable" class of functions. In general this requires the imposition of *boundary conditions*. Suppose that I is a bounded interval (a, b) and suppose that w, w', p, p', q, and r extend as continuous functions on the closed interval $[a, b]$. Suppose that u and v are twice continuously differentiable on (a, b), belong to L_w^2, and suppose that u, u', v, v' are continuous on the closed interval. Suppose also that L is symmetric. A previous calculation shows that

$$(Lu, v) - (u, Lv) = \left(p\, w\, u'\, v - p\, w\, u\, v' \right) \Big|_a^b.$$

If pw vanishes at both endpoints then we do not need additional constraints at the boundary; otherwise additional conditions must be imposed on the functions u, v. Similarly, if I is a semi-infinite interval (a, ∞), conditions must be imposed at $x = a$ unless $pw = 0$ at $x = a$.

Suppose that we have symmetry for such a maximal allowable class of functions. An allowable function u that is not identically zero is an *eigenfunction* for L with *eigenvalue* $-\lambda$ if $Lu + \lambda u = 0$.

If u_1 and u_2 are eigenfunctions with different eigenvalues $-\lambda_1$ and $-\lambda_2$, then

$$-\lambda_1(u_1, u_2) = (Lu_1, u_2) = (u_1, Lu_2) = -\lambda_2(u_1, u_2),$$

so $(u_1, u_2) = 0$: u_1 and u_2 are *orthogonal*.

In a variation on a question of Routh [242] and Bochner [33], we ask under what conditions it is the case that the set of eigenfunctions of L includes polynomials of all degrees, or at least of degrees 0, 1, and 2. The symmetry condition implies that L has the form (3.1.9):

$$p\, \frac{d^2}{dx^2} + \frac{(pw)'}{w}\, \frac{d}{dx} + r.$$

Suppose that there are polynomials of degrees 0, 1, and 2 that are eigenfunctions of L. This is equivalent to assuming that the space of polynomials of degree $\leq k$ is in the domain of L and is taken into itself by L, $k = 0, 1, 2$. In particular, constant functions belong to L_w^2, so w has finite integral:

$$\int_a^b w(x)\, dx < \infty. \tag{3.4.1}$$

Applying L to the constant function $u_0(x) \equiv 1$ gives $Lu_0 = r$, so r must be constant, and (up to translating the eigenvalues by $-r$) we may take $r = 0$. Applying L to $u_1(x) = x$ gives

$$Lu_1 = \frac{(pw)'}{w} = p' + p\, \frac{w'}{w}, \tag{3.4.2}$$

so the last expression must be a polynomial of degree at most 1. Taking $u_2(x) = \frac{1}{2}x^2$,

$$Lu_2 = p + x \left(p' + p\,\frac{w'}{w} \right).$$

Since Lu_2 must be a polynomial of degree at most 2, it follows that p is a polynomial of degree at most 2.

The symmetry condition requires that

$$0 = (Lu, v) - (u, Lv) = \int_a^b \left[pw(u'v - uv') \right]'$$

for every u, v in the domain. As noted above, a necessary condition is that $pw \to 0$ at each endpoint of the interval.

By normalizations (affine maps of the line, multiplication of the weight, the operator, and/or the polynomials by constants), we reduce to five cases, of which two turn out to be vacuous.

Case I: p constant. We take $p(x) \equiv 1$. It follows from (3.4.2) that w'/w has degree at most 1, so we take $w = e^h$ where h is a real polynomial of degree at most 2. After another normalization, $w(x) = e^{-x}$ or $w(x) = e^{\pm x^2}$. In the former case, the condition (3.4.1) requires that I be a proper subinterval, but then the condition that $pw = w$ vanish at finite boundary points cannot be met. In the latter case the endpoint condition forces $I = \mathbf{R} = (-\infty, \infty)$ and the condition (3.4.1) forces the sign choice $w(x) = e^{-x^2}$. Thus, in this case (3.1.9) is the operator

$$L = \frac{d^2}{dx^2} - 2x\,\frac{d}{dx} \quad \text{in} \quad L_w^2(\mathbf{R}); \quad w(x) = e^{-x^2}. \tag{3.4.3}$$

This operator takes the space of polynomials of degree $\le n$ to itself for each n, and it follows that for each n there is a polynomial ψ_n of degree n that is an eigenfunction. Consideration of the degree-n terms of ψ_n and of $L\psi_n$ shows that the eigenvalue is $-2n$:

$$L\,\psi_n + 2n\,\psi_n = 0.$$

Since the eigenvalues are distinct, the ψ_n are orthogonal in $L^2(\mathbf{R}, e^{-x^2}\,dx)$. Up to normalization, the associated orthogonal polynomials ψ_n are the *Hermite polynomials*.

The functions

$$\psi_n(x)\,e^{-\frac{1}{2}x^2}$$

are an orthogonal basis for $L^2(\mathbf{R})$. They are eigenfunctions for the operator

$$\frac{d^2}{dx^2} + (1 - x^2). \tag{3.4.4}$$

Case II: p linear. We may normalize to $p(x) = x$. Then (3.4.2) implies that $w'/w = b + \alpha/x$, so up to scaling, $w = x^\alpha e^{bx}$. The endpoint condition implies that the interval is either $\mathbf{R}_- = (-\infty, 0)$ or $\mathbf{R}_+ = (0, \infty)$ and we may assume $I = \mathbf{R}_+$. Condition (3.4.1) implies $b < 0$ and $\alpha > -1$. We may rescale to have $b = -1$. Thus in this case (3.1.9) is the operator

$$L = x\frac{d^2}{dx^2} + [(\alpha + 1) - x]\frac{d}{dx} \quad \text{in} \quad L_w^2(\mathbf{R}_+); \quad w(x) = x^\alpha e^{-x}, \quad \alpha > -1. \tag{3.4.5}$$

Once again, the space of polynomials of degree $\leq n$ is mapped to itself, so there is a polynomial ψ_n of degree n that is an eigenfunction. Consideration of the degree-n term shows that

$$L\,\psi_n + n\,\psi_n = 0.$$

Therefore the ψ_n are orthogonal. Up to normalization, the ψ_n are the *Laguerre polynomials*. The functions

$$\psi_n(x)\, x^{\frac{1}{2}\alpha} e^{-\frac{1}{2}x}$$

are an orthogonal basis for $L^2(\mathbf{R}_+)$. They are eigenfunctions of the operator

$$x\frac{d^2}{dx^2} + \frac{d}{dx} - \frac{x}{4} - \frac{\alpha^2}{4x} + \frac{\alpha + 1}{2}. \tag{3.4.6}$$

Case III: p quadratic, distinct real roots. We normalize to $p(x) = 1 - x^2$. Then (3.4.2) implies $w'/w = \beta(1 + x)^{-1} - \alpha(1 - x)^{-1}$, for some constants α and β, so the weight function $w(x) = (1 - x)^\alpha(1 + x)^\beta$. The endpoint condition forces $I = (-1, 1)$ and condition (3.4.1) forces $\alpha, \beta > -1$. Thus in this case (3.1.9) is the operator

$$L = (1 - x^2)\frac{d^2}{dx^2} + [\beta - \alpha - (\alpha + \beta + 2)x]\frac{d}{dx} \tag{3.4.7}$$

$$\text{in} \quad L_w^2\big((-1, 1)\big); \quad w(x) = (1 - x)^\alpha(1 + x)^\beta, \quad \alpha, \beta > -1.$$

Again the space of polynomials of degree $\leq n$ is mapped to itself, so there is a polynomial ψ_n of degree n that is an eigenfunction. As before, consideration of the degree-n term shows that

$$L\,\psi_n + n(n + \alpha + \beta + 1)\,\psi_n = 0.$$

Therefore the ψ_n are orthogonal.

Up to normalization, the associated orthogonal polynomials are the *Jacobi polynomials*.

We may rescale the interval by taking $\frac{1}{2}(1-x)$ as the new x variable so that the interval is $(0, 1)$. In the new coordinates, up to a constant factor, the weight is $x^{\alpha}(1-x)^{\beta}$ and the operator is

$$L = x(1-x)\frac{d^2}{dx^2} + [\alpha + 1 - (\alpha + \beta + 2)x]\frac{d}{dx}$$

$$\text{in} \quad L_w^2\big((0,1)\big), \quad w(x) = x^{\alpha}(1-x)^{\beta} \quad \alpha, \beta > -1.$$

This is the hypergeometric operator corresponding to equation (3.0.2) with indices $(a, b, c) = (\alpha + \beta + 1, 0, \alpha + 1)$.

Case IV: p quadratic, distinct complex roots. We normalize to $x^2 + 1$. Then (3.4.2) and the assumption that $w > 0$ implies that w would have the form $w(x) = (1 + x^2)^a$. The endpoint condition rules out bounded intervals I, and condition (3.4.1) rules out unbounded intervals.

Case V: p quadratic, double root. We may take $p(x) = x^2$. Then (3.4.2) implies that w would have the form $w(x) = x^a \exp(b/x)$, and once again the endpoint condition and the condition (3.4.1) cannot both be satisfied.

We have proved the following (somewhat informally stated) result.

Theorem 3.4.1 *Up to normalization, the classical orthogonal polynomials (Hermite, Laguerre, Jacobi) are the only ones that occur as the eigenfunctions for second-order differential operators symmetric with respect to a positive weight.*

(To account for other names: up to certain normalizations, *Gegenbauer* or *ultraspherical polynomials* are Jacobi polynomials with $\alpha = \beta$; *Legendre polynomials* are Jacobi polynomials with $\alpha = \beta = 0$; and *Chebyshev polynomials* are Jacobi polynomials in the two cases $\alpha = \beta = -\frac{1}{2}, \alpha = \beta = \frac{1}{2}$.)

There is a sense in which Case I reduces to two instances of Case II. The operator (3.4.3) and the weight $w(x) = e^{-x^2}$ are left unchanged by the reflection $x \to -x$. Therefore, the even and odd parts of a function in L_w^2 also belong to L_w^2, and (3.4.3) maps even functions to even functions and odd functions to odd functions. An even function f can be written as $f(x) = g(x^2)$ and f belongs to L_w^2 if and only if g^2 is integrable on $(0, \infty)$ with respect to the weight $x^{-\frac{1}{2}} e^{-x}$. An odd function f can be written as $f(x) = xg(x^2)$, and f belongs to L^2 if and only if g^2 is integrable on $(0, \infty)$ with respect to the weight $x^{\frac{1}{2}} e^{-x}$. It follows from these considerations that, up to multiplicative constants, Hermite polynomials of even degree must be Laguerre polynomials in x^2, with index $\alpha = -\frac{1}{2}$, and Hermite polynomials of odd degree, when divided by x, must be Laguerre polynomials in x^2 with index $\alpha = \frac{1}{2}$.

These polynomials are related also by certain limiting relations. In Case III, we may normalize the weight function by taking

$$w_{\alpha,\beta}(x) = \frac{2^{-(\alpha+\beta+1)}}{B(\alpha+1, \beta+1)} (1-x)^{\alpha}(1+x)^{\beta}$$

$$= 2^{-(\alpha+\beta+1)} \frac{\Gamma(\alpha+\beta+2)}{\Gamma(\alpha+1)\Gamma(\beta+1)} (1-x)^{\alpha}(1+x)^{\beta}. \qquad (3.4.8)$$

The change of variables $x = 1 - 2y$ shows that

$$\int_{-1}^{1} w_{\alpha,\beta}(x)\, dx = 1.$$

Now take $\beta = \alpha > 0$ and let $x = y/\sqrt{\alpha}$, so that

$$w_{\alpha,\alpha}(x)\, dx = 2^{-(2\alpha+1)} \frac{\Gamma(2\alpha+2)}{\Gamma(\alpha+1)^2} \left(1 - \frac{y^2}{\alpha}\right)^{\alpha} \frac{dy}{\sqrt{\alpha}},$$

and the interval $-1 < x < 1$ corresponds to the interval $-\sqrt{\alpha} < y < \sqrt{\alpha}$. Taking into account the duplication formula (2.3.1) and (2.1.9), we see that as $\alpha \to +\infty$, the rescaled weight on the right converges to the normalized version of Case I:

$$w_H(x) = \frac{1}{\sqrt{\pi}} e^{-x^2}. \qquad (3.4.9)$$

It follows from a compactness argument, which we omit, that the orthonormal polynomials for the weight (3.4.8) with $\alpha = \beta$ (normalized to have positive leading coefficient) converge, under the change of variables $x = y/\sqrt{\alpha}$, to the orthonormal polynomials for the weight (3.4.9). Thus Hermite polynomials are limits of rescaled equal-index Jacobi polynomials.

To obtain Laguerre polynomials as limits of Jacobi polynomials, we take the version of Case III transferred to the interval $(0, 1)$:

$$\widetilde{w}_{\alpha,\beta} = \frac{1}{B(\alpha+1, \beta+1)} x^{\alpha}(1-x)^{\beta}. \qquad (3.4.10)$$

Assume $\beta > 0$ and make the change of variables $x = y/\beta$, so

$$\widetilde{w}_{\alpha,\beta}(x)\, dx = \frac{\Gamma(\alpha+\beta+2)}{\Gamma(\alpha+1)\Gamma(\beta+1)\beta^{\alpha+1}} y^{\alpha} \left(1 - \frac{y}{\beta}\right)^{\beta} dy.$$

Taking into account (2.1.9), as $\beta \to +\infty$, the rescaled weight on the right converges to the normalized version of Case II:

$$w_\alpha(x) = \frac{1}{\Gamma(\alpha + 1)} \, x^\alpha \, e^{-x}. \tag{3.4.11}$$

Again, the orthonormal polynomials for weight (3.4.10) converge to the orthonormal polynomials for weight (3.4.11) under the change of variables $x = y/\beta$.

3.5 Maxima, minima, estimates

In order to study solutions of the eigenvalue equations

$$p(x)u''(x) + q(x)u'(x) + \lambda u(x) = 0, \quad \lambda > 0, \tag{3.5.1}$$

it is convenient to introduce the auxiliary function

$$V(x) = u(x)^2 + \frac{p(x)}{\lambda} \, u'(x)^2. \tag{3.5.2}$$

Proposition 3.5.1 *The relative maxima of $|u(x)|$ are increasing as x increases on any interval where $p' - 2q > 0$ and decreasing on any interval where $p' - 2q < 0$.*

Proof It follows from earlier results that zeros of u' are simple (u' satisfies an equation of the same form), so they determine relative extrema of u. At a relative extremum of $u(x)$, $V(x) = u(x)^2$. Equation (3.5.1) implies that

$$V'(x) = \frac{p'(x) - 2q(x)}{\lambda} \, u'(x)^2.$$

Thus V is increasing where the function $p' - 2q$ is positive and decreasing where it is negative. $\qquad\square$

A similar idea applies to equations in a somewhat different form.

Proposition 3.5.2 *Suppose that $w(x) > 0$, $r(x) > 0$, and*

$$[w \, u']'(x) + r(x) \, u(x) = 0.$$

Then as x increases, the relative maxima of $|u(x)|$ are increasing in any interval where $(w \, r)' < 0$ and decreasing in any interval where $(w \, r)' > 0$.

Proof Here let

$$W(x) = u(x)^2 + \frac{\left[w(x)\,u'(x)\right]^2}{w(x)\,r(x)}.$$

Then

$$W'(x) = -[wr]'(x)\,\frac{u'(x)^2}{r(x)^2}.$$

At the relative extrema, $u(x)^2 = W(x)$. □

This argument proves the following.

Proposition 3.5.3 *Suppose that $w(x) > 0$, $r(x) > 0$, and*

$$[w\,u']'(x) + r(x)\,u(x) = 0$$

in an interval $a < x \le b$. If $(wr)' \le 0$ in this interval, then

$$u(x)^2 \le u(b)^2 + \frac{w(b)}{r(b)}\,u'(b)^2, \quad a < x \le b.$$

Let us apply these observations to the three cases in Section 3.4. In the Hermite case, $p' - 2q = 4x$, so the relative maxima of $|H_n(x)|$ increase as $|x|$ increases. In the Laguerre case, $p' - 2q = 1 - 2(\alpha + 1 - x) = 2x - (2\alpha + 1)$, so the relative maxima of $|L^{(\alpha)}(x)|$ decrease as x increases so long as $x \le \alpha + \frac{1}{2}$, and increase with x for $x \ge \alpha + \frac{1}{2}$. In the Jacobi case,

$$p'(x) - 2q(x) = -2x - 2[\beta - \alpha - (\alpha + \beta + 2)x]$$

$$= 2[\alpha - \beta + (\alpha + \beta + 1)x].$$

It follows that the relative maxima of $|P_n^{(\alpha,\beta)}|$ are either monotone, if one of α, β is $\ge -\frac{1}{2}$ and the other is $\le -\frac{1}{2}$, or decrease from $x = -1$ until $(\alpha + \beta + 1)x = \beta - \alpha$ and then increase to $x = 1$ if both α and β exceed $-\frac{1}{2}$. Thus

$$\sup_{|x| \le 1} \left| P_n^{(\alpha,\beta)}(x) \right| = \max \left\{ \left| P_n^{(\alpha,\beta)}(-1) \right|, \left| P_n^{(\alpha,\beta)}(1) \right| \right\} \quad \text{if} \quad \max\{\alpha, \beta\} \ge -\frac{1}{2}.$$

The results on relative maxima for Hermite and Laguerre polynomials can be sharpened by using Proposition 3.5.2 with $w(x) \equiv 1$. As noted in Section 3.4, the gauge transformations

$$H_n(x) = e^{x^2/2}\,u_n(x);$$

$$L_n^{(\alpha)}(x) = e^{x/2} x^{-(\alpha+1)/2}\,v_n(x)$$

lead to the equations

$$u_n''(x) + (2n + 1 - x^2) u_n(x) = 0;$$

$$v_n''(x) + \left[\frac{1 - \alpha^2}{4x^2} + \frac{2n + \alpha + 1}{2x} - \frac{1}{4} \right] v_n(x) = 0.$$

It follows from Proposition 3.5.2 that the relative maxima of $|u_n(x)|$ are increasing away from the origin, and that the relative maxima of $|v_n(x)|$ are increasing on the interval

$$x \geq \max\left\{ 0, \frac{\alpha^2 - 1}{2n + \alpha + 1} \right\}.$$

Moreover, Theorem 3.3.6 gives information about the zeros of u_n and v_n, which are the same as those of H_n and $L_n^{(\alpha)}$. Since H_n and $L_n^{(\alpha)}$ are polynomials, u_n and v_n have limit zero as $x \to \infty$. The coefficient $r(x)$ for $u_n(x)$ is negative for $x > \sqrt{2n + 1}$, so $H_n(x) = (-1)^n H_n(-x)$ has no zeros x with $x^2 > 2n + 1$. Similarly, the coefficient $r(x)$ for $v_n(x)$ is negative for

$$x > 2n + \alpha + 1 + \sqrt{(2n + \alpha + 1)^2 + 1 - \alpha^2},$$

so $L_n^{(\alpha)}$ has no zeros in this interval.

3.6 Some equations of mathematical physics

A number of second-order ordinary differential equations arise from partial differential equations of mathematical physics, by separation of variables in special coordinate systems. We illustrate this here for problems involving the *Laplacian* in $\mathbf{R}^3$. In Cartesian coordinates $\mathbf{x} = (x_1, x_2, x_3)$ the Laplacian has the form

$$\Delta = \frac{\partial^2}{\partial x_1{}^2} + \frac{\partial^2}{\partial x_2{}^2} + \frac{\partial^2}{\partial x_3{}^2}.$$

Because it is invariant under translations and rotations, it arises in many physical problems for isotropic media. Four examples are the *heat equation*, or *diffusion equation*

$$v_t(\mathbf{x}, t) = \Delta v(\mathbf{x}, t),$$

the *wave equation*

$$v_{tt}(\mathbf{x}, t) = \Delta v(\mathbf{x}, t),$$

and the *Schrödinger equations* for the quantized harmonic oscillator

$$i v_t = \Delta v - |\mathbf{x}|^2 v \tag{3.6.1}$$

and for the Coulomb potential

$$i v_t = \Delta v - \frac{a}{|\mathbf{x}|} v. \tag{3.6.2}$$

Separating variables, i.e. looking for a solution in the form $v(\mathbf{x}, t) = \varphi(t) u(\mathbf{x})$ leads, after dividing by v, to the four equations

$$\frac{\varphi'(t)}{\varphi(t)} = \frac{\Delta u(\mathbf{x})}{u(\mathbf{x})};$$

$$\frac{\varphi''(t)}{\varphi(t)} = \frac{\Delta u(\mathbf{x})}{u(\mathbf{x})};$$

$$i \frac{\varphi'(t)}{\varphi(t)} = \frac{\Delta u(\mathbf{x})}{u(\mathbf{x})} - |\mathbf{x}|^2;$$

$$i \frac{\varphi'(t)}{\varphi(t)} = \frac{\Delta u(\mathbf{x})}{u(\mathbf{x})} - \frac{a}{|\mathbf{x}|}.$$

In each of these equations the left-hand side is a function of t alone and the right-hand side is a function of $\mathbf{x}$ alone, so each side is constant, say $-\lambda$. Thus we are led to three equations involving the Laplacian, the first of which is known as the *Helmholtz equation*:

$$\Delta u(\mathbf{x}) + \lambda u(\mathbf{x}) = 0; \tag{3.6.3}$$

$$\Delta u(\mathbf{x}) - |\mathbf{x}|^2 u(\mathbf{x}) + \lambda u(\mathbf{x}) = 0; \tag{3.6.4}$$

$$\Delta u(\mathbf{x}) - a \frac{u(\mathbf{x})}{|\mathbf{x}|} + \lambda u(\mathbf{x}) = 0. \tag{3.6.5}$$

The case $\lambda = 0$ of the Helmholtz equation is the *Laplace equation*:

$$\Delta u(\mathbf{x}) = 0. \tag{3.6.6}$$

One approach to these equations is to choose a coordinate system and separate variables once more to find special solutions. One may then try to reconstruct general solutions from the special solutions. For this purpose, in addition to Cartesian coordinates we consider the following.

Spherical coordinates (r, φ, θ):

$$(x_1, x_2, x_3) = (r \cos \varphi \sin \theta, r \sin \varphi \sin \theta, r \cos \theta), \tag{3.6.7}$$

in which the operator Δ takes the form

$$\Delta = \frac{\partial^2}{\partial r^2} + \frac{2}{r}\frac{\partial}{\partial r} + \frac{1}{r^2\sin^2\theta}\frac{\partial^2}{\partial\varphi^2} + \frac{1}{r^2\sin\theta}\frac{\partial}{\partial\theta}\sin\theta\frac{\partial}{\partial\theta}.$$

(This is the notation for spherical coordinates commonly used in physics and often used in applied mathematics. It is common in the mathematical literature to reverse the roles of θ and φ here.)

Cylindrical coordinates (r, θ, z):

$$(x_1, x_2, x_3) = (r\cos\theta, r\sin\theta, z), \tag{3.6.8}$$

in which the operator Δ takes the form

$$\Delta = \frac{\partial^2}{\partial r^2} + \frac{1}{r}\frac{\partial}{\partial r} + \frac{1}{r^2}\frac{\partial^2}{\partial\theta^2} + \frac{\partial^2}{\partial z^2}.$$

Parabolic cylindrical coordinates (ξ, ζ, z):

$$(x_1, x_2, x_3) = \left(\frac{1}{2}[\xi^2 - \zeta^2], \xi\zeta, z\right), \tag{3.6.9}$$

in which the operator Δ takes the form

$$\Delta = \frac{1}{\xi^2 + \zeta^2}\left(\frac{\partial^2}{\partial\xi^2} + \frac{\partial^2}{\partial\zeta^2}\right) + \frac{\partial^2}{\partial z^2}.$$

Separating variables in the Helmholtz equation (3.6.3) in Cartesian coordinates, by setting $u(\mathbf{x}) = u_1(x_1)u_2(x_2)u_3(x_3)$, leads to

$$\frac{u_1''}{u_1} + \frac{u_2''}{u_2} + \frac{u_3''}{u_3} + \lambda = 0.$$

If we rule out solutions with exponential growth, it follows that

$$\frac{u_j''}{u_j} = -k_j^2, \quad k_1^2 + k_2^2 + k_3^2 = \lambda.$$

Therefore the solution is a linear combination of the complex exponentials

$$u(\mathbf{x}) = e^{i\mathbf{k}\cdot\mathbf{x}}, \quad |\mathbf{k}|^2 = \lambda.$$

Separating variables in the Laplace equation (3.6.6) in spherical coordinates by setting $u(\mathbf{x}) = R(r)U(\varphi)V(\theta)$ leads to

$$\frac{r^2R'' + 2rR'}{R} + \left\{\frac{U''}{\sin^2\theta U} + \frac{[\sin\theta V']'}{\sin\theta V}\right\} = 0. \tag{3.6.10}$$

Each of the summands must be constant. It follows that R is a linear combination of powers r^ν and $r^{-1-\nu}$ and

$$\frac{U''}{U} + \left\{\sin\theta \frac{[\sin\theta V']'}{V} + \nu(\nu+1)\sin^2\theta\right\} = 0.$$

Once again each summand is constant, and $U'' = -\mu^2 U$ leads to

$$\frac{1}{\sin\theta}\frac{d}{d\theta}\left\{\sin\theta\frac{dV}{d\theta}\right\} + \left[\nu(\nu+1) - \frac{\mu^2}{\sin^2\theta}\right]V = 0.$$

The change of variables $x = \cos\theta$ converts the preceding equation to the *spherical harmonic equation*:

$$\left[(1-x^2)u'\right]' + \left[\nu(\nu+1) - \frac{\mu^2}{1-x^2}\right]u = 0, \quad 0 < x < 1. \quad (3.6.11)$$

The case $\mu = 0$ (solution-invariant under rotation around the vertical axis) is *Legendre's equation*; the solutions are known as *Legendre functions*.

Separating variables in the Helmholtz equation (3.6.3) in cylindrical coordinates, $u(\mathbf{x}) = R(r)T(\theta)Z(z)$, leads to

$$\frac{r^2R'' + rR'}{R} + \frac{T''}{T} + \frac{r^2Z''}{Z} + \lambda r^2 = 0.$$

It follows that $Z''/Z + \lambda = \mu$ and T''/T are constant. Since T is periodic, $T''/T = -n^2$ for some integer n and

$$r^2R''(r) + rR'(r) + \mu r^2R(r) - n^2R(r) = 0.$$

Assuming that $\mu = k^2$ is positive, we may set $R(r) = u(k^{-1}r)$ and obtain *Bessel's equation*

$$x^2u''(x) + xu'(x) + \left[x^2 - n^2\right]u(x) = 0. \quad (3.6.12)$$

Solutions of (3.6.12) are known as *cylinder functions*.

Separating variables in the Helmholtz equation (3.6.3) in parabolic cylindrical coordinates, $u(\mathbf{x}) = X(\xi)Y(\zeta)Z(z)$, leads to the conclusion that $Z''/Z + \lambda = \mu$ is constant and

$$\frac{X''}{X} + \frac{Y''}{Y} + \mu(\xi^2 + \zeta^2) = 0.$$

It follows that there is a constant ν such that

$$X'' + (\mu\xi^2 - \nu)X = 0 = Y'' + (\mu\zeta^2 + \nu)Y.$$

Assuming that $\mu = -k^2$ is negative, the changes of variables $\xi = k^{-1}x$ and $\zeta = k^{-1}x$ convert these to the standard forms

$$u''(x) - x^2 u(x) \pm \nu\, u(x) = 0. \tag{3.6.13}$$

Solutions are known as *parabolic cylinder functions* or *Weber functions*.

Separating variables in (3.6.4) in Cartesian coordinates leads again to (3.6.13). Separating variables in spherical or cylindrical coordinates leads to equations in the radial variable r which are not of classical type; in parabolic cylinder coordinates equation (3.6.4) does not separate.

Equation (3.6.5) separates in spherical coordinates, leading to an equation in the radial variable:

$$v''(r) + \frac{2}{r}\, v'(r) + \left(\lambda - \frac{a}{r} + \frac{\mu}{r^2}\right) v(r) = 0.$$

Taking $v(r) = r^{-1}w(r)$, followed by a change of scale, converts this to the *Coulomb wave equation*

$$u''(\rho) + \left[1 - \frac{2\eta}{\rho} - \frac{l(l+1)}{\rho^2}\right] u(\rho) = 0. \tag{3.6.14}$$

Solutions are known as *Coulomb wave functions*.

3.7 Equations and transformations

Let us begin with a list of the second-order equations encountered in this chapter. We claimed at the outset that all are related to the pair consisting of the confluent hypergeometric equation

$$xu''(x) + (c - x)\, u'(x) - a\, u(x) = 0 \tag{3.7.1}$$

and the hypergeometric equation

$$x(1 - x)u''(x) + \big[c - (a + b + 1)x\big]\, u'(x) - ab\, u(x) = 0, \tag{3.7.2}$$

in the sense that they can be reduced to one of these equations by a gauge transformation and changes of variables.

Each of the equations has, up to multiplication by a function, the form

$$u'' + \frac{q_0}{p}\, u' + \frac{r_0}{p^2}\, u = 0, \tag{3.7.3}$$

where p and r_0 are polynomials of degree at most 2 and q_0 is a polynomial of degree at most 1. This general form is preserved under a gauge transformation $u(x) = \varphi(x)\, v(x)$,

$$\frac{\varphi'}{\varphi} = \frac{q_1}{p},$$

(3.7.4)

where q_1 is any polynomial of degree at most 1. The polynomial q_1 can be chosen (not necessarily in a unique way) so that the equation for v has the canonical form

$$p\,v'' + q\,v' + \lambda\,v = 0,$$

(3.7.5)

where q has degree at most 1 and λ is constant. To accomplish this one is led to three equations for the two coefficients of q_1 and the constant λ.

Classifying symmetric problems with polynomials as eigenfunctions led to equations of the form

$$u'' - 2xu' + 2\lambda u = 0;$$

(3.7.6)

$$xu'' + (\alpha + 1 - x)u' + \lambda u = 0;$$

(3.7.7)

$$(1 - x^2)u'' + \left[\beta - \alpha - (\alpha + \beta + 2)x\right]u' + \lambda u = 0.$$

(3.7.8)

A unitary equivalence (to a symmetric problem with weight 1) mapped the first two of these three equations to

$$v'' + (2\lambda + 1 - x^2)v = 0;$$

(3.7.9)

$$x\,v'' + v' - \frac{x^2 + \alpha^2}{4x}v + \left(\lambda + \frac{\alpha + 1}{2}\right)v = 0.$$

(3.7.10)

As noted in Section 3.4, letting $x = 1 - 2y$ takes (3.7.8) to the form (3.7.2).

Separating variables in problems involving the Laplacian in special coordinate systems led to the equations

$$(1 - x^2)u'' - 2x\,u' + \lambda u - \mu^2(1 - x^2)^{-1}u = 0;$$

(3.7.11)

$$x^2u'' + xu' + (x^2 - v^2)\,u = 0;$$

(3.7.12)

$$u'' - x^2u \pm v\,u = 0;$$

(3.7.13)

$$x^2u'' + \left[x^2 - 2\eta x - l(l + 1)\right]u = 0.$$

(3.7.14)

The spherical harmonic equation (3.7.11) is in the general form (3.7.3). As noted above, it can be reduced to the canonical form by a gauge transformation, in this case

$$u(x) = \left(1 - x^2\right)^{\frac{1}{2}\mu} v(x),$$

leading to the equation

$$\left(1 - x^2\right) v''(x) - 2(\mu + 1)xv'(x) + \left[\lambda - \mu^2 - \mu\right] v(x) = 0,$$

which is the special case $\alpha = \beta = \mu$ of (3.7.8) and thus a particular case of (3.7.2).

Equation (3.7.7) is a particular case of the confluent hypergeometric equation (3.7.1), so the gauge-equivalent equation (3.7.10) is also.

Bessel's equation (3.7.12) is in the general form (3.7.3) with $p(x) = x$. Corresponding gauge transformations are

$$u(x) = e^{\pm ix} x^\nu v(x),$$

leading to the canonical form

$$x v'' + (2\nu + 1 \pm 2ix) v' \pm i(2\nu + 1) v = 0.$$

Letting $y = \mp 2ix$ converts this, in turn, to

$$y w'' + (2\nu + 1 - y) w' - \left(\nu + \frac{1}{2}\right) w = 0.$$

This is (3.7.1) with $c = 2\nu + 1$, $a = \nu + \frac{1}{2}$.

Up to the sign of the parameter, (3.7.9) and (3.7.13) are identical. Moreover, (3.7.9) is related to (3.7.6) by a gauge transformation. We can relate them to the confluent hypergeometric equation (3.7.1) by noting first that the even and odd parts of a solution $u(x)$ of (3.7.6) are also solutions. Writing an even solution as $u(x) = v(x^2)$ converts (3.7.6) to the form

$$x v'' + \left(\frac{1}{2} - x\right) v' + \frac{1}{2}\lambda v = 0, \qquad (3.7.15)$$

the particular case of (3.7.1) with $c = \frac{1}{2}$, $a = -\frac{1}{2}\lambda$. Writing an odd solution as $u(x) = xv(x^2)$ converts (3.7.6) to the form

$$x v'' + \left(\frac{3}{2} - x\right) v' + \frac{1}{2}(\lambda - 1) v = 0, \qquad (3.7.16)$$

the particular case of (3.7.1) with $c = \frac{3}{2}$, $a = \frac{1}{2}(1 - \lambda)$.

Finally, the gauge transformation $u(x) = e^{ix} x^{l+1} v(x)$ converts (3.7.14) to

$$x v''(x) + (2l + 2 + 2ix) v'(x) + \left[(2l + 2)i - 2\eta\right] v(x) = 0,$$

and the change of variables $v(x) = w(-2ix)$ converts this equation to

$$y w''(y) + (2l + 2 - y)w'(y) - (l + 1 + i\eta) w(y) = 0,$$

which is (3.7.1) with $c = 2l + 2$, $a = l + 1 + i\eta$.

3.8 Exercises

3.1 Verify that equations (3.1.1) and (3.1.5) are equivalent under the gauge transformation (3.1.4).

3.2 Show that if φ and ψ are two solutions of (3.1.7) and $\varphi(x) \neq 0$, $x \in I$, then $\psi(x) \equiv C\,\varphi(x)$, C constant.

3.3 Prove Proposition 3.1.1 by showing first that the symmetry condition is equivalent to

$$0 = (Lu, v) - (u, Lv) = \int_a^b \left[p\big(u'v - uv'\big)' + q\big(u'v - uv'\big) \right] w\,dx$$

$$= \int_a^b \big(u'v - uv'\big)\big[qw - (pw)'\big]\,dx. \tag{3.8.1}$$

In particular, if $u \equiv 1$ wherever $v \neq 0$, then

$$0 = -\int_a^b v'\,[qw - (pw)']\,dx = \int_a^b v\,[qw - (pw)']'\,dx;$$

conclude that $qw - (pw)' = c$, constant. If $u(x) = x$ wherever $v \neq 0$, then

$$0 = c\int_a^b \big(v - xv'\big)\,dx = 2c\int_a^b v\,dx$$

for all such v, so $c = 0$. Therefore symmetry implies $qw = (pw)'$. Conversely, show that the condition $qw = (pw)'$ implies symmetry.

3.4 Prove Proposition 3.1.2 by finding a first-order equation that characterizes w up to a constant.

3.5 Prove Proposition 3.1.3 by finding a first-order equation that characterizes φ up to a constant.

3.6 Prove Proposition 3.1.4 by finding a unitary map that has the form $Tf(x) = h(x)\,f(x)$.

3.7 Prove (3.1.11).

3.8 Complete the Liouville transformation, the reduction of (3.1.1) to an equation of the form

$$w''(y) + r_1(y)\,w(y) + s(y)\,w(y) = f_2(y),$$

by applying a gauge transformation to (3.1.11).

3.9 Deduce Theorem 3.2.2 from the proof of Theorem 3.2.1.

3.10 Prove the estimates (3.2.6) and (3.2.7).

3.11 Find an integral formula for the solution of the equation

$$u''(x) + \lambda^2 u(x) = f(x), \quad -\infty < x < \infty$$

that satisfies the conditions $u(0) = u'(0) = 0$, where λ is a positive constant.

3.12 Find an integral formula for the solution of the equation

$$u''(x) + u(x) = f(x), \quad -\pi < x < \pi$$

that satisfies the conditions $u(-\pi) = 0 = u(\pi)$.

3.13 Suppose that $I = (a, b)$ is a bounded interval of length $L = b - a$, suppose that $r(x)$ is continuous on I and $|r(x)| \leq C$, all $x \in I$. Suppose that $\lambda > 0$ and u is a nonzero real solution of

$$u''(x) + \left[r(x) + \lambda\right] u(x) = 0$$

on I. Let $N(\lambda)$ denote the number of zeros of u in the interval. Use Theorem 3.3.3 to prove that for sufficiently large λ,

$$\left| N(\lambda) - \frac{\sqrt{\lambda}\, L}{\pi} \right| < 2.$$

3.14 Prove Theorem 3.3.4.

3.15 Prove Corollary 3.3.5.

3.16 Verify that the gauge transformation

$$H_n(x) = e^{\frac{1}{2}x^2} h_n(x)$$

converts the equation for the Hermite function H_n to the equation

$$h_n''(x) + (2n + 1)h_n(x) = x^2 h_n(x).$$

Deduce from this that h_n is the solution of the integral equation

$$h_n(x) = A_n \cos\left(\sqrt{2n+1}\, x + b_n\right)$$

$$+ y^2 \int_0^x \frac{\sin\left[\sqrt{2n+1}\,(x-y)\right]}{\sqrt{2n+1}} h_n(y)\, dy$$

for some choice of the constants A_n and b_n; determine these constants. As shown in Chapter 10, this implies the asymptotic result

$$h_n(x) = A_n \left[\cos\left(\sqrt{2n+1}\, x + b_n\right) + O\left(n^{-\frac{1}{2}}\right)\right]$$

as $n \to \infty$.

3.17 Determine the Liouville transformation that reduces the equation (3.7.7) for Laguerre polynomials to an equation of the form

$$v''(y) + \lambda \, v(y) = r(y) \, v(y).$$

Find the form of an integral equation for the corresponding modified Laguerre functions $l_n^{(\alpha)}(y)$ analogous to that for the modified Hermite functions $h_n(x)$. Can this be used to obtain an asymptotic result? (Specifically, can the constants A_n, b_n be determined?)

3.18 Determine the Liouville transformation that reduces the equation (3.7.8) for Jacobi polynomials to an equation of the form

$$v''(y) + \lambda \, v(y) = r(y) \, v(y).$$

3.19 Verify that an equation in the general form described in connection with (3.7.3) can be reduced to the canonical form (3.7.5) by a gauge transformation as described in connection with (3.7.4).

3.20 Show that Riccati's equation [238]

$$u'(x) + u(x)^2 + x^m = 0$$

can be converted to a second-order linear equation by setting $u(x) = u_1'(x)/u_1(x)$. Show that the change of variables $u_1(x) = u_2(y)$ with

$$y = \frac{2 \, x^{(m+2)/2}}{m+2}$$

leads to the equation

$$u_2''(y) + \frac{m}{(m+2)\,y} \, u_2'(y) + u_2(y) = 0.$$

3.21 Show that eliminating the first-order term in the last equation in Exercise 3.20 leads to the equation

$$u_3''(y) + \left[1 + \frac{r - r^2}{y^2} \right] u_3(y) = 0, \qquad r = \frac{m}{2m+4}.$$

Show that the gauge transformation $u_3(y) = y^{\frac{1}{2}} v(y)$ converts this to Bessel's equation (3.6.12) with $n = 1/(m+2)$. Combined with results from Section 7.1, this proves a result of Daniel Bernoulli [24]: solutions of Riccati's equation can be expressed in terms of elementary functions when $2/(m+2)$ is an odd integer.

3.9 Summary

Various real functions defined on a finite or infinite real interval $I = (a, b)$ are denoted by p, q, r, f, g, w, ... The functions p and w are assumed to be positive on I.

3.9.1 Transformations, symmetry

The general linear second-order differential equation on the interval I is

$$p(x) u''(x) + q(x) u'(x) + r(x) u(x) = f(x).$$

The corresponding homogeneous equation is the equation with right-hand side $f \equiv 0$:

$$p(x) u''(x) + q(x) u'(x) + r(x) u(x) = 0.$$

A gauge transformation is a transformation of the form

$$u(x) = \varphi(x) v(x), \quad \varphi(x) \neq 0.$$

Then u satisfies the equation above if and only if v satisfies

$$p(x) v''(x) + \left[2p(x) \frac{\varphi'(x)}{\varphi(x)} + q(x) \right] v'(x)$$

$$+ \left[p(x) \frac{\varphi''(x)}{\varphi(x)} + q(x) \frac{\varphi'(x)}{\varphi(x)} + r(x) \right] v(x) = \frac{f(x)}{\varphi(x)}.$$

The homogeneous linear first-order differential equation $\varphi'(x) = h(x) \varphi(x)$ has a solution, unique up to a multiplicative constant,

$$\varphi(x) = \exp \left\{ \int_{x_0}^{x} h(y) \, dy \right\}, \quad x_0 \in I.$$

This solution has no zeros in the interval. Therefore a gauge transformation can be used to eliminate the first-order term $q u'$ or to symmetrize the operator L.

The weighted L^2 space L_w^2 has inner product

$$(f, g) = (f, g)_w = \int_{a}^{b} f(x) g(x) w(x) \, dx.$$

The operator L is symmetric with respect to the weight w if $(Lu, v) = (u, Lv)$ for every pair of twice continuously differentiable functions u, v that vanish outside some closed subinterval of I. An equivalent condition is that

$$L = p\frac{d^2}{dx^2} + \frac{(pw)'}{w}\frac{d}{dx} + r = \frac{1}{w}\frac{d}{dx}\left(pw\frac{d}{dx}\right) + r.$$

Given L, there is always a weight function, unique up to a multiplicative constant, such that L is symmetric with respect to w.

The coefficient p can be eliminated by a change of variables

$$y(x) = \int_{x_0}^{x} \frac{dt}{\sqrt{p(t)}}.$$

Then the original equation becomes

$$v'' + \left[\frac{q}{\sqrt{p}} - \frac{p'}{2\sqrt{p}}\right]v' + r\,v = f.$$

The first-order term can be eliminated by a gauge transformation; the resulting composite transformation is the Liouville transformation.

3.9.2 Existence and uniqueness

The set of solutions of the homogeneous equation

$$p(x)\,u''(x) + q(x)\,u'(x) + r(x)\,u(x) = 0$$

is a vector space of dimension two.

Given a point x_0 in the interval I and two constants c_0, c_1, there is a unique solution of the homogeneous equation that satisfies the conditions

$$u(x_0) = c_0; \quad u'(x_0) = c_1.$$

It follows that if u is a solution that does not vanish identically, then any zero of u in I is a simple zero.

3.9.3 Wronskians, Green's functions, comparison

The Wronskian of two functions is

$$W(u_1, u_2)(x) = \begin{vmatrix} u_1(x) & u_2(x) \\ u_1'(x) & u_2'(x) \end{vmatrix} = u_1(x)\,u_2'(x) - u_1'(x)\,u_2(x).$$

If u_1 and u_2 are two solutions of the homogeneous equation, the Wronskian is either identically zero, if the solutions are linearly dependent, or never zero, if they are independent.

The equation $pu'' + qu' + ru = f$ has a solution

$$u(x) = \int_{x_0}^{x} G(x, y) f(y) \, dy,$$

with the Green's function G given by

$$G(x, y) = \frac{u_1(y)u_2(x) - u_2(y)u_1(x)}{p(y) \, W(y)},$$

where u_1 and u_2 are independent solutions of the homogeneous equation and $W = W(u_1, u_2)$ is the Wronskian. A consequence is the general existence and uniqueness theorem for the inhomogeneous equation $pu'' + qu' + ru = f$ with conditions

$$u(x_0) = c_0, \quad u'(x_0) = c_1.$$

To satisfy more general boundary conditions, we use the form

$$u(x) = \int_{y<x} G_+(x, y) f(y) \, dy + \int_{y>x} G_-(x, y) f(y) \, dy,$$

where

$$G_+(x, y) = \frac{u_+(y) \, u_-(x)}{p(y) \, W(y)}, \quad G_-(x, y) = \frac{u_-(y) \, u_+(x)}{p(y) \, W(y)}.$$

Zeros of solutions of the homogeneous equation with different zero-order coefficients can be compared using Sturm's theorem: if u_1 and u_2 are solutions of the equations

$$p(x) \, u_j''(x) + q(x) \, u_j'(x) + r_j(x) \, u_j(x) = 0, \quad j = 1, 2$$

on the interval I, neither is identically zero, and

$$r_1(x) < r_2(x), \quad \text{all} \quad x \in I,$$

then between any two zeros of u_1 in I there is a zero of u_2.

Zeros of solutions of equations in the form

$$[w \, u']'(x) + r(x) \, u(x) = 0$$

can be located by noting that if $r < 0$ on an interval (c, d) and $u(x) \to 0$ at d and u is not identically zero, or if $u(x)u'(x) > 0$ for x close to c, then u has no zeros in (c, d). Otherwise there is at most one zero in (c, d).

3.9.4 Polynomials as eigenfunctions

To extend the symmetry condition for a second-order differential operator L to a larger class of functions in the weighted L^2 space L_w^2 requires imposing conditions such as $pw \to 0$ at the boundary points. An allowable function $u \not\equiv 0$ is an eigenfunction for L with eigenvalue $-\lambda$ if $Lu + \lambda u = 0$. Eigenfunctions corresponding to different eigenvalues are orthogonal.

Up to certain normalizations (linear transformations, translations, multiplications by constants) there are only three cases of weights and symmetric operators for which there are polynomials of degrees 0, 1, and 2 that are eigenfunctions. In each case the set of eigenfunctions consists of polynomials of all degrees:

Case I: $I = (-\infty, +\infty)$, $w(x) = \exp\left(-x^2\right)$,

$$L = \frac{d^2}{dx^2} - 2x \frac{d}{dx}, \quad \text{in} \quad L^2\left(\mathbf{R}, e^{-x^2} dx\right).$$

The eigenfunctions $\{\psi_n\}$ have eigenvalues $\{-2n\}$. Up to normalization, they are the Hermite polynomials. The functions

$$\psi_n(x) e^{-\frac{1}{2}x^2}$$

are an orthogonal basis for $L^2(\mathbf{R}, dx)$. They are eigenfunctions for the operator

$$\frac{d^2}{dx^2} + \left(1 - x^2\right). \tag{3.9.1}$$

Case II: $I = (0, +\infty)$, $w(x) = x^\alpha \exp(-x)$, $\alpha > -1$,

$$L = x \frac{d^2}{dx^2} + [(\alpha + 1) - x] \frac{d}{dx} \quad \text{in} \quad L^2\left(\mathbf{R}_+, x^\alpha e^{-x} dx\right).$$

The eigenfunctions $\{\psi_n\}$ have eigenvalues $\{-n\}$. Up to normalization, they are the Laguerre polynomials. The functions

$$\psi_n(x) x^{\frac{1}{2}\alpha} e^{-\frac{1}{2}x}$$

are an orthogonal basis for $L^2(\mathbf{R}_+, dx)$. They are eigenfunctions of the operator

$$x \frac{d^2}{dx^2} + \frac{d}{dx} - \frac{x}{4} - \frac{\alpha^2}{4x} + \frac{\alpha + 1}{2}.$$

Case III: $I = (-1, 1)$, $w(x) = (1-x)^{\alpha}(1+x)^{\beta}$, $\alpha, \beta > -1$,

$$L = (1-x^2)\frac{d^2}{dx^2} + [\beta - \alpha - (\alpha + \beta + 2)x]\frac{d}{dx}.$$

The eigenfunctions $\{\psi_n\}$ have eigenvalues $\{-n(n + \alpha + \beta + 1)\}$. Up to normalization, they are the Jacobi polynomials. We may rescale the interval by taking $\frac{1}{2}(1-x)$ as the new x-variable so that the interval is $(0, 1)$. In the new coordinates, up to a constant factor, the weight is $x^{\alpha}(1-x)^{\beta}$ and the operator is

$$L = x(1-x)\frac{d^2}{dx^2} + [\alpha + 1 - (\alpha + \beta + 2)x]\frac{d}{dx}.$$

This is the hypergeometric operator corresponding to equation (3.0.2) with indices $(a, b, c) = (\alpha + \beta + 1, 0, \alpha + 1)$.

Up to certain normalizations, Gegenbauer or ultraspherical polynomials are Jacobi polynomials with $\alpha = \beta$, Legendre polynomials are Jacobi polynomials with $\alpha = \beta = 0$, and Chebyshev polynomials are Jacobi polynomials in the two cases $\alpha = \beta = -\frac{1}{2}$, $\alpha = \beta = \frac{1}{2}$.

A Hermite polynomial of even degree is a multiple of a Laguerre polynomial of index $\alpha = -\frac{1}{2}$ in x^2; a Hermite polynomial of odd degree is the product of x and a Laguerre polynomial of index $\alpha = \frac{1}{2}$ in x^2.

If the weight function in Case III is normalized to have mass 1, then the limit $\alpha = \beta \to +\infty$, $x = y/\sqrt{\alpha}$ is the normalized weight for Case I. A consequence is that the Hermite polynomial H_n is a limit of rescaled Jacobi polynomials $P_n^{(\alpha,\alpha)}$. If Case III is transferred to the interval $(0, 1)$ and normalized, the limit $\beta \to +\infty$, $x = y/\beta$ is the normalized weight for Case II. A consequence is that the Laguerre polynomial $L_n^{(\alpha)}$ is a limit of rescaled Jacobi polynomials $P_n^{(\alpha,\beta)}$. For specifics, see Section 4.2.

3.9.5 Maxima, minima, estimates

The relative maxima of $|u(x)|$, where

$$p(x)u''(x) + q(x)u'(x) + \lambda u(x) = 0, \quad \lambda > 0,$$

are increasing on any interval where $p' - 2q$ is positive and decreasing on any interval where $p' - 2q$ is negative.

If $w(x)$ is positive and

$$\left[w\,u'\right]'(x) + r(x)\,u(x) = 0,$$

then the relative maxima of $|u(x)|$ increase with x where $(wr)' < 0$ and decrease as x increases where $(wr)' > 0$.

Consequences of these results and Theorem 3.3.6 for the three cases in Section 3.4:

$$\sup_{|x|\leq 1} \left|P_n^{(\alpha,\beta)}(x)\right| = \max\left\{\left|P_n^{(\alpha,\beta)}(-1)\right|, \left|P_n^{(\alpha,\beta)}(1)\right|\right\} \quad \text{if } \alpha \geq -\frac{1}{2} \text{ or } \beta \geq -\frac{1}{2}.$$

The relative maxima of $|e^{-x^2/2}H_n(x)|$ increase as $|x|$ increases. The relative maxima of $|e^{-x/2}x^{(\alpha+1)/2}L_n^{(\alpha)}(x)|$ increase with x on the interval

$$x \geq \max\left\{0, \frac{\alpha^2 - 1}{2n + \alpha + 1}\right\}.$$

The zeros of $H_n(x)$ lie in the interval $x^2 \leq 2n + 1$, and the zeros of $L_n^{(\alpha)}(x)$ lie in the interval

$$x \leq 2n + \alpha + 1 + \sqrt{(2n + \alpha + 1)^2 + 1 - \alpha^2}.$$

3.9.6 Some equations of mathematical physics

The Laplacian in Cartesian coordinates is

$$\Delta = \frac{\partial^2}{\partial x_1{}^2} + \frac{\partial^2}{\partial x_2{}^2} + \frac{\partial^2}{\partial x_3{}^2}.$$

It occurs in Laplace's equation and the Helmholtz equation:

$$\Delta u(\mathbf{x}) = 0,$$

$$\Delta u(\mathbf{x}) + \lambda u(\mathbf{x}) = 0;$$

the heat (diffusion) equation and the wave equation:

$$v_t(\mathbf{x}, t) = \Delta v(\mathbf{x}, t),$$

$$v_{tt}(\mathbf{x}, t) = \Delta v(\mathbf{x}, t);$$

and the Schrödinger equations for the quantized harmonic oscillator and for the Coulomb potential:

$$iv_t = \Delta v - |\mathbf{x}|^2 v,$$

$$iv_t = \Delta v - \frac{a}{|\mathbf{x}|} v.$$

Various linear second-order ordinary differential equations are obtained by writing such equations in special coordinate systems and separating variables (looking for solutions that are products of functions of a single coordinate).

In spherical coordinates (r, φ, θ), cylindrical coordinates (r, θ, z), and parabolic cylindrical coordinates (ξ, ζ, z), respectively,

$$\Delta = \frac{\partial^2}{\partial r^2} + \frac{2}{r}\frac{\partial}{\partial r} + \frac{1}{r^2 \sin^2 \theta}\frac{\partial^2}{\partial \varphi^2} + \frac{1}{r^2 \sin \theta}\frac{\partial}{\partial \theta}\sin\theta\frac{\partial}{\partial \theta};$$

$$= \frac{\partial^2}{\partial r^2} + \frac{1}{r}\frac{\partial}{\partial r} + \frac{1}{r^2}\frac{\partial^2}{\partial \theta^2} + \frac{\partial^2}{\partial z^2};$$

$$= \frac{1}{\xi^2 + \zeta^2}\left(\frac{\partial^2}{\partial \xi^2} + \frac{\partial^2}{\partial \zeta^2}\right) + \frac{\partial^2}{\partial z^2}.$$

Separating variables in the Helmholtz equation in Cartesian coordinates leads to the plane-wave solutions

$$u(\mathbf{x}) = e^{i\mathbf{k}\cdot\mathbf{x}}, \quad |\mathbf{k}|^2 = \lambda.$$

Separating variables in the Laplace equation in spherical coordinates leads to the spherical harmonic equation

$$\left[(1 - x^2)u'\right]' + \left[\nu(\nu + 1) - \frac{\mu^2}{1 - x^2}\right]u = 0, \quad 0 < x < 1.$$

The case $\mu = 0$ is Legendre's equation; the solutions are known as Legendre functions.

Separating variables in the Helmholtz equation in cylindrical coordinates leads to Bessel's equation

$$x^2 u''(x) + x u'(x) + \left[x^2 - n^2\right]u(x) = 0.$$

Solutions are known as cylinder functions.

Separating variables in the Helmholtz equation (3.6.3) in parabolic cylindrical coordinates leads to

$$u''(x) - x^2 u(x) \pm \nu u(x) = 0.$$

Solutions are known as parabolic cylinder functions or Weber functions.

Separating variables in the Coulomb equation (3.6.5) leads to

$$u''(\rho) + \left[1 - \frac{2\eta}{\rho} - \frac{l(l + 1)}{\rho^2}\right]u(\rho) = 0.$$

Solutions are known as Coulomb wave functions.

3.9.7 Equations and transformations

Classifying symmetric problems with polynomials as eigenfunctions led to equations of the form

$$u'' - 2xu' + 2\lambda u = 0;$$

$$xu'' + (\alpha + 1 - x)u' + \lambda u = 0;$$

$$(1 - x^2)u'' + \left[\beta - \alpha - (\alpha + \beta + 2)x\right]u' + \lambda u = 0.$$

The second of these equations is the confluent hypergeometric equation. Gauge transformations take the first of these three equations to

$$v'' + \left(2\lambda + 1 - x^2\right)v = 0$$

and the second, which is the confluent hypergeometric equation, to

$$x v'' + v' - \frac{x^2 + \alpha^2}{4x} v + \left(\lambda + \frac{\alpha + 1}{2}\right) v = 0.$$

As noted in Section 3.4, letting $x = 1 - 2y$ takes the third equation to the hypergeometric equation.

Separating variables in problems involving the Laplacian in special coordinate systems led to

$$\left(1 - x^2\right)u'' - 2x\,u' + \lambda\,u - \mu^2\left(1 - x^2\right)^{-1}u = 0;$$

$$x^2 u'' + xu' + \left(x^2 - \nu^2\right)u = 0;$$

$$u'' - x^2 u \pm \nu\,u = 0;$$

$$x^2\,u'' + \left[x^2 - 2\eta x - l(l + 1)\right]u = 0.$$

A gauge transformation takes the first of these to

$$\left(1 - x^2\right)v''(x) - 2(\mu + 1)xv'(x) + \left[\lambda - \mu^2 - \mu\right]v(x) = 0,$$

and taking $x = 1 - 2y$ gives a hypergeometric equation. The second is Bessel's equation. A gauge transformation and an imaginary coordinate change convert it to a special case of the confluent hypergeometric equation. The third equation is the parabolic cylinder equation, which is gauge equivalent to the first equation in the first set of three equations above. These two can also be related to special cases of the confluent hypergeometric equation by considering even and odd solutions. A gauge transformation and an imaginary coordinate transformation convert the fourth equation, the Coulomb wave equation, to the confluent hypergeometric equation.

3.10 Remarks

The general theory of second-order equations is covered in most textbooks of ordinary differential equations. Three classic texts are Forsyth [99], Ince [135], and Coddington and Levinson [55]. The book by Ince has an extensive discussion of the classification of second-order linear equations with rational coefficients. The book by Hille [129] is concerned specifically with equations in the complex domain. For some indication of the modern ramifications of the study of equations in the complex domain, see the survey article by Varadarajan [296]. Explicit solutions for many second-order equations are collected in the handbooks by Kamke [150], Sachdev [244], and Zwillinger [322].

The idea of solving partial differential equations by separation of variables developed throughout the 18th century, e.g. in the work of D. Bernoulli, D'Alembert, and Euler. Fourier [100] was the first to put all the ingredients of the method in place. For a discussion of this, Sturm–Liouville theory, and other developments, see Painlevé's survey [225] and Lützen [191].

Separation of variables for the Laplace, Helmholtz, and other equations is treated in detail by Miller [205]; see also Müller [209]. There are other coordinate systems in which one can separate variables for the Helmholtz or Laplace equations, but the functions that arise are not among those treated here: Mathieu fuctions, modified Mathieu functions, prolate spheroidal functions, etc.

The Wronskian appears in [134]. The concept of linear dependence of solutions and the connection with the Wronskian goes back to Christoffel [52]. The idea of the Green's function goes back to Green in 1828 [119]. The use of the method of successive approximation to prove existence of solutions originated with Liouville [186] and was developed in full generality by Picard [230]. The Liouville transform was introduced in [187]. Sturm's comparison theorem appeared in [276]. The systematic study of orthogonal functions is due to Murphy [211]. Techniques for estimating relative extrema were developed by Stieltjes [272] for Legendre polynomials, Sonine for Bessel functions [265], and others.

For extensive coverage of the 19th-century history, see [225] and Part 2 of [198].

4

Orthogonal polynomials

It was shown in Chapter 3 that there are three cases in which the eigenfunctions of a second-order ordinary differential operator that is symmetric with respect to a weight are polynomials. The polynomials in the three cases are the classical orthogonal polynomials: Hermite polynomials, Laguerre polynomials, and Jacobi polynomials.

Each of these sets of polynomials is an example of a family of polynomials that are orthogonal with respect to an inner product that is induced by a positive weight function on an interval of the real line. The basic theory of general orthogonal polynomials is covered in the first section: expressions as determinants, three-term recurrence relations, properties of the zeros, and so on. It is shown that under a certain condition on the weight, which is satisfied in each of the three classical cases, each element of the L^2 space can be expanded in a series using the orthogonal polynomials, analogous to the Fourier series expansion.

We then examine some features common to the three classical cases, including Rodrigues formulas and representations as integrals. In succeeding sections each of the three classical cases is considered in more detail, as well as some special cases of Jacobi polynomials (Legendre and Chebyshev polynomials). The question of pointwise convergence of the expansion in orthogonal polynomials is addressed.

Finally we return to integral representations and the construction of a second solution of each of the differential equations.

4.1 General orthogonal polynomials

We begin with some properties common to any sequence of orthogonal polynomials. Let $w(x)$ be a positive weight function on an open interval $I = (a, b)$ and assume that the moments

$$A_n = \int_a^b x^n \, w(x) \, dx, \quad n = 0, 1, 2, \ldots$$

are finite.

Let $\Delta_{-1} = 1$ and let Δ_n, $n \geq 0$, be the determinant

$$\Delta_n = \begin{vmatrix} A_0 & A_1 & \ldots & A_n \\ A_1 & A_2 & \ldots & A_{n+1} \\ & & \ddots & \\ A_n & A_{n+1} & \ldots & A_{2n} \end{vmatrix}. \tag{4.1.1}$$

The associated quadratic form

$$\sum_{j,k=0}^{n} A_{j+k} \, a_j \, a_k = \int_a^b \left[\sum_{j=0}^{n} a_j x^j \right]^2 w(x) \, dx$$

is positive definite, so the determinant Δ_n is positive.

Consider the Hilbert space L_w^2, with inner product

$$(f, g) = (f, g)_w = \int_a^b f(x) \, g(x) \, w(x) \, dx.$$

The polynomial

$$Q_n(x) = \begin{vmatrix} A_0 & A_1 & \ldots & A_{n-1} & 1 \\ A_1 & A_2 & \ldots & A_n & x \\ & & \ddots & & \\ A_n & A_{n+1} & \ldots & A_{2n-1} & x^n \end{vmatrix} \tag{4.1.2}$$

is orthogonal to x^m, $m < n$, while $(Q_n, x^n) = \Delta_n$. To see this, expand the determinant (4.1.2) along the last column. Computing the inner product of Q_n with x^m results in a determinant in which the last column of the determinant (4.1.1) has been replaced by column $m + 1$ of the same determinant. Thus if $m < n$ there is a repeated column, while $m = n$ gives Δ_n. Now $Q_n(x) = \Delta_{n-1} x^n$ modulo terms of lower degree, so

$$(Q_n, Q_n) = \left(Q_n, \Delta_{n-1} x^n \right) = \Delta_{n-1} \left(Q_n, x^n \right) = \Delta_{n-1} \, \Delta_n.$$

Therefore the polynomials

$$P_n(x) = \frac{1}{\sqrt{\Delta_{n-1} \Delta_n}} \, Q_n(x)$$

are orthonormal. They are uniquely determined by the requirement that the leading coefficient be positive. The leading coefficient of P_n is $h_n = \sqrt{\Delta_{n-1}/\Delta_n}$.

Note that $x P_n(x)$ has degree $n + 1$ and is orthogonal to x^m, $m < n - 1$. It follows that for some constants a_n, b_n, c_n,

$$x\, P_n(x) = a_n P_{n+1}(x) + b_n P_n(x) + c_n P_{n-1}(x). \tag{4.1.3}$$

Comparing coefficients of x^{n+1}, we see that $a_n = h_n/h_{n+1}$. On the other hand, taking the inner product with P_{n-1}, we have

$$c_n = (x P_n, P_{n-1}) = (P_n, x P_{n-1}) = \frac{h_{n-1}}{h_n} = a_{n-1}.$$

For later use, we note that the existence of a three-term recurrence formula of the form (4.1.3) depends only on the orthogonality properties of the P_n, not on the fact that they have norm one (so long as we do not require that $c_n = a_{n-1}$, as we do in the following calculation).

It follows from the previous two equations that

$$(x - y)\, P_n(x)\, P_n(y) = a_n \big[P_{n+1}(x) P_n(y) - P_n(x) P_{n+1}(y) \big]$$
$$- a_{n-1} \big[P_n(x) P_{n-1}(y) - P_{n-1}(x) P_n(y) \big].$$

Iterating, summing, and dividing by $x - y$, we get the *Christoffel–Darboux formula*

$$a_n \left[\frac{P_{n+1}(x) P_n(y) - P_n(x) P_{n+1}(y)}{x - y} \right] = \sum_{j=0}^{n} P_j(x) P_j(y). \tag{4.1.4}$$

Recall that the coefficient on the left is the ratio of the leading coefficient of P_n to the leading coefficient of P_{n+1}.

This has an interesting consequence. Suppose that q is any polynomial of degree $\leq n$. Then q is a linear combination of the P_k, $k \leq n$, and orthonormality implies that

$$q = \sum_{j=0}^{n} (q, P_j)\, P_j.$$

Thus (4.1.4) implies the following.

Proposition 4.1.1 *If q is a polynomial of degree $\leq n$, then*

$$q(x) = \int_a^b K_n(x, y)\, q(y)\, w(y)\, dy, \tag{4.1.5}$$

where

$$K_n(x, y) = a_n \left[\frac{P_{n+1}(x)P_n(y) - P_n(x)P_{n+1}(y)}{x - y} \right]$$

and a_n is the ratio of the leading coefficient of P_n to the leading coefficient of P_{n+1}.

The kernel function K_n plays the same role with respect to expansions in orthogonal polynomials as the classical Dirichlet kernel plays with respect to the classical Fourier expansion, so we refer to it as the *Dirichlet kernel* for the polynomials $\{P_n\}$.

The kernel function K_n can be realized as a determinant:

$$K_n(x, y) = -\frac{1}{\Delta_n} \begin{vmatrix} 0 & 1 & x & x^2 & \dots & x^n \\ 1 & A_0 & A_1 & A_2 & \dots & A_n \\ y & A_1 & A_2 & A_3 & \dots & A_{n+1} \\ & & & \ddots & \\ y^n & A_n & A_{n+1} & A_{n+2} & \dots & A_{2n} \end{vmatrix}. \qquad (4.1.6)$$

See the exercises.

Taking the limit as $y \to x$ in (4.1.4) gives

$$a_n \left[P'_{n+1}(x)P_n(x) - P_{n+1}(x)P'_n(x) \right] = \sum_{j=0}^{n} P_j(x)^2. \qquad (4.1.7)$$

A first consequence of (4.1.7) is that the real roots of the P_n are simple: if x_0 were a double root, the left side of (4.1.7) would vanish at $x = x_0$, but P_0 is a nonzero constant, so the right side of (4.1.7) is positive.

The next result locates the zeros.

Proposition 4.1.2 *P_n has n real zeros, all lying in the interval I.*

Proof This is trivial for $n = 0$. Suppose $n \geq 1$ and let $x_1, \dots, x_m$ be the real roots of P_n lying in I. Set $q(x) = \prod(x - x_j)$. The sign changes of P_n and q in the interval I occur precisely at the x_j. Therefore the product $q P_n$ has fixed sign in I, so $(P_n, q) \neq 0$. This implies that q has degree at least n, so $m = n$. $\qquad \square$

A second consequence of (4.1.7) is that the roots of successive polynomials P_{n-1} and P_n interlace.

Proposition 4.1.3 *Between each pair of zeros of P_n is a zero of P_{n-1}.*

Proof For $n < 2$ there is nothing to prove. For $n \geq 2$, (4.1.7) implies that

$$P_n'(x)\, P_{n-1}(x) - P_n(x)\, P_{n-1}'(x) > 0.$$

Suppose that $x_1 < x_2$ are two successive zeros of P_n. Then the previous inequality implies that

$$P_n'(x_j)\, P_{n-1}(x_j) > 0, \quad j = 1, 2.$$

Since $P_n'(x_1)$ and $P_n'(x_2)$ must have different signs, it follows from the preceding inequality that the $P_{n-1}(x_j)$ have different signs. Therefore, P_{n-1} has a zero in the interval (x_1, x_2). $\qquad\square$

We turn to the question of *completeness*: can every element of the Hilbert space L_w^2 be written as a linear combination of the P_n? Suppose that f belongs to L_w^2. Consider the question of finding an element in the span of $\{P_j\}_{j \leq n}$ that is closest to f, with respect to the L^2 distance

$$d(f, g) = \|f - g\| = (f - g, f - g)^{\frac{1}{2}}.$$

Proposition 4.1.4 *Let*

$$f_n = \sum_{j=0}^{n} (f, P_j) P_j = \int_a^b K_n(x, y)\, f(y)\, w(y)\, dy. \qquad (4.1.8)$$

Then f_n is the closest function to f in the span of $\{P_0, P_1, \ldots, P_n\}$.

Proof Write

$$f - g = (f - f_n) + (f_n - g).$$

Computing inner products shows that $f - f_n$ is orthogonal to every element of the span of $\{P_j\}_{j \leq n}$, so if g is also in the span, then $f - f_n$ and $f_n - g$ are orthogonal, so

$$\|f - g\|^2 = \|f - f_n\|^2 + \|f_n - g\|^2. \qquad (4.1.9)$$

The left-hand side is minimal exactly when $g = f_n$. $\qquad\square$

Taking $g = 0$ in (4.1.9) and using orthonormality of the P_n, we obtain Bessel's inequality

$$\|f\|^2 = \sum_{j=0}^{n} (f, P_n)^2 + \|f - f_n\|^2 \geq \sum_{j=0}^{n} (f, P_n)^2. \qquad (4.1.10)$$

The following completeness theorem applies to all the cases we shall consider.

Theorem 4.1.5 *Suppose that w is a positive weight on the interval (a, b) and suppose that for some $c > 0$,*

$$\int_a^b e^{2c|x|} w(x)\, dx < \infty. \tag{4.1.11}$$

Let $\{P_n\}$ be the orthonormal polynomials for w. For any $f \in L_w^2$,

$$f = \sum_{n=0}^{\infty} \left(f, P_n \right) P_n$$

in the sense that the partial sums of the series on the right converge to f in norm in L_w^2. Moreover, one has Parseval's equality

$$\|f\|^2 = \sum_{n=0}^{\infty} \left(f, P_n \right)^2. \tag{4.1.12}$$

If the interval (a, b) is bounded, then the condition (4.1.11) is redundant and one may use the Weierstrass polynomial approximation theorem to show that polynomials are dense in L_w^2. A proof that remains valid in the general case is given in Appendix B.

4.2 Classical polynomials: general properties, I

We return to the three cases corresponding to the classical polynomials, with interval I, weight w, and eigenvalue equation of the form

$$p(x)\, \psi_n''(x) + q(x)\, \psi_n'(x) + \lambda_n \psi_n(x) = 0, \quad q = \frac{(pw)'}{w}. \tag{4.2.1}$$

Or, equivalently,

$$\left(pw\psi_n' \right)' + \lambda_n w\psi_n = 0.$$

The cases are

$$I = \mathbf{R} = (-\infty, \infty), \quad w(x) = e^{-x^2}, \quad p(x) = 1, \ q(x) = -2x;$$

$$I = \mathbf{R}_+ = (0, \infty), \quad w(x) = x^\alpha e^{-x}, \ \alpha > -1, \quad p(x) = x,$$

$$q(x) = \alpha + 1 - x;$$

$$I = (-1, 1), \quad w(x) = (1 - x)^\alpha (1 + x)^\beta, \ \alpha > -1, \ \beta > -1,$$

$$p(x) = 1 - x^2, \quad q(x) = \beta - \alpha - (\alpha + \beta + 2)x.$$

The derivative of a solution ψ_n of (4.2.1) satisfies a similar equation: differentiating (4.2.1) gives

$$p \left[\psi_n' \right]'' + (q + p') \left[\psi_n' \right]' + (q' + \lambda_n) \psi_n' = 0.$$

Now

$$q + p' = \frac{(pw)'}{w} + p' = \frac{p^2 w' + 2pp'w}{pw} = \frac{[p(pw)]'}{pw}.$$

Thus the function pw is also a weight. Since q' is a constant in each of the three cases, ψ_n' is an orthogonal polynomial of degree $n - 1$ for the weight pw. Continuing, ψ_n'' is an orthogonal polynomial of degree $n - 2$ for the weight $p^2 w$, with eigenvalue

$$-\lambda_n - q' - (p' + q)' = -\lambda_n - 2q' - p''.$$

By induction, the mth derivative $\psi_n^{(m)}$ corresponds to weight $p^m w$, with eigenvalue

$$-\lambda_n - mq' - \frac{1}{2} m(m - 1) p''.$$

Since $\psi_n^{(n)}$ is constant, the corresponding eigenvalue is zero and we have the general formula

$$\lambda_n = -nq' - \frac{1}{2} n(n - 1) p'',$$

which corresponds to the results obtained in the three cases above:

$$\lambda_n = 2n, \quad \lambda_n = n, \quad \lambda_n = n(n + \alpha + \beta + 1), \qquad (4.2.2)$$

respectively.

Equation (4.2.1) can be rewritten as

$$w\psi_n = -\lambda_n^{-1} \left(pw\psi_n' \right)'. \qquad (4.2.3)$$

Since pw is the weight corresponding to ψ_n', this leads to

$$w\psi_n = \left[\lambda_n (\lambda_n + q') \right]^{-1} \left(p^2 w \psi_n'' \right)'',$$

and finally to

$$w\psi_n = (-1)^n \prod_{m=0}^{n-1} \left[\lambda_n + mq' + \frac{1}{2} m(m - 1) p'' \right]^{-1} \left(p^n w \psi_n^{(n)} \right)^{(n)}. \qquad (4.2.4)$$

Since $\psi_n^{(n)}$ is constant, we may normalize by taking

$$\psi_n(x) = w(x)^{-1} \frac{d^n}{dx^n} \{ p(x)^n w(x) \}. \tag{4.2.5}$$

This is known as the *Rodrigues formula*. In view of (4.2.4), with this choice of ψ_n, we have

$$\psi_n(x) = a_n x^n + \text{lower order}, \tag{4.2.6}$$

$$n! \, a_n = (-1)^n \prod_{m=0}^{n-1} \left[\lambda_n + mq' + \frac{1}{2}m(m-1)p'' \right].$$

In our three cases the product on the right is, respectively,

$$(-2)^n n!, \quad (-1)^n n!, \quad (-1)^n n! (\alpha + \beta + n + 1)_n,$$

so the leading coefficient a_n is, respectively,

$$(-2)^n, \quad (-1)^n, \quad (-1)^n (\alpha + \beta + n + 1)_n. \tag{4.2.7}$$

As a first application of the Rodrigues formula, we consider the calculation of weighted inner products of the form

$$(f, \psi_n) = (f, \psi_n)_w = \int_a^b f(x) \, \psi_n(x) \, w(x) \, dx.$$

By (4.2.5), $\psi_n w = (p^n w)^{(n)}$. The function $p^n w$ vanishes fairly rapidly at the endpoints of the interval I. Therefore, under rather mild conditions on the function f, we may integrate by parts n times without acquiring boundary terms, to obtain

$$\int_a^b \psi_n(x) \, f(x) \, w(x) \, dx = (-1)^n \int_a^b p(x)^n f^{(n)}(x) \, w(x) \, dx. \tag{4.2.8}$$

This idea can be used in conjunction with (4.2.6) to calculate the weighted L^2 norms:

$$\int_a^b \psi_n^2(x) w(x) \, dx = (-1)^n \int_a^b \psi_n^{(n)} p(x)^n w(x) \, dx$$

$$= \prod_{m=0}^{n-1} \left[\lambda_n + mq' + \frac{1}{2}m(m-1)p'' \right] \int_a^b p(x)^n w(x) \, dx.$$

$$\tag{4.2.9}$$

The preceding integral, in the three cases considered, is respectively

$$\int_{-\infty}^{\infty} e^{-x^2} \, dx = \sqrt{\pi};$$

$$\int_{0}^{\infty} x^{n+\alpha} e^{-x} \, dx = \Gamma(n + \alpha + 1);$$

$$\int_{-1}^{1} (1-x)^{n+\alpha} (1+x)^{n+\beta} \, dx = 2^{2n+\alpha+\beta+1} \int_{0}^{1} s^{n+\alpha} (1-s)^{n+\beta} \, ds$$

$$= 2^{2n+\alpha+\beta+1} \mathrm{B}(n + \alpha + 1, n + \beta + 1).$$

Thus the square of the weighted L^2 norm of ψ_n in the three cases is, respectively,

$$||\psi_n||^2 = 2^n \, n! \sqrt{\pi},$$

$$||\psi_n||^2 = n! \, \Gamma(n + \alpha + 1),$$

$$||\psi_n||^2 = 2^{2n+\alpha+\beta+1} \, n! \, \frac{\Gamma(n + \alpha + 1) \, \Gamma(n + \beta + 1)}{(2n + \alpha + \beta + 1) \, \Gamma(n + \alpha + \beta + 1)}.$$

The standard normalizations of the classical polynomials differ from the choice given by the Rodrigues formula (4.2.5). The Rodrigues formula for the Hermite, Laguerre, and Jacobi polynomials, respectively, is taken to be

$$H_n(x) = (-1)^n e^{x^2} \frac{d^n}{dx^n} \left(e^{-x^2} \right); \tag{4.2.10}$$

$$L_n^{(\alpha)}(x) = \frac{1}{n!} x^{-\alpha} e^{x} \frac{d^n}{dx^n} \left(e^{-x} x^{n+\alpha} \right); \tag{4.2.11}$$

$$P_n^{(\alpha,\beta)}(x) = \frac{(-1)^n}{n! \, 2^n} (1-x)^{-\alpha} (1+x)^{-\beta}$$

$$\times \frac{d^n}{dx^n} \left\{ (1-x)^{n+\alpha} (1+x)^{n+\beta} \right\}. \tag{4.2.12}$$

In view of these normalizations, the previous calculation of weighted L^2 norms gives

$$||H_n||^2 = 2^n \, n! \sqrt{\pi}, \tag{4.2.13}$$

$$||L_n^{(\alpha)}||^2 = \frac{\Gamma(n + \alpha + 1)}{n!}, \tag{4.2.14}$$

$$||P_n^{(\alpha,\beta)}||^2 = \frac{2^{\alpha+\beta+1} \, \Gamma(n + \alpha + 1) \, \Gamma(n + \beta + 1)}{n! \, (2n + \alpha + \beta + 1) \, \Gamma(n + \alpha + \beta + 1)}. \tag{4.2.15}$$

The normalizations, together with (4.2.7), imply that the leading coefficients of H_n, $L_n^{(\alpha)}$, and $P_n^{(\alpha,\beta)}$ are

$$2^n, \quad \frac{(-1)^n}{n!}, \quad \frac{(\alpha + \beta + n + 1)_n}{2^n\, n!}, \tag{4.2.16}$$

respectively. The discussion at the end of Section 3.4 shows that Laguerre and Hermite polynomials can be obtained as certain limits of Jacobi polynomials. In view of that discussion and this calculation of leading coefficients, it follows that

$$H_n(x) = \lim_{\alpha \to +\infty} \frac{2^n\, n!}{\alpha^{n/2}} P_n^{(\alpha,\alpha)}\left(\frac{x}{\sqrt{\alpha}}\right); \tag{4.2.17}$$

$$L_n^{(\alpha)}(x) = \lim_{\beta \to +\infty} P_n^{(\alpha,\beta)}\left(1 - \frac{2x}{\beta}\right). \tag{4.2.18}$$

4.3 Classical polynomials: general properties, II

We may take advantage of the Cauchy integral formula for derivatives of wp^n to derive an integral formula from (4.2.5); see Appendix A. Let Γ be a curve that encloses $x \in I$ but excludes the endpoints of I. Then (4.2.5) shows that

$$\frac{\psi_n(x)}{n!} = \frac{1}{2\pi i} \int_\Gamma \frac{w(z)}{w(x)} \frac{p(z)^n}{(z-x)^n} \frac{dz}{z-x}. \tag{4.3.1}$$

The *generating function* for the orthogonal polynomials $\{\psi_n/n!\}$ is defined as

$$G(x, s) = \sum_{n=0}^{\infty} \frac{\psi_n(x)}{n!} s^n.$$

The integral formula (4.3.1) allows the evaluation of G:

$$G(x, s) = \frac{1}{2\pi i} \int_\Gamma \sum_{n=0}^{\infty} \frac{s^n p(z)^n}{(z-x)^n} \cdot \frac{w(z)}{w(x)} \cdot \frac{dz}{z-x}$$

$$= \frac{1}{2\pi i} \int_\Gamma \frac{w(z)}{w(x)} \cdot \frac{dz}{z - x - s\, p(z)}.$$

We may assume that Γ encloses a single solution $z = \zeta(x, s)$ of $z - x = sp(z)$. Since the residue of the integrand at this point is $w(\zeta)w(x)^{-1}[1 - sp'(\zeta)]^{-1}$, we obtain

$$G(x, s) = \frac{w(\zeta)}{w(x)} \cdot \frac{1}{1 - sp'(\zeta)}, \quad \zeta - sp(\zeta) = x. \tag{4.3.2}$$

In the following sections we give the explicit evaluation in the case of the Jacobi, Hermite, and Laguerre functions. In each of the latter two cases we give a second derivation of the formula for the generating function.

The integral formula (4.3.1) can also be used to obtain recurrence relations of the type (4.1.3) for the polynomials ψ_n. It will be slightly more convenient to work with

$$\varphi_n(x) = \frac{\psi_n(x)}{n!} \, w(x) = \frac{1}{2\pi i} \int_C \frac{p(z)^n w(z) \, dz}{(z - x)^{n+1}}$$

and look for a three-term recurrence relation

$$a_n \varphi_{n+1}(x) = b_n(x) \, \varphi_n(x) + c_n \varphi_{n-1}(x), \quad b_n(x) = b_{n0} + b_{n1}x. \quad (4.3.3)$$

If a is constant then an integration by parts gives

$$\int_C \left[\frac{ap(z)^{n+1} w(z)}{(z - x)^{n+2}} - \frac{b(z)p(z)^n w(z)}{(z - x)^{n+1}} - \frac{cp^{n-1}(z)w(z)}{(z - x)^n} \right] dz$$

$$= \int_C \left[\frac{\widetilde{a}[p^{n+1} w]'(z) - b(z)p(z)^n w(z) - (z - x)cp(z)^{n-1} w(z)}{(z - x)^{n+1}} \right] dz,$$

$$\widetilde{a} = \frac{a}{n + 1}.$$

Constants a, b_0, b_1, and c can be chosen so that for fixed x the last integrand is a derivative:

$$\frac{d}{dz} \left\{ \frac{Q(z) \, p(z)^n \, w(z)}{(z - x)^n} \right\}, \quad Q(z) = Q_0 + Q_1(z - x),$$

where Q_0 and Q_1 are constants. Since $(pw)' = qw$ it follows that

$$\left[p^m w \right]' = \left[p^{m-1}(pw) \right]' = \left[(m - 1)p' + q \right] p^{m-1} w.$$

Applying this to both previous expressions we find that the condition on $\widetilde{a}$, b, c and Q is:

$$\widetilde{a}(np' + q)p - bp - c(z - x) = -nQp + \left\{ Q'p + Q[(n - 1)p' + q] \right\}(z - x). \quad (4.3.4)$$

Expanding p, q, and Q in powers of $z - x$ leads to a system of four linear equations for the four unknowns $b/\widetilde{a}$, $c/\widetilde{a}$, Q_0, and Q_1. The results for the three cases we have been considering are the following, respectively:

$$a_n \varphi_{n+1}(x) = b_n(x)\varphi_n(x) + c_n \varphi_{n-1}(x); \tag{4.3.5}$$

$$\begin{cases} \qquad p(x) = 1, \quad w(x) = e^{-x^2}: \\ a_n = n+1, \quad b_n(x) = -2x, \quad c_n = -2; \end{cases}$$

$$\begin{cases} \qquad p(x) = x, \quad w(x) = x^\alpha e^{-x}: \\ a_n = n+1, \quad b_n(x) = (2n+\alpha+1) - x, \quad c_n = -(n+\alpha); \end{cases}$$

$$\begin{cases} \qquad p(x) = 1 - x^2, \quad w(x) = (1-x)^\alpha (1+x)^\beta: \\ \qquad a_n = (n+1)(n+\alpha+\beta+1)(2n+\alpha+\beta); \\ b_n(x) = (2n+\alpha+\beta+1)\left[\beta^2 - \alpha^2 - (2n+\alpha+\beta)(2n+\alpha+\beta+2)x\right]; \\ \qquad c_n = -4(2n+\alpha+\beta+2)(n+\alpha)(n+\beta). \end{cases}$$

Taking into account the normalizations above, the three-term recurrence relations in the three classical cases are

$$H_{n+1}(x) = 2x\, H_n(x) - 2n\, H_{n-1}(x); \tag{4.3.6}$$

$$(n+1) L_{n+1}^{(\alpha)}(x) = (2n+\alpha+1-x) L_n^{(\alpha)}(x) - (n+\alpha) L_{n-1}^{(\alpha)}(x); \tag{4.3.7}$$

$$\frac{(2n+2)(n+\alpha+\beta+1)}{2n+\alpha+\beta+1} P_{n+1}^{(\alpha,\beta)}(x)$$

$$= \left[\frac{\alpha^2 - \beta^2}{2n+\alpha+\beta} + (2n+\alpha+\beta+2)x\right] P_n^{(\alpha,\beta)}(x)$$

$$- \frac{2(2n+\alpha+\beta+2)(n+\alpha)(n+\beta)}{(2n+\alpha+\beta)(2n+\alpha+\beta+1)} P_{n-1}^{(\alpha,\beta)}(x). \tag{4.3.8}$$

The starting point in the derivation of the Rodrigues formula (4.2.5) was that if ψ_n is the degree-n polynomial for weight w, then the derivative ψ_n' is a multiple of the degree-$(n-1)$ polynomial for weight pw. In the Hermite case, $pw = w$; in the other two cases going to pw raises the index or indices by 1. Taking into account the leading coefficients of the ψ_n given in (4.2.7) and the normalizations, we obtain

$$H_n'(x) = 2n\, H_{n-1}(x); \tag{4.3.9}$$

$$\left[L_n^{(\alpha)}\right]'(x) = -L_{n-1}^{(\alpha+1)}(x); \tag{4.3.10}$$

$$\left[P_n^{(\alpha,\beta)}\right]'(x) = \frac{1}{2}(n+\alpha+\beta+1) P_{n-1}^{(\alpha+1,\beta+1)}(x). \tag{4.3.11}$$

The method used to derive (4.3.3) can be used to obtain derivative formulas of a different form:

$$p(x)\,\varphi_n'(x) = a_n(x)\,\varphi_n(x) + b_n\,\varphi_{n-1}(x).$$

In fact,

$$p(x)\,\varphi_n'(x) = \frac{n+1}{2\pi i}\int_C \frac{p(x)\,p(z)^n\,w(z)\,dz}{(z-x)^{n+2}}.$$

For fixed x we may expand $p(x)$ in powers of $(z-x)$ and integrate one summand by parts to put the integrand into the form

$$\frac{\tilde{p}_0\big(p^n w\big)'(z) + p_1 p(z)^n w(z) + p_2(z-x)p(z)^n w(z)}{(z-x)^{n+1}}, \quad \tilde{p}_0 = \frac{p_0}{n+1},$$

where $p_0 = p(x)$, $p_1 = -p'(x)$, $p_2 = \frac{1}{2}p''$. Using the integral forms of φ_n and φ_{n-1} as well, we may treat the equation

$$p(x)\varphi_n'(x) - a(x)\varphi_n(x) - b\varphi_{n-1}(x) = 0 \tag{4.3.12}$$

in the same way as we treated the equation (4.3.3). The resulting formulas are

$$H_n'(x) = 2n\,H_{n-1}(x); \tag{4.3.13}$$

$$x\left[L_n^{(\alpha)}\right]'(x) = n\,L_n^{(\alpha)}(x) - (n+\alpha)\,L_{n-1}^{(\alpha)}(x); \tag{4.3.14}$$

$$\left(1-x^2\right)\left[P_n^{(\alpha,\beta)}\right]'(x) = \left[\frac{n(\alpha-\beta)}{2n+\alpha+\beta} - nx\right]P_n^{(\alpha,\beta)}(x)$$
$$+ \frac{2(n+\alpha)(n+\beta)}{2n+\alpha+\beta}\,P_{n-1}^{(\alpha,\beta)}(x). \tag{4.3.15}$$

The recurrence formulas and the calculation of the norms allow us to compute the associated Dirichlet kernels. To see this, suppose that we have the identities

$$x\varphi_n(x) = a_n\varphi_{n+1}(x) + b_n\varphi_n(x) + c_n\varphi_{n-1}(x) \tag{4.3.16}$$

for a sequence of orthogonal polynomials $\{\varphi_n\}$. The associated orthonormal polynomials are

$$\tilde{\varphi}_n = \|\varphi_n\|^{-1}\varphi_n.$$

The Christoffel–Darboux formula (4.1.4) implies that the Dirichlet kernel

$$K_n(x, y) = \sum_{j=0}^{n} \widetilde{\varphi}_j(x)\widetilde{\varphi}_j(y)$$

$$= ||\varphi_n||^{-2}\varphi_n(x)\varphi_n(y) + K_{n-1}(x, y)$$

is given by

$$\alpha_n \left[\frac{\widetilde{\varphi}_{n+1}(x)\widetilde{\varphi}_n(y) - \widetilde{\varphi}_n(x)\widetilde{\varphi}_{n+1}(y)}{x - y} \right] = \beta_n \left[\frac{\varphi_{n+1}(x)\varphi_n(y) - \varphi_n(x)\varphi_{n+1}(y)}{x - y} \right]$$

for some constants α_n, β_n. It follows from these equations together with (4.3.5) that the constant β_n is $a_n/||\varphi_n||^2$. These observations lead to the following evaluations of the Dirichlet kernels associated with the Hermite, Laguerre, and Jacobi polynomials respectively:

$$K_n^H(x, y) = \frac{1}{2^{n+1}n!\sqrt{\pi}}$$

$$\times \left[\frac{H_{n+1}(x)H_n(y) - H_n(x)H_{n+1}(y)}{x - y} \right]; \qquad (4.3.17)$$

$$K_n^{(\alpha)}(x, y) = -\frac{(n+1)!}{\Gamma(n + \alpha + 1)}$$

$$\times \left[\frac{L_{n+1}^{(\alpha)}(x)L_n^{(\alpha)}(y) - L_n^{(\alpha)}(x)L_{n+1}^{(\alpha)}(y)}{x - y} \right]; \qquad (4.3.18)$$

$$K_n^{(\alpha,\beta)}(x, y) = \frac{2^{-\alpha-\beta}(n+1)!\,\Gamma(n + \alpha + \beta + 2)}{(2n + \alpha + \beta + 2)\Gamma(n + \alpha + 1)\,\Gamma(n + \beta + 1)}$$

$$\times \left[\frac{P_{n+1}^{(\alpha,\beta)}(x)P_n^{(\alpha,\beta)}(y) - P_n^{(\alpha,\beta)}(x)P_{n+1}^{(\alpha,\beta)}(y)}{x - y} \right]. \qquad (4.3.19)$$

The *discriminant* of a polynomial $P(z) = a \prod_{j=1}^{n}(z - z_j)$ is the polynomial

$$D(z) = a^{2n-2} \prod_{1 \leq j < k \leq n} (z_j - z_k)^2.$$

The discriminants of the Hermite, Laguerre, and Jacobi polynomials are, respectively:

$$D_n^H = 2^{3n(n-1)/2} \prod_{j=1}^{n} j^j; \tag{4.3.20}$$

$$D_n^{(\alpha)} = \prod_{j=1}^{n} j^{j-2n+2}(j+\alpha)^{j-1};$$

$$D_n^{(\alpha,\beta)} = \frac{1}{2^{n(n-1)}} \prod_{j=1}^{n} \frac{j^{j-2n+2}(j+\alpha)^{j-1}(j+\beta)^{j-1}}{(n+j+\alpha+\beta)^{j-n}}.$$

For a proof, see for example Section 6.71 of [279].

The next three sections contain additional results and some alternate derivations for these three classical cases.

4.4 Hermite polynomials

The Hermite polynomials $\{H_n\}$ are orthogonal polynomials associated with the weight e^{-x^2} on the line $\mathbf{R} = (-\infty, \infty)$. They are eigenfunctions

$$H_n''(x) - 2x\, H_n'(x) + 2n\, H_n(x) = 0, \tag{4.4.1}$$

satisfy the derivative relation

$$H_n'(x) = 2n\, H_{n-1}(x),$$

and can be defined by the Rodrigues formula

$$H_n(x) = (-1)^n e^{x^2} \frac{d^n}{dx^n}\{e^{-x^2}\} = \left(2x - \frac{d}{dx}\right)^n \{1\}.$$

It follows that the leading coefficient is 2^n and that

$$H_n'(x) - 2x\, H_n(x) = -H_{n+1}(x). \tag{4.4.2}$$

They are limits

$$H_n(x) = \lim_{\alpha \to +\infty} \frac{2^n n!}{\alpha^{n/2}} P_n^{(\alpha,\alpha)}\left(\frac{x}{\sqrt{\alpha}}\right).$$

The three-term recurrence relation (4.3.6) may also be derived as follows. It is easily shown by induction that H_n is even if n is even and odd if n is odd:

$$H_n(-x) = (-1)^n H_n(x).$$

Therefore the relation must have the form

$$x \, H_n(x) = a_n \, H_{n+1}(x) + b_n \, H_{n-1}(x).\tag{4.4.3}$$

Identities (4.4.2) and (4.4.3) imply that $H'_n = 2b_n \, H_{n-1}$. Comparing leading coefficients, we see that $a_n = \frac{1}{2}$ and $b_n = n$:

$$x \, H_n(x) = \frac{1}{2} H_{n+1}(x) + n \, H_{n-1}(x).\tag{4.4.4}$$

If we write $H_n(x) = \sum_{k=0}^{n} a_k x^k$, then equation (4.4.1) implies the relation

$$(k+2)(k+1) \, a_{k+2} = 2(k-n) \, a_k.$$

Since $a_n = 2^n$ and $a_{n-1} = 0$, this recursion gives

$$H_n(x) = \sum_{2j \le n} (-1)^j \frac{n!}{j!\,(n-2j)!} (2x)^{n-2j}.\tag{4.4.5}$$

The first six of the H_n are

$$H_0(x) = 1;$$
$$H_1(x) = 2x;$$
$$H_2(x) = 4x^2 - 2;$$
$$H_3(x) = 8x^3 - 12x;$$
$$H_4(x) = 16x^4 - 48x^2 + 12;$$
$$H_5(x) = 32x^5 - 160x^3 + 120x.$$

Taking into account the factor $(-1)^n$, the generating function

$$G(x, s) = \sum_{n=0}^{\infty} \frac{H_n(x)}{n!} s^n$$

is calculated from (4.3.2) with $p(x) = 1$, and s replaced by $-s$, so $\zeta(x, s) = x - s$ and

$$\sum_{n=0}^{\infty} \frac{H_n(x)}{n!} s^n = \frac{e^{-(x-s)^2}}{e^{-x^2}} = e^{2xs - s^2}.\tag{4.4.6}$$

This can also be calculated from the three-term recurrence relation (4.4.4), which is equivalent to

$$2x \, G(x, s) = \frac{\partial G}{\partial s}(x, s) + 2s \, G(x, s).$$

Therefore

$$G(x, s) = c(x) e^{-(s-x)^2}.$$

Since $G(x, 0) = 1$, we obtain (4.4.6).

The generating function (4.4.6) can be used to obtain two *addition formulas* for the Hermite polynomials:

$$H_n(x + y) = \sum_{j+k+2l=n} \frac{n!}{j!\,k!\,l!}\, H_j(x)\, H_k(y) \qquad (4.4.7)$$

and

$$H_n(x + y) = 2^{-\frac{1}{2}n} \sum_{m=0}^{n} \binom{n}{m} H_m\left(\sqrt{2}\,x\right) H_{n-m}\left(\sqrt{2}\,y\right); \qquad (4.4.8)$$

see the exercises.

The generating function may also be used to give an alternative calculation of the weighted L^2 norms (4.2.13):

$$\sum_{m,n=0}^{\infty} \frac{s^m\, t^n}{m!\,n!} \int_{-\infty}^{\infty} H_m(x)\, H_n(x)\, e^{-x^2}\, dx = \int_{-\infty}^{\infty} G(x, s)\, G(x, t)\, e^{-x^2}\, dx$$

$$= \int_{-\infty}^{\infty} e^{2st}\, e^{-(x-s-t)^2}\, dx$$

$$= e^{2st}\, \sqrt{\pi} = \sqrt{\pi} \sum_{n=0}^{\infty} \frac{(2st)^n}{n!}.$$

This confirms that the H_n are mutually orthogonal and that

$$\int_{-\infty}^{\infty} H_n(x)^2\, e^{-x^2}\, dx = n!\, 2^n\, \sqrt{\pi}. \qquad (4.4.9)$$

Therefore the normalized polynomials are

$$\widetilde{H}_n(x) = \frac{1}{\pi^{\frac{1}{4}} \sqrt{n!\, 2^n}}\, H_n(x). \qquad (4.4.10)$$

According to Theorem (4.1.5), a given function $f \in L^2\left(\mathbf{R}, e^{-x^2} dx\right)$ can be approximated in L^2 by the sequence

$$f_n(x) = \sum_{m=0}^{n} \left(f, \widetilde{H}_m\right) \widetilde{H}_m = \int K_n(x, y)\, f(y)\, e^{-y^2}\, dy. \qquad (4.4.11)$$

To compute the coefficients $\left(f, \widetilde{H}_n\right)$ or, equivalently, (f, H_n), we may use the Rodrigues formula, as in (4.2.8). If the function f and its derivatives to order n are of at most exponential growth as $|x| \to \infty$, then

$$(f, H_n) = \int_{-\infty}^{\infty} f(x)\, H_n(x)\, e^{-x^2}\, dx = \int_{-\infty}^{\infty} e^{-x^2}\, f^{(n)}(x)\, dx.$$

For example, $(x^m, H_n) = 0$ unless m and n are both even or both odd, and unless $m \geq n$. If $m = n + 2k$, the previous calculation and a change of variable give

$$(x^m, H_n) = \frac{m!}{(m-n)!} \int_{-\infty}^{\infty} e^{-x^2} x^{m-n}\, dx$$

$$= \frac{m!}{(m-n)!} \int_0^{\infty} e^{-t} t^{\frac{1}{2}(m-n-1)}\, dt = \frac{m!}{(m-n)!}\, \Gamma\left(\frac{1}{2}[m-n+1]\right).$$

$$\tag{4.4.12}$$

Similarly,

$$(e^{ax}, H_n) = a^n \int_{-\infty}^{\infty} e^{ax - x^2}\, dx$$

$$= a^n \int_{-\infty}^{\infty} e^{-\left(x - \frac{1}{2}a\right)^2} e^{\frac{1}{4}a^2}\, dx = a^n e^{\frac{1}{4}a^2} \int_{-\infty}^{\infty} e^{-x^2}\, dx$$

$$= a^n e^{\frac{1}{4}a^2} \sqrt{\pi}.$$

$$\tag{4.4.13}$$

It follows from Cauchy's theorem or by analytic continuation that the identity (4.4.13) remains valid for all complex a. In particular, we may take $a = \pm ib$ to calculate

$$(\cos bx, H_n) = \begin{cases} \sqrt{\pi}\,(ib)^n e^{-\frac{1}{4}b^2}, & n \text{ even,} \\ 0, & n \text{ odd;} \end{cases} \tag{4.4.14}$$

$$(\sin bx, H_n) = \begin{cases} -i\sqrt{\pi}\,(ib)^n e^{-\frac{1}{4}b^2}, & n \text{ odd,} \\ 0, & n \text{ even.} \end{cases} \tag{4.4.15}$$

A particular case of the calculation in (4.4.13) is the identity

$$e^{-x^2} = \frac{1}{\sqrt{\pi}} \int_{-\infty}^{\infty} e^{-2ixt - t^2}\, dt.$$

This identity and the Rodrigues formula imply the integral formula

$$H_n(x) = (-1)^n \frac{e^{x^2}}{\sqrt{\pi}} \int_{-\infty}^{\infty} (-2it)^n e^{-2ixt - t^2}\, dt. \tag{4.4.16}$$

This formula can be used to find a generating function for the products

$$H_n(x)H_n(y) = \frac{e^{x^2+y^2}}{\pi} \int_{-\infty}^{\infty} \int_{-\infty}^{\infty} (-4tu)^n e^{-2ixt-2iyu-t^2-u^2} \, dt \, du.$$

Indeed, for $|s| < 1$,

$$\sum_{n=0}^{\infty} \frac{H_n(x)H_n(y)}{2^n n!} s^n = \frac{e^{x^2+y^2}}{\pi} \int_{-\infty}^{\infty} \int_{-\infty}^{\infty} e^{-2ixt-2iyu-2tus-t^2-u^2} \, dt \, du$$

$$= \frac{e^{x^2+y^2}}{\pi} \int_{-\infty}^{\infty} \left\{ \int_{-\infty}^{\infty} e^{2i(-x+ius)t-t^2} \, dt \right\} e^{-2iyu-u^2} \, du$$

$$= \frac{e^{x^2+y^2}}{\pi} \int_{-\infty}^{\infty} e^{-(x-ius)^2-2iyu-u^2} \, du$$

$$= \frac{e^{y^2}}{\sqrt{\pi}} \int_{-\infty}^{\infty} e^{-2i(y-xs)u-(1-s^2)u^2} \, du.$$

Taking $v = u\sqrt{1 - s^2}$ as a new variable of integration gives

$$\sum_{n=0}^{\infty} \frac{H_n(x)H_n(y)s^n}{2^n n!} = \frac{1}{\sqrt{1-s^2}} \exp\left(\frac{2xys - s^2x^2 - s^2y^2}{1-s^2} \right). \qquad (4.4.17)$$

The general results about zeros of orthogonal polynomials, together with the results proved in Section 3.5, give the following.

Theorem 4.4.1 *The Hermite polynomial $H_n(x)$ has n simple roots, lying in the interval*

$$-\sqrt{2n+1} < x < \sqrt{2n+1}.$$

The relative maxima of

$$\left| e^{-x^2/2} H_n(x) \right|$$

increase as $|x|$ increases.

Results in Section 3.5 can be used to give more detailed information about the zeros of $H_n(x)$: see the exercises.

Theorem 4.4.2 *The positive zeros $x_{1n} < x_{2n} < \ldots$ of $H_n(x)$ satisfy the following estimates. If $n = 2m$ is even, then*

$$\frac{(2k-1)\pi}{2\sqrt{2n+1}} < x_{kn} < \frac{4k+1}{\sqrt{2n+1}}, \qquad k = 1, 2, \ldots, m. \qquad (4.4.18)$$

If $n = 2m + 1$ is odd, then

$$\frac{k\pi}{\sqrt{2n+1}} < x_{kn} < \frac{4k+3}{\sqrt{2n+1}}, \quad k = 1, 2, \ldots, m. \quad (4.4.19)$$

The following asymptotic result will be proved in Chapter 10:

$$H_n(x) = 2^{\frac{1}{2}n} \frac{2^{\frac{1}{4}}(n!)^{\frac{1}{2}}}{(n\pi)^{\frac{1}{4}}} e^{\frac{1}{2}x^2} \left[\cos\left(\sqrt{2n+1}\, x - \frac{1}{2}n\pi \right) + O\left(n^{-\frac{1}{2}} \right) \right],$$

$$(4.4.20)$$

as $n \to \infty$, uniformly on any bounded interval. In view of (4.4.22) and (4.4.23) below, (4.4.20) also follows from Fejér's result for Laguerre polynomials [94].

A different normalization $w(x) = e^{-\frac{1}{2}x^2}$ is sometimes used for the weight function. The corresponding orthogonal polynomials, denoted by $\{He_n\}$, are eigenfunctions

$$He_n''(x) - x\, He_n'(x) + n\, He_n(x) = 0$$

and are given by the Rodrigues formula

$$He_n(x) = (-1)^n e^{\frac{1}{2}x^2} \frac{d^n}{dx^n} \left\{ e^{-\frac{1}{2}x^2} \right\} = \left[x - \frac{d}{dx} \right]^n \{1\}.$$

Setting $y = x/\sqrt{2}$, it is clear that $H_n(y)$ must be a multiple of $He_n(x)$, and consideration of the leading coefficients shows that

$$He_n(x) = 2^{-n/2} H_n\left(\frac{x}{\sqrt{2}} \right). \quad (4.4.21)$$

As noted in Section 3.4, there is a close relationship between Hermite polynomials and certain Laguerre polynomials. Since the Hermite polynomials H_{2n} of even order are even functions, they are orthogonal with respect to the weight $w(x) = e^{-x^2}$ on the half line $x > 0$. Let $y = x^2$; then the measure $w(x)\, dx$ becomes

$$\frac{1}{2} y^{-\frac{1}{2}} e^{-y}\, dy.$$

The polynomials $\{H_{2n}(\sqrt{y})\}$ are, therefore, multiples of the Laguerre polynomials $L_n^{\left(-\frac{1}{2}\right)}$. Consideration of the leading coefficients (see the next section) shows that the relationship is

$$H_{2n}(x) = (-1)^n 2^{2n} n!\, L_n^{\left(-\frac{1}{2}\right)}(x^2), \quad n = 0, 1, 2, \ldots \quad (4.4.22)$$

Similarly, the polynomials $x^{-1} H_{2n+1}$ are even functions that are orthogonal with respect to the weight $x^2 e^{-x^2}$ on the half line, so that they must be multiples of the Laguerre polynomials $L_n^{\left(\frac{1}{2}\right)}(x^2)$:

$$H_{2n+1}(x) = (-1)^n 2^{2n+1} n! \, x \, L_n^{\left(\frac{1}{2}\right)}(x^2), \quad n = 0, 1, 2, \ldots \qquad (4.4.23)$$

4.5 Laguerre polynomials

The Laguerre polynomials $\left\{L_n^{(\alpha)}\right\}$ are orthogonal polynomials associated with the weight $w(x) = x^\alpha e^{-x}$ on the half line $\mathbf{R}_+ = (0, \infty)$. For a given $\alpha > -1$ they are eigenfunctions

$$x\left[L_n^{(\alpha)}\right]''(x) + (\alpha + 1 - x)\left[L_n^{(\alpha)}\right]'(x) + n L_n^{(\alpha)}(x) = 0; \qquad (4.5.1)$$

see Case II of 3.4. They satisfy the derivative relation

$$\left[L_n^{(\alpha)}\right]'(x) = -L_{n-1}^{(\alpha+1)}(x)$$

and are given by the Rodrigues formula (4.2.11)

$$L_n^{(\alpha)}(x) = \frac{1}{n!} x^{-\alpha} e^x \frac{d^n}{dx^n}\left\{x^\alpha e^{-x} x^n\right\}$$

$$= \frac{1}{n!}\left[\frac{d}{dx} + \frac{\alpha}{x} - 1\right]^n \{x^n\}, \qquad (4.5.2)$$

where the second version is obtained by using the gauge transformation $u = \varphi v$ with $\varphi = x^\alpha e^{-x}$. It follows that the leading coefficient is $(-1)^n/n!$. The Laguerre polynomials for $\alpha = 0$ are also denoted by L_n:

$$L_n(x) = L_n^{(0)}(x).$$

The Laguerre polynomials are limits

$$L_n^{(\alpha)}(x) = \lim_{\beta \to +\infty} P_n^{(\alpha, \beta)}\left(1 - \frac{2x}{\beta}\right).$$

Writing $L_n^{(\alpha)}(x) = \sum_{k=0}^n b_k x^k$, equation (4.5.1) gives the recurrence relation

$$(k+1)(k+\alpha+1)\, b_{k+1} = -(n-k)\, b_k,$$

so

$$L_n^{(\alpha)}(x) = \sum_{k=0}^n (-1)^k \frac{(\alpha+1)_n}{k!\,(n-k)!\,(\alpha+1)_k} x^k \qquad (4.5.3)$$

$$= \frac{(\alpha+1)_n}{n!} \sum_{k=0}^n \frac{(-n)_k}{(\alpha+1)_k\, k!} x^k. \qquad (4.5.4)$$

The first four of the $L_n^{(\alpha)}$ are

$$L_0^{(\alpha)}(x) = 1;$$

$$L_1^{(\alpha)}(x) = \alpha + 1 - x;$$

$$L_2^{(\alpha)}(x) = \frac{(\alpha + 1)(\alpha + 2)}{2} - (\alpha + 2)x + \frac{1}{2}x^2;$$

$$L_3^{(\alpha)}(x) = \frac{(\alpha + 1)(\alpha + 2)(\alpha + 3)}{6} - \frac{(\alpha + 2)(\alpha + 3)}{2}x + \frac{(\alpha + 3)}{2}x^2 - \frac{1}{6}x^3.$$

In particular,

$$L_n(x) = \sum_{k=0}^{n} (-1)^k \frac{n!}{k!(n-k)!k!} x^k.$$

Comparing coefficients, the general three-term recurrence relation (4.1.3) is

$$x\,L_n^{(\alpha)}(x) = -(n+1)\,L_{n+1}^{(\alpha)}(x) + (2n + \alpha + 1)\,L_n^{(\alpha)}(x) - (n + \alpha)\,L_{n-1}^{(\alpha)}(x). \tag{4.5.5}$$

By induction

$$\frac{d^n}{dx^n}\{x\,f(x)\} = x\frac{d^n}{dx^n}\{f(x)\} + n\frac{d^{n-1}}{dx^{n-1}}\{f(x)\},$$

so the Rodrigues formula gives the recurrence relation (4.3.7):

$$(n+1)L_{n+1}^{(\alpha)}(x) = \left[x\frac{d}{dx} + \alpha + n + 1 - x\right]\{L_n^{(\alpha)}(x)\}. \tag{4.5.6}$$

Taking into account the normalization (4.5.2), the generating function

$$G(x, s) = \sum_{n=0}^{\infty} L_n^{(\alpha)}(x)\,s^n$$

can be calculated from (4.3.2). Here, $p(x) = x$, so $\zeta = x/(1 - s)$ and therefore

$$G(x, s) = \frac{e^{-xs/(1-s)}}{(1 - s)^{\alpha+1}}. \tag{4.5.7}$$

This can also be calculated from the three-term recurrence relation (4.5.5), which is equivalent to

$$\frac{\partial G}{\partial s}(x, s) = \frac{\alpha + 1}{1 - s}\,G(x, s) - \frac{x}{(1 - s)^2}\,G(x, s).$$

Since $G(x, 0) \equiv 1$, this implies (4.5.7).

As for Hermite polynomials, the generating function can be used to obtain an addition formula:

$$L_n^{(\alpha)}(x + y) = \sum_{j+k+l=n} (-1)^j \frac{(\alpha + 2 - j)_j}{j!} L_k^{(\alpha)}(x) L_l^{(\alpha)}(y); \qquad (4.5.8)$$

see the exercises.

The generating function can also be used to calculate L^2 norms:

$$\sum_{m,n=0}^{\infty} s^m t^n \int_0^{\infty} L_m^{(\alpha)}(x) L_n^{(\alpha)}(x) x^\alpha e^{-x} dx$$

$$= \int_0^{\infty} G(x, s) G(x, t) x^\alpha e^{-x} dx$$

$$= \frac{1}{(1 - s)^{\alpha+1} (1 - t)^{\alpha+1}} \int_0^{\infty} e^{-x(1-st)/(1-s)(1-t)} x^{\alpha+1} \frac{dx}{x}.$$

Letting $y = x(1 - st)/(1 - s)(1 - t)$, the last integral is

$$\frac{\Gamma(\alpha + 1)}{(1 - st)^{\alpha+1}} = \sum_{n=0}^{\infty} \frac{\Gamma(\alpha + 1 + n)}{n!} (st)^n.$$

This confirms that the $L_n^{(\alpha)}$ are mutually orthogonal, and

$$\int_0^{\infty} \left[L_n^{(\alpha)}(x)\right]^2 x^\alpha e^{-x} dx = \frac{\Gamma(\alpha + n + 1)}{n!}.$$

Therefore the normalized polynomials are

$$\widetilde{L}_n^{(\alpha)}(x) = \frac{\sqrt{n!}}{\sqrt{\Gamma(\alpha + n + 1)}} L_n^{(\alpha)}(x).$$

To compute the coefficients of the expansion

$$f = \sum_{n=0}^{\infty} \left(f, \widetilde{L}_n^{(\alpha)}\right) \widetilde{L}_n^{(\alpha)},$$

we may use (4.2.8). If f and its derivatives to order n are bounded as $x \to 0$ and of at most polynomial growth as $x \to +\infty$,

$$(f, L_n^{(\alpha)}) = \int_0^{\infty} f(x) L_n^{(\alpha)}(x) x^\alpha e^{-x} dx = \frac{(-1)^n}{n!} \int_0^{\infty} e^{-x} f^{(n)}(x) x^{n+\alpha} dx.$$

In particular, if $m \geq n$, then

$$\int_0^{\infty} x^m L_n^{(\alpha)}(x) x^\alpha e^{-x} dx = (-1)^n \binom{m}{n} \Gamma(\alpha + m + 1) \qquad (4.5.9)$$

and for $\text{Re}\, a > -1$,

$$\left(e^{-ax}, L_n^{(\alpha)}\right) = \frac{a^n\, \Gamma(n + \alpha + 1)}{n!\, (a + 1)^{n+\alpha+1}}. \tag{4.5.10}$$

For general values of λ and for $c > 1$, the equation

$$xu''(x) + (c - x)u'(x) - \lambda u(x) = 0$$

has a unique solution that is regular at $x = 0$ with $u(0) = 1$. It is known as the *confluent hypergeometric function* or *Kummer function* $_1F_1(\lambda, c; x) = M(\lambda, c; x)$; see Chapter 6. In view of this and (4.5.3),

$$L_n^{(\alpha)}(x) = \frac{(\alpha + 1)_n}{n!}\, {}_1F_1(-n, \alpha + 1; x) = \frac{(\alpha + 1)_n}{n!}\, M(-n, \alpha + 1; x). \tag{4.5.11}$$

The general results about zeros of orthogonal polynomials, together with the results proved in Section 3.5, give the following.

Theorem 4.5.1 *The Laguerre polynomial $L_n^{(\alpha)}(x)$ has n simple roots in the interval*

$$0 < x < 2n + \alpha + 1 + \sqrt{(2n + \alpha + 1)^2 + (1 - \alpha^2)}.$$

The relative maxima of

$$\left| e^{-x/2} x^{(\alpha+1)/2} L_n^{(\alpha)}(x) \right|$$

increase as x increases.

The following asymptotic result of Fejér [94, 95] will be proved in Chapter 10:

$$L_n^{(\alpha)}(x) = \frac{e^{\frac{1}{2}x} n^{\frac{1}{2}\alpha - \frac{1}{4}}}{\sqrt{\pi}\, x^{\frac{1}{2}\alpha + \frac{1}{4}}} \left[\cos\left(2\sqrt{nx} - \frac{1}{2}\left[\alpha + \frac{1}{2}\right]\pi\right) + O\left(n^{-\frac{1}{2}}\right) \right], \tag{4.5.12}$$

as $n \to \infty$, uniformly on any subinterval $0 < \delta \le x \le \delta^{-1}$.

4.6 Jacobi polynomials

The Jacobi polynomials $\left\{P_n^{(\alpha,\beta)}\right\}$ with indices $\alpha, \beta > -1$ are orthogonal with respect to the weight $w(x) = (1 - x)^\alpha (1 + x)^\beta$ on the interval $(-1, 1)$. The norms are

$$\|P_n^{(\alpha,\beta)}\|^2 = \frac{2^{\alpha+\beta+1}\, \Gamma(n + \alpha + 1)\, \Gamma(n + \beta + 1)}{n!\, (2n + \alpha + \beta + 1)\, \Gamma(n + \alpha + \beta + 1)}.$$

Changing the sign of x,

$$P_n^{(\alpha,\beta)}(-x) = (-1)^n P_n^{(\beta,\alpha)}(x). \tag{4.6.1}$$

The $P_n^{(\alpha,\beta)}$ are eigenfunctions:

$$(1 - x^2) \left[P_n^{(\alpha,\beta)} \right]'' + \left[\beta - \alpha - (\alpha + \beta + 2)x \right] \left[P_n^{(\alpha,\beta)} \right]'$$
$$+ n(n + \alpha + \beta + 1) P_n^{(\alpha,\beta)} = 0. \tag{4.6.2}$$

They satisfy the derivative relation

$$\left[P_n^{(\alpha,\beta)} \right]'(x) = \frac{1}{2}(n + \alpha + \beta + 1) P_{n-1}^{(\alpha+1,\beta+1)}(x)$$

and can be defined by the Rodrigues formula (4.2.12)

$$P_n^{(\alpha,\beta)}(x) = \frac{(-1)^n}{n!\, 2^n} (1 - x)^{-\alpha}(1 + x)^{-\beta} \frac{d^n}{dx^n} \left\{ (1 - x)^{\alpha+n}(1 + x)^{\beta+n} \right\}.$$

It follows from the extended form of Leibniz's rule that

$$P_n^{(\alpha,\beta)}(x) = \frac{(-1)^n}{2^n} \sum_{k=0}^{n} (-1)^k \frac{1}{k!\,(n - k)!}$$
$$\times \frac{(\alpha + 1)_n\,(\beta + 1)_n}{(\alpha + 1)_{n-k}\,(\beta + 1)_k} (1 - x)^{n-k}(1 + x)^k, \tag{4.6.3}$$

which gives the endpoint values

$$P_n^{(\alpha,\beta)}(1) = \frac{(\alpha + 1)_n}{n!}, \qquad P_n^{(\alpha,\beta)}(-1) = (-1)^n \frac{(\beta + 1)_n}{n!}. \tag{4.6.4}$$

In view of the discussion in Section 3.5, we have the estimate

$$\sup_{|x|\leq 1} \left| P_n^{(\alpha,\beta)}(x) \right| = \max \left\{ \frac{(\alpha + 1)_n}{n!}, \frac{(\beta + 1)_n}{n!} \right\} \quad \text{if } \alpha \text{ or } \beta \geq -\frac{1}{2}. \tag{4.6.5}$$

Jacobi polynomials with either $\alpha = \pm\frac{1}{2}$ or $\beta = \pm\frac{1}{2}$ can be reduced to those with equal indices by using the following two identities and (4.6.1):

$$P_{2n}^{(\alpha,\alpha)}(x) = \frac{(n + \alpha + 1)_n}{(n + 1)_n} P_n^{\left(\alpha,-\frac{1}{2}\right)}(2x^2 - 1); \tag{4.6.6}$$

$$P_{2n+1}^{(\alpha,\alpha)}(x) = \frac{(n + \alpha + 1)_{n+1}}{(n + 1)_{n+1}} x\, P_n^{\left(\alpha,\frac{1}{2}\right)}(2x^2 - 1).$$

These identities follow from the relationship with hypergeometric functions, (4.6.12), together with two of the quadratic transformations (8.6.2) and (8.6.22). For a direct proof, see the exercises.

The generating function can be calculated from (4.3.2). Taking into account the factor $\left(-\frac{1}{2}\right)^n$ in the normalization (4.2.12), we replace s by $-\frac{1}{2}s$. Since $p(x) = 1 - x^2$, $y(x, s)$ is the solution of

$$y = x - \frac{s}{2}\left(1 - y^2\right)$$

so

$$y = s^{-1}\left[1 - \sqrt{1 - 2xs + s^2}\right];$$

$$\frac{1-y}{1+x} = \frac{2}{1 - s + \sqrt{1 - 2xs + s^2}};$$

$$\frac{1+y}{1+x} = \frac{2}{1 + s + \sqrt{1 - 2xs + s^2}}.$$

Thus

$$\sum_{n=0}^{\infty} P_n^{(\alpha,\beta)}(x)\, s^n = \frac{2^{\alpha+\beta}}{R\,(1 - s + R)^{\alpha}(1 + s + R)^{\beta}}, \quad R = \sqrt{1 - 2xs + s^2}.$$

$$(4.6.7)$$

The Liouville transformation for the Jacobi case starts with the change of variable

$$\theta(x) = \int_1^x \frac{dy}{\sqrt{p(y)}} = \int_1^x \frac{dy}{\sqrt{1 - y^2}} = \cos^{-1} x.$$

In the variable θ the operator in (4.6.2) takes the form

$$\frac{d^2}{d\theta^2} + \frac{\alpha - \beta + (\alpha + \beta + 1)\cos\theta}{\sin\theta}\,\frac{d}{d\theta}.$$

The coefficient of the first-order term can be rewritten as

$$(2\alpha + 1)\frac{\cos\frac{1}{2}\theta}{2\sin\frac{1}{2}\theta} - (2\beta + 1)\frac{\sin\frac{1}{2}\theta}{2\cos\frac{1}{2}\theta}.$$

Therefore this coefficient can be eliminated by the gauge transformation $u = \varphi v$ with

$$\varphi(\theta) = \left(\sin\frac{1}{2}\theta\right)^{-\alpha-\frac{1}{2}}\left(\cos\frac{1}{2}\theta\right)^{-\beta-\frac{1}{2}}.$$

After this gauge transformation, the operator acting on the function v is

$$\frac{d^2}{d\theta^2} + \frac{(\alpha + \beta + 1)^2}{4} + \frac{(1 + 2\alpha)(1 - 2\alpha)}{16\sin^2\frac{1}{2}\theta} + \frac{(1 + 2\beta)(1 - 2\beta)}{16\cos^2\frac{1}{2}\theta} \quad (4.6.8)$$

and the eigenvalue equation with $\lambda_n = n(n + \alpha + \beta + 1)$ is

$$v''(\theta) + \left[\frac{(2n + \alpha + \beta + 1)^2}{4} + \frac{(1 + 2\alpha)(1 - 2\alpha)}{16 \sin^2 \frac{1}{2}\theta} \right.$$

$$\left. + \frac{(1 + 2\beta)(1 - 2\beta)}{16 \cos^2 \frac{1}{2}\theta} \right] v(\theta) = 0. \tag{4.6.9}$$

In particular, if $2\alpha = \pm 1$ and $2\beta = \pm 1$, then (4.6.9) can be solved explicitly. Equation (4.6.9) leads to estimates for the zeros of $P_n^{(\alpha,\beta)}$ for certain α, β: see the exercises.

Theorem 4.6.1 *Suppose $\alpha^2 \le \frac{1}{4}$ and $\beta^2 \le \frac{1}{4}$, and let $\cos\theta_{1n}, \ldots, \cos\theta_{nn}$ be the zeros of $P_n^{(\alpha,\beta)}$, $\theta_{1n} < \cdots < \theta_{nn}$. Then*

$$\frac{(k - 1 + \gamma)\pi}{n + \gamma} \le \theta_{kn} \le \frac{k\pi}{n + \gamma}, \quad \gamma = \frac{1}{2}(\alpha + \beta + 1). \tag{4.6.10}$$

The inequalities are strict unless $\alpha^2 = \beta^2 = \frac{1}{4}$.

The following asymptotic result of Darboux [62] will be proved in Chapter 10:

$$P_n^{(\alpha,\beta)}(\cos\theta) = \frac{\cos\left(n\theta + \frac{1}{2}[\alpha + \beta + 1]\theta - \frac{1}{2}\alpha\pi - \frac{1}{4}\pi\right) + O\left(n^{-\frac{1}{2}}\right)}{\sqrt{n\pi}\left(\sin\frac{1}{2}\theta\right)^{\alpha + \frac{1}{2}}\left(\cos\frac{1}{2}\theta\right)^{\beta + \frac{1}{2}}}$$

$$\tag{4.6.11}$$

as $n \to \infty$, uniformly on any subinterval $\delta \le \theta \le \pi - \delta$, $\delta > 0$.

It is convenient for some purposes, such as making the connection to hypergeometric functions, to rescale the x-interval to $(0, 1)$. Let $y = \frac{1}{2}(1 - x)$. Up to a constant factor the corresponding weight function is $w(y) = y^\alpha(1 - y)^\beta$, while the rescaled polynomials are eigenfunctions for the operator

$$y(1 - y)\frac{d^2}{dy^2} + \left[\alpha + 1 - (\alpha + \beta + 2)y\right]\frac{d}{dy}$$

with eigenvalues $-n(n + \alpha + \beta + 1)$; see Case III in Section 3.4. If we set

$$P_n^{(\alpha,\beta)}(1 - 2y) = \sum_{k=0}^{n} c_k\, y^k$$

then the eigenvalue equation implies the identities

$$c_{k+1} = \frac{(n + \alpha + \beta + 1 + k)(-n + k)}{k(\alpha + 1 + k)} c_k.$$

By (4.6.4), $c_0 = (\alpha + 1)_n / n !$, so

$$P_n^{(\alpha,\beta)}(x) = \frac{(\alpha + 1)_n}{n !} \sum_{k=0}^{n} \frac{(\alpha + \beta + 1 + n)_k (-n)_k}{(\alpha + 1)_k \, k !} y^k$$

$$= \frac{(\alpha + 1)_n}{n !} F\left(\alpha + \beta + 1 + n, -n, \alpha + 1; \frac{1}{2}(1 - x)\right), \quad (4.6.12)$$

where F is the hypergeometric function associated with the equation (1.0.2), (3.7.2) with indices $\alpha + \beta + 1 + n, -n, \alpha + 1$: the solution that has value 1 at $y = 0$.

4.7 Legendre and Chebyshev polynomials

Up to normalization, these are Jacobi polynomials $P_n^{(\alpha,\alpha)}$ with a repeated index $\alpha = \beta$. Note that in any such case the weight function $(1 - x^2)^\alpha$ is an even function. It follows by induction that orthogonal polynomials of even degree are even functions, those of odd degree are odd functions.

The Legendre polynomials $\{P_n\}$ are the case $\alpha = \beta = 0$:

$$P_n(x) = P_n^{(0,0)}(x).$$

The associated weight function is $w(x) \equiv 1$ and the eigenvalue equation is

$$\left(1 - x^2\right) P_n''(x) - 2x \, P_n'(x) + n(n + 1) P_n(x) = 0. \quad (4.7.1)$$

The generating function is

$$\sum_{n=0}^{\infty} P_n(x) s^n = \left(1 - 2xs + s^2\right)^{-\frac{1}{2}}. \quad (4.7.2)$$

The recurrence and derivative formulas (4.3.8) and (4.3.15) specialize to

$$(n + 1) P_{n+1}(x) = (2n + 1)x \, P_n(x) - n P_{n-1}(x); \quad (4.7.3)$$

$$\left(1 - x^2\right) P_n'(x) = -nx P_n(x) + n P_{n-1}(x). \quad (4.7.4)$$

The recurrence relation (4.7.3) and the derivative identity

$$P_n'(x) - 2x P_{n-1}'(x) + P_{n-2}'(x) = P_{n-1}(x) \quad (4.7.5)$$

can be derived from the generating function; see the exercises.

In Sections 9.2 and 8.7 we establish two integral representations:

$$P_n(\cos\theta) = \frac{1}{2\pi} \int_0^{2\pi} (\cos\theta + i\sin\theta\sin\alpha)^n \, d\alpha \qquad (4.7.6)$$

$$= \frac{1}{\pi} \int_0^1 \frac{\cos\left(s\left(n + \frac{1}{2}\right)\theta\right)}{\cos\frac{1}{2}s\theta} \frac{ds}{\sqrt{s(1-s)}}. \qquad (4.7.7)$$

The general formula (4.3.19) for the Dirichlet kernel specializes to

$$K_n^{(0,0)}(x, y) = \frac{n+1}{2} \left[\frac{P_{n+1}(x)P_n(y) - P_n(x)P_{n+1}(y)}{x - y} \right]. \qquad (4.7.8)$$

The Liouville transformation takes the eigenvalue equation to the following equation for $u_n(\theta) = (\sin\theta)^{\frac{1}{2}} P_n(\cos\theta)$, the case $\alpha = \beta = 0$ of (4.6.9):

$$u_n''(\theta) + \left[\left(n + \frac{1}{2}\right)^2 + \frac{1}{4\sin^2\theta} \right] u_n(\theta) = 0. \qquad (4.7.9)$$

The first part of the following result is the specialization to $\alpha = \beta = 0$ of (4.6.5). The second part is the specialization of a result in Section 3.5. The remaining two statements follow from (4.7.9), together with Propositions 3.5.2 and 3.5.3.

Theorem 4.7.1 *The Legendre polynomials satisfy*

$$\sup_{|x|\leq 1} |P_n(x)| = 1.$$

The relative maxima of $|P_n(\cos\theta)|$ *decrease as θ increases for* $0 \leq \theta \leq \frac{1}{2}\pi$ *and increase as θ increases for* $\frac{1}{2}\pi \leq \theta \leq \pi$.

The relative maxima of $|(\sin\theta)^{\frac{1}{2}} P_n(\cos\theta)|$ *increase with θ for* $0 \leq \theta \leq \frac{1}{2}\pi$ *and decrease as θ increases for* $\frac{1}{2}\pi \leq \theta \leq \pi$. *Moreover, for* $0 \leq \theta \leq \pi$,

$$\left[(\sin\theta)^{\frac{1}{2}} P_n(\cos\theta) \right]^2 \leq P_n(0)^2 + \frac{P_n'(0)^2}{\left(n + \frac{1}{2}\right)^2 + \frac{1}{4}}. \qquad (4.7.10)$$

A more explicit form of this last estimate can be derived:

$$\left| (\sin\theta)^{\frac{1}{2}} P_n(\cos\theta) \right| < \left(\frac{2}{n\pi} \right)^{\frac{1}{2}}; \qquad (4.7.11)$$

see the exercises. This is a sharp version due to Bernstein [29] of an earlier result of Stieltjes [272].

Theorem 4.6.1 specializes to the following: let $\cos\theta_{1n}, \ldots, \cos\theta_{nn}$ be the zeros of $P_n(x), \theta_{1n} < \cdots < \theta_{nn}$. Then

$$\frac{\left(k - 1 + \frac{1}{2}\right)\pi}{n + \frac{1}{2}} < \theta_{kn} < \frac{k\pi}{n + \frac{1}{2}}. \tag{4.7.12}$$

The following inequality is due to Turán [278, 290]:

$$P_n(x)^2 - P_{n-1}(x)\, P_{n+1}(x) \geq 0. \tag{4.7.13}$$

The Chebyshev (or Tchebycheff) polynomials $\{T_n\}$ and $\{U_n\}$ are the cases $\alpha = \beta = -\frac{1}{2}$ and $\alpha = \beta = \frac{1}{2}$, respectively:

$$T_n(x) = \frac{2 \cdot 4 \cdot 6 \cdots (2n)}{1 \cdot 3 \cdot 5 \cdots (2n-1)}\, P_n^{\left(-\frac{1}{2}, -\frac{1}{2}\right)}(x) = \frac{n!}{\left(\frac{1}{2}\right)_n}\, P_n^{\left(-\frac{1}{2}, -\frac{1}{2}\right)}(x);$$

$$U_n(x) = \frac{4 \cdot 6 \cdots (2n+2)}{3 \cdot 5 \cdot \cdots (2n+1)}\, P_n^{\left(\frac{1}{2}, \frac{1}{2}\right)}(x) = \frac{(n+1)!}{\left(\frac{3}{2}\right)_n}\, P_n^{\left(\frac{1}{2}, \frac{1}{2}\right)}(x).$$

Thus $T_n(1) = 1, U_n(1) = n + 1$.

The Gegenbauer polynomials or ultraspherical polynomials $\left\{C_n^\lambda\right\}$ are the general case $\alpha = \beta$, normalized as follows:

$$C_n^\lambda(x) = \frac{(2\lambda)_n}{\left(\lambda + \frac{1}{2}\right)_n}\, P_n^{\left(\lambda - \frac{1}{2}, \lambda - \frac{1}{2}\right)}(x).$$

In particular,

$$C_n^0(x) = T_n(x), \quad C_n^{\frac{1}{2}}(x) = P_n(x), \quad C_n^1(x) = U_n(x).$$

The Chebyshev polynomials simplify considerably under the Liouville transformation and gauge transformation considered above. With $\alpha = \beta = -\frac{1}{2}$ the operator (4.6.8) has zero-order term and the eigenvalue $-\lambda_n$ is $-n^2$. Therefore the solutions of (4.6.2) that are even (resp. odd) functions of $\cos\theta$ when n is even (resp. odd) are multiples of $\cos(n\theta)$. The normalization gives the Chebyshev polynomials of the first kind

$$T_n(\cos\theta) = \cos n\theta. \tag{4.7.14}$$

Since

$$\frac{dx}{\sqrt{1 - x^2}} = d\theta,$$

the square of the L^2 norm is π if $n = 0$, and

$$\int_0^\pi \cos^2 n\theta\, d\theta = \frac{\pi}{2}, \quad n > 0.$$

The recurrence and derivative formulas are easily derived from the trigonometric identities

$$\cos(n\theta \pm \theta) = \cos n\theta \cos \theta \mp \sin n\theta \sin \theta.$$

These imply that

$$T_{n+1}(x) + T_{n-1}(x) = 2x\, T_n(x), \quad n \geq 1. \tag{4.7.15}$$

Also

$$\left(1 - x^2\right) T_n'(x)\Big|_{x=\cos\theta} = -\sin\theta \frac{d}{d\theta} \cos n\theta = n \sin n\theta \sin\theta,$$

so

$$\left(1 - x^2\right) T_n'(x) = -nx T_n(x) + n T_{n-1}(x). \tag{4.7.16}$$

It follows from (4.7.15) and the norm calculation that the associated Dirichlet kernel is

$$
\begin{aligned}
K_n^T(x, y) &= \frac{2}{\pi} \sum_{k=0}^{n} T_k(x) T_k(y) \\
&= \frac{1}{\pi} \left[\frac{T_{n+1}(x) T_n(y) - T_n(x) T_{n+1}(y)}{x - y} \right]
\end{aligned}
\tag{4.7.17}
$$

for $n > 0$.

Note that $T_0 = 1$ and $T_1(x) = x$. Together with (4.7.15), this allows us to calculate an alternative generating function

$$G_T(x, s) = \sum_{n=0}^{\infty} T_n(x)\, s^n.$$

In fact,

$$s^{-1} \left[G_T(x, s) - xs - 1 \right] + s\, G_T(x, s) = 2x \left[G_T(x, s) - 1 \right]$$

so

$$\sum_{n=0}^{\infty} T_n(x)\, s^n = \frac{1 - xs}{1 - 2xs + s^2}. \tag{4.7.18}$$

With $\alpha = \beta = \frac{1}{2}$ the gauge transformation $u(\theta) = (\sin\theta)^{-1} v(\theta)$ reduces the eigenvalue equation to

$$v''(x) + [1 + n(n+2)]v(x) = v''(x) + (n+1)^2 v(x) = 0,$$

so the solutions u that are regular at $\theta = 0$ are multiples of $\sin(n + 1)\theta / \sin \theta$. The normalization gives the Chebyshev polynomials of the second kind

$$U_n(\cos \theta) = \frac{\sin(n + 1)\theta}{\sin \theta}. \tag{4.7.19}$$

The weight function here is $\sin \theta$, so the square of the L^2 norm is

$$\int_0^\pi \sin^2 (n + 1)\theta \, d\theta = \frac{\pi}{2}.$$

The recurrence and derivation formulas are easily derived from the trigonometric identities

$$\sin(n\theta \pm \theta) = \sin n\theta \cos \theta \pm \cos n\theta \sin \theta.$$

These imply that

$$U_{n+1}(x) + U_{n-1}(x) = 2x \, U_n(x). \tag{4.7.20}$$

Also

$$\left(1 - x^2\right) U_n'(x)\Big|_{x=\cos\theta} = -\sin\theta \frac{d}{d\theta} \left\{ \frac{\sin (n + 1)\theta}{\sin \theta} \right\}$$

$$= \frac{-(n + 1)\cos (n + 1)\theta \sin \theta + \sin (n + 1)\theta \cos \theta}{\sin \theta},$$

so

$$\left(1 - x^2\right) U_n'(x) = -nx \, U_n(x) + (n + 1)U_{n-1}(x). \tag{4.7.21}$$

It follows from the formulas following (4.3.16) and the norm calculation that the associated Dirichlet kernel is

$$K_n^U(x, y) = \frac{2}{\pi} \sum_{k=0}^n U_k(x)U_k(y)$$

$$= \frac{1}{\pi} \left[\frac{U_{n+1}(x)U_n(y) - U_n(x)U_{n+1}(y)}{x - y} \right]. \tag{4.7.22}$$

Note that $U_0 = 1$ and $U_1(x) = 2x$. These facts and the recurrence relation allow us to compute an alternative generating function in analogy with G_T above:

$$\sum_{n=0}^\infty U_n(x) s^n = \frac{1}{1 - 2xs + s^2}. \tag{4.7.23}$$

The remaining cases when the eigenvalue equation (4.6.9) can be solved immediately are $\alpha = -\frac{1}{2}$, $\beta = \frac{1}{2}$ and $\alpha = \frac{1}{2}$, $\beta = -\frac{1}{2}$. In each case the eigenvalue parameter for degree n is $\lambda_n = n(n + 1)$, so after the gauge transformation the constant term is $\lambda_n + \frac{1}{4} = \left(n + \frac{1}{2}\right)^2$. In the first case the gauge function is $\left(\cos \frac{1}{2}\theta\right)^{-1}$ and the value at $\theta = 0$ should be $\left(\frac{1}{2}\right)_n / n!$, so

$$P_n^{\left(-\frac{1}{2}, \frac{1}{2}\right)}(\cos\theta) = \frac{\left(\frac{1}{2}\right)_n}{n!} \frac{\cos\left(n + \frac{1}{2}\right)\theta}{\cos \frac{1}{2}\theta}. \tag{4.7.24}$$

In the second case the gauge function is $\left(\sin \frac{1}{2}\theta\right)^{-1}$ and the value at $\theta = 0$ should be $\left(\frac{3}{2}\right)_n / n!$, so

$$P_n^{\left(\frac{1}{2}, -\frac{1}{2}\right)}(\cos\theta) = \frac{\left(\frac{1}{2}\right)_n}{n!} \frac{\sin\left(n + \frac{1}{2}\right)\theta}{\sin \frac{1}{2}\theta}. \tag{4.7.25}$$

Combining the results of this section with (4.6.12) gives explicit evaluations of hypergeometric functions associated with the Jacobi indices $\alpha = \pm\frac{1}{2}$, $\beta = \pm\frac{1}{2}$:

$$F\left(n, -n, \frac{1}{2}; \frac{1}{2}[1 - \cos\theta]\right) = \cos(n\theta); \tag{4.7.26}$$

$$F\left(n + 2, -n, \frac{3}{2}; \frac{1}{2}[1 - \cos\theta]\right) = \frac{\sin(n + 1)\theta}{(n + 1)\sin\theta}; \tag{4.7.27}$$

$$F\left(n + 1, -n, \frac{1}{2}; \frac{1}{2}[1 - \cos\theta]\right) = \frac{\cos\left(n + \frac{1}{2}\right)\theta}{\cos \frac{1}{2}\theta}; \tag{4.7.28}$$

$$F\left(n + 1, -n, \frac{3}{2}; \frac{1}{2}[1 - \cos\theta]\right) = \frac{\sin\left(n + \frac{1}{2}\right)\theta}{(2n + 1)\sin \frac{1}{2}\theta}. \tag{4.7.29}$$

The previous arguments show that these identities are valid for all values of the parameter n. It will be shown in Section 8.7 that the integral representation (4.7.7) is a consequence of (4.7.28).

4.8 Expansion theorems

Suppose that w is one of the weights associated with the classical orthogonal polynomials:

$$w_H(x) = e^{-x^2}, \quad -\infty < x < \infty;$$

$$w_\alpha(x) = x^\alpha e^{-x}, \quad 0 < x < \infty;$$

$$w_{\alpha\beta}(x) = (1 - x)^\alpha(1 + x)^\beta, \quad -1 < x < 1,$$

and suppose that $\{\varphi_n\}_{n=0}^\infty$ is the corresponding set of orthonormal polynomials: the normalized version of the Hermite, Laguerre, or Jacobi polynomials associated with w. It follows from Theorem 4.1.5 that if f is a function in L_w^2, then the series

$$\sum_{n=0}^\infty (f, \varphi_n)\, \varphi_n(x) \qquad (4.8.1)$$

converges to f in the L^2 sense:

$$\lim_{n\to\infty} ||f_n - f|| \to 0, \quad f_n(x) = \sum_{k=0}^n (f, \varphi_k)\, \varphi_k(x).$$

The partial sums are given by integration against the associated Dirichlet kernel:

$$f_n(x) = \int_I K_n(x, y)\, f(y)\, w(y)\, dy, \quad K_n(x, y) = \sum_{k=0}^n \varphi_k(x)\, \varphi_k(y).$$

The kernel K_n can be written in more compact form by using the Christoffel–Darboux formula (4.1.4). Taking $f \equiv 1$ and using orthogonality shows that

$$1 = \int_I K_n(x, y)\, w(y)\, dy. \qquad (4.8.2)$$

In this section we consider pointwise convergence. In each case, if f belongs to L_w^2, then the series (4.8.1) converges to $f(x)$ at each point where f is differentiable. In fact, we may replace differentiability at x with the weaker condition that for some $\delta > 0$,

$$\left| \frac{f(y) - f(x)}{y - x} \right| \le C \quad \text{for} \quad 0 < |y - x| < \delta. \qquad (4.8.3)$$

If f is piecewise continuously differentiable, then this condition is satisfied at every point of continuity.

We begin with the Hermite case. According to (4.2.13), the normalized polynomials can be taken to be

$$\varphi_n(x) = a_n\, H_n(x), \quad a_n = \frac{1}{\sqrt{2^n\, n!}\, \pi^{\frac{1}{4}}}.$$

Thus in this case the expansion (4.8.1) is

$$\sum_{n=0}^\infty c_n\, H_n(x), \quad c_n = \frac{1}{2^n\, n!\, \sqrt{\pi}} \int_{-\infty}^\infty f(x)\, H_n(x)\, e^{-x^2}\, dx. \qquad (4.8.4)$$

Theorem 4.8.1 *Suppose that $f(x)$ is a real-valued function that satisfies*

$$\int_{-\infty}^{\infty} f(x)^2 e^{-x^2} dx < \infty, \tag{4.8.5}$$

and suppose that f satisfies the condition (4.8.3) at the point x. Then the series (4.8.4) converges to $f(x)$.

It follows from (4.8.2) that

$$f(x) - f_n(x) = \int_{-\infty}^{\infty} K_n^H(x, y) \left[f(y) - f(x) \right] e^{-y^2} dy.$$

The Dirichlet kernel here is given by (4.3.17):

$$K_n^H(x, y) = \frac{1}{2^{n+1} n! \sqrt{\pi}} \left[\frac{H_{n+1}(x) H_n(y) - H_n(x) H_{n+1}(y)}{x - y} \right].$$

Therefore

$$f(x) - f_n(x) = \frac{H_{n+1}(x)}{2^{n+1} n! \sqrt{\pi}} \int_{-\infty}^{\infty} H_n(y) g(x, y) e^{-y^2} dy$$

$$- \frac{H_n(x)}{2^{n+1} n! \sqrt{\pi}} \int_{-\infty}^{\infty} H_{n+1}(y) g(x, y) e^{-y^2} dy, \tag{4.8.6}$$

where

$$g(x, y) = \frac{f(y) - f(x)}{y - x}.$$

For $|y| \geq 2|x|$,

$$y^2 g(x, y)^2 \leq 8 \left[f(x)^2 + g(y)^2 \right].$$

Together with assumptions (4.8.3) and (4.8.5), this implies that

$$\int_{-\infty}^{\infty} (1 + y^2) g(x, y)^2 e^{-y^2} dy < \infty. \tag{4.8.7}$$

From (4.4.20) and Stirling's formula (2.5.1), we have the estimates

$$|H_n(x)| \leq A(x) n^{-\frac{1}{4}} (2^n n!)^{\frac{1}{2}}.$$

Thus it is enough to prove that both the integrals

$$\frac{n^{\frac{1}{4}}}{(2^n n!)^{\frac{1}{2}}} \int_{-\infty}^{\infty} H_n(y) g(x, y) e^{-y^2} dy, \quad \frac{n^{-\frac{1}{4}}}{(2^n n!)^{\frac{1}{2}}} \int_{-\infty}^{\infty} H_{n+1}(y) g(x, y) e^{-y^2} dy$$

have limit zero as $n \to \infty$. Changing n to $n - 1$ in the second expression shows that the two expressions are essentially the same. Thus it is enough to prove the following lemma.

Lemma 4.8.2 *If*

$$\int_{-\infty}^{\infty} \left(1 + x^2\right) g(x)^2 e^{-x^2} \, dx < \infty,$$

then

$$\lim_{n \to \infty} \frac{n^{\frac{1}{4}}}{(2^n \, n!)^{\frac{1}{2}}} \int_{-\infty}^{\infty} H_n(x) \, g(x) \, e^{-x^2} \, dx = 0. \tag{4.8.8}$$

We begin with an estimate for an integral involving H_n.

Lemma 4.8.3 *There is a constant B such that*

$$\int_{-\infty}^{\infty} \frac{H_n(x)^2}{1 + x^2} e^{-x^2} \, dx \le B \, 2^n \, n! \, n^{-\frac{1}{2}} \tag{4.8.9}$$

for all n.

Proof First,

$$J_n \equiv \frac{1}{2^n \, n!} \int_{-\infty}^{\infty} \frac{H_n(x)^2}{1 + x^2} e^{-x^2} \, dx = \int_{-\infty}^{\infty} \left(\frac{1 - x^2}{1 + x^2}\right)^n \frac{e^{-x^2}}{1 + x^2} \, dx$$

$$= 2 \int_{0}^{\infty} \left(\frac{1 - x^2}{1 + x^2}\right)^n \frac{e^{-x^2}}{1 + x^2} \, dx; \tag{4.8.10}$$

see Exercise 4.16. Making the change of variable $x \to 1/x$ in the integral over the interval $[1, \infty)$ converts the integral to

$$J_n = 2 \int_{0}^{1} \left(\frac{1 - x^2}{1 + x^2}\right)^n \frac{e^{-x^2} + (-1)^n \, e^{-1/x^2}}{1 + x^2} \, dx$$

$$\le 4 \int_{0}^{1} \left(\frac{1 - x^2}{1 + x^2}\right)^n \frac{dx}{1 + x^2}.$$

Let $t = 1 - \left(1 - x^2\right)^2/(1 + x^2)^2 = 4x^2/\left(1 + x^2\right)^2$, so that

$$\sqrt{1 - t} = \frac{2}{1 + x^2} - 1, \qquad \frac{4dx}{1 + x^2} = \frac{dt}{\sqrt{t(1 - t)}}.$$

Then

$$4 \int_{0}^{1} \left(\frac{1 - x^2}{1 + x^2}\right)^n \frac{dx}{1 + x^2} = \int_{0}^{1} (1 - t)^{\frac{1}{2}n} \frac{dt}{\sqrt{t(1 - t)}} = B\left(\frac{1}{2}, \frac{1}{2}n + \frac{1}{2}\right).$$

By (2.1.9), the last expression is $O\left(n^{-\frac{1}{2}}\right)$, which gives (4.8.9).   □

It follows from this result and the Cauchy–Schwarz inequality that

$$\left(\int_{-\infty}^{\infty} H_n(x) \, g(x) \, e^{-x^2} \, dx \right)^2 \leq B \, 2^n \, n! \, n^{-\frac{1}{2}} \int_{-\infty}^{\infty} g(x)^2 \big(1 + x^2\big) e^{-x^2} \, dx \tag{4.8.11}$$

for all n, if the integral on the right is finite; here B is the constant in (4.8.9).

We can now prove the first lemma, and thus complete the proof of Theorem 4.8.1. Given $\varepsilon > 0$ we can choose $N = N(\varepsilon)$ so large that

$$\int_{|x| > N} g(x)^2 \big(1 + x^2\big) e^{-x^2} \, dx < \frac{\varepsilon^2}{B}.$$

In view of (4.8.11), up to ε we only need to consider the integral (4.8.8) over the bounded interval $[-N, N]$. The asymptotic estimate (4.4.20) is uniform over such an interval. The product of the constant in (4.4.20) and the constant in (4.8.8) is independent of n. Thus it is enough to consider the integrals

$$\int_{-N}^{N} \cos\left(\sqrt{2n+1}\, x - \frac{1}{2} n\pi \right) g(x) \, e^{-x^2/2} \, dx, \qquad \int_{-N}^{N} n^{-\frac{1}{2}} g(x) \, e^{-x^2/2} \, dx.$$

Since N is fixed, the second integral is $O\big(n^{-\frac{1}{2}}\big)$. The first integral tends to zero as $n \to \infty$ by the Riemann–Lebesgue lemma: see Appendix B. (Consider separately the cases n even, n odd.)

This result, and the ones to follow, can be extended to points of discontinuity: if $f \in L_w^2$ has one-sided limits $f(x\pm)$ at x, and one-sided estimates

$$\frac{|f(y) - f(x\pm)|}{|y - x|} \leq C \quad \text{for} \quad 0 < \pm(y - x) < \delta, \tag{4.8.12}$$

then

$$\lim_{n \to \infty} \sum_{k=0}^{n} (f, \varphi_k)\, \varphi_k(x) = \frac{1}{2}\big[f(x+) + f(x-) \big]; \tag{4.8.13}$$

see the exercises.

The Laguerre case is quite similar. According to (4.2.14), the normalized polynomials for the weight w_α on the interval $(0, \infty)$ can be taken to be

$$\varphi_n(x) = a_n \, L_n^{(\alpha)}(x), \qquad a_n = \left[\frac{n!}{\Gamma(n + \alpha + 1)} \right]^{1/2}.$$

Therefore the corresponding expansion of a function $f \in L_{w_\alpha}^2$ is

$$\sum_{n=0}^{\infty} c_n \, L_n^{(\alpha)}(x), \qquad c_n = \frac{n!}{\Gamma(n + \alpha + 1)} \int_0^{\infty} f(x) \, L_n^{(\alpha)}(x) \, x^\alpha \, e^{-x} \, dx. \tag{4.8.14}$$

Theorem 4.8.4 *Suppose that $f(x)$ is a real-valued function that satisfies*

$$\int_{-\infty}^{\infty} f(x)^2 x^{\alpha} e^{-x} \, dx < \infty, \tag{4.8.15}$$

and suppose that f satisfies the condition (4.8.3) at the point x, $0 < x < \infty$. Then the series (4.8.14) converges to $f(x)$.

The argument here is similar to, but somewhat more complicated than, the proof of Theorem 4.8.1. We refer to Uspensky's paper [291] for the details.

The proof of the corresponding result for Jacobi polynomials is simpler, because we have already established the necessary estimates. According to (4.2.15), the orthonormal polynomials for the weight $w_{\alpha\beta}$ on the interval $(-1, 1)$ can be taken to be

$$\varphi_n(x) = a_n \, P_n^{(\alpha,\beta)}(x), \quad a_n = \left[\frac{n! \, (2n + \alpha + \beta + 1) \, \Gamma(n + \alpha + \beta + 1)}{2^{\alpha+\beta+1} \Gamma(n + \alpha + 1) \, \Gamma(n + \beta + 1)} \right]^{1/2}.$$

Therefore the corresponding expansion of a function f in $L^2_{w_{\alpha\beta}}$ is

$$\sum_{n=0}^{\infty} c_n \, P_n^{(\alpha,\beta)}(x), \quad c_n = \frac{n! \, (2n + \alpha + \beta + 1) \, \Gamma(n + \alpha + \beta + 1)}{2^{\alpha+\beta+1} \Gamma(n + \alpha + 1) \, \Gamma(n + \beta + 1)}$$

$$\times \int_{-1}^{1} f(x) \, P_n^{(\alpha,\beta)}(x) \, (1 - x)^{\alpha} (1 + x)^{\beta} \, dx.$$

$$\tag{4.8.16}$$

Theorem 4.8.5 *Suppose that $f(x)$ is a real–valued function that satisfies*

$$\int_{-1}^{1} f(x)^2 \, (1 - x)^{\alpha} (1 + x)^{\beta} \, dx < \infty, \tag{4.8.17}$$

and suppose that f satisfies the condition (4.8.3) at the point x, $-1 < x < 1$. Then the series (4.8.16) converges to $f(x)$.

Proof   It follows from (4.8.2) that the partial sums f_n of the series (4.8.16) satisfy

$$f_n(x) - f(x) = \int_{-1}^{1} K_n^{(\alpha,\beta)}(x, y) \left[f(y) - f(x) \right] (1 - y)^{\alpha} (1 + y)^{\beta} \, dy,$$

where the Dirichlet kernel here is given by (4.3.19)

$$K_n^{(\alpha,\beta)}(x, y) = \frac{2^{-\alpha-\beta} \, (n + 1)! \, \Gamma(n + \alpha + \beta + 2)}{(2n + \alpha + \beta + 2) \Gamma(n + \alpha + 1) \, \Gamma(n + \beta + 1)}$$

$$\times \left[\frac{P_{n+1}^{(\alpha,\beta)}(x) P_n^{(\alpha,\beta)}(y) - P_n^{(\alpha,\beta)}(x) P_{n+1}^{(\alpha,\beta)}(y)}{x - y} \right].$$

It follows from (2.1.9) that the coefficient here is $O(n)$ as $n \to \infty$. The asymptotic result (4.6.11) implies that $P_n^{(\alpha,\beta)}(x)$ is $O(n^{-\frac{1}{2}})$ as $n \to \infty$. Thus to prove the result we only need to show that the inner product

$$(P_n^{(\alpha,\beta)}, h) \equiv \int_{-1}^{1} P_n^{(\alpha,\beta)}(y)\, h(y)\, (1-y)^\alpha (1+y)^\beta \, dy = o\left(n^{-\frac{1}{2}}\right)$$

as $n \to \infty$, where $h(y) = [f(y) - f(x)]/(y - x)$.

Let $\{Q_n\}$ be the normalized polynomials

$$Q_n(x) = \|P_n^{(\alpha,\beta)}\|^{-1} P_n^{(\alpha,\beta)}(x).$$

By (4.2.15) and (2.1.9), $\|P_n^{(\alpha,\beta)}\|$ is $O(n^{-1/2})$. Therefore it is enough to show that

$$(Q_n, h) = o(1). \tag{4.8.18}$$

The assumptions (4.8.3) and (4.8.15) imply that h is square-integrable, so (4.8.18) follows from (4.1.12). □

4.9 Functions of second kind

A principal result of Section 4.3 was that, in the three cases considered, the function

$$u(x) = \frac{1}{w(x)} \int_C \frac{p^\nu(z)\, w(z)\, dz}{(z-x)^{\nu+1}} \tag{4.9.1}$$

is a solution of the equation

$$\left(pwu'\right)' + \lambda_\nu wu = 0, \qquad \lambda_\nu = -\nu q' - \frac{1}{2}\nu(\nu-1)p'' \tag{4.9.2}$$

for non-negative integer values of ν. In each case p was a polynomial of degree at most 2, q a polynomial of degree at most 1, $qw = (pw)'$, and the contour C enclosed x but excluded the zeros of p. This was derived as a consequence of the Rodrigues equation (4.2.5). We give now, under the assumptions just stated, a direct proof that (4.9.1) implies (4.9.2). We then note how similar integral representations of solutions can be obtained for arbitrary ν.

First,

$$p(x)w(x)u'(x) = (\nu+1)p(x) \int_C \frac{p(z)^\nu\, w(z)\, dz}{(z-x)^{\nu+2}}$$

$$- \frac{p(x)w'(x)}{w(x)} \int_C \frac{p(z)^\nu\, w(z)\, dz}{(z-x)^{\nu+1}}.$$

By assumption

$$-pw' = -(pw)' + p'w = (p' - q)w,$$

and since p and q have degree at most 2 and 1, respectively,

$$p(x) = p(z) - p'(z)(z - x) + \frac{1}{2}p''(z)(z - x)^2,$$

$$p'(x) - q(x) = p'(z) - q(z) + [q'(z) - p''(z)](z - x)$$

$$= -\frac{p(z)w'(z)}{w(z)} + [q'(z) - p''(z)](z - x).$$

Combining the three preceding equations gives

$$(pwu')(x) = (v + 1) \int_C \frac{p(z)^{v+1}w(z)\,dz}{(z - x)^{v+2}}$$

$$- \int_C \frac{(v + 1)p(z)^v p'(z)w(z) + p(z)^{v+1}w'(z)\,dz}{(z - x)^{v+1}}$$

$$+ \left[q' + \frac{1}{2}(v - 1)p'' \right] \int_C \frac{p(z)^v w(z)\,dz}{(z - x)^v}$$

$$= (v + 1) \int_C \frac{p(z)^{v+1}w(z)\,dz}{(z - x)^{v+2}} - \int_C \frac{\left[p^{v+1}w \right]'(z)\,dz}{(z - x)^{v+1}}$$

$$+ \left[q' + \frac{1}{2}(v - 1)p'' \right] \int_C \frac{p(z)^v w(z)\,dz}{(z - x)^v}.$$

Differentiating,

$$(pwu')'(x) = (v + 1)(v + 2) \int_C \frac{p(z)^{v+1}w(z)\,dz}{(z - x)^{v+3}}$$

$$- (v + 1) \int_C \frac{\left[p^{v+1}w \right]'(z)\,dz}{(z - x)^{v+2}} + \left[vq' + \frac{1}{2}v(v - 1)p'' \right] u(x)w(x).$$

$$(4.9.3)$$

Since

$$\frac{v + 2}{(z - x)^{v+3}} = -\frac{d}{dz}\left\{ \frac{1}{(z - x)^{v+2}} \right\},$$

an integration by parts gives

$$(v + 2) \int_C \frac{p(z)^{v+1}\,w(z)\,dz}{(z - x)^{v+3}} = \int_C \frac{(p^{v+1}w)'(z)\,dz}{(z - x)^{v+2}}. \qquad (4.9.4)$$

Therefore (4.9.3) reduces to (4.9.2)

For the contour just discussed, the assumption that v is an integer was necessary in order for the contour to lie in one branch of $(s - x)^{-v-1}$: otherwise the integration by parts results in contributions from the points of C where one crosses from one branch to another. On the other hand, the argument would apply to a contour that lies (except possibly for its endpoints) in one branch of the power, provided that the endpoint contributions vanish. This provides a way to obtain a solution of (4.9.2) for more general values of v.

For each of three cases considered above, let $I = (a, b)$ be the associated real interval, and define a function of *second kind*

$$u_v(x) = \frac{c_v}{w(x)} \int_a^b \frac{p(s)^v \, w(s) \, ds}{(s - x)^{v+1}}, \quad \text{Re } v \geq 0, \; x \notin I. \tag{4.9.5}$$

In each case $p^{v+1} w$ vanishes at a finite endpoint and vanishes exponentially at an infinite endpoint of the interval so long as $\text{Re } v \geq 0$, so that once again (4.9.3) leads to (4.9.4). In addition, the argument that led to the recurrence relations and derivative formulas also carries over to the functions of second kind. Summarizing, we have the following result.

Theorem 4.9.1 *For* $\text{Re } v \geq 0$ *and* x *not real, the functions*

$$u_{v1}(x) = (-1)^n \, n! \, e^{x^2} \int_{-\infty}^{\infty} \frac{e^{-s^2} \, ds}{(s - x)^{v+1}}; \tag{4.9.6}$$

$$u_{v2}(x) = x^{-\alpha} e^x \int_0^{\infty} \frac{s^{v+\alpha} e^{-s} \, ds}{(s - x)^{v+1}}; \tag{4.9.7}$$

$$u_{v3}(x) = \frac{1}{2^v} (1 - x)^{-\alpha} (1 + x)^{-\beta} \int_{-1}^1 \frac{(1 - s)^{v+\alpha} (1 + s)^{v+\beta} \, ds}{(s - x)^{v+1}} \tag{4.9.8}$$

satisfy the equations

$$u_{v1}''(x) - 2x \, u_{v1}'(x) + 2v u_{v1}(x) = 0;$$

$$x \, u_{v2}''(x) + (\alpha + 1 - x) \, u_{v2}'(x) + v u_{v2}(x) = 0;$$

$$\left(1 - x^2\right) u_{v3}''(x) + \left[\beta - \alpha - (\alpha + \beta + 2)x\right] u_{v3}'(x)$$
$$+ v(v + \alpha + \beta + 1) u_{v3}(x) = 0$$

respectively. Moreover, for $\text{Re } v \geq 1$ *they satisfy the corresponding recurrence and derivative equations* (4.3.6), (4.3.9) *for* u_{v1}, (4.3.7), (4.3.10), (4.3.14) *for* u_{v2}, *and* (4.3.8), (4.3.11), (4.3.14) *for* u_{v3}, *with n replaced by v.*

As the preceding proof shows, the basic result here holds in greater generality than the three specific cases considered above.

Theorem 4.9.2 *Suppose that* p *is a polynomial of degree at most 2,* q *a polynomial of degree at most 1,* $pw'/w = q - p'$, *and* ν, x_0 *are complex numbers. Suppose that* C *is an oriented contour in a region where* p^ν *and* $(z - x_0)^\nu$ *are holomorphic, and that the functions*

$$\frac{p(z)^\nu \, w(z)}{(z - x_0)^\nu}, \quad \frac{p(z)^{\nu+1} \, w(z)}{(z - x_0)^{\nu+1}}$$

are integrable on C, *while the limit along the curve of*

$$\frac{p(z)^{\nu+1} \, w(z)}{(z - x_0)^{\nu+2}}$$

as z *approaches the (finite or infinite) endpoints* a *and* b *is zero. Then the function*

$$u_\nu(x) = \frac{1}{w(x)} \int_C \frac{p(z)^\nu \, w(z) \, dz}{(z - x)^{\nu+1}}$$

is a solution of the equation

$$(pwu')' + \lambda_\nu w \, u = 0, \quad \lambda_\nu = -\nu q' - \frac{1}{2}\nu(\nu - 1)p''$$

in any region containing x_0 *in which the assumptions continue to hold.*

4.10 Exercises

4.1 Show that the kernel K_n in Proposition 4.1.1 is uniquely determined by the conditions (i) it is a polynomial of degree n in x and in y and (ii) the identity (4.1.5) holds for every polynomial of degree $\leq n$.

4.2 Prove (4.1.6). Hint: show that the function on the right-hand side satisfies the conditions in Exercise 4.1.

4.3 Verify that the normalizations (4.2.10), (4.2.11), (4.2.12) lead to (4.2.13), (4.2.14), and (4.2.15), respectively.

4.4 Verify the limit (4.2.17), at least for the leading coefficients.

4.5 Verify the limit (4.2.18), at least for the leading coefficients.

4.6 Verify one or more of the identities (4.3.5).

4.7 Use (4.3.5) and the normalizations (4.2.10), (4.2.11), (4.2.12) to verify the identities (4.3.6), (4.3.7), (4.3.8).

4.8 Use (4.2.7) and the normalizations (4.2.10), (4.2.11), (4.2.12) to verify the identities (4.3.9), (4.3.10), (4.3.11).

4.9 Use the method for (4.3.3) applied to (4.3.12) to derive one or more of the identities (4.3.13), (4.3.14), (4.3.15).

4.10 Prove a converse to the result in Section 4.4: the recurrence relation
(4.4.4) can be derived from the generating function formula (4.4.6).

4.11 Prove the addition formula (4.4.7).

4.12 Prove the addition formula (4.4.8).

4.13 Prove

$$\lim_{n \to \infty} \left(\frac{x}{n}\right)^n H_n\left(\frac{n}{2x}\right) = e^{-x^2}.$$

4.14 Prove that if $a^2 + b^2 = 1$, then

$$H_n(ax + by) = \sum_{k=0}^{n} \binom{n}{k} H_{n-k}(x) H_k(y) a^{n-k} b^k.$$

4.15 Let $u_n(x) = e^{-x^2/2} H_n(x)$. Prove that the Fourier transform of u_n,

$$\frac{1}{\sqrt{2\pi}} \int_{-\infty}^{\infty} e^{-ix\xi} u_n(x) \, dx,$$

is $(-i)^n u_n(\xi)$. Hint: use the identity $\exp(-ix\xi + \frac{1}{2}x^2) = e^{\xi^2/2} \exp\left(\frac{1}{2}[x - i\xi]^2\right)$.

4.16 Prove that

$$\frac{1}{2^n n!} \int_{-\infty}^{\infty} \frac{H_n(x)^2}{1 + x^2} e^{-x^2} \, dx = \int_{-\infty}^{\infty} \left(\frac{1 - x^2}{1 + x^2}\right)^n \frac{e^{-x^2}}{1 + x^2} \, dx.$$

Hint: (i) set $y = x$ in (4.4.17) and multiply both sides by $e^{-x^2}/(1 + x^2)$.
(ii) Integrate the resulting equation on both sides with respect to x and
evaluate the integral on the right by making the change of variables
$x = \sqrt{(1 + s)/(1 - s)} \, y$.

4.17 Prove the lower bounds in Theorem 4.4.2 by using the gauge
transformation $H_n(x) = e^{x^2/2} h_n(x)$ and noting that the zeros of H_n and
h_n coincide.

4.18 Prove the upper bounds in Theorem 4.4.2 by using the gauge
transformation $H_n(x) = e^{x^2/2} h_n(x)$ and then writing $h_n(x) = u_n(y)$,
$y = x\sqrt{2n + 1}$. Use Corollary 3.3.5 to relate the kth positive zero of u_n
to that of u_{2k+1} (n even) or u_{2k+1} (n odd), and thus relate x_{kn} to
$x_{k,2k} < \sqrt{4k + 1}$ or to $x_{k,2k+1} < \sqrt{4k + 3}$.

4.19 Prove

$$\lim_{\alpha \to \infty} \alpha^{-n} L_n^{(\alpha)}(\alpha x) = \frac{(1 - x)^n}{n!}.$$

4.20 Verify (4.5.5) by comparing coefficients, as in the derivation of (4.4.4).

4.21 Show that (4.5.5) can be derived from the generating function formula
(4.5.7).

4.22 Prove the addition formula (4.5.8).

4.23 Prove the addition formula

$$L_n^{(\alpha+\beta+1)}(x+y) = \sum_{k=0}^{n} L_{n-k}^{(\alpha)}(x) L_k^{(\beta)}(y).$$

4.24 Expand the integrand in series to prove Koshlyakov's formula [163]:

$$L_n^{(\alpha+\beta)}(x) = \frac{\Gamma(n+\alpha+\beta+1)}{\Gamma(\beta)\,\Gamma(n+\alpha+1)} \int_{-1}^{1} t^\alpha (1-t)^{\beta-1} L_n^{(\alpha)}(xt)\,dt.$$

4.25 The *Laplace transform* of a function $f(x)$ defined for $x \geq 0$ is the function $\mathcal{L}f$ defined by

$$[\mathcal{L}f](s) = \int_0^\infty e^{-sx} f(x)\,dx$$

for all values of s for which the integral converges. Show that the Laplace transform of $x^\alpha L_n^{(\alpha)}$ is

$$\frac{\Gamma(n+\alpha+1)}{n!} \frac{(s-1)^n}{s^{n+\alpha+1}}, \quad \mathrm{Re}\,s > 1.$$

4.26 Prove

$$\lim_{\alpha \to \infty} \alpha^{-n} P_n^{(\alpha,\beta)}(x) = \frac{1}{n!} \left(\frac{1+x}{2} \right)^n.$$

4.27 Prove (4.6.6) by showing that the functions on the right-hand side are orthogonal to polynomials in x of lower degree, with respect to the weight function $(1-x^2)^\alpha$ and comparing coefficients at $x = -1$.

4.28 Prove Theorem 4.6.1: use (4.6.9) and Theorem 3.3.3 to prove the lower bounds, then use (4.6.1) and the lower bounds to obtain the upper bounds. (Note that $x \to -x$ corresponds to $\theta \to \pi - \theta$.)

4.29 Suppose that f and its derivatives of order $\leq n$ are bounded on the interval $(-1, 1)$. Show that

$$\int_0^1 f(x)\, P_n^{(\alpha,\beta)}(x)\, (1-x)^\alpha (1+x)^\beta\, dx$$

$$= \frac{1}{2^n\, n!} \int_0^1 f^{(n)}(x)(1-x)^{n+\alpha}(1+x)^{n+\beta}\, dx.$$

4.30 Show that for integers $m \geq n$,

$$\int_0^1 (1+x)^m\, P_n^{(\alpha,\beta)}(x)\, (1-x)^\alpha (1+x)^\beta\, dx$$

$$= \binom{m}{n} \frac{\Gamma(n+\alpha+1)\,\Gamma(m+\beta+1)}{\Gamma(n+m+\alpha+\beta+2)}\, 2^{n+m+\alpha+\beta+1}.$$

4.31 Derive the recurrence relation (4.7.3) from the generating function (4.7.2) by differentiating both sides of (4.7.2) with respect to s, multiplying by $\left(1 - 2xs + x^2\right)$, and equating coefficients of s.

4.32 Derive the identity (4.7.5) from (4.7.2).

4.33 Derive the generating function (4.7.2) from the recurrence relation (4.7.3).

4.34 Prove

$$\int_{-1}^{1} x^{n+2k} P_n(x) \, dx = \frac{(n + 2k)!}{2^n (2k)!} \frac{\Gamma\left(k + \frac{1}{2}\right)}{\Gamma\left(n + k + \frac{3}{2}\right)}, \quad k = 0, 1, 2, \dots$$

4.35 Derive the recurrence relation (4.7.15) and the identity (4.7.16) from the generating function (4.7.18).

4.36 Derive (4.7.23) from (4.7.20) and the special cases U_0, U_1.

4.37 Derive (4.7.20) and (4.7.21) from (4.7.23).

4.38 Use the generating function to show that the Legendre polynomials satisfy the following: for even $n = 2m$, $P_n'(0) = 0$ and

$$P_n(0) = (-1)^m \frac{\left(\frac{1}{2}\right)_m}{m!} = (-1)^m \frac{\Gamma\left(m + \frac{1}{2}\right)}{\sqrt{\pi}\, \Gamma(m + 1)},$$

while for odd $n = 2m + 1$, $P_n(0) = 0$ and

$$P_n'(0) = (-1)^m \frac{\left(\frac{3}{2}\right)_m}{m!} = (-1)^m \frac{2\,\Gamma\left(m + \frac{3}{2}\right)}{\sqrt{\pi}\, \Gamma(m + 1)}.$$

4.39 Use the functional equation for the gamma function to show that the sequence

$$\frac{\sqrt{m}\, \Gamma\left(m + \frac{1}{2}\right)}{\Gamma(m + 1)}$$

is increasing, and use (2.1.9) to find the limit. Deduce from this and Exercise 4.38 that (4.7.11) is true for n even.

4.40 Prove (4.7.11) for n odd.

4.41 Use Exercise 4.30 to show that for any integer $m \geq 0$,

$$(1 + x)^m = \sum_{n=0}^{m} c_n P_n^{(\alpha, \beta)}(x),$$

where

$$c_n = \frac{\Gamma(n + \alpha + \beta + 1)}{\Gamma(2n + \alpha + \beta + 1)} \frac{m!\,(n + \beta + 1)_{m-n}}{(2n + \alpha + \beta + 1)_{m-n}} 2^m.$$

4.42 Prove that if the assumption (4.8.3) in any of the theorems of 4.8 is replaced by the conditions (4.8.12), then the corresponding series converges to $[f(x_+) + f(x_-)]/2$. Hint: by subtracting a function that satisfies (4.8.3) at x one can essentially reduce this to the fact that the integral, over an interval centered at x, of a function that is odd around x, is zero.

4.11 Summary

4.11.1 General orthogonal polynomials

Suppose that $w(x) > 0$ on an interval (a, b), with finite moments

$$A_n = \int_a^b x^n \, w(x) \, dx.$$

Let $\Delta_{-1} = 1$ and let $\Delta_n, n \geq 0$, be the $(n + 1) \times (n + 1)$ determinant

$$\Delta_n = \begin{vmatrix} A_0 & A_1 & \ldots & A_n \\ A_1 & A_2 & \ldots & A_{n+1} \\ & & \ddots & \\ A_n & A_{n+1} & \ldots & A_{2n} \end{vmatrix} > 0.$$

The polynomials

$$Q_n(x) = \begin{vmatrix} A_0 & A_1 & \ldots & A_{n-1} & 1 \\ A_1 & A_2 & \ldots & A_n & x \\ & & \ddots & & \\ A_n & A_{n+1} & \ldots & A_{2n-1} & x^n \end{vmatrix}$$

are orthogonal with respect to the inner product

$$(f, g) = \int_a^b f(x) \, g(x) \, w(x) \, dx$$

and

$$(Q_n, Q_n) = \Delta_{n-1} \Delta_n,$$

so the polynomials

$$P_n(x) = \frac{1}{\sqrt{\Delta_{n-1} \Delta_n}} Q_n(x)$$

are orthonormal. They satisfy the recurrence relation

$$x P_n(x) = a_n P_{n+1}(x) + b_n P_n(x) + a_{n-1} P_{n-1}(x),$$

which implies the Christoffel–Darboux formula

$$a_n \left[\frac{P_{n+1}(x) P_n(y) - P_n(x) P_{n+1}(y)}{x - y} \right] = \sum_{j=0}^{n} P_j(x) P_j(y)$$

and the limiting form

$$a_n \left[P'_{n+1}(x) P_n(x) - P_{n+1}(x) P'_n(x) \right] = \sum_{j=0}^{n} P_j(x)^2.$$

The polynomial P_n has n distinct real roots in the interval (a, b); each root of P_{n-1} lies between consecutive roots of P_n.

If for some $c > 0$

$$\int_a^b e^{2c|x|} w(x) \, dx < \infty,$$

then for any $f \in L_w^2$

$$f = \sum_{n=0}^{\infty} (f, P_n) \, P_n.$$

The partial sums of this series are

$$\sum_{j=0}^{n} (f, P_j) P_j(x) = \int_a^b K_n(x, y) \, f(y) \, w(y) \, dy,$$

where the Dirichlet kernel K_n for $\{P_m\}$ is

$$K_n(x, y) = a_n \left[\frac{P_{n+1}(x) P_n(y) - P_n(x) P_{n+1}(y)}{x - y} \right].$$

4.11.2 Classical polynomials: general properties, I

Rodrigues formula:

$$\psi_n(x) = w(x)^{-1} \frac{d^n}{dx^n} \{ p(x)^n w(x) \}.$$

Limiting relations:

$$H_n(x) = \lim_{\alpha \to +\infty} \frac{2^n \, n!}{\alpha^{n/2}} \, P_n^{(\alpha,\alpha)} \left(\frac{x}{\sqrt{\alpha}} \right);$$

$$L_n^{(\alpha)}(x) = \lim_{\beta \to +\infty} P_n^{(\alpha,\beta)} \left(1 - \frac{2x}{\beta} \right).$$

4.11.3 Classical polynomials: general properties, II

Integral version of the Rodrigues formula:

$$\frac{\psi_n(x)}{n!} = \frac{1}{2\pi i} \int_\Gamma \frac{w(z)}{w(x)} \frac{p(z)^n}{(z-x)^n} \frac{dz}{z-x}.$$

This can be used to calculate the exponential generating function

$$G(x,s) = \sum_{n=0}^\infty \frac{\psi_n(x)}{n!} s^n$$

$$= \frac{w(\zeta)}{w(x)} \cdot \frac{1}{1 - s \, p'(\zeta)}, \quad \zeta - s \, p(\zeta) = x,$$

coefficients of the three-term recurrence relation

$$a_n \varphi_{n+1}(x) = [b_{n0} + b_{n1}x]\varphi_n(x) + c_n \varphi_{n-1}(x), \quad \varphi_n(x) = \frac{\psi_n(x) \, w(x)}{n!},$$

and derivative formula

$$p(x) \varphi_n'(x) = c_n(x) \, \varphi_n(x) + d_n \, \varphi_{n-1}(x).$$

The three-term recurrence formula allows computation of the Dirichlet kernel $K_n(x, y)$.

Results of these calculations are given in the summaries for Hermite, Laguerre, and Jacobi polynomials.

4.11.4 Hermite polynomials

The Hermite polynomials $\{H_n\}$ are orthogonal polynomials associated with the weight $w(x) = e^{-x^2}$ on the line. They are eigenfunctions

$$H_n''(x) - 2x \, H_n'(x) + 2n \, H_n(x) = 0,$$

and can be defined by the Rodrigues formula

$$H_n(x) = (-1)^n e^{x^2} \frac{d^n}{dx^n} \{e^{-x^2}\} = \left(2x - \frac{d}{dx} \right)^n \{1\}.$$

They satisfy identities

$$H_n(-x) = (-1)^n H_n(x);$$

$$H_{n+1}(x) = 2x\, H_n(x) - 2n\, H_{n-1}(x);$$

$$H_n'(x) = 2n\, H_{n-1}(x);$$

$$H_n(x) = \sum_{2j \le n} (-1)^j \frac{n!}{j!\,(n-2j)!}\,(2x)^{n-2j}.$$

First six of the H_n:

$$H_0(x) = 1;$$

$$H_1(x) = 2x;$$

$$H_2(x) = 4x^2 - 2;$$

$$H_3(x) = 8x^3 - 12x;$$

$$H_4(x) = 16x^4 - 48x^2 + 12;$$

$$H_5(x) = 32x^5 - 160x^3 + 120x.$$

Generating function:

$$G(x, s) = \sum_{n=0}^{\infty} \frac{H_n(x)}{n!}\, s^n = e^{2xs - s^2}.$$

Addition formulas:

$$H_n(x + y) = \sum_{j+k+2l=n} \frac{n!}{j!\,k!\,l!}\, H_j(x)\, H_k(y)$$

$$= 2^{-\frac{1}{2}n} \sum_{m=0}^{n} \binom{n}{m} H_m(\sqrt{2}\,x)\, H_{n-m}(\sqrt{2}\,y).$$

Norms:

$$\|H_n\|^2 \equiv \int_{-\infty}^{\infty} H_n(x)^2\, e^{-x^2}\, dx = 2^n n!\, \sqrt{\pi}.$$

Inner products:

$$(f, H_n) \equiv \int_{-\infty}^{\infty} f(x)\, H_n(x)\, e^{-x^2}\, dx = \int_{-\infty}^{\infty} e^{-x^2}\, f^{(n)}(x)\, dx;$$

$$\left(x^{n+2k}, H_n\right) = \frac{(n+2k)!}{(2k)!}\, \Gamma\left(k + \frac{1}{2}\right);$$

$$\left(e^{ax}, H_n\right) = a^n e^{\frac{1}{4}a^2} \sqrt{\pi};$$

$$(\cos bx, H_n) = \begin{cases} \sqrt{\pi}(ib)^n e^{-\frac{1}{4}b^2}, & n \text{ even,} \\ 0, & n \text{ odd;} \end{cases}$$

$$(\sin bx, H_n) = \begin{cases} -i\sqrt{\pi}(ib)^n e^{-\frac{1}{4}b^2}, & n \text{ odd,} \\ 0, & n \text{ even.} \end{cases}$$

Integral representation:

$$H_n(x) = (-1)^n \frac{e^{x^2}}{\sqrt{\pi}} \int_{-\infty}^{\infty} (-2it)^n e^{-2ixt-t^2} \, dt.$$

Generating function for products:

$$\sum_{n=0}^{\infty} \frac{H_n(x)H_n(y)}{2^n n!} s^n = \frac{1}{\sqrt{1-s^2}} \exp\left(\frac{2xys - s^2x^2 - s^2y^2}{1-s^2}\right).$$

The relative maxima of $|e^{-x^2/2} H_n(x)|$ increase as $|x|$ increases.

$H_n(x)$ has n simple roots in the interval $-\sqrt{2n+1} < x < \sqrt{2n+1}$; the positive roots $0 < x_{1n} < x_{2n} < \ldots$ satisfy the inequalities

$$\frac{(2k-1)\pi}{2\sqrt{2n+1}} < x_{kn} < \frac{4k+1}{\sqrt{2n+1}}, \quad k = 1, 2, \ldots, m, \quad n = 2m;$$

$$\frac{k\pi}{\sqrt{2k+1}} < x_{kn} < \frac{4k+3}{\sqrt{2n+1}}, \quad k = 1, 2, \ldots, m, \quad n = 2m+1.$$

Dirichlet kernel:

$$K_n^H(x, y) = \frac{1}{2^{n+1}n!\sqrt{\pi}} \left[\frac{H_{n+1}(x)H_n(y) - H_n(x)H_{n+1}(y)}{x - y}\right].$$

Discriminant:

$$D_n^H = 2^{3n(n-1)/2} \prod_{j=1}^{n} j^j.$$

Asymptotic behavior as $n \to \infty$:

$$H_n(x) = 2^{\frac{1}{2}n} \frac{2^{\frac{1}{4}}(n!)^{\frac{1}{2}}}{(n\pi)^{\frac{1}{4}}} e^{\frac{1}{2}x^2} \left[\cos\left(\sqrt{2n+1}\,x - \frac{1}{2}n\pi\right) + O\left(n^{-\frac{1}{2}}\right)\right].$$

If the weight is taken instead as $w(x) = e^{-\frac{1}{2}x^2}$, the orthogonal polynomials, denoted by $\{He_n\}$, satisfy

$$He_n''(x) - x\,He_n'(x) + n\,He_n(x) = 0;$$

$$He_n(x) = \left[x - \frac{d}{dx}\right]^n \{1\} = 2^{-n/2}H_n\left(\frac{x}{\sqrt{2}}\right).$$

Relation to Laguerre polynomials:

$$H_{2n}(x) = (-1)^n 2^{2n} n!\, L_n^{\left(-\frac{1}{2}\right)}(x^2), \quad n = 0, 1, 2, \dots,$$

$$H_{2n+1}(x) = (-1)^n 2^{2n+1} n!\, x\, L_n^{\left(\frac{1}{2}\right)}(x^2), \quad n = 0, 1, 2, \dots$$

4.11.5 Laguerre polynomials

The Laguerre polynomials $\left\{L_n^{(\alpha)}\right\}$ are orthogonal polynomials associated with the weight $x^\alpha e^{-x}\,dx$ on the half line $x > 0$. For a given $\alpha > -1$ they are eigenfunctions

$$x\left[L_n^{(\alpha)}\right]''(x) + (\alpha + 1 - x)\left[L_n^{(\alpha)}\right]'(x) + n\,L_n^{(\alpha)}(x) = 0;$$

and can be defined by the Rodrigues formula

$$L_n^{(\alpha)}(x) = \frac{1}{n!}x^{-\alpha}e^x\frac{d^n}{dx^n}\{x^\alpha e^{-x}x^n\} = \frac{1}{n!}\left[\frac{d}{dx} + \frac{\alpha}{x} - 1\right]^n\{x^n\}.$$

They satisfy the identities

$$x\,L_n^{(\alpha)}(x) = -(n + 1)L_{n+1}^{(\alpha)}(x) + (2n + \alpha + 1)L_n^{(\alpha)}(x) - (n + \alpha)\,L_{n-1}^{(\alpha)}(x);$$

$$\left[L_n^{(\alpha)}\right]'(x) = -L_{n-1}^{(\alpha+1)}(x);$$

$$x\left[L_n^{(\alpha)}\right]'(x) = n\,L_n^{(\alpha)}(x) - (n + \alpha)\,L_{n-1}^{(\alpha)}(x);$$

$$L_n^{(\alpha)}(x) = \sum_{k=0}^{n}(-1)^k\frac{(\alpha + 1)_n}{k!\,(n - k)!(\alpha + 1)_k}x^k.$$

First four Laguerre polynomials:

$$L_0^{(\alpha)}(x) = 1;$$

$$L_1^{(\alpha)}(x) = \alpha + 1 - x;$$

$$L_2^{(\alpha)}(x) = \frac{(\alpha + 1)(\alpha + 2)}{2} - (\alpha + 2)\,x + \frac{1}{2}\,x^2;$$

$$L_3^{(\alpha)}(x) = \frac{(\alpha + 1)(\alpha + 2)(\alpha + 3)}{6} - \frac{(\alpha + 2)(\alpha + 3)}{2}\,x + \frac{(\alpha + 3)}{2}\,x^2 - \frac{1}{6}\,x^3.$$

Alternate notation in the case $\alpha = 0$:

$$L_n^{(0)}(x) = L_n(x).$$

Generating function:

$$G(x, s) \equiv \sum_{n=0}^{\infty} L_n^{(\alpha)}(x)\,s^n = \frac{e^{-xs/(1-s)}}{(1 - s)^{\alpha+1}}.$$

Addition formula:

$$L_n^{(\alpha)}(x + y) = \sum_{j+k+l=n} (-1)^j \frac{(\alpha + 2 - j)_j}{j\,!}\,L_k^{(\alpha)}(x)\,L_l^{(\alpha)}(y).$$

Norms:

$$\left\|L_n^{(\alpha)}\right\|^2 \equiv \int_{-\infty}^{\infty} \left[L_n^{(\alpha)}(x)\right]^2 x^\alpha\,e^{-x}\,dx = \frac{\Gamma(\alpha + n + 1)}{n!}.$$

Inner products:

$$\left(f, L_n^{(\alpha)}\right) \equiv \int_0^{\infty} L_n^{(\alpha)}(x)\,f(x)\,x^\alpha e^{-x}\,dx = \frac{(-1)^n}{n\,!} \int_0^{\infty} e^{-x} f^{(n)}(x) x^{n+\alpha}\,dx;$$

$$\left(x^m, L_n^{(\alpha)}\right) = (-1)^n \binom{m}{n}\Gamma(\alpha + m + 1);$$

$$\left(e^{-ax}, L_n^{(\alpha)}\right) = \frac{a^n\,\Gamma(n + \alpha + 1)}{n\,!\,(a + 1)^{n+\alpha+1}}, \quad \mathrm{Re}\,a > -1.$$

Relation to confluent hypergeometric functions (Kummer functions):

$$L_n^{(\alpha)}(x) = \frac{(\alpha + 1)_n}{n\,!}\,{}_1F_1(-n, \alpha + 1; x) = \frac{(\alpha + 1)_n}{n\,!}\,M(-n, \alpha + 1; x).$$

$L_n^{(\alpha)}(x)$ has n simple roots in the interval

$$0 < x < 2n + \alpha + 1 + \sqrt{(2n + \alpha + 1)^2 + (1 - \alpha^2)}.$$

The relative maxima of $|e^{-x/2}x^{(\alpha+1)/2}L_n^{(\alpha)}(x)|$ increase as x increases.

Dirichlet kernel:

$$K_n^{(\alpha)}(x, y) = -\frac{(n + 1)\,!}{\Gamma(n + \alpha + 1)} \left[\frac{L_{n+1}^{(\alpha)}(x)L_n^{(\alpha)}(y) - L_n^{(\alpha)}(x)L_{n+1}^{(\alpha)}(y)}{x - y}\right].$$

Discriminant:

$$D_n^{(\alpha)} = \prod_{j=1}^{n} j^{j-2n+2}(j+\alpha)^{j-1}.$$

Asymptotic behavior as $n \to \infty$:

$$L_n^{(\alpha)}(x) = \frac{e^{\frac{1}{2}x} n^{\frac{1}{2}\alpha - \frac{1}{4}}}{\sqrt{\pi}\, x^{\frac{1}{2}\alpha + \frac{1}{4}}} \left[\cos\left(2\sqrt{nx} - \frac{1}{2}\left[\alpha + \frac{1}{2}\right]\pi\right) + O\left(n^{-\frac{1}{2}}\right) \right],$$

uniformly on any subinterval $0 < \delta \le x \le \delta^{-1}$.

4.11.6 Jacobi polynomials

Jacobi polynomials $\left\{ P_n^{(\alpha,\beta)} \right\}$ with indices $\alpha, \beta > -1$ are orthogonal with respect to the weight $(1-x)^\alpha (1+x)^\beta$ on the interval $(-1, 1)$; the norms are

$$\| P_n^{(\alpha,\beta)} \|^2 = \frac{2^{\alpha+\beta+1}\,\Gamma(n+\alpha+1)\,\Gamma(n+\beta+1)}{n!\,(2n+\alpha+\beta+1)\,\Gamma(n+\alpha+\beta+1)}.$$

They are eigenfunctions

$$\left(1 - x^2\right) \left[P_n^{(\alpha,\beta)} \right]''(x) + \left[\beta - \alpha - (\alpha+\beta+2)x \right] \left[P_n^{(\alpha,\beta)} \right]'(x)$$
$$+ n(n+\alpha+\beta+1)\, P_n^{(\alpha,\beta)}(x) = 0$$

and can be defined by the Rodrigues formula

$$P_n^{(\alpha,\beta)}(x) = \frac{(-1)^n}{n!\,2^n}(1-x)^{-\alpha}(1+x)^{-\beta} \frac{d^n}{dx^n}\left\{ (1-x)^{\alpha+n}(1+x)^{\beta+n} \right\}.$$

They satisfy the identities

$$P_n^{(\alpha,\beta)}(-x) = (-1)^n\, P_n^{(\beta,\alpha)}(x);$$

$$\frac{(2n+2)(n+\alpha+\beta+1)}{2n+\alpha+\beta+1}\, P_{n+1}^{(\alpha,\beta)}(x)$$

$$= \left[\frac{\alpha^2 - \beta^2}{2n+\alpha+\beta} + (2n+\alpha+\beta+2)\,x \right] P_n^{(\alpha,\beta)}(x)$$

$$- \frac{2(2n+\alpha+\beta+2)(n+\alpha)(n+\beta)}{(2n+\alpha+\beta)(2n+\alpha+\beta+1)}\, P_{n-1}^{(\alpha,\beta)}(x);$$

$$\left[P_n^{(\alpha,\beta)}\right]'(x) = \frac{1}{2}(n + \alpha + \beta + 1)\, P_{n-1}^{(\alpha+1,\beta+1)}(x);$$

$$\left(1 - x^2\right)\left[P_n^{(\alpha,\beta)}\right]'(x) = \left[\frac{n(\alpha - \beta)}{2n + \alpha + \beta} - nx\right] P_n^{(\alpha,\beta)}(x)$$

$$+ \frac{2(n + \alpha)(n + \beta)}{2n + \alpha + \beta}\, P_{n-1}^{(\alpha,\beta)}(x);$$

$$P_{2n}^{(\alpha,\alpha)}(x) = \frac{(n + \alpha + 1)_n}{(n + 1)_n}\, P_n^{\left(\alpha,-\frac{1}{2}\right)}\left(2x^2 - 1\right);$$

$$P_{2n+1}^{(\alpha,\alpha)}(x) = \frac{(n + \alpha + 1)_{n+1}}{(n + 1)_{n+1}}\, x\, P_n^{\left(\alpha,\frac{1}{2}\right)}\left(2x^2 - 1\right).$$

The endpoint values are

$$P_n^{(\alpha,\beta)}(1) = \frac{(\alpha + 1)_n}{n!}, \quad P_n^{(\alpha,\beta)}(-1) = (-1)^n\, \frac{(\beta + 1)_n}{n!}.$$

The Jacobi polynomials satisfy

$$\sup_{|x|\leq 1} |P_n^{(\alpha,\beta)}(x)| = \max\left\{\frac{(\alpha + 1)_n}{n!}, \frac{(\beta + 1)_n}{n!}\right\}, \quad \text{if } \alpha \text{ or } \beta \geq -\frac{1}{2}.$$

Generating function:

$$G(x, s) \equiv \sum_{n=0}^{\infty} P_n^{(\alpha,\beta)}(x)\, s^n = \frac{2^{\alpha+\beta}}{R\,(1 - s + R)^{\alpha}(1 + s + R)^{\beta}},$$

$$R = \sqrt{1 - 2xs + s^2}.$$

Dirichlet kernel:

$$K_n^{(\alpha,\beta)}(x, y) = \frac{2^{-\alpha-\beta}\,(n + 1)!\,\Gamma(n + \alpha + \beta + 2)}{(2n + \alpha + \beta + 2)\Gamma(n + \alpha + 1)\,\Gamma(n + \beta + 1)}$$

$$\times \left[\frac{P_{n+1}^{(\alpha,\beta)}(x)P_n^{(\alpha,\beta)}(y) - P_n^{(\alpha,\beta)}(x)P_{n+1}^{(\alpha,\beta)}(y)}{x - y}\right].$$

Discriminant:

$$D_n^{(\alpha,\beta)} = \frac{1}{2^{n(n-1)}} \prod_{j=1}^{n} \frac{j^{j-2n+2}(j + \alpha)^{j-1}(j + \beta)^{j-1}}{(n + j + \alpha + \beta)^{j-n}}.$$

The Liouville transformation converts the equation for Jacobi polynomials to $u''(\theta)$

$$+ \left[\frac{(2n + \alpha + \beta + 1)^2}{4} + \frac{(1 + 2\alpha)(1 - 2\alpha)}{16 \sin^2 \frac{1}{2}\theta} + \frac{(1 + 2\beta)(1 - 2\beta)}{16 \cos^2 \frac{1}{2}\theta} \right] u(x) = 0.$$

It follows that, for $\alpha^2 \leq \frac{1}{4}$ and $\beta^2 \leq \frac{1}{4}$, if $\cos \theta_{1n} < \ldots < \cos \theta_{nn}$ are the zeros of $P_n^{(\alpha,\beta)}$,

$$\frac{(k - 1 + \gamma)\pi}{n + \gamma} \leq \theta_{kn} \leq \frac{k\pi}{n + \gamma}, \quad \gamma = \frac{1}{2}(\alpha + \beta + 1).$$

Asymptotics as $n \to \infty$:

$$P_n^{(\alpha,\beta)}(\cos\theta) = \frac{\cos\left(n\theta + \frac{1}{2}(\alpha + \beta + 1)\theta - \frac{1}{2}\alpha\pi - \frac{1}{4}\pi\right) + O\left(n^{-\frac{1}{2}}\right)}{\sqrt{n\pi} \left(\sin\frac{1}{2}\theta\right)^{\alpha + \frac{1}{2}} \left(\cos\frac{1}{2}\theta\right)^{\beta + \frac{1}{2}}},$$

uniformly on any subinterval $0 < \delta \leq \theta \leq \pi - \delta$.

Relation between Jacobi polynomials and hypergeometric functions:

$$P_n^{(\alpha,\beta)}(x) = \frac{(\alpha + 1)_n}{n!} F\left(\alpha + \beta + 1 + n, -n, \alpha + 1; \frac{1}{2}(1 - x)\right).$$

4.11.7 Legendre and Chebyshev polynomials

Legendre polynomials: $w(x) = 1$,

$$P_n(x) = P_n^{(0,0)}(x).$$

Eigenvalue equation:

$$\left(1 - x^2\right) P_n''(x) - 2x\, P_n'(x) + n(n + 1)\, P_n(x) = 0.$$

Generating function:

$$\sum_{n=0}^{\infty} P_n(x)\, s^n = \left(1 - 2xs + s^2\right)^{-\frac{1}{2}}.$$

The general recurrence and derivative formulas specialize to

$$(n + 1)P_{n+1}(x) = (2n + 1)x\, P_n(x) - n P_{n-1}(x);$$

$$\left(1 - x^2\right)P_n'(x) = -nx P_n(x) + n P_{n-1}(x);$$

$$P_n'(x) = 2x P_{n-1}'(x) - P_{n-2}'(x) + P_{n-1}(x).$$

Two integral representations:

$$P_n(\cos\theta) = \frac{1}{2\pi} \int_0^{2\pi} (\cos\theta + i\sin\theta\sin\alpha)^n \, d\alpha$$

$$= \frac{1}{\pi} \int_0^1 \frac{\cos\left(s\left(n+\frac{1}{2}\right)\theta\right)}{\cos\frac{1}{2}s\theta} \frac{ds}{\sqrt{s(1-s)}}.$$

Dirichlet kernel:

$$K_n^{(0,0)}(x, y) = \frac{n+1}{2} \left[\frac{P_{n+1}(x)P_n(y) - P_n(x)P_{n+1}(y)}{x - y} \right].$$

Legendre polynomials satisfy

$$\sup_{|x|\le 1} \left| P_n(x) \right| = 1.$$

Relative maxima of $|P_n(\cos\theta)|$ decrease as θ increases for $0 \le \theta \le \frac{1}{2}\pi$ and increase as θ increases for $\frac{1}{2}\pi \le \theta \le \pi$.

Relative maxima of $|\sin\theta P_n(\cos\theta)|$ increase with θ for $0 \le \theta \le \frac{1}{2}\pi$ and decrease as θ increases for $\frac{1}{2}\pi \le \theta \le \pi$. Moreover,

$$\left| (\sin\theta)^{\frac{1}{2}} P_n(\cos\theta) \right| < \left(\frac{2}{n\pi} \right)^{\frac{1}{2}}.$$

Let $\cos\theta_{1n}, \ldots, \cos\theta_{nn}$ be the zeros of $P_n(x)$, $\theta_{1n} < \cdots < \theta_{nn}$. Then

$$\frac{\left(k-1+\frac{1}{2}\right)\pi}{n+\frac{1}{2}} < \theta_{kn} < \frac{k\pi}{n+\frac{1}{2}}.$$

Turán's inequality:

$$P_n(x)^2 - P_{n-1}(x)\,P_{n+1}(x) \ge 0.$$

Chebyshev (or *Tchebycheff*) *polynomials* $\{T_n\}$ and $\{U_n\}$: the cases $\alpha = \beta = -\frac{1}{2}$ and $\alpha = \beta = \frac{1}{2}$, respectively. *Gegenbauer polynomials* or *ultraspherical polynomials* $\{C_n^\lambda\}$: the general case $\alpha = \beta$:

$$C_n^\lambda(x) = \frac{(2\lambda)_n}{\left(\lambda + \frac{1}{2}\right)_n} P_n^{\left(\lambda-\frac{1}{2}, \lambda-\frac{1}{2}\right)}(x);$$

$$C_n^0(x) = T_n(x), \quad C_n^{\frac{1}{2}}(x) = P_n(x), \quad C_n^1(x) = U_n(x).$$

Chebyshev polynomials of the first kind:

$$T_n(\cos\theta) = \cos n\theta.$$

The square of the L^2 norm is π if $n = 0$, and

$$\int_0^\pi \cos^2 n\theta \, d\theta = \frac{\pi}{2}, \quad n > 0.$$

Recurrence and derivative formulas:

$$T_{n+1}(x) + T_{n-1}(x) = 2x \, T_n(x), \quad n \geq 1;$$

$$\left. \left(1 - x^2\right) T_n'(x) \right|_{x=\cos\theta} = -\sin\theta \frac{d}{d\theta} \cos n\theta = -nx T_n(x) + n T_{n-1}(x).$$

Dirichlet kernel:

$$K_n^T(x, y) = \frac{1}{\pi} \left[\frac{T_{n+1}(x)T_n(y) - T_n(x)T_{n+1}(y)}{x - y} \right].$$

Generating function:

$$G_T(x, s) = \sum_{n=0}^{\infty} T_n(x) s^n = \frac{1 - xs}{1 - 2xs + s^2}.$$

Chebyshev polynomials of the second kind:

$$U_n(\cos\theta) = \frac{\sin(n+1)\theta}{\sin\theta}.$$

The square of the L^2 norm is

$$\int_0^\pi \sin^2(n+1)\theta \, d\theta = \frac{\pi}{2}.$$

Recurrence and derivative formulas:

$$U_{n+1}(x) + U_{n-1}(x) = 2x U_n(x);$$

$$(1 - x^2) U_n'(x) = -nx \, U_n(x) + (n+1) U_{n-1}(x).$$

Dirichlet kernel:

$$K_n^U(x, y) = \frac{1}{\pi} \left[\frac{U_{n+1}(x)U_n(y) - U_n(x)U_{n+1}(y)}{x - y} \right].$$

Generating function:

$$G_U(x, s) = \sum_{n=0}^{\infty} U_n(x) s^n = \frac{1}{1 - 2xs + s^2}.$$

Other cases:

$$P_n^{\left(-\frac{1}{2},\frac{1}{2}\right)}(\cos\theta) = \frac{\left(\frac{1}{2}\right)_n}{n!}\frac{\cos\left(n+\frac{1}{2}\theta\right)}{\cos\frac{1}{2}\theta};$$

$$P_n^{\left(\frac{1}{2},-\frac{1}{2}\right)}(\cos\theta) = \frac{\left(\frac{1}{2}\right)_n}{n!}\frac{\sin\left(n+\frac{1}{2}\theta\right)}{\sin\frac{1}{2}\theta}.$$

Special cases of hypergeometric functions:

$$F\left(n, -n, \frac{1}{2}; \frac{1}{2}[1-\cos\theta]\right) = \cos(n\theta);$$

$$F\left(n+2, -n, \frac{3}{2}; \frac{1}{2}[1-\cos\theta]\right) = \frac{\sin(n+1)\theta}{(n+1)\sin\theta};$$

$$F\left(n+1, -n, \frac{1}{2}; \frac{1}{2}[1-\cos\theta]\right) = \frac{\cos\left(n+\frac{1}{2}\right)\theta}{\cos\frac{1}{2}\theta};$$

$$F\left(n+1, -n, \frac{3}{2}; \frac{1}{2}[1-\cos\theta]\right) = \frac{\sin\left(n+\frac{1}{2}\right)\theta}{(2n+1)\sin\frac{1}{2}\theta}.$$

4.11.8 Expansion theorems

Suppose that f is a piecewise continuously differentiable function that is square integrable with respect to one of the weights associated with the classical orthogonal polynomials. The corresponding series expansion of f takes one of the forms

$$\sum_{n=0}^{\infty} c_n H_n(x), \quad c_n = \frac{1}{2^n\sqrt{\pi}\,n!}\int_{-\infty}^{\infty} f(x)\,H_n(x)\,e^{-x^2}\,dx$$

in the Hermite case,

$$\sum_{n=0}^{\infty} c_n L_n^{(\alpha)}(x), \quad c_n = \frac{n!}{\Gamma(n+\alpha+1)}\int_0^{\infty} f(x)\,L_n^{(\alpha)}(x)\,x^\alpha\,e^{-x}\,dx$$

in the Laguerre case, or

$$\sum_{n=0}^{\infty} c_n P_n^{(\alpha,\beta)}(x), \quad c_n = \frac{n!\,(2n+\alpha+\beta+1)\,\Gamma(n+\alpha+\beta+1)}{2^{\alpha+\beta+1}\Gamma(n+\alpha+1)\,\Gamma(n+\beta+1)}$$

$$\times \int_{-1}^{1} f(x)\,P_n^{(\alpha,\beta)}(x)\,(1-x)^\alpha(1+x)^\beta\,dx$$

in the Jacobi case. The series converges to $f(x)$ at every point x of continuity, and converges to $\left[f(x+) + f(x-)\right]/2$ at every point x of discontinuity, in the open interval. (For weaker, pointwise conditions on f, see Section 4.8.)

4.11.9 Functions of second kind

The complex integral representation for the classical orthogonal polynomials can be adapted to give an integral representation for other values of the parameter ν. In each of the three cases let $I = (a, b)$ denote the associated real interval. The associated function of second kind

$$u_\nu(x) = \frac{c_\nu}{w(x)} \int_a^b \frac{p(s)^\nu \, w(s) \, ds}{(s - x)^{\nu+1}}, \quad \operatorname{Re} \nu \geq 0, \quad x \notin I,$$

is a solution of the corresponding equation

$$p(x) \, u''(x) + q(x) \, u'(x) + \lambda_\nu \, u(x) = 0, \quad \lambda_\nu = -\nu \, q' - \frac{\nu(\nu - 1)}{2} \, p'',$$

for $\operatorname{Re} \nu \geq 0$, x not real. This result also holds in somewhat greater generality: Theorem 4.9.2.

4.12 Remarks

General orthogonal polynomials are an integral part of the theory of moments, continued fractions, and spectral theory; see, for example, Akhiezer [5] and the various books cited below.

Legendre and Hermite polynomials were the first to be studied, and most of the general theory was first worked out in these cases. Legendre found recurrence relations and the generating function for Legendre polynomials in 1784–5 [180, 181]; Legendre and Laplace [175] found the orthogonality relation. Rodrigues proved the Rodrigues formula for the Legendre polynomials in 1816 [241]. Schläfli [250] gave the integral formula (4.3.1) for the Legendre polynomials in 1881. The series expansion (4.6.12) for Legendre polynomials was given by Murphy in 1835 [211].

Hermite polynomials occur in the work of Laplace on celestial mechanics [175] and on probability [176, 177], and in Chebyshev's 1859 paper [47], as well as in Hermite's 1864 paper [126]. Laguerre polynomials for $\alpha = 0$ were considered by Lagrange [170], Abel [2], Chebyshev [47], Laguerre [172] and, for general α, by Sonine [265].

Jacobi polynomials in full generality were introduced by Jacobi in 1859 [143]. The special case of Chebyshev polynomials was studied by Chebyshev in 1854 [44]. Gegenbauer polynomials were studied by Gegenbauer in 1874 [111, 112]. The Christoffel–Darboux formula was found by Chebyshev [45] in 1855, then rediscovered in the case of Legendre polynomials by Christoffel [51] in 1858 and in the general case by Christoffel [53] and Darboux [62] in 1877–8. Further remarks on the history are contained in several of the books cited below, and in Szegő's article and Askey's addendum [280].

A common approach to the classical orthogonal polynomials is to use the generating function formulas (4.4.6), (4.5.7), and (4.6.7) as definitions. The three-term recurrence relations and some other identities are easy consequences, as is orthogonality (once one has selected the correct weight) in the Hermite and Laguerre cases. The fact that the polynomials are the eigenfunctions of a symmetric second-order operator is easily established, but in the generating function approach this fact appears as something of a (fortunate) accident. It does not seem clear, from the generating function approach, why it is these polynomials and no others that have arisen as the "classical" orthogonal polynomials. In our view the characterization theorem, Theorem 3.4.1, the connection with the basic equations of mathematical physics in special coordinates (Section 3.6), and the characterization of "recursive" second-order equations (Section 1.1) provide natural explanations for the significance of precisely this set of polynomials.

The approach used in this chapter, deriving basic properties directly from the differential equation via the Rodrigues formula and the resulting complex integral representation, is the one used by Nikiforov and Uvarov [219].

The classical orthogonal polynomials are treated in every treatise or handbook of special functions. Some more comprehensive sources for the classical and general orthogonal polynomials are Szegő's classic treatise [279], as well as Askey [14], Chihara [50], Freud [101], Gautschi [108], Geronimus [114], Ismail [136], Krall [164], Khrushchev [167], Macdonald [193], Nevai [213], Stahl and Totik [266], Simon [259] and Tricomi [286].

The addition formulas for Hermite and Laguerre polynomials are easily derived from the generating function representation. There is a classical addition formula for Legendre polynomials that follows from the connection of these polynomials with spherical harmonics; see Section 9.1. Gegenbauer [113] obtained a generalization to all ultraspherical (Gegenbauer) polynomials ($\alpha = \beta$). More recently, Šapiro [247] found a formula valid for $\beta = 0$ and Koornwinder [160, 161] found a generalization valid for all Jacobi polynomials.

Discriminants of Jacobi polynomials were calculated by Hilbert [128] and Stieltjes [270, 271]. Zeros of Hermite and Laguerre polynomials are treated in the report by Hahn [121]. Hille, Shohat, and Walsh compiled an exhaustive bibliography to 1940 [130].

The three-term recurrence relation for orthogonal polynomials connects them closely to certain types of continued fractions. The book by Khrushchev [167] approaches the classical orthogonal polynomials from this point of view (which was Euler's).

Expansion in orthogonal polynomials is treated in Sansone [246]. The book by Van Assche [292] is devoted to asymptotics of various classes of orthogonal polynomials.

The subject of orthogonal polynomials continues to be a very active area of research, as evidenced, for example, by the books of Ismail [136] and Stahl and Totik [266]. Zeros of orthogonal polynomials are related to stationary points for electrostatic potentials, a fact that was used by Stieltjes in his calculation of the discriminants of Jacobi polynomials. These ideas have been extended considerably; see the books of Levin and Lubinsky [185] and Saff and Totik [245]. For asymptotic results, see the survey article by Wong [319].

We have not touched here on the topic of "q-orthogonal polynomials," which are related to the q-difference operator

$$D_q f(x) = \frac{f(x) - f(qx)}{(1-q)x}$$

in ways similar to the ways in which the classical orthogonal polynomials are related to the derivative. For extensive treatments, see the books by Andrews, Askey, and Roy [7], Bailey [18], Gasper and Rahman [102], Ismail [136], and Slater [261].

5

Discrete orthogonal polynomials

In Chapter 4 we discussed the question of polynomials orthogonal with respect to a weight function, which was assumed to be a positive continuous function on a real interval. This is an instance of a measure. Another example is a discrete measure, for example, one supported on the integers with masses w_m, $m = 0, \pm1, \pm2, \ldots$ Most of the results of Section 4.1 carry over to this case, although if w_m is positive at only a finite number $N + 1$ of points, the associated function space has dimension $N + 1$ and will be spanned by orthogonal polynomials of degrees zero through N.

In this context the role of differential operators is played by difference operators. An analogue of the characterization in Theorem 3.4.1 is valid: up to normalization, the orthogonal polynomials that are eigenfunctions of a symmetric second-order difference operator are the "classical discrete polynomials," associated with the names Charlier, Krawtchouk, Meixner, and Hahn.

The theory of the classical discrete polynomials can be developed in a way that parallels the treatment of the classical polynomials in Chapter 4, using a discrete analogue of the formula of Rodrigues.

5.1 Discrete weights and difference operators

Suppose that $w = \{w_n\}_{n=-\infty}^{\infty}$ is a two-sided sequence of non-negative numbers. The corresponding inner product

$$(f, g) = (f, g)_w = \sum_{m=-\infty}^{\infty} f(m) \, g(m) \, w_m$$

is well-defined for all real functions f and g for which the norms $||f||_w$ and $||g||_w$ are finite, where

154

$$\|f\|_w^2 = (f, f)_w = \sum_{m=-\infty}^{\infty} f(m)^2 \, w_m.$$

The norm and inner product depend only on the values taken by functions on the integers, although it is convenient to continue to regard polynomials, for example, as being defined on the line. Polynomials have finite norm if and only if the even moments

$$\sum_{m=-\infty}^{\infty} m^{2n} \, w_m$$

are finite. If so, then orthogonal polynomials ψ_n can be constructed exactly as in the case of a continuous weight function. Usually we normalize so that w is a probability distribution:

$$\sum_{m=-\infty}^{\infty} w_m = 1.$$

Again, there is a three-term recurrence relation. Suppose that

$$\psi_n(x) = a_n x^n + b_n x^{n-1} + \dots$$

The polynomial

$$x \, \psi_n(x) - \frac{a_n}{a_{n+1}} \psi_{n+1} + \left(\frac{b_{n+1}}{a_{n+1}} - \frac{b_n}{a_n} \right) \psi_n(x)$$

has degree $n - 1$ and is orthogonal to polynomials of degree $< n - 1$, and is therefore a multiple $\gamma_n \psi_{n-1}$. Then

$$\gamma_n (\psi_{n-1}, \psi_{n-1})_w = (x\psi_n, \psi_{n-1})_w = (\psi_n, x\psi_{n-1})_w = \frac{a_{n-1}}{a_n} (\psi_n, \psi_n)_w.$$

Thus

$$x\psi_n(x) = \alpha_n \psi_{n+1}(x) + \beta_n \psi_n(x) + \gamma_n \psi_{n-1}(x), \tag{5.1.1}$$

$$\alpha_n = \frac{a_n}{a_{n+1}}, \quad \beta_n = \frac{b_n}{a_n} - \frac{b_{n+1}}{a_{n+1}}, \quad \gamma_n = \frac{a_{n-1}}{a_n} \frac{(\psi_n, \psi_n)_w}{(\psi_{n-1}, \psi_{n-1})_w}.$$

As before, the three-term recurrence implies a Christoffel-Darboux formula.

If the weight w_m is positive at only finitely many integers, say at $m = 0$, $1, \dots, N$, then the L^2 space has dimension $N + 1$. The orthogonal polynomials have degrees $0, 1, \dots, N$ and are a basis. In general there is a completeness result analogous to Theorem 4.1.5, applicable to all the cases to be considered in this chapter.

Theorem 5.1.1 *Suppose that w is a positive weight on the integers and suppose that for some $c > 0$,*

$$\sum_{m=-\infty}^{\infty} e^{2c|m|} w_m < \infty. \tag{5.1.2}$$

Let $\{P_n\}$ be the orthonormal polynomials for w. Then $\{P_n\}$ is complete: for any $f \in L_w^2$,

$$\lim_{n \to \infty} \left\| f - \sum_{k=0}^{n} (f, P_k) P_k \right\|_w = 0. \tag{5.1.3}$$

In the discrete case, it is easy to see that convergence in norm implies pointwise convergence at the points m where $w_m > 0$.

Corollary 5.1.2 *Under the assumptions of Theorem 5.1.1, for any $f \in L_w^2$ and any m such that $w_m > 0$,*

$$f(m) = \sum_{n=0}^{\infty} (f, P_n) P_n(m). \tag{5.1.4}$$

Like an integrable function, a finite measure is determined by its Fourier transform. Therefore the proof of Theorem 4.1.5 in Appendix B also proves Theorem 5.1.1.

For functions defined on the integers, differentiation can be replaced by either the forward or backward difference operators

$$\Delta_+ f(m) = f(m+1) - f(m), \quad \Delta_- f(m) = f(m) - f(m-1).$$

Each operator maps polynomials of degree d to polynomials of degree $d - 1$, so the product

$$\left[\Delta_+ \Delta_-\right] f(m) = \left[\Delta_- \Delta_+\right] f(m) = f(m+1) + f(m-1) - 2f(m)$$

decreases the degree of a polynomial by two and plays the role of a second-order derivative. It is convenient to express these operators in terms of the shift operators

$$S_\pm f(m) = f(m \pm 1).$$

Thus

$$\Delta_+ = S_+ - I, \quad \Delta_- = I - S_-,$$

and

$$\Delta_+ \Delta_- = \Delta_- \Delta_+ = S_+ + S_- - 2I = \Delta_+ - \Delta_-. \tag{5.1.5}$$

Therefore the general real second-order difference operator can be written as

$$L = p_+ S_+ + p_- S_- + r,$$

where p_+, p_-, and r are real-valued functions. The condition for L to be symmetric with respect to w is that

$$0 = (Lf, g) - (f, Lg)$$

$$= (p_+ S_+ f, g) - (f, p_- S_- g) + (p_- S_- f, g) - (f, p_+ S_+ g)$$

$$= \sum_{m=-\infty}^{\infty} \left[p_+ w - S_+ (p_- w) \right] (m) f(m+1) g(m)$$

$$+ \sum_{m=-\infty}^{\infty} \left[p_- w - S_- (p_+ w) \right] (m) f(m-1) g(m)$$

for every f and g that vanish for all but finitely many integers. By choosing g to vanish except at one value m and f to vanish except at $m+1$ or at $m-1$, we conclude that symmetry is equivalent to

$$S_-(p_+ w) = p_- w, \quad S_+(p_- w) = p_+ w. \tag{5.1.6}$$

Note that S_+ and S_- are inverses, so these two conditions are mutually equivalent.

As for differential equations, symmetry implies that eigenfunctions that correspond to distinct eigenvalues are orthogonal: if $Lu_j = \lambda_j u_j$, $j = 1, 2$, then

$$\lambda_1 (u_1, u_2)_w = (Lu_1, u_2)_w = (u_1, Lu_2)_w = \lambda_2 (u_1, u_2)_w.$$

We now ask: for which discrete weights w (with finite moments) and which symmetric operators L with coefficients $p_\pm(m)$ positive where $w_m > 0$ (with exceptions at endpoints) are the eigenfunctions of L polynomials? More precisely, when do the eigenfunctions of L include polynomials of degrees 0, 1, and 2? We assume that $w_m > 0$ for integers m in a certain interval that is either infinite or has $N + 1$ points, and is zero otherwise, so we normalize to the cases

(a) $w_m > 0$ if and only if $0 \le m \le N$;
(b) $w_m > 0$ if and only if $m \ge 0$;
(c) $w_m > 0$ all m.

We shall see that case (c) does not occur.

The symmetry condition (5.1.6) implies that $p_-(0) = 0$ in cases (a) and (b) and that $p_+(N) = 0$ in case (a). We shall assume that otherwise $p_\pm$ is positive where w is positive.

With a change of notation for the zero-order term, we may write L as

$$L = p_+\Delta_+ - p_-\Delta_- + r. \tag{5.1.7}$$

Then $L(1) = r$ must be constant, and we may assume $r = 0$. Since $\Delta_\pm(x) = 1$, we have

$$L(x) = p_+ - p_-,$$

so $p_+ - p_-$ is a polynomial of degree 1. Next, $\Delta_\pm(x^2) = 2x \pm 1$, so

$$L(x^2) = 2x\,(p_+ - p_-) + (p_+ + p_-),$$

and it follows that $p_+ + p_-$ is a polynomial of degree at most 2. Therefore p_+ and p_- are both polynomials of degree at most 2, and at least one has positive degree. Moreover, if either has degree 2 then both do and they have the same leading coefficient.

The symmetry condition (5.1.6) and our positivity assumptions imply that

$$w_{m+1} = \frac{p_+(m)}{p_-(m+1)}\,w_m = \varphi(m)\,w_m, \tag{5.1.8}$$

wherever $w_m > 0$. This allows us to compute all values w_m from w_0.

Suppose first that one of $p_\pm$ has degree 0, so the other has degree 1. Then (5.1.8) implies that $\varphi(m)$ is $O(|m|)$ in one direction or the other, so the moment condition rules out case (c). It follows that $p_-(0) = 0$, and p_+ is constant, which rules out case (a). We normalize by taking $p_+(x) = 1$. Then $p_-(x) = x/a, a > 0$, so

$$\frac{p_+(m)}{p_-(m+1)} = \frac{a}{m+1}.$$

To normalize w we introduce the factor e^{-a}:

$$w_m = e^{-a}\,\frac{a^m}{m!}, \quad m = 0, 1, 2, 3, \ldots \tag{5.1.9}$$

This is the probability distribution known as the Poisson distribution. The associated polynomials, suitably normalized, are the *Charlier polynomials*, also called the Poisson–Charlier polynomials.

Suppose next that both of $p_\pm$ have degree 1. If the leading coefficients are not the same, then $\varphi(m)$ grows in one direction, which rules out case (c). If the leading coefficients are the same, then asymptotically $\varphi(m) - 1 \sim b/m$ for some constant b, which implies that the products

$$\prod_{j=0}^{m} \varphi(j), \quad \prod_{j=0}^{m} \varphi(j)^{-1} \tag{5.1.10}$$

are either identically 1 ($p_+ = p_-$), or one grows like m and the other decays like $1/m$ as $m \to \infty$. This rules out case (c). Thus if both have degree 1, then we have case (a) or (b) and $p_-(0) = 0$.

Continuing to assume that both of $p_\pm$ have degree 1, in the case of a finite interval we normalize by taking $p_+(x) = p(N - x)$ and $p_-(x) = qx$, where $p, q > 0$, $p + q = 1$. Then the normalized weight is

$$w_m = \binom{N}{m} p^m q^{N-m}, \quad m = 0, 1, 2, 3, \ldots, N, \tag{5.1.11}$$

the binomial probability distribution. Up to normalization, the associated polynomials are the *Krawtchouk polynomials*.

Suppose now that both $p_\pm$ have degree 1 and the interval is infinite. We may normalize by taking $p_-(x) = x$ and $p_+(x) = c(x + b)$. Positivity implies $b, c > 0$ and finiteness implies $c < 1$. Then

$$\frac{p_+(m)}{p_-(m+1)} = \frac{c(m+b)}{(m+1)},$$

and the normalized weight is

$$w_m = (1 - c)^b \frac{(b)_m}{m!} c^m, \quad m = 0, 1, 2, 3, \ldots \tag{5.1.12}$$

The associated polynomials are the *Meixner polynomials*.

Suppose finally that one of $p_\pm$ has degree 2. Then both do, and the leading coefficients are the same, so

$$\frac{p_+(m)}{p_-(m+1)} = \frac{1 + am^{-1} + bm^{-2}}{1 + cm^{-1} + dm^{-2}} = 1 + \frac{a-c}{m} + O\left(\frac{1}{m^2}\right).$$

Arguing as before, we see that the moment condition rules out the possibility of an infinite interval. With the weight supported on $0 \le m \le N$ we have $p_-(0) = 0 = p_+(N)$ and there are two cases, which can be normalized to

$$p_-(x) = x(N + 1 + \beta - x), \quad p_+(x) = (N - x)(x + \alpha + 1), \quad \alpha, \beta > -1$$

or

$$p_-(x) = x(x - \beta - N - 1), \quad p_+(x) = (N - x)(-\alpha - 1 - x), \quad \alpha, \beta < -N.$$

It is convenient to give up the positivity assumption for $p_\pm$ and use the first formula for $p_\pm$ in either case. The weight function is

$$w_m = C \, \frac{(N - m + 1)_m (\alpha + 1)_m}{m! \, (N + \beta + 1 - m)_m} = C \binom{N}{m} \frac{(\alpha + 1)_m \, (\beta + 1)_{N-m}}{(\beta + 1)_N}.$$
(5.1.13)

The associated normalized polynomials are commonly known as the *Hahn polynomials*. We refer to them here as the *Chebyshev–Hahn polynomials*; see the remarks at the end of the chapter. In the case $\alpha = \beta = 0$, the polynomials are commonly known as the *discrete Chebyshev polynomials*.

We have proved the discrete analogue of Theorem 3.4.1, somewhat loosely stated.

Theorem 5.1.3 *Up to normalization, the Charlier, Krawtchouk, Meixner, and Chebyshev–Hahn polynomials are the only ones that occur as eigenfunctions of a second-order difference operator that is symmetric with respect to a positive weight.*

5.2 The discrete Rodrigues formula

Suppose that w is a weight on the integers and L is symmetric with respect to w and has polynomials as eigenfunctions. The eigenvalue equation for a polynomial ψ_n of degree n is

$$p_+ \, \Delta_+ \psi_n - p_- \, \Delta_- \psi_n + \lambda_n \psi_n = 0. \qquad (5.2.1)$$

Applying Δ_+ to this equation gives an equation for the "derivative" $\psi_n^{(1)} = \Delta_+ \psi_n$. Using the discrete Leibniz identity

$$\Delta_+ (fg) = (S_+ f) \Delta_+ g + g \, \Delta_+ f,$$

we obtain

$$p_+^{(1)} \, \Delta_+ \psi_n^{(1)} - p_-^{(1)} \, \Delta_- \psi_n^{(1)} + \lambda_n^{(1)} \psi_n^{(1)} = 0,$$

with

$$p_+^{(1)} = S_+ p_+, \quad p_-^{(1)} = p_-, \quad \lambda_n^{(1)} = \lambda_n + \Delta_+ (p_+ - p_-).$$

The new operator here is symmetric with respect to the weight $w^{(1)} = p_+ w$:

$$S_+ \left(w^{(1)} p_-^{(1)} \right) = (S_+ p_+) \, S_+ (w p_-) = p_+^{(1)} w p_+ = w^{(1)} \, p_+^{(1)}.$$

Continuing, we see that the successive differences $\psi_n^{(k)} = (\Delta_+)^k \psi_n$ satisfy

$$p_+^{(k)} \, \Delta_+ \psi_n^{(k)} - p_- \, \Delta_- \psi_n^{(k)} + \lambda_n^{(k)} \psi_n^{(k)} = 0,$$

with

$$p_+^{(k)} = S_+^k p_+, \tag{5.2.2}$$

$$\lambda_n^{(k)} = \lambda_n + \left(I + S_+ + \cdots + S_+^{k-1}\right)\Delta_+ p_+ - k\Delta_+ p_-,$$

and the corresponding operator is symmetric with respect to the weight

$$w^{(k)} = w \prod_{j=0}^{k-1} S_+^j p_+. \tag{5.2.3}$$

Now $\psi_n^{(k)}$ has degree $n - k$. In particular, $\psi_n^{(n)}$ is constant, so $\lambda_n^{(n)} = 0$, and we have proved that

$$\lambda_n = -\left(I + S_+ + \cdots + S_+^{n-1}\right)\Delta_+ p_+ + n\Delta_+ p_-$$

$$= n\Delta_+(p_- - p_+) + \sum_{j=1}^{n-1} \left(I - S_+^j\right)\Delta_+ p_+. \tag{5.2.4}$$

The eigenvalue equation can be rewritten to obtain $\psi_n^{(k-1)}$ from $\psi_n^{(k)}$, which leads to a discrete Rodrigues formula for ψ_n. At the first stage we rewrite the operator L, using the identities (5.1.6) and $S_- \Delta_+ = \Delta_-$,

$$wL = wp_+ \Delta_+ - S_-(w \, p_+) \, \Delta_- = wp_+ \Delta_+ - S_-(wp_+ \Delta_+)$$

$$= \Delta_-(wp_+ \Delta_+) = \Delta_-(w^{(1)} \Delta_+).$$

Therefore the eigenvalue equation can be solved to obtain

$$\psi_n = -\frac{1}{\lambda_n w} \, \Delta_-\left(w^{(1)} \psi_n^{(1)}\right)$$

on the points where w is positive. We continue this process, and take the constant $\psi_n^{(n)}$ to be

$$\psi_n^{(n)} = \prod_{k=0}^{n-1} \left[\lambda_n + \left(S_+^k - I\right)p_+ - k\Delta_+ p_-\right] \equiv A_n. \tag{5.2.5}$$

The result is the discrete Rodrigues formula: where $w > 0$,

$$\psi_n = (-1)^n \frac{1}{w} \Delta_-^n \left(w^{(n)}\right), \quad w^{(n)} = w \prod_{k=0}^{n-1} S_+^k p_+. \tag{5.2.6}$$

Since $\Delta_- = I - S_-$, we may expand $\Delta_-^n = (I - S_-)^n$ and rewrite (5.2.6) as

$$\psi_n = \frac{1}{w} \sum_{k=1}^{n} (-1)^{n-k} \binom{n}{k} S_-^k \left(w^{(n)} \right). \qquad (5.2.7)$$

As an application, we obtain a formula for the norm of ψ_n. Note that whenever the sums are absolutely convergent,

$$\sum_k \Delta_- f(k) \, g(k) = -\sum_k f(k) \, \Delta_+ g(k).$$

Therefore

$$\lambda_n ||\psi_n||_w^2 = -(L\psi_n, \psi_n)_w = -\sum_k (wL\psi_n)(k) \, \psi_n(k)$$

$$= -\sum_k \left[\Delta_- \left(w p_+ \psi_n^{(1)} \right) \right](k) \, \psi_n(k) = \sum_k \left[w p_+ \psi_n^{(1)} \right](k) \, \psi_n^{(1)}(k)$$

$$= ||\psi_n^{(1)}||_{w^{(1)}}^2.$$

Continuing, we obtain

$$A_n ||\psi_n||_w^2 = ||\psi_n^{(n)}||_{w^{(n)}}^2 = A_n^2 ||1||_{w^{(n)}}^2.$$

Therefore

$$||\psi_n||_w^2 = A_n \sum_k w^{(n)}(k). \qquad (5.2.8)$$

It is useful to rewrite the operator in (5.2.6) as

$$\frac{1}{w} \Delta_-^n w^{(n)} = \left(\frac{1}{w} \Delta_- w^{(1)} \right) \left(\frac{1}{w^{(1)}} \Delta_- w^{(2)} \right) \cdots \left(\frac{1}{w^{(n-1)}} \Delta_- w^{(n)} \right).$$

It follows from the symmetry condition (5.1.6) that

$$\frac{1}{w} \Delta_- \left(w^{(1)} f \right) = p_+ f - \frac{S_-(p_+ w)}{w} S_- f = p_+ f - p_- S_- f.$$

Then

$$\frac{1}{w} \Delta_- w^{(1)} \cdot \frac{1}{w^{(1)}} \Delta_- w^{(2)} = (p_+ - p_- S_-)\left(p_+^{(1)} - p_- S_- \right)$$

$$= p_+ p_+^{(1)} - 2 p_+ p_- S_- + p_- (S_- p_-) S_-^2$$

and by induction

$$(-1)^n \frac{1}{w} \Delta_-^n w^{(n)} = \sum_{k=0}^n (-1)^{n-k} \binom{n}{k} \prod_{j=0}^{n-k-1} p_+^{(j)} \prod_{j=0}^{k-1} p_-^{(j)} S_-^k,$$

where $p_-^{(j)} = S_-^j p_-$. Applying this operator to the constant function 1 gives

$$\psi_n = \sum_{k=0}^n (-1)^{n-k} \binom{n}{k} \prod_{j=0}^{n-k-1} \left(S_+^j p_+\right) \prod_{j=0}^{k-1} \left(S_-^j p_-\right). \tag{5.2.9}$$

A second approach to computing ψ_n is to use a discrete analogue of the series expansion, with the monomials x^k replaced by the polynomials

$$e_k(x) = (x - k + 1)_k = x(x - 1)(x - 2) \cdots (x - k + 1) = (-1)^k (-x)_k.$$

Then

$$\Delta_+ e_k(x) = k\, e_{k-1}(x), \quad x\, \Delta_- e_k(x) = k\, e_k(x), \quad x\, e_k = e_{k+1} + k\, e_k.$$

In each of our cases, $p_-(x)$ is divisible by x, so applying the operator $L = p_+\Delta_+ - p_-\Delta_-$ to the expansion

$$\psi_n(x) = \sum_{k=0}^n a_{nk}\, e_k(x) \tag{5.2.10}$$

leads to recurrence relations for the coefficients a_{nk} that identify a_{nk} as a certain multiple of $a_{n,k-1}$.

Finally, we remark that both the coefficient β_n of the three-term recurrence relation (5.1.1) and the eigenvalue λ_n can be recovered directly from equation (5.2.1) by computing the coefficients of x^n and x^{n-1}:

$$0 = L\psi_n(x) + \lambda_n \psi_n(x)$$

$$= p_+\Delta_+\left(a_n x^n + b_n x^{n-1}\right) - p_-\Delta_-\left(a_n x^n + b_n x^{n-1}\right)$$

$$+ \lambda_n a_n x^n + \lambda_n b_n x^{n-1} + \cdots$$

$$= (p_+ - p_-)\left[n a_n x^{n-1} + (n-1)b_n x^{n-2}\right]$$

$$+ (p_+ + p_-)\left[\binom{n}{2} a_n x^{n-2} + \binom{n-1}{2} b_n x^{n-3}\right]$$

$$+ \lambda_n a_n x^n + \lambda_n b_n x^{n-1} + \cdots \tag{5.2.11}$$

The coefficient of x^n on the right must vanish, and this determines λ_n. Using this value of λ_n in the coefficient of x^{n-1} determines the ratio b_n/a_n and therefore the term $\beta_n = b_n/a_n - b_{n+1}/a_{n+1}$ in the three-term recurrence.

5.3 Charlier polynomials

The interval is infinite, and

$$p_+(x) = 1, \quad p_-(x) = \frac{x}{a}, \quad w_m = e^{-a}\frac{a^m}{m!}.$$

Then $p_+^{(k)} = p_+$, $w^{(k)} = w$, and

$$\lambda_n = \frac{n}{a}.$$

Therefore the constant A_n in (5.2.5) is $n!/a^n$. From (5.2.8) we obtain the norm:

$$\|\psi_n\|_w^2 = \frac{n!}{a^n}. \tag{5.3.1}$$

A standard normalization is $C_n(x; a) = (-1)^n \psi_n(x)$. Equation (5.2.9) gives

$$C_n(x; a) = \sum_{k=0}^{n} (-1)^k \binom{n}{k} a^{-k}(x - k + 1)_k$$

$$= \sum_{k=0}^{n} \frac{(-n)_k\,(-x)_k}{k!}\left(-\frac{1}{a}\right)^k$$

$$= {}_2F_0\left(-n, -x; -\frac{1}{a}\right), \tag{5.3.2}$$

where ${}_2F_0$ is a generalized hypergeometric series; see Chapter 8. The leading coefficient is $(-a)^{-n}$. In this case $\Delta_+ C_n$ is an orthogonal polynomial with respect to the same weight, and comparison of leading coefficients gives

$$\Delta_+ C_n(x; a) = -\frac{n}{a}\,C_{n-1}(x; a). \tag{5.3.3}$$

The associated difference equation is

$$\Delta_+ C_n(x; a) - \frac{x}{a}\,\Delta_- C_n(x; a) + \frac{n}{a}\,C_n(x; a) = 0. \tag{5.3.4}$$

The first four polynomials are

$$C_0(x; a) = 1;$$

$$C_1(x; a) = -\frac{x}{a} + 1;$$

$$C_2(x; a) = \frac{x(x-1)}{a^2} - \frac{2x}{a} + 1 = \frac{x^2}{a^2} - (1 + 2a)\frac{x}{a^2} + 1;$$

$$C_3(x; a) = -\frac{x(x-1)(x-2)}{a^3} + \frac{3x(x-1)}{a^2} - \frac{3x}{a} + 1$$

$$= -\frac{x^3}{a^3} + 3(a+1)\frac{x^2}{a^3} - (3a^2 + 3a + 2)\frac{x}{a^3} + 1.$$

It is a simple matter to compute the generating function

$$G(x, t; a) \equiv \sum_{n=0}^{\infty} \frac{C_n(x; a)}{n!} t^n.$$

Note that the constant term of $C_n(x; a)$ is 1, so

$$G(x, 0; a) = 1, \quad G(0, t; a) = e^t. \tag{5.3.5}$$

In general,

$$G(x, t; a) = \sum_{n=0}^{\infty} \sum_{k=0}^{n} t^n \frac{(-x)_k}{(n-k)!\, k!} \left(\frac{1}{a}\right)^k$$

$$= \sum_{m=0}^{\infty} \sum_{k=0}^{\infty} \frac{t^m}{m!} \frac{(-x)_k}{k!} \left(\frac{t}{a}\right)^k$$

$$= e^t \left(1 - \frac{t}{a}\right)^x. \tag{5.3.6}$$

The generating function allows another computation of the norms and inner products:

$$\sum_{n=0}^{\infty} \sum_{m=0}^{\infty} \frac{(C_n, C_m)_w}{n!\, m!} t^n s^m = \sum_{k=0}^{\infty} G(k, t; a)\, G(k, s; a)\, w_k = e^{st/a}. \tag{5.3.7}$$

The identity

$$\sum_{n=0}^{\infty} \frac{C_{n+1}(x; a)}{n!} t^n = \frac{\partial G}{\partial t}(x, t; a)$$

leads to the recurrence relation

$$C_{n+1}(x; a) = C_n(x; a) - \frac{x}{a} C_n(x - 1; a). \tag{5.3.8}$$

The identity

$$G(x + 1, t; a) = \left(1 - \frac{t}{a}\right) G(x, t; a)$$

leads to (5.3.3).

The general three-term recurrence (5.1.1) is easily computed. It follows from (5.3.2) that

$$(-1)^n C_n(x; a) = \frac{1}{a^n} x^n - \frac{\binom{n}{2} + na}{a^n} x^{n-1} + \cdots$$

Therefore

$$x\, C_n(x; a) = -a\, C_{n+1}(x; a) + (n + a)\, C_n(x; a) - n\, C_{n-1}(x; a). \tag{5.3.9}$$

The generating function can be used to obtain an addition formula

$$C_n(x + y; a) = \sum_{j+k+l=n} (-1)^l \frac{n!}{j!\,k!\,l!} C_j(x; a)\, C_k(x; a); \tag{5.3.10}$$

see the exercises. The proof of the following addition formula is also left as an exercise:

$$C_n(x + y; a) = \sum_{k=0}^{n} \binom{n}{k} C_{n-k}(x; a) \frac{(-y)_k}{a^k}. \tag{5.3.11}$$

Charlier polynomials are connected with the Laguerre polynomials:

$$C_n(x; a) = (-1)^n \frac{n!}{a^n} L_n^{(x-n)}(a);$$

see [74]. Therefore $u(a) = a^n C_n(x; a)$ satisfies the confluent hypergeometric equation

$$a\, u''(a) + (1 + x - n - a)\, u'(a) + n\, u(a) = 0.$$

Dunster [74] derived uniform asymptotic expansions in n for x in each of three intervals whose union is the real line $-\infty < x < \infty$. His results imply that for fixed real x,

$$C_n(x; a) \sim -\frac{n!\, e^a}{a^n} \frac{\sin \pi x}{\pi} \frac{\Gamma(1 + x)}{n^{x+1}} = \frac{n!\, e^a}{a^n\, \Gamma(-x)\, n^{x+1}} \tag{5.3.12}$$

as $n \to \infty$. Therefore the zeros are asymptotically close to the positive integers. See Exercise 10.22 in Chapter 10.

5.4 Krawtchouk polynomials

The interval is finite and

$$
p_+(x) = p\,(N - x), \quad p_-(x) = q\,x, \quad w_m = \binom{N}{m} p^m q^{N-m}, \quad (5.4.1)
$$

where $p, q > 0$, $p + q = 1$. Then $\lambda_n = n$ and the constant A_n in (5.2.5) is $n\,!$.
Two standard normalizations here are

$$
k_n^{(p)}(x, N) = \frac{1}{n\,!}\,\psi_n(x);
$$

$$
K_n(x; p, N) = (-1)^n \frac{(N - n)\,!}{N\,!\,p^n}\,\psi_n(x) = (-1)^n \binom{N}{n}^{-1} \frac{1}{p^n}\,k_n^{(p)}(x; N).
$$

From (5.2.8) we obtain the norms:

$$
\|k_n^{(p)}\|_w^2 = \binom{N}{n}(pq)^n; \quad (5.4.2)
$$

$$
\|K_n\|_w^2 = \binom{N}{n}^{-1} \left(\frac{q}{p}\right)^n.
$$

Most of the identities to follow have a simpler form in the version $k_n^{(p)}$.
Equation (5.2.9) gives

$$
k_n^{(p)}(x; N) = \sum_{k=0}^{n} p^{n-k} q^k \frac{(x - N)_{n-k}(x - k + 1)_k}{(n - k)\,!\,k\,!}. \quad (5.4.3)
$$

The leading coefficient is

$$
\frac{1}{n\,!} \sum_{k=0}^{n} \binom{n}{k} p^{n-k} q^k = \frac{(p + q)^n}{n\,!} = \frac{1}{n\,!}.
$$

The polynomial $\Delta_+ k_n^{(p)}$ is an eigenfunction for the weight $w^{(1)}$, which, after normalization, is the weight associated with p and $N - 1$. Taking into account the leading coefficients, it follows that

$$
\Delta_+ k_n^{(p)}(x; N) = k_{n-1}^{(p)}(x; N - 1). \quad (5.4.4)
$$

The associated difference equation is

$$
p\,(N - x)\,\Delta_+ k_n^{(p)}(x; N) - q\,x\,\Delta_- k_n^{(p)}(x; N) + n\,k_n^{(p)}(x; N) = 0. \quad (5.4.5)
$$

Using this equation and the expansion (5.2.10), we may derive a second form for the Krawtchouk polynomials:

$$k_n^{(p)}(x; N) = (-p)^n \binom{N}{n} \sum_{k=0}^{n} \frac{(-n)_k(-x)_k}{(-N)_k \, k!} \, p^{-k}$$

$$= (-p)^n \binom{N}{n} F\left(-n, -x, -N; \frac{1}{p}\right), \qquad (5.4.6)$$

where F is the hypergeometric function (Chapter 8). The normalization of the alternate form K_n is chosen so that

$$K_n(x; p, N) = F\left(-n, -x, -N; \frac{1}{p}\right).$$

The first four polynomials are:

$$k_0^{(p)}(x; N) = 1;$$

$$k_1^{(p)}(x; N) = x - Np;$$

$$k_2^{(p)}(x; N) = \frac{1}{2}\left[x^2 + (2p - 1 - 2Np)\,x + N(N-1)\,p^2\right];$$

$$k_3^{(p)}(x; N) = \frac{1}{6}\left\{x^3 + (6p - 3 - 3Np)\,x^2\right.$$

$$+ \left[3Np(Np + 1 - 3p) + 2(3p^2 - 3p + 1)\right]x$$

$$\left. - N(N-1)(N-2)p^3\right\}.$$

We may consider $k_n^{(p)}(x; N)$ as being defined by (5.4.3) for all $n = 0, 1, 2, \ldots$ Note that for $n > N$ these polynomials vanish at the points $m = 0, 1, 2, \ldots, N$. The generating function is

$$G(x, t; N, p) \equiv \sum_{n=0}^{\infty} k_n^{(p)}(x; N)\,t^n = (1 + qt)^x (1 - pt)^{N-x}; \qquad (5.4.7)$$

see the exercises. The identity

$$\sum_{n=0}^{\infty} (n+1)k_{n+1}^{(p)}(x; N)\,t^n = \frac{\partial G}{\partial t}(x, t; N, p)$$

leads to the recurrence relation

$$(n+1)\,k_{n+1}^{(p)}(x; N) = x\,q\,k_n^{(p)}(x-1; N-1) - (N-x)\,p\,k_n^{(p)}(x; N-1). \tag{5.4.8}$$

The identity

$$(1 - pt) G(x + 1, t; N, p) = (1 + qt) G(x, t; N, p) \tag{5.4.9}$$

leads to

$$k_n^{(p)}(x + 1; N) - k_n^{(p)}(x; N) = k_{n-1}^{(p)}(x; N). \tag{5.4.10}$$

The three-term recurrence

$$x \, k_n^{(p)}(x; N) = (n + 1) \, k_{n+1}^{(p)}(x; N)$$

$$+ (pN + n - 2pn) \, k_n^{(p)}(x; N) + p \, q \, (N - n + 1) \, k_{n-1}^{(p)}(x; N) \tag{5.4.11}$$

can be computed using (5.2.11); see the exercises.

The generating function can be used to prove the addition formula

$$k_n^{(p)}(x + y; N) = \sum_{j+l+m=n} \frac{(N)_j \, p^j}{j \,!} \, k_l^{(p)}(x; N) \, k_m^{(p)}(y; N). \tag{5.4.12}$$

To describe the asymptotic behavior of $k_n^{(p)}(x; N)$ for fixed $x > 0$ and $p > 0$, we first let $N = n\mu$ for fixed $\mu \geq 1$, and set $q - 1 = p$. Following Qiu and Wong [235] we note first that there is a unique $\eta = \eta(\mu)$ such that

$$\eta - (\mu - 1) \log \eta = (\mu - 1)(1 - \log q) - \mu \log \mu - \log p.$$

Let

$$t_0 = \frac{\mu - 1}{\mu}, \quad s_0 = \frac{\mu - 1}{\eta}, \quad \lambda = n \, \eta,$$

$$\gamma = -\mu \log \mu + (\mu - 1)(1 + \log \eta), \quad \zeta = \pm \sqrt{2(1 - s_0 + s_0 \log s_0)},$$

where the positive sign is taken if and only if $s_0 \geq 1$. Then let

$$g_0(s_0) = \begin{cases} -\dfrac{1}{\sqrt{1 - t_0}} \left(\dfrac{t_0 - q}{s_0 - 1} \right)^x, & s_0 \neq 1, \\[2ex] -q^x p^{(x-1)/2}, & s_0 = 1. \end{cases}$$

Using an extension of the steepest descent method (see Chapter 10), Qiu and Wong showed that as $n \to \infty$,

(i) for $\mu \geq \frac{1}{p} + \varepsilon$,

$$k_n^{(p)}(x; n\mu) \sim (-1)^{n+1} \, e^{\lambda(s_0 - s_0 \log s_0)} \, \frac{(s_0 - 1)^x g(s_0)}{\sqrt{2\pi \lambda s_0}} \, \frac{p^{n-x}}{e^{n\gamma}};$$

(ii) for $\frac{1}{p} - \varepsilon < \mu < \frac{1}{p} + \varepsilon$,

$$k_n^{(p)}(x; n\mu) \sim (-1)^{n+1} a_0 \, W\left(x, \sqrt{\lambda}\,\zeta\right) \frac{e^\lambda \, p^{n-x}}{e^{n\gamma} \, \lambda^{(x+1)/2}},$$

where

$$W(x, \zeta) = \frac{D_x(\zeta)}{\sqrt{2\pi} \, e^{\zeta^2/4}}, \qquad a_0 = \begin{cases} g_0(1) \left(\dfrac{\zeta}{s_0 - 1}\right)^{x+1}, & s_0 \neq 1; \\ g_0(1), & s_0 = 1, \end{cases}$$

and D_x is the parabolic cylinder function of Section 6.6;

(iii) for $1 \leq \mu \leq \frac{1}{p} - \varepsilon$,

$$k_n^{(p)}(x; n\mu) \sim (-1)^{n+1} \frac{g_0(1) \, e^\lambda \, p^{n-x}}{\Gamma(-x) \, (\lambda - \lambda s_0)^{x+1} e^{n\gamma}}.$$

5.5 Meixner polynomials

The interval is infinite and

$$p_+(x) = c(x + b), \qquad p_-(x) = x, \qquad w_m = (1 - c)^b \, \frac{(b)_m}{m!} \, c^m, \qquad (5.5.1)$$

where $b > 0$ and $0 < c < 1$. Therefore

$$\lambda_n = (1 - c)\, n, \qquad A_n = (1 - c)^n \, n!.$$

Equation (5.2.9) gives

$$\psi_n(x) = \sum_{k=0}^{n} (-1)^{n-k} \binom{n}{k} (x + b)_{n-k} \, (x - k + 1)_k \, c^{n-k}$$

with leading coefficient $(1 - c)^n$. Two standard normalizations are:

$$m_n(x; b, c) = (-c)^{-n} \psi_n(x) = \sum_{k=0}^{n} \binom{n}{k} (x + b)_{n-k} \, (x - k + 1)_k \, (-c)^{-k};$$

$$(5.5.2)$$

$$M_n(x; b, c) = \frac{(-1)^n}{c^n \, (b)_n} \, \psi_n(x) = \frac{1}{(b)_n} \, m_n(x; b, c).$$

It follows that

$$||m_n||_w^2 = n! \frac{(b)_n}{c^n}; \tag{5.5.3}$$

$$||M_n||_w^2 = \frac{n!}{(b)_n c^n}.$$

Most of the identities to follow have a simpler form in the version m_n.

The polynomial $\Delta_+ m_n$ is an eigenfunction for the weight $w^{(1)}$, which, after normalization, is the weight associated with $b + 1$ and c. The leading coefficient of m_n is $(1 - 1/c)^n$, so

$$\Delta_+ m_n(x; b, c) = n \left(1 - \frac{1}{c}\right) m_{n-1}(x; b + 1, c). \tag{5.5.4}$$

The associated difference equation is

$$c(x + b) \Delta_+ m_n(x; b, c) - x \Delta_- m_n(x; b, c) + (1 - c) n m_n(x; b, c) = 0. \tag{5.5.5}$$

Using this equation and the expansion (5.2.10) leads to a second form (normalized by taking into account the leading coefficient):

$$m_n(x; b, c) = (b)_n \sum_{k=0}^{n} \frac{(-n)_k (-x)_k}{(b)_k k!} \left(1 - \frac{1}{c}\right)^k$$

$$= (b)_n F\left(-n, -x, b; 1 - \frac{1}{c}\right), \tag{5.5.6}$$

where again F is the hypergeometric function. The normalization of M_n is chosen so that

$$M_n(x; b, c) = F\left(-n, -x, b; 1 - \frac{1}{c}\right).$$

The first four polynomials are:

$$m_0(x; b, c) = 1;$$

$$m_1(x; b, c) = \left(1 - \frac{1}{c}\right) x + b;$$

$$m_2(x; b, c) = \left(1 - \frac{1}{c}\right)^2 x^2 + \left(2b + 1 - \frac{2b}{c} - \frac{1}{c^2}\right) x + b(b + 1);$$

$$m_3(x; b, c) = \left(1 - \frac{1}{c}\right)^3 x^3 + \left(3b + 3 - \frac{6b + 3}{c} + \frac{3b - 3}{c^2} + \frac{3}{c^3}\right) x^2$$

$$+ \left(3b^2 + 6b + 2 - \frac{3b^2 + 3b}{c} - \frac{3b}{c^2} - \frac{2}{c^3}\right) x + b(b + 1)(b + 2).$$

The generating function is

$$G(x; b, c) \equiv \sum_{n=0}^{\infty} \frac{m_n(x; b, c)}{n!} t^n = (1 - t)^{-x-b} \left(1 - \frac{t}{c}\right)^x; \qquad (5.5.7)$$

see the exercises.

The identity

$$\sum_{n=0}^{\infty} \frac{m_{n+1}(x; b, c)}{n!} t^n = \frac{\partial G}{\partial t}(x; b, c)$$

implies the recurrence relation

$$m_{n+1}(x; b, c) = (x + b)\, m_n(x; b + 1, c) - \frac{x}{c}\, m_n(x - 1; b + 1, c). \quad (5.5.8)$$

The identity

$$(1 - t)\, G(x + 1; b, c) = \left(1 - \frac{t}{c}\right) G(x; b, c)$$

implies

$$m_n(x + 1; b, c) - m_n(x; b, c) = n\, m_{n-1}(x + 1; b, c) - \frac{n}{c}\, m_{n-1}(x; b, c). \quad (5.5.9)$$

The three-term recurrence

$$(c - 1)\, x\, m_n(x; b, c) = c\, m_{n+1}(x; b, c) - (bc + nc + n)\, m_n(x; b, c)$$

$$+ n(b + n - 1)\, m_{n-1}(x; b, c) \qquad (5.5.10)$$

can be computed using (5.2.11); see the exercises.

The generating function can be used to prove the addition formula

$$m_n(x + y; b, c) = \sum_{j+k+l=n} \frac{n!\, (-b)_j}{j!\, k!\, l!}\, m_l(x; b, c)\, m_k(y; b, c); \qquad (5.5.11)$$

see the exercises.

It follows from the expansion (5.5.6) and the identity (4.5.11) that Laguerre polynomials are limits of Meixner polynomials: for $x \neq 0$,

$$m_n \left(\frac{cx}{1-c}; \alpha + 1, c \right) \sim (\alpha + 1)_n \, M(-n, \alpha + 1; x) = n! \, L_n^{(\alpha)}(x) \quad (5.5.12)$$

as $c \to 1-$; see the exercises.

Jin and Wong [147] used a modification of the steepest descent method due to Chester, Friedman, and Ursell [49, 318] to derive an asymptotic expansion for $m_n(n\alpha; b, c)$ for $\alpha > 0$. When $n\alpha$ is bounded, they gave the simplified result

$$m_n(n\alpha; b, c) \sim - \frac{n! \, \Gamma(\alpha n + 1)}{c^n \, (1-c)^{n\alpha+b} \, n^{n\alpha+1}} \frac{\sin \pi n\alpha}{\pi} \quad (5.5.13)$$

as $n \to \infty$: see (3.13), (3.14), and (4.1) of [148] and Exercise 10.21 of Chapter 10.

Fix x and take $\alpha = x/n$. It follows from (5.5.13) that

$$m_n(x; b, c) \sim - \frac{\Gamma(b+n) \, \Gamma(x+1)}{c^n (1-c)^{x+b} \, n^{b+x}} \frac{\sin \pi x}{\pi}. \quad (5.5.14)$$

Thus the zeros of the Meixner polynomials are asymptotically close to the positive integers as $n \to \infty$.

5.6 Chebyshev–Hahn polynomials

The interval is finite and we take

$$p_+(x) = (N - x)(x + \alpha + 1), \quad p_-(x) = x(N + \beta + 1 - x),$$

$$\alpha, \beta > -1 \quad \text{or} \quad \alpha, \beta < -N. \quad (5.6.1)$$

In the case $\alpha, \beta < -N$ we have violated our condition that $p_\pm$ be positive at the interior points $\{1, 2, \ldots, N - 1\}$. In the following formulas this will only show up in the appearance of absolute values in the formulas for norms.

The weight is

$$w_m = C_0 \binom{N}{m} (\alpha + 1)_m \, (\beta + 1)_{N-m}, \quad (5.6.2)$$

where

$$C_0 = \frac{\Gamma(N + \alpha + \beta + 2)}{\Gamma(\alpha + \beta + 2)}.$$

With this choice of C_0 the total mass is

$$\sum_{m=0}^{N} w_m = 1; \tag{5.6.3}$$

see the exercises.

According to (5.2.9),

$$\psi_n(x) = \sum_{k=0}^{n} (-1)^{n-k} \binom{n}{k} \prod_{j=0}^{n-k-1} p_+(x+j) \prod_{j=0}^{k-1} p_-(x-j). \tag{5.6.4}$$

This appears to have degree $2n$ rather than n, but there is considerable cancellation. A more useful form can be obtained using (5.2.10), which leads to

$$\psi_n(x) = C \sum_{k=0}^{n} \frac{(-n)_k (-x)_k (n + \alpha + \beta + 1)_k}{(-N)_k (\alpha + 1)_k \, k!}$$

$$= C \, {}_3F_2(-n, -x, n + \alpha + \beta + 1; -N, \alpha + 1; 1),$$

where ${}_3F_2$ denotes the generalized hypergeometric series; see Chapter 8. To determine the constant C we note that the constant term in (5.6.4) comes from the summand with $k = 0$ and is therefore

$$(-1)^n \prod_{j=0}^{n-1} p_+(j) = (-1)^n (N + 1 - n)_n (\alpha + 1)_n.$$

It follows that

$$\psi_n(x) = (-1)^n (N + 1 - n)_n (\alpha + 1)_n$$

$$\times \, {}_3F_2(-n, -x, n + \alpha + \beta + 1; -N, \alpha + 1; 1).$$

One normalization is

$$Q_n(x; \alpha, \beta, N) = {}_3F_2(-n, -x, n + \alpha + \beta + 1; -N, \alpha + 1; 1)$$

$$= \sum_{k=0}^{n} \frac{(-n)_k (-x)_k (n + \alpha + \beta + 1)_k}{(-N)_k (\alpha + 1)_k \, k!}. \tag{5.6.5}$$

A second is

$$h_n^{(\alpha,\beta)}(x, N) = (-1)^n \frac{(N - n)_n (\beta + 1)_n}{n!} Q_n(x; \beta, \alpha, N - 1). \tag{5.6.6}$$

Neither normalization results in particularly simple forms for the identities that follow; we use Q_n.

It follows from (5.2.4) that $\lambda_n = n(n + \alpha + \beta + 1)$, so

$$\|\psi_n\|_w^2 = (n!)^2 \binom{N}{n} \left| \frac{(\alpha+1)_n(\beta+1)_n(\alpha+\beta+n+1)_{N+1}}{(\alpha+\beta+2)_N(\alpha+\beta+2n+1)} \right|.$$

Therefore

$$\|Q_n(x; \alpha, \beta, N)\|^2$$
$$= \left| \frac{n!(N-n)!(\beta+1)_n(\alpha+\beta+n+1)_{N+1}}{N!(\alpha+1)_n(\alpha+\beta+2)_N(\alpha+\beta+2n+1)} \right|. \tag{5.6.7}$$

The weight $w^{(1)} = p_+ w$ is a constant multiple of the weight associated with the indices $(\alpha+1, \beta+1, N-1)$. It follows from (5.6.5) that the leading coefficient of $Q_n(x; \alpha, \beta, N)$ is

$$\frac{(n+\alpha+\beta+1)_n}{(-N)_n(\alpha+1)_n},$$

so

$$\Delta_+ Q_n(x; \alpha, \beta, N)$$
$$= -\frac{n(\alpha+\beta+n+1)}{N(\alpha+1)} Q_{n-1}(x; \alpha+1, \beta+1, N-1). \tag{5.6.8}$$

The difference equation is

$$(N-x)(x+\alpha+1)\Delta_+ Q_n(x; \alpha, \beta, N)$$
$$- x(N+\beta+1-x)\Delta_- Q_n(x; \alpha, \beta, N)$$
$$+ n(n+\alpha+\beta+1)Q_n(x; \alpha, \beta, N) = 0. \tag{5.6.9}$$

The first three polynomials are:

$$Q_0(x; \alpha, \beta, N) = 1;$$

$$Q_1(x; \alpha, \beta, N) = -\frac{\alpha+\beta+2}{N(\alpha+1)} x + 1;$$

$$Q_2(x; \alpha, \beta, N) = \frac{(\alpha+\beta+3)(\alpha+\beta+4)}{N(N-1)(\alpha+1)(\alpha+2)} x^2$$
$$- \frac{(\alpha+\beta+3)\left[\alpha+\beta+4+2(N-1)(\alpha+2)\right]}{N(N-1)(\alpha+1)(\alpha+2)} x + 1.$$

A straightforward (but tedious) application of (5.2.11) yields

$$Q_n(x; \alpha, \beta, N) = a_n x^n + b_n x^{n-1} + \cdots ,$$

$$\frac{b_n}{a_n} = -\frac{n \left[2N (\alpha + 1) + (2N + \beta - \alpha)(n - 1) \right]}{2(\alpha + \beta + 2n)}.$$

The ratio of leading coefficients is

$$\frac{a_n}{a_{n+1}} = -\frac{(n + \alpha + \beta + 1)(\alpha + n + 1)(N - n)}{(2n + \alpha + \beta + 1)(2n + \alpha + \beta + 2)}.$$

Therefore, by (5.1.1), the three-term recurrence is

$$x \, Q_n(x; \alpha, \beta, N) = \alpha_n \, Q_{n+1}(x; \alpha, \beta, N) + \beta_n \, Q_n(x; \alpha, \beta, N)$$

$$+ \gamma_n \, Q_{n-1}(x; \alpha, \beta, N); \qquad (5.6.10)$$

$$\alpha_n = -\frac{(\alpha + \beta + n + 1)(\alpha + n + 1)(N - n)}{(\alpha + \beta + 2n + 1)(\alpha + \beta + 2n + 2)},$$

$$\gamma_n = -\frac{n(n + \beta)(\alpha + \beta + n + N + 1)}{(\alpha + \beta + 2n)(\alpha + \beta + 2n + 1)},$$

$$\beta_n = -(\alpha_n + \gamma_n).$$

It follows from the expansion (5.6.5) and the identity (4.6.12) that Jacobi polynomials are limits of Chebyshev–Hahn polynomials: for $x \neq 0$,

$$Q_n(Nx; \alpha, \beta, N) \sim F(\alpha + \beta + 1 + n, -n, \alpha + 1; x)$$

$$= \frac{n!}{(\alpha + 1)_n} P_n^{(\alpha, \beta)}(1 - 2x) \qquad (5.6.11)$$

as $N \to \infty$; see the exercises. For refinements of this result, see Sharapudinov [258].

The polynomials that are commonly called "discrete Chebyshev polynomials" are the case $\alpha = \beta = 0$.

$$t_n(x, N) = (-1)^n (N - n)_n \, Q_n(x; 0, 0, N - 1)$$

$$= (-1)^n \sum_{k=0}^{n} \binom{n + k}{k} \frac{(N - n)_{n-k}(-x)_k(n - k + 1)_k}{k!}. \qquad (5.6.12)$$

5.7 Exercises

5.1 Use the expansion (5.2.10) and the difference equation for the Charlier polynomials to give another derivation of (5.3.2), assuming that the leading coefficient is $(-a)^{-n}$.

5.2 Verify (5.3.7) and show that it implies that the C_n are orthogonal and have norm given by (5.3.1).

5.3 Verify the recurrence relation (5.3.8).

5.4 Show that (5.3.3) and the identity $G(0, t; a) = e^t$ determine the Charlier generating function $G(x, t; a)$ uniquely (for integer x).

5.5 Show that the recurrence relation (5.3.8) and the identity $G(x, 0; a) = 1$ determine $G(x, t; a)$ uniquely (for integer x).

5.6 Prove the addition formula (5.3.10).

5.7 Let $p_m(x) = (x - m + 1)_m = (x - m + 1)(x - m + 2) \cdots (x - 1)x$. Show that $\Delta_+ p_m = m p_{m-1}$ and conclude that $(\Delta_+)^k p_m(0) = m!$ if $k = m$ and 0 otherwise. Use this to conclude that if f is any polynomial, it has a discrete Taylor expansion

$$f(x + y) = \sum_{k \geq 0} \frac{(\Delta_+)^k f(x)}{k!} (y - k + 1)_k.$$

5.8 Use Exercise 5.7 to prove (5.3.11).

5.9 Show that

$$\lim_{a \to \infty} C_n(ax; a) = (1 - x)^n.$$

5.10 Derive (5.4.6).

5.11 Use the binomial expansion to verify (5.4.7).

5.12 Compute the constant term of $k_n^{(p)}(x; N)$, and deduce from this and the computation of the leading coefficient that the generating function must satisfy

$$G(x, 0) = 1, \quad G(0, t) = (1 - pt)^N.$$

5.13 Use the result of Exercise 5.12 to show that the recurrence relation (5.4.8) determines the generating function (5.4.7) for integer x.

5.14 Verify (5.4.8) and (5.4.10).

5.15 Verify (5.4.11) by using (5.2.11).

5.16 Show that the coefficient b_n of x^{n-1} in $k_n^{(p)}(x; N)$ in (5.4.3) is

$$b_n = \frac{1}{n!}\left[-nNp + \binom{n}{2}(p-q) \right] = \frac{-Np + (n-1)\left(p - \frac{1}{2}\right)}{(n-1)!}.$$

Hint: compute the coefficient of the kth summand in (5.4.3) and reduce the problem to computing sums like

$$\sum_{k=0}^{n}\binom{n}{k}p^k q^{n-k}k, \quad \sum_{k=0}^{n}\binom{n}{k}p^k q^{n-k}k(k-1).$$

Compare these sums to the partial derivatives F_p and F_{pp} of the function of two variables $F(p,q) = (p+q)^n$, evaluated at $q = 1 - p$.

5.17 Use Exercise 5.16 to give another proof of (5.4.11).

5.18 Prove the addition formula (5.4.12).

5.19 Show that as $N \to \infty$,

$$k_n^{(p)}(Nx; N) \sim \frac{N^n}{n!}(x-p)^n.$$

5.20 Use (5.5.5) and the expansion in (5.2.10) to prove (5.5.6).

5.21 Verify (5.5.7) using (5.5.2).

5.22 Verify (5.5.7) using (5.5.6).

5.23 Verify (5.5.8) and (5.5.9).

5.24 Verify (5.5.10) by using (5.2.11).

5.25 Prove the addition formula (5.5.11).

5.26 Verify (5.5.12).

5.27 Verify (5.6.3): show that

$$(\alpha + 1)_m (\beta + 1)_{N-m}$$

$$= \frac{\Gamma(N + \alpha + \beta + 2)}{\Gamma(\alpha + 1)\Gamma(\beta + 1)} B(\alpha + 1 + m, \beta + 1 + N - m)$$

and use the integral representation (2.1.7) of the beta function to show that

$$\sum_{m=0}^{N}\binom{N}{m}B(\alpha + 1 + m, \beta + 1 + n - m) = B(\alpha + 1, \beta + 1). \quad (5.7.1)$$

5.28 Verify (5.6.10).

5.29 Verify (5.6.11).

5.8 Summary

5.8.1 Discrete weights and difference operators

Suppose that $w = \{w_n\}_{n=-\infty}^{\infty}$ is a two-sided sequence of non-negative numbers. Corresponding inner product:

$$(f, g) = (f, g)_w = \sum_{m=-\infty}^{\infty} f(m)\, g(m)\, w_m.$$

Assume

$$\sum_{m=-\infty}^{\infty} w_m = 1; \qquad \sum_{m=-\infty}^{\infty} m^{2n}\, w_m < \infty, \quad n = 0, 1, 2, \ldots$$

Orthogonal polynomials ψ_n can be constructed as in Chapter 4.

Suppose

$$\psi_n(x) = a_n x^n + b_n x^{n-1} + \cdots$$

Then

$$x\psi_n(x) = \alpha_n \psi_{n+1}(x) + \beta_n \psi_n(x) + \gamma_n \psi_{n-1}(x),$$

$$\alpha_n = \frac{a_n}{a_{n+1}}, \quad \beta_n = \frac{b_n}{a_n} - \frac{b_{n+1}}{a_{n+1}}, \quad \gamma_n = \frac{a_{n-1}}{a_n} \frac{(\psi_n, \psi_n)_w}{(\psi_{n-1}, \psi_{n-1})_w}.$$

Forward and backward difference operators:

$$\Delta_+ f(m) = f(m+1) - f(m), \quad \Delta_- f(m) = f(m) - f(m-1).$$

In terms of shift operators

$$S_\pm f(m) = f(m \pm 1),$$

we have

$$\Delta_+ = S_+ - I, \quad \Delta_- = I - S_-,$$

$$\Delta_+ \Delta_- = \Delta_- \Delta_+ = S_+ + S_- - 2I = \Delta_+ - \Delta_-.$$

General second-order difference operator:

$$L = p_+ S_+ + p_- S_- + r.$$

Operator L is symmetric with respect to w if either of the two equivalent conditions

$$S_-(p_+ w) = p_- w, \quad S_+(p_- w) = p_+ w.$$

hold. Eigenfunctions that correspond to distinct eigenvalues are orthogonal.

Symmetric operators L with coefficients $p_\pm(m)$ positive where $w_m > 0$ (with exceptions at endpoints), whose eigenfunctions are polynomials, normalize to

(a) $w_m > 0$ if and only if $0 \le m \le N$;
(b) $w_m > 0$ if and only if $m \ge 0$.

Coefficients are polynomials of degree at most 2. After further normalization the possibilities are, first,

$$p_+(x) = 1, \quad p_-(x) = \frac{x}{a}, \quad a > 0;$$

$$w_m = e^{-a}\frac{a^m}{m!}, \quad m = 0, 1, 2, 3, \ldots,$$

leading to the Charlier polynomials.
Second,

$$p_+(x) = p(N - x), \quad p_-(x) = qx, \quad p, q > 0, \quad p + q = 1;$$

$$w_m = \binom{N}{m} p^m q^{N-m}, \quad m = 0, 1, 2, 3, \ldots, N,$$

leading to the Krawtchouk polynomials.
Third,

$$p_+(x) = c(x + b), \quad p_-(x) = x, \quad b > 0, \ 0 < c < 1;$$

$$w_m = (1 - c)^b \frac{(b)_m}{m!} c^m, \quad m = 0, 1, 2, 3, \ldots,$$

leading to the Meixner polynomials.
Fourth, with $\alpha, \beta > -1$ or $\alpha, \beta < -N$,

$$p_-(x) = x(N + \beta + 1 - x), \quad p_+(x) = (N - x)(x + \alpha + 1);$$

$$w_m = C_0 \binom{N}{m} (\alpha + 1)_m (\beta + 1)_{N-m},$$

$$C_0 = \frac{\Gamma(N + \alpha + \beta + 2)}{\Gamma(\alpha + \beta + 2)},$$

leading to the Chebyshev–Hahn polynomials.
Thus the Charlier, Krawtchouk, Meixner, and Chebyshev–Hahn polynomials are the only ones that occur as the eigenfunctions of a second-order difference operator that is symmetric with respect to a positive weight.

5.8.2 The discrete Rodrigues formula

Suppose w is a weight on the integers, L is symmetric with respect to w and has polynomials ψ_n as eigenfunctions.

Corresponding discrete Rodrigues formula: where $w > 0$,

$$\psi_n = (-1)^n \frac{1}{w} \Delta_-^n \left(w^{(n)} \right) = \frac{1}{w} \sum_{k=1}^n (-1)^{n-k} \binom{n}{k} S_-^k \left(w^{(n)} \right)$$

$$= \sum_{k=0}^n (-1)^{n-k} \binom{n}{k} \prod_{j=0}^{n-k-1} \left(S_+^j p_+ \right) \prod_{j=0}^{k-1} \left(S_-^j p_- \right),$$

where

$$w^{(n)} = w \prod_{k=0}^{n-1} S_+^k p_+.$$

It follows that

$$\|\psi_n\|_w^2 = A_n \sum_k w^{(n)}(k).$$

The eigenfunction ψ_n can also be computed using the expansion

$$\psi_n(x) = \sum_{k=0}^n a_{nk} e_k(x),$$

where

$$e_k(x) = (x - k + 1)_k = x(x-1)(x-2) \cdots (x - k + 1) = (-1)^k (-x)_k.$$

The eigenvalue equation leads to recurrence relations for the coefficients a_{nk} that identify a_{nk} as a certain multiple of $a_{n,k-1}$.

The coefficient β_n of the three-term recurrence relation (5.1.1) and the eigenvalue λ_n can be obtained directly from equation (5.2.1) by computing the coefficients of x^n and x^{n-1}.

5.8.3 Charlier polynomials

The interval is infinite,

$$p_+(x) = 1, \quad p_-(x) = \frac{x}{a}, \quad w_m = e^{-a} \frac{a^m}{m!};$$

$$\lambda_n = \frac{n}{a}, \quad \|\psi_n\|_w^2 = \frac{n!}{a^n}.$$

A standard normalization is

$$C_n(x; a) = (-1)^n \psi_n(x) = \sum_{k=0}^{n} \frac{(-n)_k \, (-x)_k}{k!} \left(-\frac{1}{a}\right)^k$$

$$= {}_2F_0\left(-n, -x; -\frac{1}{a}\right).$$

Difference relation and eigenvalue equation:

$$\Delta_+ C_n(x; a) = -\frac{n}{a} C_{n-1}(x; a);$$

$$\Delta_+ C_n(x; a) - \frac{x}{a} \Delta_- C_n(x; a) + \frac{n}{a} C_n(x; a) = 0.$$

First four polynomials:

$$C_0(x; a) = 1;$$

$$C_1(x; a) = -\frac{x}{a} + 1;$$

$$C_2(x; a) = \frac{x(x-1)}{a^2} - \frac{2x}{a} + 1 = \frac{x^2}{a^2} - (1 + 2a)\frac{x}{a^2} + 1;$$

$$C_3(x; a) = -\frac{x(x-1)(x-2)}{a^3} + \frac{3x(x-1)}{a^2} - \frac{3x}{a} + 1$$

$$= -\frac{x^3}{a^3} + 3(a+1)\frac{x^2}{a^3} - (3a^2 + 3a + 2)\frac{x}{a^3} + 1.$$

Generating function:

$$G(x, t; a) \equiv \sum_{n=0}^{\infty} \frac{C_n(x; a)}{n!} t^n = e^t \left(1 - \frac{t}{a}\right)^x.$$

Recurrence relations:

$$C_{n+1}(x; a) = C_n(x; a) - \frac{x}{a} C_n(x - 1; a);$$

$$x \, C_n(x; a) = -a \, C_{n+1}(x; a) + (n + a) \, C_n(x; a) - n \, C_{n-1}(x; a).$$

Addition formulas:

$$C_n(x + y; a) = \sum_{j+k+l=n} (-1)^l \frac{n!}{j! \, k! \, l!} C_j(x; a) \, C_k(x; a);$$

$$= \sum_{k=0}^{n} \binom{n}{k} C_{n-k}(x; a) \frac{(-y)_k}{a^k}.$$

Connection with Laguerre polynomials:

$$C_n(x; a) = (-1)^n \frac{n!}{a^n} L_n^{(x-n)}(a).$$

Asymptotics, x fixed, $n \to \infty$:

$$C_n(x; a) \sim -\frac{n! \, e^a}{a^n} \frac{\sin \pi x}{\pi} \frac{\Gamma(1 + x)}{n^{x+1}} = \frac{n! \, e^a}{a^n \, \Gamma(-x) \, n^{x+1}}.$$

5.8.4 Krawtchouk polynomials

The interval is finite,

$$p_+(x) = p\,(N - x), \quad p_-(x) = q\,x, \quad p, q > 0, \ p + q = 1;$$
$$w_m = \binom{N}{m} p^m q^{N-m}, \quad \lambda_n = n.$$

Standard normalizations:

$$k_n^{(p)}(x, N) = \frac{1}{n!} \, \psi_n(x);$$

$$K_n(x; p, N) = (-1)^n \frac{(N - n)!}{N! \, p^n} \, \psi_n(x) = (-1)^n \binom{N}{n}^{-1} \frac{1}{p^n} k_n^{(p)}(x, N).$$

Norms:

$$\|k_n^{(p)}\|_w^2 = \binom{N}{n} (pq)^n;$$

$$\|K_n\|^2 = \binom{N}{n}^{-1} \left(\frac{q}{p}\right)^n.$$

Expansions:

$$k_n^{(p)}(x; N) = \sum_{k=0}^{n} p^{n-k} q^k \frac{(x - N)_{n-k}(x - k + 1)_k}{(n - k)! \, k!}$$

$$= (-p)^n \binom{N}{n} \sum_{k=0}^{n} \frac{(-n)_k(-x)_k}{(-N)_k \, k!} \, p^{-k};$$

$$K_n(x; p, N) = F\left(-n, -x, -N; \frac{1}{p}\right).$$

Difference relation and eigenvalue equation:

$$\Delta_+ k_n^{(p)}(x; N) = k_{n-1}^{(p)}(x; N-1);$$

$$p(N-x)\,\Delta_+ k_n^{(p)}(x; N) - q\,x\,\Delta_- k_n^{(p)}(x, N) + n\,k_n^{(p)}(x; N) = 0.$$

First four polynomials:

$$k_0^{(p)}(x; N) = 1;$$

$$k_1^{(p)}(x; N) = x - Np;$$

$$k_2^{(p)}(x; N) = \frac{1}{2}\left[x^2 + (2p - 1 - 2Np)\,x + N(N-1)\,p^2\right];$$

$$k_3^{(p)}(x; N) = \frac{1}{6}\Big\{x^3 + (6p - 3 - 3Np)\,x^2$$

$$+ \left[3Np(Np + 1 - 3p) + 2(3p^2 - 3p + 1)\right]x$$

$$- N(N-1)(N-2)p^3\Big\}.$$

Generating function:

$$G(x, t; N, p) \equiv \sum_{n=0}^{\infty} k_n^{(p)}(x; N)\,t^n = (1 + qt)^x (1 - pt)^{N-x}.$$

Recurrence relations:

$$(n+1)\,k_{n+1}^{(p)}(x; N) = x\,q\,k_n^{(p)}(x-1; N-1) - (N-x)\,p\,k_n^{(p)}(x; N-1);$$

$$k_n^{(p)}(x+1; N) = k_n^{(p)}(x; N) + k_{n-1}^{(p)}(x; N);$$

$$x\,k_n^{(p)}(x; N) = (n+1)\,k_{n+1}^{(p)}(x; N)$$

$$+ (pN + n - 2pn)\,k_n^{(p)}(x, N)$$

$$+ p\,q\,(N - n + 1)\,k_{n-1}^{(p)}(x; N).$$

Addition formula:

$$k_n^{(p)}(x + y; N) = \sum_{j+l+m=n} \frac{(N)_j\,p^j}{j!}\,k_l^{(p)}(x; N)\,k_m^{(p)}(y; N).$$

For the asymptotic behavior, see above.

5.8.5 Meixner polynomials

The interval is infinite,

$$p_+(x) = c(x+b), \quad p_-(x) = x, \quad b > 0, \quad 0 < c < 1;$$

$$w_m = (1-c)^b \frac{(b)_m}{m!} c^m, \quad \lambda_n = (1-c)n, \quad A_n = (1-c)^n n!.$$

Standard normalizations:

$$m_n(x; b, c) = (-c)^{-n} \psi_n(x) = \sum_{k=0}^{n} \binom{n}{k} (x+b)_{n-k} (x-k+1)_k (-c)^{-k}$$

$$= (b)_n \sum_{k=0}^{n} \frac{(-n)_k (-x)_k}{(b)_k k!} \left(1 - \frac{1}{c}\right)^k;$$

$$M_n(x; b, c) = \frac{1}{(b)_n} m_n(x; b, c) = F\left(-n, -x, b; 1 - \frac{1}{c}\right).$$

Norms:

$$\|m_n\|_w^2 = n! \frac{(b)_n}{c^n};$$

$$\|M_n\|_w^2 = \frac{n!}{(b)_n c^n}.$$

Difference relation and eigenvalue equation:

$$\Delta_+ m_n(x; b, c) = n\left(1 - \frac{1}{c}\right) m_{n-1}(x; b+1, c);$$

$$c(x+b)\,\Delta_+ m_n(x; b, c) - x\,\Delta_- m_n(x; b, c) + (1-c)\,n\,m_n(x; b, c) = 0.$$

First four polynomials:

$$m_0(x; b, c) = 1;$$

$$m_1(x; b, c) = \left(1 - \frac{1}{c}\right) x + b;$$

$$m_2(x; b, c) = \left(1 - \frac{1}{c}\right)^2 x^2 + \left(2b + 1 - \frac{2b}{c} - \frac{1}{c^2}\right) x + b(b+1);$$

$$m_3(x; b, c) = \left(1 - \frac{1}{c}\right)^3 x^3 + \left(3b + 3 - \frac{6b+3}{c} + \frac{3b-3}{c^2} + \frac{3}{c^3}\right) x^2$$

$$+ \left(3b^2 + 6b + 2 - \frac{3b^2+3b}{c} - \frac{3b}{c^2} - \frac{2}{c^3}\right) x + b(b+1)(b+2).$$

Generating function:

$$G(x; b, c) \equiv \sum_{n=0}^{\infty} \frac{m_n(x; b, c)}{n!} t^n = (1 - t)^{-x-b} \left(1 - \frac{t}{c}\right)^x.$$

Recurrence relations:

$$m_{n+1}(x; b, c) = (x + b) m_n(x; b + 1, c) - \frac{x}{c} m_n(x - 1; b + 1, c);$$

$$m_n(x + 1; b, c) = m_n(x; b, c) + n m_{n-1}(x + 1; b, c) - \frac{n}{c} m_{n-1}(x; b, c);$$

$$(c - 1) x m_n(x; b, c) = c m_{n+1}(x; b, c)$$
$$- (bc + nc + n) m_n(x; b, c)$$
$$+ n(b + n - 1) m_{n-1}(x; b, c).$$

Addition formula:

$$m_n(x + y; b, c) = \sum_{j+k+l=n} \frac{n! (-b)_j}{j! k! l!} m_l(x; b, c) m_k(y; b, c).$$

Asymptotics as $c \to 1-$, $x \neq 0$:

$$m_n \left(\frac{cx}{1 - c}; \alpha + 1, c\right) \sim n! L_n^{(\alpha)}(x).$$

Asymptotics as $n \to \infty$, x fixed:

$$m_n(x; b, c) \sim -\frac{\Gamma(b + n) \Gamma(x + 1)}{c^n (1 - c)^{x+b} n^{b+x}} \frac{\sin \pi x}{\pi}.$$

5.8.6 Chebyshev–Hahn polynomials

The interval is finite; with $\alpha, \beta > -1$ or $\alpha, \beta < -N$, set

$$p_+(x) = (N - x)(x + \alpha + 1), \quad p_-(x) = x(N + \beta + 1 - x);$$

$$w_m = C_0 \binom{N}{m} (\alpha + 1)_m (\beta + 1)_{N-m},$$

$$C_0 = \frac{\Gamma(N + \alpha + \beta + 2)}{\Gamma(\alpha + \beta + 2)};$$

$$\psi_n(x) = (-1)^n (N + 1 - n)_n (\alpha + 1)_n \sum_{k=0}^{n} \frac{(-n)_k (-x)_k (n + \alpha + \beta + 1)_k}{(-N)_k (\alpha + 1)_k k!}$$

$$= (-1)^n (N + 1 - n)_n (\alpha + 1)_n$$
$$\times {}_3F_2(-n, -x, n + \alpha + \beta + 1; -N, \alpha + 1; 1).$$

Standard normalizations:

$$Q_n(x; \alpha, \beta, N) = {}_3F_2(-n, -x, n + \alpha + \beta + 1; -N, \alpha + 1; 1);$$

$$h_n^{(\alpha,\beta)}(x, N) = (-1)^n \frac{(N-n)_n (\beta+1)_n}{n!} Q_n(x; \beta, \alpha, N - 1).$$

Norm:

$$\|Q_n(x; \alpha, \beta, N)\|^2 = \left| \frac{n! (N-n)! (\beta+1)_n (\alpha+\beta+n+1)_{N+1}}{N! (\alpha+1)_n (\alpha+\beta+2)_N (\alpha+\beta+2n+1)} \right|.$$

Difference relation and eigenvalue equation:

$$\Delta_+ Q_n(x; \alpha, \beta, N)$$

$$= -\frac{n(\alpha+\beta+n+1)}{N(\alpha+1)} Q_{n-1}(x; \alpha+1, \beta+1, N-1);$$

$$(N-x)(x+\alpha+1)\Delta_+ Q_n(x; \alpha, \beta, N)$$

$$- x(N+\beta+1-x)\Delta_- Q_n(x; \alpha, \beta, N)$$

$$+ n(n+\alpha+\beta+1)Q_n(x; \alpha, \beta, N) = 0.$$

First three polynomials:

$$Q_0(x; \alpha, \beta, N) = 1;$$

$$Q_1(x; \alpha, \beta, N) = -\frac{\alpha+\beta+2}{N(\alpha+1)} x + 1;$$

$$Q_2(x; \alpha, \beta, N) = \frac{(\alpha+\beta+3)(\alpha+\beta+4)}{N(N-1)(\alpha+1)(\alpha+2)} x^2$$

$$- \frac{(\alpha+\beta+3)[\alpha+\beta+4+2(N-1)(\alpha+2)]}{N(N-1)(\alpha+1)(\alpha+2)} x + 1.$$

Three-term recurrence:

$$x Q_n(x; \alpha, \beta, N) = \alpha_n Q_{n+1}(x; \alpha, \beta, N) + \beta_n Q_n(x; \alpha, \beta, N)$$

$$+ \gamma_n Q_{n-1}(x; \alpha, \beta), N);$$

$$\alpha_n = -\frac{(\alpha+\beta+n+1)(\alpha+n+1)(N-n)}{(\alpha+\beta+2n+1)(\alpha+\beta+2n+2)};$$

$$\gamma_n = -\frac{n(n+\beta)(\alpha+\beta+n+N+1)}{(\alpha+\beta+2n)(\alpha+\beta+2n+1)},$$

$$\beta_n = -(\alpha_n + \gamma_n).$$

Asymptotics as $N \to \infty$, $x \neq 0$:

$$Q_n(Nx; a, b, n) \sim \frac{n!}{(\alpha + 1)_n} P_n^{(\alpha, \beta)}(1 - 2x).$$

"Discrete Chebyshev polynomials" are the case $\alpha = \beta = 0$:

$$t_n(x, N) = (-1)^n (N - n)_n Q_n(x; 0, 0, N - 1)$$

$$= (-1)^n \sum_{k=0}^n \binom{n+k}{k} \frac{(N - n)_{n-k}(-x)_k(n - k + 1)_k}{k!}.$$

5.9 Remarks

Discrete orthogonal polynomials are treated in the books by Chihara [50] and Ismail [136], who discuss the history and some of the classification results. Nikiforov, Suslov, and Uvarov [218] present the subject from the point of view of difference equations. Asymptotics are studied in the book by Baik *et al.* [17], using the Riemann–Hilbert method. Notation has not been completely standardized. The notation selected for [223; 224] is such that each of these polynomials is a generalized hypergeometric series. This choice does not necessarily yield the simplest formulas.

The terminology here is more faithful to history than is often the case, in part because much of the history is relatively recent. Nevertheless, Chebyshev introduced the version of the "Hahn polynomials" treated above in 1858 [46]; see also [48]. Charlier introduced the Charlier polynomials in 1905 [43], Krawtchouk introduced the Krawtchouk polynomials in 1929 [166], Meixner introduced the Meixner polynomials in 1934 [203]. Explicit formulas for the polynomials $h_n^{(\alpha, \beta)}$ were obtained by Weber and Erdélyi [310]. Dunkl [73] obtained addition formulas for Krawtchouk polynomials, analogous to those for Legendre and Jacobi polynomials, by group theoretic methods.

In 1949 Hahn [122] introduced a large class of "q-polynomials" that contain as a limiting case the discrete polynomials that had been introduced by Chebyshev. For q-polynomials in general, see the remarks and references at the end of Chapter 4.

Stanton [267] generalized Dunkl's addition formula to q-Krawtchouk polynomials. For other recent extensions of classical results to discrete and q-polynomials, see the book by Ismail [136] and the detailed report by Koekoek and Swarttouw [158]. For results on asymptotics, see the survey article by Wong [319].

6

Confluent hypergeometric functions

The confluent hypergeometric equation

$$x\,u''(x) + (c - x)\,u'(x) - a\,u(x) = 0$$

has one solution, the Kummer function $M(a, c; x)$, with value 1 at the origin, and a second solution, $x^{1-c}M(a + 1 - c, 2 - c; x)$, which is $\sim x^{1-c}$ at the origin, provided that c is not an integer. A particular linear combination of the two gives a solution $U(a, c; x) \sim x^{-a}$ as $x \to +\infty$. The Laguerre polynomials are particular cases, corresponding to particular values of the parameters. Like the Laguerre polynomials, the general solutions satisfy a number of linear relations involving derivatives and different values of the parameters a and c. Special consideration is required when c is an integer.

In addition to the Laguerre polynomials, functions that can be expressed in terms of Kummer functions include the exponential function, error function, incomplete gamma function, complementary incomplete gamma function, Fresnel integrals, the exponential integral, and the sine integral and cosine integral functions.

The closely related parabolic cylinder functions are solutions of

$$u''(x) + \left[\mp\frac{x^2}{4} + \nu + \frac{1}{2}\right]u(x) = 0.$$

Three solutions are obtained by utilizing the three solutions of the confluent hypergeometric equation mentioned above.

A gauge transformation removes the first-order term of the confluent hypergeometric equation and converts it to Whittaker's equation

$$u''(x) + \left[-\frac{1}{4} + \frac{\kappa}{x} + \frac{1 - 4\mu^2}{4x^2}\right]u(x) = 0.$$

Again there are three solutions, related to the three solutions of the confluent hypergeometric equation, which satisfy a number of relations. The Coulomb wave functions are special cases.

6.1 Kummer functions

The confluent hypergeometric equation is the equation

$$x u''(x) + (c - x) u'(x) - a u(x) = 0. \tag{6.1.1}$$

As noted in Chapter 1, it has a solution represented by the power series

$$M(a, c; x) = \sum_{n=0}^{\infty} \frac{(a)_n}{(c)_n \, n!} x^n. \tag{6.1.2}$$

This is only defined for $c \neq 0, -1, -2, \ldots$, since otherwise the denominators vanish for $n > -c$. The function (6.1.2) is known as the *Kummer function*. The notation $\Phi(a, c; x)$ is common in the Russian literature. Another notation is $_1F_1(a, c; x)$, a special case of the class of functions $_pF_q$ introduced in Chapter 8. Thus

$$M(a, c; x) = \Phi(a, c; x) = {}_1F_1(a, c; x), \quad c \neq 0, -1, -2, \ldots$$

If $a = -m$ is an integer ≤ 0, then $(a)_n = 0$ for $n > m$: $M(-m, c; x)$ is a polynomial of degree m. If $c > 0$, it is a multiple of the Laguerre polynomial $L_m^{(c-1)}(x)$; see the next section. In general, the ratio test shows that $M(a, c; x)$ is an entire function of x. The adjective "confluent" here will be explained in Chapter 8.

Assuming that $\operatorname{Re} c > \operatorname{Re} a > 0$, we obtain an integral representation, using the identity

$$\frac{(a)_n}{(c)_n} = \frac{\Gamma(a + n)}{\Gamma(a)} \frac{\Gamma(c)}{\Gamma(c + n)} = \frac{\Gamma(c)}{\Gamma(a) \Gamma(c - a)} B(a + n, c - a)$$

$$= \frac{\Gamma(c)}{\Gamma(a) \Gamma(c - a)} \int_0^1 s^{n+a-1} (1 - s)^{c-a-1} \, ds.$$

As in Section 1.2, this leads to the computation

$$M(a, c; x) = \frac{\Gamma(c)}{\Gamma(a) \Gamma(c - a)} \int_0^1 \left\{ s^{a-1} (1 - s)^{c-a-1} \sum_{n=0}^{\infty} \frac{(sx)^n}{n!} \right\} ds$$

$$= \frac{\Gamma(c)}{\Gamma(a) \Gamma(c - a)} \int_0^1 s^{a-1} (1 - s)^{c-a-1} e^{sx} \, ds, \quad \operatorname{Re} c > \operatorname{Re} a > 0.$$

$$\tag{6.1.3}$$

Let D denote the operator

$$D = x \frac{d}{dx}.$$

After multiplication by x, the confluent hypergeometric equation (6.1.1) has the form

$$D(D + c - 1)u - x(D + a)u = 0. \tag{6.1.4}$$

Under the gauge transformation $u(x) = x^b v(x)$, the operator D acting on the function u becomes the operator $D + b$ acting on v:

$$x^{-b} D\{x^b v(x)\} = (D + b)v(x). \tag{6.1.5}$$

Letting $b = 1 - c$, equation (6.1.4) becomes

$$D(D + 1 - c)v - x(D + a + 1 - c)v = 0.$$

Therefore, if $c \neq 2, 3, 4, \ldots,$

$$x^{1-c} M(a + 1 - c, 2 - c; x) \tag{6.1.6}$$

is a second solution of (6.1.1). (This is easily checked directly.) If c is an integer $\neq 1$, then one of the two functions (6.1.2), (6.1.6) is not defined. If $c = 1$, the functions coincide. We consider these cases in Section 6.3.

Denote the solution (6.1.2) by $M_1(x)$ and the solution (6.1.6) by $M_2(x)$. Equation (6.1.1) implies that the Wronskian

$$W(x) = W(M_1, M_2)(x) \equiv M_1(x)M_2'(x) - M_2(x)M_1'(x)$$

satisfies the first-order equation

$$x W'(x) = (x - c) W(x),$$

so $W(x) = Ax^{-c}e^x$ for some constant A. We may evaluate A by looking at the behavior of M_1 and M_2 as $x \to 0$:

$$M_1(x) = 1 + O(x), \quad M_2(x) = x^{1-c}\left[1 + O(x)\right].$$

The result is

$$W(M_1, M_2)(x) = (1 - c) x^{-c} e^x. \tag{6.1.7}$$

Write (6.1.1) for $x \neq 0$ as

$$u''(x) - u'(x) + \frac{c}{x} u'(x) - \frac{a}{x} u(x) = 0.$$

As $x \to \infty$ we may expect any solution of (6.1.1) to resemble a solution of $v'' - v' = 0$: a linear combination of e^x and 1. In fact, the change of variables $(1 - s)x = t$ converts the integral in (6.1.3) to

$$x^{a-c} e^x \int_0^x e^{-t} \left(1 - \frac{t}{x}\right)^{a-1} t^{c-a-1} \, dt.$$

This last integral has limit

$$\int_0^\infty e^{-t} t^{c-a-1} \, dt = \Gamma(c - a)$$

as $\mathrm{Re}\, x \to +\infty$, giving the asymptotics

$$M(a, c; x) \sim \frac{\Gamma(c)}{\Gamma(a)} x^{a-c} e^x \quad \text{as} \quad \mathrm{Re}\, x \to +\infty, \quad \mathrm{Re}\, c > \mathrm{Re}\, a > 0.$$
$$(6.1.8)$$

Expanding $(1 - t/x)^{a-1}$ gives the full asymptotic expansion

$$M(a, c; x) \sim \frac{\Gamma(c)}{\Gamma(a) \Gamma(c - a)} x^{a-c} e^x \sum_{n=0}^\infty \int_0^\infty e^{-t} \frac{(1 - a)_n}{n!} \left(\frac{t}{x}\right)^n t^{c-a-1} \, dt$$

$$= \frac{\Gamma(c)}{\Gamma(a)} x^{a-c} e^x \sum_{n=0}^\infty \frac{(1 - a)_n (c - a)_n}{n!} \frac{1}{x^n}, \quad \mathrm{Re}\, c > \mathrm{Re}\, a > 0.$$
$$(6.1.9)$$

The results in Section 6.5 can be used to show that this asymptotic result is valid for all indices (a, c) with $c \neq 0, -1, -2, \ldots$

It is easily seen that if $v(x)$ is a solution of (6.1.1) with the index a replaced by $c - a$,

$$x \, v''(x) + (c - x) \, v'(x) - (c - a) \, v(x) = 0,$$

then $u(x) = e^x v(-x)$ is a solution of (6.1.1). Comparing values at $x = 0$ establishes *Kummer's identity*

$$e^x M(c - a, c; -x) = M(a, c; x). \qquad (6.1.10)$$

A second identity due to Kummer is

$$M(a, 2a; 4x) = e^{2x} {}_0F_1\left(a + \frac{1}{2}; x^2\right) \equiv e^{2x} \sum_{n=0}^\infty \frac{x^{2n}}{\left(a + \frac{1}{2}\right)_n n!}; \qquad (6.1.11)$$

see Exercise 8.19 in Chapter 8.

It will be shown in Chapter 10 that $M(a, c; x)$ has the following asymptotic behavior as $a \to -\infty$:

$$M(a, c; x) = \frac{\Gamma(c)}{\sqrt{\pi}} \left(\frac{1}{2}cx - ax \right)^{\frac{1}{4} - \frac{1}{2}c} e^{\frac{1}{2}x}$$

$$\times \left[\cos\left(\sqrt{2cx - 4ax} - \frac{1}{2}c\pi + \frac{1}{4}\pi \right) + O\left(|a|^{-\frac{1}{2}} \right) \right].$$

This was proved by Erdélyi [80] and Schmidt [253], generalizing Fejér's result for Laguerre polynomials (4.5.12).

6.2 Kummer functions of the second kind

To find a solution u of the confluent hypergeometric equation (6.1.1) that does not have exponential growth as $\operatorname{Re} x \to +\infty$, we note that such a solution should be expressible as a Laplace transform (see Exercise 4.25 in Chapter 4). Thus we look for an integral representation

$$u(x) = \left[\mathcal{L}\varphi \right](x) = \int_0^\infty e^{-xt} \varphi(t) \, dt$$

for some integrable function $\varphi(t)$. If a function $u(x)$ has this form, then

$$xu''(x) + (c - x) u'(x) - a u(x) = \int_0^\infty e^{-xt} \left[xt^2 + (x - c)t - a \right] \varphi(t) \, dt$$

$$= \int_0^\infty \left[xe^{-xt} \right] (t^2 + t)\varphi(t) \, dt - \int_0^\infty e^{-xt} (a + ct) \varphi(t) \, dt.$$

Suppose that $t \varphi(t)$ has limit zero as $t \to 0$ and that φ' is integrable. Then integration by parts of the first integral in the last line leads to

$$xu''(x) + (c - x) u'(x) - a u(x)$$

$$= \int_0^\infty e^{-xt} \left\{ \left[(t^2 + t)\varphi(t) \right]' - (a + ct) \varphi(t) \right\} dt.$$

It follows that u is a solution of (6.1.1) if the expression in braces vanishes. This condition is equivalent to

$$\frac{\varphi'(t)}{\varphi(t)} = \frac{(c - 2)t + a - 1}{t^2 + t} = \frac{c - a - 1}{1 + t} + \frac{a - 1}{t},$$

or $\varphi(t) = A t^{a-1}(1 + t)^{c-a-1}$. Then $t \varphi(t)$ has limit zero as $t \to 0$ if $\operatorname{Re} a > 0$. This argument shows that we may obtain a solution $U(a, c; \cdot)$ of (3.0.1) by taking

$$U(a, c; x) = \frac{1}{\Gamma(a)} \int_0^\infty e^{-xt} t^{a-1} (1+t)^{c-a-1} \, dt$$

$$= \frac{x^{-a}}{\Gamma(a)} \int_0^\infty e^{-s} s^{a-1} \left(1 + \frac{s}{x}\right)^{c-a-1} ds, \quad \mathrm{Re}\, a > 0. \quad (6.2.1)$$

The second form leads to a full asymptotic series expansion for U as $\mathrm{Re}\, x \to +\infty$. The first term gives

$$U(a, c; x) \sim x^{-a} \quad \text{as} \quad \mathrm{Re}\, x \to +\infty, \quad \mathrm{Re}\, a > 0. \quad (6.2.2)$$

The full expansion is

$$U(a, c; x) \sim x^{-a} \sum_{n=0}^\infty \frac{(a)_n (a+1-c)_n}{n!} \frac{(-1)^n}{x^n}, \quad \mathrm{Re}\, a > 0. \quad (6.2.3)$$

As noted below, U can be extended to all index pairs (a, c) with c not an integer. The results in Section 6.5, together with (6.2.6) below, make it possible to show that the asymptotic results (6.2.2) and (6.2.3) extend to all such index pairs.

The function U is called the *confluent hypergeometric function of the second kind*. In the Russian literature it is denoted by Ψ:

$$\Psi(a, c; x) = U(a, c; x). \quad (6.2.4)$$

The solution (6.2.1) must be a linear combination of the solutions (6.1.2) and (6.1.6),

$$U(a, c; x) = A(a, c) \, M(a, c; x) + B(a, c) \, x^{1-c} M(a+1-c, 2-c; x) \quad (6.2.5)$$

with coefficients $A(a, c)$ and $B(a, c)$ that are meromorphic functions of a and c. The coefficients can be determined by considering the integral representation in two special cases.

If $1 - c > 0$ then the second summand on the right in (6.2.5) vanishes at $x = 0$, while the value of (6.2.1) at $x = 0$ is

$$\frac{1}{\Gamma(a)} \int_0^\infty t^a (1+t)^{c-a-1} \frac{dt}{t} = \frac{B(a, 1-c)}{\Gamma(a)} = \frac{\Gamma(1-c)}{\Gamma(a+1-c)}.$$

Therefore $A(a, c) = \Gamma(1-c)/\Gamma(a+1-c)$ for $\mathrm{Re}\, a > 0$, $1 - c > 0$, and, by analytic continuation, for all non-integer values of c.

If $c - 1 > 0$ then $x^{c-1} M(a, c; x)$ vanishes at $x = 0$, while

$$x^{c-1} U(a, c; x) = x^{c-1} \frac{x^{-a}}{\Gamma(a)} \int_0^\infty e^{-t} t^{a-1} \left(1 + \frac{t}{x}\right)^{c-a-1} dt$$

$$= \frac{1}{\Gamma(a)} \int_0^\infty e^{-t} t^{a-1} (x + t)^{c-a-1} dt$$

$$\to \frac{\Gamma(c - 1)}{\Gamma(a)} \quad \text{as} \quad x \to 0+.$$

Therefore $B(a, c) = \Gamma(c - 1)/\Gamma(a)$ for $\text{Re}\, a > 0$ and all non-integer values of c.

We have proved

$$U(a, c; x) = \frac{\Gamma(1 - c)}{\Gamma(a + 1 - c)} M(a, c; x) + \frac{\Gamma(c - 1)}{\Gamma(a)} x^{1-c} M(a + 1 - c, 2 - c; x)$$

$$(6.2.6)$$

for all values of $\text{Re}\, a > 0$ and c not an integer. Conversely, we may use (6.2.6) to remove the limitation $\text{Re}\, a > 0$ and *define* U for all a and all non-integer values of c. It follows from (6.2.6) and (6.1.10) that

$$U(a, c; x) = x^{1-c} U(a + 1 - c, 2 - c; x); \qquad (6.2.7)$$

see the exercises.

The Wronskian of the solutions M and U may be computed from (6.1.8) and (6.2.2), or from (6.2.6) and (6.1.7). The result is

$$W\big(M(a, c; \cdot), U(a, c; \cdot)\big)(x) = -\frac{\Gamma(c)}{\Gamma(a)} x^{-c} e^x; \qquad (6.2.8)$$

see the exercises.

It follows from (6.1.10) that a solution of (6.1.1) that decays exponentially as $x \to -\infty$ is

$$\widetilde{U}(a, c; x) = e^x U(c - a, c; -x)$$

$$= \frac{1}{\Gamma(c - a)} \int_0^\infty e^{x(1+t)} t^{c-a-1} (1 + t)^{a-1} dt$$

$$= \frac{(-x)^{a-c} e^x}{\Gamma(c - a)} \int_0^\infty e^{-s} s^{c-a-1} \left(1 - \frac{s}{x}\right)^{a-1} ds, \quad \text{Re}\, c > \text{Re}\, a > 0.$$

$$(6.2.9)$$

Then $\widetilde{U}(a, c; x) = e^x (-x)^{a-c}(1 + O(-1/x))$ as $x \to -\infty$. Again, the results in Section 6.5 show that this asymptotic result extends to all index pairs (a, c), c not an integer.

In terms of the solutions (6.1.2) and (6.1.6), we use (6.2.6) and (6.1.10) to obtain

$$\tilde{U}(a, c; x) = \frac{\Gamma(1-c)}{\Gamma(1-a)} e^x M(c-a, c; -x)$$

$$+ \frac{\Gamma(c-1)}{\Gamma(c-a)} e^x (-x)^{1-c} M(1-a, 2-c; -x)$$

$$= \frac{\Gamma(1-c)}{\Gamma(1-a)} M(a, c; x) + \frac{\Gamma(c-1)}{\Gamma(c-a)} (-x)^{1-c} M(1+a-c, 2-c; x).$$

$$(6.2.10)$$

It will be shown in Chapter 10 that a consequence of (6.1.12) and (6.2.6) is that U has the following asymptotic behavior as $a \to -\infty$:

$$U(a, c; x) = \frac{\Gamma\left(\frac{1}{2}c - a + \frac{1}{4}\right)}{\sqrt{\pi}} x^{\frac{1}{4} - \frac{1}{2}c} e^{\frac{1}{2}x}$$

$$\times \left[\cos\left(\sqrt{2cx - 4ax} - \frac{1}{2}c\pi + a\pi + \frac{1}{4}\pi \right) + O\left(|a|^{-\frac{1}{2}} \right) \right].$$

$$(6.2.11)$$

6.3 Solutions when c is an integer

As noted above, for $c \neq 1$ an integer, only one of the two functions (6.1.2) and (6.1.6) is defined, while if $c = 1$ they coincide. When c is an integer, a second solution can be obtained from the solution of the second kind U. In view of (6.2.7), it is enough to consider the case $c = m$ a positive integer.

Assume first that a is not an integer. We begin by modifying the solution M so that it is defined for all c. Assuming first that $c \neq 0, -1, -2, \ldots$, let

$$N(a, c; x) \equiv \frac{\Gamma(a)}{\Gamma(c)} M(a, c; x) = \sum_{n=0}^{\infty} \frac{\Gamma(a+n)}{\Gamma(c+n) n!} x^n.$$

The series expansion is well-defined for all values of the parameters except for a a non-positive integer. Note that if $c = -k$ is an integer ≤ 0, the first $k + 1$ terms of the series vanish. In particular, if $c = m$ is a positive integer,

$$N(a+1-m, 2-m; x) = \sum_{n=m-1}^{\infty} \frac{\Gamma(a+1-m+n)}{\Gamma(2-m+n) n!} x^n$$

$$= x^{m-1} \sum_{k=0}^{\infty} \frac{\Gamma(a+k)}{\Gamma(m+k) k!} x^k$$

$$= x^{m-1} N(a, m; x). \qquad (6.3.1)$$

For non-integer c and a we use the reflection formula (2.2.7) to rewrite (6.2.6) as

$$U(a, c; x) = \frac{\pi}{\sin \pi c \, \Gamma(a) \, \Gamma(a + 1 - c)}$$
$$\times \left[N(a, c; x) - x^{1-c} N(a + 1 - c; 2 - c, x) \right]. \quad (6.3.2)$$

In view of (6.3.1), the difference in brackets has limit zero as $c \to m$, m a positive integer. It follows that $U(a, c; x)$ has a limit as $c \to m$, and the limit is given by l'Hôpital's rule. For a not an integer,

$$U(a, m; x) = \frac{(-1)^m}{\Gamma(a) \, \Gamma(a + 1 - m)}$$
$$\times \left. \frac{\partial}{\partial c} \right|_{c=m} \left[N(a, c; x) - x^{1-c} N(a + 1 - c, 2 - c; x) \right].$$

Therefore, for non-integer values of a and positive integer values of m, calculating the derivative shows that

$$U(a, m; x) = \frac{(-1)^m}{\Gamma(a + 1 - m)(m - 1)!} \left\{ \log x \, M(a, m; x) \right.$$
$$+ \sum_{n=0}^{\infty} \frac{(a)_n}{(m)_n \, n!} \left[\psi(a + n) - \psi(n + 1) - \psi(m + n) \right] x^n \right\}$$
$$+ \frac{(m - 2)!}{\Gamma(a)} x^{1-m} \sum_{n=0}^{m-2} \frac{(a + 1 - m)_n}{(2 - m)_n \, n!} x^n, \quad (6.3.3)$$

where $\psi(b) = \Gamma'(b)/\Gamma(b)$ and the last sum is taken to be zero if $m = 1$.

The function in (6.3.3) is well-defined for all values of a. By a continuity argument, it is a solution of (6.1.1) for all values of a and c and all values of x not in the interval $(-\infty, 0]$. If a is not an integer less than m, $U(a, m; x)$ has a logarithmic singularity at $x = 0$ and is therefore independent of the solution $M(a, c; x)$. If a is a positive integer less than m, then the coefficient of the term in braces vanishes and $U(a, c; x)$ is the finite sum, which is a rational function that is again independent of $M(a, c; x)$.

If a is a non-positive integer, then $U(a, m; x) \equiv 0$. To obtain a solution in this case we start with non-integer a and multiply (6.3.3) by $\Gamma(a)$. Since

$$\frac{\Gamma(a)}{\Gamma(a + 1 - m)} = (a + 1 - m)_{m-1} \neq 0$$

for $a = 0, -1, -2, \ldots$, the limiting value of $\Gamma(a) U(a, c; x)$ is a solution of (6.1.1) that has a logarithmic singularity at $x = 0$.

6.4 Special cases

The exponential function

$$e^x = \sum_{n=0}^{\infty} \frac{(c)_n}{(c)_n \, n\,!} \, x^n = M(c, c; x), \quad c \neq 0, -1, -2, \ldots, \qquad (6.4.1)$$

is one example of a confluent hypergeometric function.

The Laguerre polynomial $L_n^{(\alpha)}(x)$ satisfies equation (6.1.1) with $c = \alpha + 1$ and $a = -n$. Since the constant term is $(\alpha + 1)_n / n\,!$, it follows that

$$L_n^{(\alpha)}(x) = \frac{(\alpha + 1)_n}{n\,!} \, M(-n, \alpha + 1; x). \qquad (6.4.2)$$

Combining this with the identities (4.4.22) and (4.4.23) that relate Hermite and Laguerre polynomials gives

$$H_{2m}(x) = (-1)^m \, 2^{2m} \left(\frac{1}{2}\right)_m M\left(-m, \frac{1}{2}; x^2\right); \qquad (6.4.3)$$

$$H_{2m+1}(x) = (-1)^m \, 2^{2m+1} \left(\frac{3}{2}\right)_m x \, M\left(-m, \frac{3}{2}; x^2\right).$$

These identities, together with the identity (4.4.21) relating the Hermite polynomials $\{H_n\}$ and the modified version $\{He_n\}$, give

$$He_{2m}(x) = (-1)^m \, 2^m \left(\frac{1}{2}\right)_m M\left(-m, \frac{1}{2}; \frac{1}{2}x^2\right); \qquad (6.4.4)$$

$$He_{2m+1}(x) = (-1)^m \, 2^{m+\frac{1}{2}} \left(\frac{3}{2}\right)_m \frac{x}{\sqrt{2}} \, M\left(-m, \frac{3}{2}; \frac{1}{2}x^2\right).$$

Other special cases of Kummer functions include the error function, the incomplete gamma function, and the complementary incomplete gamma function of Section 2.6:

$$\operatorname{erf} x \equiv \frac{2}{\sqrt{\pi}} \int_0^x e^{-t^2} \, dt = \frac{2x}{\sqrt{\pi}} \, M\left(\frac{1}{2}, \frac{3}{2}; -x^2\right); \qquad (6.4.5)$$

$$\gamma(a, x) \equiv \int_0^x e^{-t} \, t^{a-1} \, dt = \frac{x^a}{a} \, M(a, a + 1; -x); \qquad (6.4.6)$$

$$\Gamma(a, x) \equiv \int_x^{\infty} e^{-t} t^{a-1} \, dt = x^a e^{-x} \, U(1, a + 1; x). \qquad (6.4.7)$$

Among other such functions are the Fresnel integrals

$$C(x) \equiv \int_0^x \cos\left(\frac{1}{2}t^2\pi\right) dt;$$

$$S(x) \equiv \int_0^x \sin\left(\frac{1}{2}t^2\pi\right) dt;$$

and the exponential integral, cosine integral, and sine integral functions

$$\mathrm{Ei}(z) \equiv \int_{-\infty}^z e^t \frac{dt}{t}, \quad z \notin [0, \infty);$$

$$\mathrm{Ci}(z) \equiv \int_\infty^z \cos t \frac{dt}{t}, \quad z \notin (-\infty, 0];$$

$$\mathrm{Si}(z) \equiv \int_0^z \sin t \frac{dt}{t}, \quad z \notin (-\infty, 0].$$

These functions are related to the Kummer functions for special values of the indices:

$$C(x) = \frac{x}{2}\left[M\left(\frac{1}{2}, \frac{3}{2}; \frac{1}{2}ix^2\pi\right) + M\left(\frac{1}{2}, \frac{3}{2}; -\frac{1}{2}ix^2\pi\right)\right]; \qquad (6.4.8)$$

$$S(x) = \frac{x}{2i}\left[M\left(\frac{1}{2}, \frac{3}{2}; \frac{1}{2}ix^2\pi\right) - M\left(\frac{1}{2}, \frac{3}{2}; -\frac{1}{2}ix^2\pi\right)\right]; \qquad (6.4.9)$$

and

$$\mathrm{Ei}(-z) = -e^{-z} U(1, 1; z); \qquad (6.4.10)$$

$$\mathrm{Ci}(x) = -\frac{1}{2}\left[e^{-ix} U(1, 1; ix) + e^{ix} U(1, 1; -ix)\right]; \qquad (6.4.11)$$

$$\mathrm{Si}(x) = \frac{1}{2i}\left[e^{-ix} U(1, 1; ix) - e^{ix} U(1, 1; -ix)\right] + \frac{\pi}{2}. \qquad (6.4.12)$$

See the exercises for the identities (6.4.5)–(6.4.12).

6.5 Contiguous functions

Two Kummer functions are said to be *contiguous* if the first of the two indices is the same for each function and the second indices differ by ± 1, or the second indices are the same and the first indices differ by ± 1. Any triple of contiguous Kummer functions satisfies a linear relationship. For convenience, fix a, c, x, and let

$$M = M(a, c; x); \quad M(a\pm) = M(a \pm 1, c; x); \quad M(c\pm) = M(a, c \pm 1; x).$$

There are six basic linear relations: the relations between M and any of the six pairs chosen from the four contiguous functions $M(a\pm)$, $M(c\pm)$.

Denote the coefficient of x^n in the expansion of M by

$$\varepsilon_n = \frac{(a)_n}{(c)_n \, n!}.$$

As above, let $D = x(d/dx)$. The coefficients of x^n in the expansions of DM, $M(a+)$ and $M(c-)$, respectively, are

$$n\,\varepsilon_n, \qquad \frac{a+n}{a}\,\varepsilon_n, \qquad \frac{c-1+n}{c-1}\,\varepsilon_n.$$

Therefore

$$DM = a\big[M(a+) - M\big] = (c-1)\big[M(c-) - M\big], \tag{6.5.1}$$

and

$$(a - c + 1)\,M = a\,M(a+) - (c - 1)\,M(c-). \tag{6.5.2}$$

The coefficient of x^n in the expansion of xM is

$$\frac{(c+n-1)\,n}{a+n-1}\,\varepsilon_n = \left[n + (c - a) - (c - a)\frac{a-1}{a+n-1}\right]\varepsilon_n,$$

so

$$xM = DM + (c - a)\,M - (c - a)\,M(a-).$$

Combining this with (6.5.1) gives

$$(2a - c + x)\,M = a\,M(a+) - (c - a)\,M(a-); \tag{6.5.3}$$

$$(a - 1 + x)\,M = (c - 1)\,M(c-) - (c - a)\,M(a-). \tag{6.5.4}$$

The coefficient of x^n in the expansion of M' is

$$\frac{a+n}{c+n}\,\varepsilon_n = \left[1 + \frac{a-c}{c}\,\frac{c}{c+n}\right]\varepsilon_n,$$

so

$$c\,M' = c\,M - (c - a)\,M(c+).$$

Multiplying by x gives

$$c\,DM = cx\,M - (c - a)x\,M(c+),$$

and combining this with (6.5.1) gives

$$c(a + x) M = ac M(a+) + (c - a)x M(c+);$$ \hfill (6.5.5)

$$c(c - 1 + x) M = c(c - 1) M(c-) + (c - a)x M(c+).$$ \hfill (6.5.6)

Eliminating $M(c-)$ from (6.5.4) and (6.5.6) gives

$$c M = c M(a-) + x M(c+).$$ \hfill (6.5.7)

The relations (6.5.2)–(6.5.7) are the six basic relations mentioned above.

Contiguous relations for the solution $U(a, c; x)$ can be derived from those for $M(a, c; x)$, using (6.2.6). However, it is simpler to start with $\operatorname{Re} a > 0$ and use the integral representation (6.2.1). The identities extend to general values of the parameter by analytic continuation. We use the same notational conventions as before:

$$U = U(a, c; x); \quad U(a\pm) = U(a \pm 1, c; x); \quad U(c\pm) = U(a, c \pm 1; x).$$

Differentiating the first integral in (6.2.1) with respect to x gives the two identities

$$U'(a, c; x) = -a\, U(a + 1, c + 1; x);$$ \hfill (6.5.8)

$$U'(a, c; x) = U(a, c; x) - U(a, c + 1; x).$$ \hfill (6.5.9)

Replacing c by $c - 1$ and combining these two identities gives

$$U = a\, U(a+) + U(c-).$$ \hfill (6.5.10)

Integrating the identity

$$\frac{d}{dt}\left[e^{-xt} t^{a-1}(1 + t)^{c-a}\right] = e^{-xt}\left[-x t^{a-1}(1 + t)^{c-a} + (a - 1)t^{a-2}(1 + t)^{c-a}\right.$$
$$\left. + (c - a)t^{a-1}(1 + t)^{c-a-1}\right]$$

for $0 < t < \infty$ gives

$$(c - a) U = x\, U(c+) - U(a-).$$ \hfill (6.5.11)

The two identities (6.5.8), (6.5.9) give

$$a\, U(a + 1, c + 1; x) = U(c+) - U.$$

Combining this with (6.5.11) with $a + 1$ in place of a gives

$$(x + a) U = x\, U(c+) + a(a + 1 - c) U(a+).$$ \hfill (6.5.12)

Eliminating $U(a+)$ from (6.5.10) and (6.5.12) gives

$$(x + c - 1) U = x U(c+) + (c - a - 1) U(c-). \qquad (6.5.13)$$

Eliminating $U(c+)$ from (6.5.11) and (6.5.12) gives

$$(x + 2a - c) U = U(a-) + a(1 + a - c) U(a+). \qquad (6.5.14)$$

Finally, eliminating $U(a+)$ from (6.5.10) and (6.5.14) gives

$$(a + x - 1) U = U(a-) + (c - a - 1) U(c-). \qquad (6.5.15)$$

6.6 Parabolic cylinder functions

A parabolic cylinder function, or Weber function, is a solution of one of the
equations

$$u''(x) - \frac{x^2}{4} u(x) + \left(v + \frac{1}{2} \right) u(x) = 0; \qquad (6.6.1)$$

$$u''(x) + \frac{x^2}{4} u(x) + \left(v + \frac{1}{2} \right) u(x) = 0. \qquad (6.6.2)$$

If u is a solution of (6.6.1), then $v(x) = u(e^{\frac{1}{4} i \pi} x)$ is a solution of (6.6.2) with
$v + \frac{1}{2}$ replaced by $i(v + \frac{1}{2})$, and conversely. We shall consider only (6.6.1).

Suppose that u is a solution of (6.6.1) that is holomorphic near $x = 0$. The
coefficients of the expansion $u(x) = \sum b_n x^n$ can be computed from the first
two terms b_0, b_1 by setting $b_{-2} = b_{-1} = 0$ and using the recursion

$$(n + 2)(n + 1)b_{n+2} = - \left(v + \frac{1}{2} \right) b_n + \frac{1}{4} b_{n-2}. \qquad (6.6.3)$$

In particular, taking $b_0 \neq 0, b_1 = 0$ determines an even solution, while $b_0 = 0$,
$b_1 \neq 0$ determines an odd solution.

The gauge transformation $u(x) = e^{-\frac{1}{4} x^2} v(x)$ converts (6.6.1) to

$$v''(x) - x v'(x) + v v(x) = 0. \qquad (6.6.4)$$

For $v = n$ a non-negative integer, one solution of this equation is the modified
Hermite polynomial He_n.

Each of the equations (6.6.1), (6.6.2), and (6.6.4) is unchanged under the
coordinate change $x \to -x$. Therefore each has an even solution and an odd
solution. In the case of (6.6.4), we know that the polynomial solutions are even
or odd according as v is an even or odd non-negative integer. Let us look for

an even solution of the modified equation (6.6.4) in the form $v(x) = w(x^2/2)$. Then (6.6.4) is equivalent to

$$y\, w''(y) + \left(\frac{1}{2} - y\right) w'(y) + \frac{1}{2}\nu\, w(y) = 0. \tag{6.6.5}$$

This is the confluent hypergeometric equation (6.1.1) with $c = \frac{1}{2}$, $a = -\frac{1}{2}\nu$. Thus the even solutions of (6.6.1) are multiples of

$$Y_{\nu1}(x) = e^{-\frac{1}{4}x^2}\, M\left(-\frac{1}{2}\nu, \frac{1}{2}; \frac{1}{2}x^2\right).$$

According to (6.1.6) a second solution of (6.6.5) is

$$\sqrt{y}\, M\left(-\frac{1}{2}\nu + \frac{1}{2}, \frac{3}{2}; y\right).$$

Therefore the odd solutions of (6.6.1) are multiples of

$$Y_{\nu2}(x) = e^{-\frac{1}{4}x^2}\, \frac{x}{\sqrt{2}}\, M\left(-\frac{1}{2}\nu + \frac{1}{2}, \frac{3}{2}; \frac{1}{2}x^2\right).$$

It follows from (6.1.8) that the solutions $Y_{\nu1}$ and $Y_{\nu2}$ of (6.6.5) grow like $e^{\frac{1}{4}x^2}$ as $|x| \to \infty$. To obtain a solution with decay at ∞ we use the Kummer function of the second kind $U(-\frac{1}{2}\nu, \frac{1}{2}; \cdot)$ instead. The standard normalized solution is

$$\begin{aligned}
D_\nu(x) &= 2^{\frac{1}{2}\nu} e^{-\frac{1}{4}x^2}\, U\left(-\frac{1}{2}\nu, \frac{1}{2}; \frac{1}{2}x^2\right) \\
&= 2^{\frac{1}{2}\nu} \left\{ \frac{\Gamma\left(\frac{1}{2}\right)}{\Gamma\left(\frac{1}{2} - \frac{1}{2}\nu\right)}\, Y_{\nu1}(x) + \frac{\Gamma\left(-\frac{1}{2}\right)}{\Gamma\left(-\frac{1}{2}\nu\right)}\, Y_{\nu2}(x) \right\}.
\end{aligned} \tag{6.6.6}$$

For Re $\nu < 0$, (6.2.1) implies the integral representation

$$D_\nu(x) = \frac{2^{\frac{1}{2}\nu} e^{-\frac{1}{4}x^2}}{\Gamma\left(-\frac{1}{2}\nu\right)} \int_0^\infty e^{-\frac{1}{2}tx^2} t^{-\frac{1}{2}\nu-1} (1+t)^{\frac{1}{2}\nu-\frac{1}{2}}\, dt. \tag{6.6.7}$$

Another integral representation valid for Re $\nu < 0$ is

$$D_\nu(x) = \frac{2^{\frac{1}{2}\nu} \Gamma\left(\frac{1}{2}\right) e^{-\frac{1}{4}x^2}}{\Gamma\left(\frac{1}{2} - \frac{1}{2}\nu\right) \Gamma\left(-\frac{1}{2}\nu\right)} \int_0^\infty e^{-t-\sqrt{2t}\, x}\, t^{-\frac{1}{2}\nu-1}\, dt; \tag{6.6.8}$$

see Exercise 6.8.

If $\nu = 2m$ is an even non-negative integer, then the second summand on the right in (6.6.6) vanishes and the coefficient of the first summand is

$$2^m \frac{\Gamma(\frac{1}{2})}{\Gamma(\frac{1}{2} - m)} = (-1)^m 2^m \left(\frac{1}{2}\right)_m.$$

If $\nu = 2m + 1$ is an odd positive integer, then the first summand on the right in (6.6.6) vanishes and the coefficient of the second summand is

$$2^{m+\frac{1}{2}} \frac{\Gamma(-\frac{1}{2})}{\Gamma(-\frac{1}{2} - m)} = (-1)^m 2^{m+\frac{1}{2}} \left(\frac{3}{2}\right)_m.$$

In view of (6.4.4), therefore,

$$D_n(x) = e^{-\frac{1}{4}x^2} He_n(x), \quad n = 0, 1, 2, \ldots \tag{6.6.9}$$

The behavior as $x \to 0$ is given by

$$D_\nu(x) = 2^{\frac{1}{2}\nu} \sqrt{\pi} \left\{ \frac{1}{\Gamma(\frac{1}{2} - \frac{1}{2}\nu)} - \frac{2^{\frac{1}{2}} x}{\Gamma(-\frac{1}{2}\nu)} \right\} + O(x^2); \tag{6.6.10}$$

the remaining coefficients can be computed from the recursion (6.6.3). The definition (6.6.6) and the identity (6.2.7) imply the identity

$$D_\nu(x) = 2^{\frac{1}{2}\nu - \frac{1}{2}} e^{-\frac{1}{4}x^2} x U\left(-\frac{1}{2}\nu + \frac{1}{2}, \frac{3}{2}; \frac{1}{2}x^2\right). \tag{6.6.11}$$

As noted above, equation (6.6.1) is unchanged under the coordinate change $x \to -x$. It is also unchanged if we replace the pair $(x, \nu + \frac{1}{2})$ with the pair $(\pm ix, -\nu - \frac{1}{2})$. Therefore four solutions of (6.6.1) are

$$D_\nu(x), \quad D_\nu(-x), \quad D_{-\nu-1}(ix), \quad D_{-\nu-1}(-ix). \tag{6.6.12}$$

The behavior of each of these solutions as $x \to 0$ follows from (6.6.10). This allows one to express any one of the solutions (6.6.12) as a linear combination of any two of these solutions. In particular, it follows from (6.6.10), using (2.2.7) and (2.3.1), that

$$D_\nu(x) = \frac{\Gamma(\nu + 1)}{\sqrt{2\pi}} \left\{ e^{\frac{1}{2}\nu\pi i} D_{-\nu-1}(ix) + e^{-\frac{1}{2}\nu\pi i} D_{-\nu-1}(-ix) \right\}; \tag{6.6.13}$$

$$D_{-\nu-1}(ix) = \frac{\Gamma(-\nu)}{\sqrt{2\pi}} \left\{ i e^{\frac{1}{2}\nu\pi i} D_\nu(x) - i e^{-\frac{1}{2}\nu\pi i} D_\nu(-x) \right\}.$$

The Wronskian of two solutions of (6.6.1) is constant. It follows from (6.6.10) and the duplication formula (2.3.1) that

$$W\big(D_\nu(x), D_\nu(-x)\big) = \frac{2^{\nu+\frac{3}{2}}\pi}{\Gamma\big(\frac{1}{2} - \frac{1}{2}\nu\big)\,\Gamma\big(-\frac{1}{2}\nu\big)} = \frac{\sqrt{2\pi}}{\Gamma(-\nu)}. \qquad (6.6.14)$$

The reflection formula (2.2.7) and (6.6.10) imply that

$$W\big(D_\nu(x), D_{-\nu-1}(ix)\big) = -i\,e^{-\frac{1}{2}\pi\nu i} = e^{-\frac{1}{2}\pi(\nu+1)i}; \qquad (6.6.15)$$

$$W\big(D_\nu(x), D_{-\nu-1}(-ix)\big) = i\,e^{\frac{1}{2}\nu\pi i} = e^{\frac{1}{2}\pi(\nu+1)i}.$$

The remaining Wronskians of the four solutions (6.6.12) can be deduced from these. In particular, $D_\nu(x)$ and $D_\nu(-x)$ are independent if and only if ν is not a non-negative integer, while $D_\nu(x)$ and $D_{-\nu-1}(ix)$ are always independent. The identities (6.6.6), (6.5.8), and (6.6.11) imply

$$D_\nu'(x) + \frac{x}{2} D_\nu(x) - \nu D_{\nu-1}(x) = 0. \qquad (6.6.16)$$

Similarly, the identities (6.6.6), (6.5.9), and (6.6.11) imply

$$D_\nu'(x) - \frac{x}{2} D_\nu(x) + D_{\nu+1}(x) = 0. \qquad (6.6.17)$$

Eliminating D_ν' from these identities gives the recurrence identity

$$D_{\nu+1}(x) - x\,D_\nu(x) + \nu\,D_{\nu-1}(x) = 0. \qquad (6.6.18)$$

The asymptotic result

$$D_\nu(x) \sim \frac{2^{\frac{1}{2}\nu}}{\sqrt{\pi}}\,\Gamma\left(\frac{1}{2}\nu + \frac{1}{2}\right) \cos\left(\sqrt{\nu + \frac{1}{2}}\,x - \frac{1}{2}\pi\nu\right) \qquad (6.6.19)$$

as $\nu \to +\infty$ is a consequence of (6.6.6) and the asymptotic result (6.2.11). A direct proof will be given in Chapter 10.

6.7 Whittaker functions

The first-order term in the confluent hypergeometric equation (6.1.1) can be eliminated by a gauge transformation $u(x) = \varphi(x)v(x)$ with $\varphi'/\varphi = \frac{1}{2}(1 - c/x)$, i.e. $\varphi(x) = e^{\frac{1}{2}x}x^{-\frac{1}{2}c}$. The resulting equation for v is *Whittaker's equation*

$$v''(x) + \left[-\frac{1}{4} + \frac{\kappa}{x} + \frac{1 - 4\mu^2}{4x^2} \right] v(x) = 0, \quad \kappa = \frac{c}{2} - a, \quad \mu = \frac{c-1}{2}.$$
(6.7.1)

Conversely, solutions of (6.7.1) have the form $x^{\frac{1}{2}c} e^{-\frac{1}{2}x} V(x)$, where V is a solution of the confluent hypergeometric equation (6.1.1) with indices a, c given by

$$a = \mu - \kappa + \frac{1}{2}, \quad c = 1 + 2\mu.$$

Since also

$$a + 1 - c = -\mu - \kappa + \frac{1}{2}, \quad 2 - c = 1 - 2\mu, \quad x^{\frac{1}{2}c} x^{1-c} = x^{1-\frac{1}{2}c} = x^{-\mu+\frac{1}{2}},$$

so long as 2μ is not an integer, there are independent solutions, the Whittaker functions

$$M_{\kappa,\mu}(x) = e^{-\frac{1}{2}x} x^{\mu+\frac{1}{2}} M\left(\mu - \kappa + \frac{1}{2}, 1 + 2\mu; x\right); \qquad (6.7.2)$$

$$M_{\kappa,-\mu}(x) = e^{-\frac{1}{2}x} x^{-\mu+\frac{1}{2}} M\left(-\mu - \kappa + \frac{1}{2}, 1 - 2\mu; x\right),$$

that correspond to (6.1.2) and (6.1.6), respectively.

The Kummer functions are entire when defined. Both the functions in (6.7.2) are defined so long as 2μ is not a nonzero integer. The Whittaker functions are multiple-valued unless 2μ is an odd integer, in which case whichever of the functions (6.7.2) is defined is a single-valued function.

The asymptotics of the solutions (6.7.2) follow from (6.1.8):

$$M_{\kappa,\mu} \sim \frac{\Gamma(1 + 2\mu)}{\Gamma\left(\mu - \kappa + \frac{1}{2}\right)} x^{-\kappa} e^{\frac{1}{2}x} \quad \text{as} \quad \operatorname{Re} x \to +\infty. \qquad (6.7.3)$$

The Wronskian of two solutions of (6.7.1) is constant. It follows from this and the behavior at zero,

$$M_{\kappa,\mu}(x) \sim x^{\mu+\frac{1}{2}},$$

that the Wronskian of the solutions (6.7.2) is

$$W\left(M_{\kappa,\mu}, M_{\kappa,-\mu}\right)(x) \equiv -2\mu. \qquad (6.7.4)$$

In view of (6.2.2) there is a solution, exponentially decreasing as $x \to +\infty$:

$$W_{\kappa,\mu}(x) = e^{-\frac{1}{2}x} x^{\mu+\frac{1}{2}} \, U\left(\mu - \kappa + \frac{1}{2}, 1 + 2\mu; x\right)$$

$$= \frac{\Gamma(-2\mu)}{\Gamma\left(-\mu - \kappa + \frac{1}{2}\right)} M_{\kappa,\mu}(x) + \frac{\Gamma(2\mu)}{\Gamma\left(\mu - \kappa + \frac{1}{2}\right)} M_{\kappa,-\mu}(x)$$

$$= W_{\kappa,-\mu}(x), \tag{6.7.5}$$

provided 2μ is not an integer. It follows from (6.2.2) that

$$W_{\kappa,\mu}(x) \sim x^{\kappa} e^{-\frac{1}{2}x} \quad \text{as} \quad \operatorname{Re} x \to +\infty. \tag{6.7.6}$$

The Wronskian of $M_{\kappa,\mu}$ and $W_{\kappa,\mu}$ can be computed from (6.7.5) and (6.7.4) or from the asymptotics (6.7.3) and (6.7.6):

$$W(M_{\kappa,\mu}, W_{\kappa,\mu})(x) \equiv -\frac{\Gamma(1 + 2\mu)}{\Gamma\left(\mu - \kappa + \frac{1}{2}\right)}. \tag{6.7.7}$$

Since (6.7.1) is unchanged under $(x, \kappa) \to (-x, -\kappa)$, it follows that a solution exponentially decreasing at $-\infty$ is

$$W_{-\kappa,\mu}(x) = \frac{\Gamma(-2\mu)}{\Gamma\left(-\mu + \kappa + \frac{1}{2}\right)} M_{-\kappa,\mu}(-x)$$

$$+ \frac{\Gamma(2\mu)}{\Gamma\left(\mu + \kappa + \frac{1}{2}\right)} M_{-\kappa,-\mu}(-x). \tag{6.7.8}$$

It follows from (6.7.5) and (6.2.11) that

$$W_{\kappa,\mu}(x) \sim \frac{\Gamma\left(\kappa + \frac{1}{4}\right) x^{\frac{1}{4}}}{\sqrt{\pi}} \cos\left(2\sqrt{\kappa x} - \kappa\pi + \frac{1}{4}\pi\right) \tag{6.7.9}$$

as $\kappa \to +\infty$.

The Coulomb wave equation (3.6.14) is

$$u''(\rho) + \left[1 - \frac{2\eta}{\rho} - \frac{l(l+1)}{\rho^2}\right] u(\rho) = 0.$$

Let $u(\rho) = v(2i\rho)$. Then the equation becomes

$$v''(x) + \left[-\frac{1}{4} + \frac{i\eta}{x} - \frac{l(l+1)}{x^2}\right] v(x) = 0.$$

This is Whittaker's equation with $\kappa = i\eta$ and $\mu = l + \frac{1}{2}$. When the equation is obtained by separating variables in spherical coordinates, the parameter l is a non-negative integer. In this case, in addition to a solution that is regular at $x = 0$, there is a solution with a logarithmic singularity. The normalization of the regular solution in [224] is

$$
\begin{aligned}
F_l(\eta, \rho) &= \frac{C_l(\eta)}{(\pm 2i)^{l+1}} \, M_{\pm i\eta, l+\frac{1}{2}}(\pm 2i\rho) \\
&= C_l(\eta) \, \rho^{l+1} e^{\mp i\rho} \, M(l + 1 \mp i\eta, 2l + 2; \pm 2i\rho),
\end{aligned} \qquad (6.7.10)
$$

where the normalizing constant is

$$
C_l(\eta) = 2^l \, e^{-\pi\eta/2} \, \frac{|\Gamma(l + 1 + i\eta)|}{\Gamma(2l + 2)}.
$$

(By (6.1.10), the choice of sign does not matter.) The reason for this choice of C_l is so that

$$
F_l(\eta, \rho) \sim \sin \theta_l(\eta, \rho) \qquad (6.7.11)
$$

as $\rho \to +\infty$, where θ_l is given by (6.7.14) below; see the exercises.

Irregular solutions are defined by

$$
\begin{aligned}
H_l^{\pm}(\eta, \rho) &= (\pm i)^l \, e^{\pm i\sigma_l(\eta)} \, e^{\frac{1}{2}\pi\eta} \, W_{\mp i\eta, l+\frac{1}{2}}(\mp 2i\rho) \\
&= e^{\pm i\theta_l(\eta, \rho)} \, (\mp 2i\rho)^{l+1\pm i\eta} \, U(l + 1 \pm i\eta, 2l + 1; \mp 2i\rho).
\end{aligned} \qquad (6.7.12)
$$

Here the normalizing phases are the *Coulomb wave shift*

$$
\sigma_l(\eta) = \arg \Gamma(l + 1 + i\eta) \qquad (6.7.13)
$$

and

$$
\theta_l(\eta, \rho) = \rho - \eta \, \log(2\rho) - \frac{1}{2} l\pi + \sigma_l(\eta). \qquad (6.7.14)
$$

The real and imaginary parts are

$$
H_l^{\pm}(\eta, \rho) = G_\lambda(\eta, \rho) \pm i \, F_l(\eta, \rho), \qquad (6.7.15)
$$

where F_l is the regular solution (6.7.10) and G_l is defined by (6.7.15):

$$
G_l(\eta, \rho) = \frac{1}{2} \left[H_l^+(\eta, \rho) + H_l^-(\eta, \rho) \right]. \qquad (6.7.16)
$$

When l is a non-negative integer, G_l has a logarithmic singularity at $\rho = 0$; see (6.3.3).

6.8 Exercises

6.1 Show that the expansion of $e^{\lambda x}$ as a sum of Jacobi polynomials is

$$e^{\lambda x} = \sum_{n=0}^{\infty} C_n \, M(n + \beta + 1, 2n + \alpha + \beta + 2; 2\lambda) \, P_n^{(\alpha,\beta)}(x),$$

where

$$C_n = \frac{\Gamma(n + \alpha + \beta + 1)}{\Gamma(2n + \alpha + \beta + 1)} (2\lambda)^n \, e^{-\lambda}.$$

Hint: write $e^{\lambda x} = e^{-\lambda} \, e^{\lambda(1+x)}$ and use Exercise 4.41 in Chapter 4.

6.2 Show that integration and summation can be interchanged to calculate the Laplace transform (see Exercise 4.25 in Chapter 4) of $f(x) = x^{b-1} M(a, c; x)$:

$$[\mathcal{L}f](s) = \frac{\Gamma(b)}{s^b} \, F\left(a, b, c; \frac{1}{s}\right),$$

where F is the hypergeometric function (see Exercise 1.7 in Chapter 1, or Chapter 8).

6.3 Verify the integral representation

$$M(a, c; x) = \frac{\Gamma(c)}{2\pi i} \int_C \left(1 - \frac{x}{t}\right)^{-a} t^{-c} e^t \, dt,$$

where C is the Hankel loop of Corollary (2.2.4).

6.4 The asymptotic result (6.1.8) implies that for most values of nonzero constants A and B the linear combination

$$u(x) = A \, M(a, c; x) + B \, x^{1-c} M(a + 1 - c, 2 - c; x)$$

will have exponential growth as $x \to +\infty$. Determine a necessary condition on the ratio A/B to prevent this from happening. Compare this to (6.2.6).

6.5 Prove the identity (6.2.7).

6.6 Verify the evaluation (6.2.8).

6.7 Prove that

$$U\left(a, \frac{1}{2}; x^2\right) = \frac{\Gamma\left(\frac{1}{2}\right)}{\Gamma\left(a + \frac{1}{2}\right) \Gamma(a)} \sum_{n=0}^{\infty} \frac{\Gamma\left(a + \frac{1}{2}n\right)}{n!} (-2x)^n.$$

6.8 Use Exercise 6.7 to prove

$$U\left(a, \frac{1}{2}; x^2\right) =$$

$$\frac{\Gamma(\frac{1}{2})}{\Gamma(a + \frac{1}{2})\Gamma(a)} \int_0^\infty \exp\left(-t - 2\sqrt{t}\,x\right) t^{a-1}\, dt, \quad \operatorname{Re} a > 0.$$

6.9 Verify the identity for the Laguerre polynomials

$$L_n^{(\alpha)}(x) = \frac{(-1)^n}{n!}\, U(-n, \alpha + 1; x).$$

6.10 Prove the identity (6.4.5) for the error function. Hint:
$1/(2n + 1) = \left(\frac{1}{2}\right)_n / \left(\frac{3}{2}\right)_n$.

6.11 Prove the identity (6.4.6) for the incomplete gamma function. Hint:
$a/(a + n) = (a)_n/(a + 1)_n$.

6.12 Use (6.2.1) to prove the identity (6.4.7) for the complementary incomplete gamma function.

6.13 Let $\omega = e^{i\pi/4} = (1 + i)/\sqrt{2}$. Show that

$$\frac{1}{1 + i}\, \operatorname{erf}\left(\omega x \sqrt{\frac{\pi}{2}}\right) = \int_0^x e^{-it^2\pi/2}\, dt = C(x) - i S(x),$$

where $C(x)$ and $S(x)$ are the Fresnel integrals of Section 6.4.

6.14 Use Exercise 6.13 and (6.4.5) to prove (6.4.8) and (6.4.9).

6.15 Use (6.2.1) to prove (6.4.10).

6.16 Use Cauchy's theorem to prove that

$$\int_0^\infty \frac{\sin t}{t}\, dt = \frac{1}{2} \int_{-\infty}^\infty \frac{\sin t}{t}\, dt$$

$$= \frac{1}{2} \operatorname{Im}\left[\lim_{\varepsilon \to 0+} \int_{|t| > \varepsilon} \frac{e^{it}}{t}\, dt\right] = \frac{\pi}{2}.$$

Hint: integrate e^{it}/t over the boundary of the region
$\{|z| > \varepsilon,\ 0 < \operatorname{Im} z < R,\ |\operatorname{Re} z| < S\}$ and let first S, then R go to ∞.
Thus for $x > 0$,

$$\int_x^\infty \frac{\sin t}{t}\, dt = \frac{\pi}{2} - \int_0^x \frac{\sin t}{t}\, dt.$$

6.17 Use (6.4.10) and Exercise 6.16 to prove (6.4.11) and (6.4.12).

6.18 Use the results of Section 6.5 to show that the asymptotic results (6.1.8) and (6.1.9) are valid for all indices (a, c), $c \neq 0, 1, 2, \ldots$

6.19 Use the results of Section 6.5 to show that the asymptotic results (6.2.2) and (6.2.3) are valid for all indices (a, c), c not an integer.

6.20 Suppose $c > 0$. The operator

$$L = x \frac{d^2}{dx^2} + (c - x) \frac{d}{dx}$$

is symmetric in L_w^2, $w(x) = x^{c-1} e^{-x}$, $x > 0$, with eigenvalues
$0, 1, 2, \ldots$; the Laguerre polynomials with index $c - 1$ are
eigenfunctions.

Given $\lambda > 0$ and $f \in L_w^2$, the equation $Lu + \lambda u = f$ has a unique
solution $u \in L_w^2$, expressible in the form

$$u(x) = \int_0^\infty G_\lambda(x, y) f(y) \, dy.$$

Compute the Green's function G_λ. Hint: see Section 3.3. The
appropriate boundary conditions here are: regular at $x = 0$, having at
most polynomial growth as $x \to +\infty$.

6.21 Prove (6.6.13).

6.22 Express $D_{-\nu-1}(ix)$ as a linear combination of $D_\nu(x)$ and $D_\nu(-x)$.

6.23 Prove the Wronskian formulas (6.6.14), (6.6.15).

6.24 Compute the Wronskians

$$W\big(D_\nu(x), D_{-\nu-1}(ix)\big);$$

$$W\big(D_\nu(x), D_{-\nu-1}(-ix)\big);$$

$$W\big(D_{-\nu-1}(ix), D_{-\nu-1}(-ix)\big).$$

6.25 Verify the identities (6.6.16), (6.6.17), (6.6.18).

6.26 Show that the asymptotic result (6.2.2) is also valid for imaginary values
of the argument x. Deduce that the Coulomb wave functions satisfy

$$H_l^\pm(\eta, \rho) \sim e^{\pm i \theta_l(\eta, \rho)}$$

as $\rho \to +\infty$, and thus verify (6.7.11).

6.9 Summary

6.9.1 Kummer functions

The confluent hypergeometric equation is

$$x u''(x) + (c - x) u'(x) - a u(x) = 0.$$

An entire solution is Kummer's function

$$M(a, c; x) = \sum_{n=0}^{\infty} \frac{(a)_n}{(c)_n \, n\,!} x^n$$

$$= \Phi(a, c; x) = {}_1F_1(a, c; x), \quad c \neq 0, -1, -2, \ldots$$

For $\operatorname{Re} c > \operatorname{Re} a > 0$,

$$M(a, c; x) = \frac{\Gamma(c)}{\Gamma(a)\,\Gamma(c - a)} \int_0^1 s^{a-1} (1 - s)^{c-a-1} e^{sx} \, ds.$$

Asymptotically,

$$M(a, c; x) \sim \frac{\Gamma(c)}{\Gamma(a)} x^{a-c} e^x \sum_{n=0}^{\infty} \frac{(1 - a)_n (c - a)_n}{n\,!} x^{-n}, \quad x \to +\infty.$$

Second solution:

$$x^{1-c} M(a + 1 - c, 2 - c; x), \quad c \neq 0, \pm 1, \pm 2, \ldots$$

Let

$$M_1(x) = M(a, c; x), \quad M_2(x) = x^{1-c} M(a + 1 - c, 2 - c; x),$$

$$c \neq 0, \pm 1, \pm 2, \ldots$$

Wronskian:

$$W(M_1, M_2)(x) = (1 - c) x^{-c} e^x.$$

Two identities due to Kummer:

$$M(a, c; x) = e^x \, M(c - a, c; -x);$$

$$M(a, 2a; 4x) = e^{2x} \, {}_0F_1\left(a + \frac{1}{2}; x^2\right) = e^{2x} \sum_{n=0}^{\infty} \frac{x^{2n}}{(a + \frac{1}{2})_n \, n!}.$$

Asymptotic behavior as $a \to -\infty$:

$$M(a, c; x) = \frac{\Gamma(c)}{\sqrt{\pi}} \left(\frac{1}{2} cx - ax\right)^{\frac{1}{4} - \frac{1}{2}c} e^{\frac{1}{2}x}$$

$$\times \left[\cos\left(\sqrt{2cx - 4ax} - \frac{1}{2}c\pi + \frac{1}{4}\pi\right) + O\left(|a|^{-\frac{1}{2}}\right)\right].$$

6.9.2 Kummer functions of the second kind

Solution with at most polynomial growth as $x \to +\infty$:

$$U(a, c; x) = \frac{\Gamma(1-c)}{\Gamma(a+1-c)} M(a, c; x)$$

$$+ \frac{\Gamma(c-1)}{\Gamma(a)} x^{1-c} M(a+1-c, 2-c; x), \quad c \neq 0, \pm 1, \pm 2, \ldots$$

Integral representation:

$$U(a, c; x) = \frac{1}{\Gamma(a)} \int_0^\infty e^{-xt} t^{a-1} (1+t)^{c-a-1} \, dt, \quad \text{Re}\, a > 0.$$

Asymptotics:

$$U(a, c; x) \sim x^{-a} \quad \text{as} \quad \text{Re}\, x \to +\infty,$$

$$U(a, c; x) \sim x^{-a} \sum_{n=0}^\infty \frac{(a)_n (a+1-c)_n}{n!} \frac{(-1)^n}{x^n}.$$

Kummer's identity:

$$U(a, c; x) = x^{1-c} U(a+1-c, 2-c; x).$$

Wronskian:

$$W\big(M(a, c; \cdot), U(a, c; \cdot)\big)(x) = -\frac{\Gamma(c)}{\Gamma(a)} x^{-c} e^x.$$

Solution exponentially decreasing as $x \to -\infty$:

$$\widetilde{U}(a, c; x) = \frac{\Gamma(1-c)}{\Gamma(1-a)} e^x M(c-a, c; -x)$$

$$+ \frac{\Gamma(c-1)}{\Gamma(c-a)} e^x (-x)^{1-c} M(1-a, 2-c; -x)$$

$$= \frac{\Gamma(1-c)}{\Gamma(1-a)} M(a, c; x) + \frac{\Gamma(c-1)}{\Gamma(c-a)} (-x)^{1-c} M(1+a-c, 2-c; x),$$

$$c \neq 0, \pm 1, \pm 2, \ldots$$

Integral representation:

$$\widetilde{U}(a, c; x) = e^x U(c-a, c; -x)$$

$$= \frac{1}{\Gamma(c-a)} \int_0^\infty e^{x(1+t)} t^{c-a-1} (1+t)^{a-1} \, dt, \quad \text{Re}\, c > \text{Re}\, a.$$

Asymptotic behavior as $a \to -\infty$:

$$U(a, c; x) = \frac{\Gamma\left(\frac{1}{2}c - a + \frac{1}{4}\right)}{\sqrt{\pi}} x^{\frac{1}{4} - \frac{1}{2}c} e^{\frac{1}{2}x}$$

$$\times \left[\cos\left(\sqrt{2cx - 4ax} - \frac{1}{2}c\pi + a\pi + \frac{1}{4}\pi\right) + O\left(|a|^{-\frac{1}{2}}\right) \right].$$

6.9.3 Solutions when c is an integer

Second solution of (6.1.1) when c is an integer: (6.2.7) allows reduction to the case $c = m$ a positive integer. Limit of $U(a, c; x)$, as $c \to m$, a not an integer:

$$U(a, m; x) = \frac{(-1)^m}{\Gamma(a + 1 - m)(m - 1)!} \left\{ \log x \, M(a, m; x) \right.$$

$$+ \sum_{n=0}^{\infty} \frac{(a)_n}{(m)_n \, n!} [\psi(a + n) - \psi(n + 1) - \psi(m + n)] x^n \Big\}$$

$$+ \frac{(m - 2)!}{\Gamma(a)} x^{1-m} \sum_{n=0}^{m-2} \frac{(a + 1 - m)_n}{(2 - m)_n \, n!} x^n, \quad \psi(b) = \frac{\Gamma'(b)}{\Gamma(b)};$$

last sum taken to be zero if $m = 1$. This solution is independent of $M(a, c; x)$ except for $a = 0, -1, -2, \ldots$ For these values, an independent solution is obtained by multiplying this expression by $\Gamma(a)$.

6.9.4 Special cases

The exponential function:

$$e^x = M(c, c; x).$$

Laguerre, Hermite, and modified Hermite polynomials:

$$L_n^{(\alpha)}(x) = \frac{(\alpha + 1)_n}{n!} M(-n, \alpha + 1; x);$$

$$H_{2m}(x) = (-1)^m 2^{2m} \left(\frac{1}{2}\right)_m M\left(-m, \frac{1}{2}; x^2\right);$$

$$H_{2m+1}(x) = (-1)^m 2^{2m+1} \left(\frac{3}{2}\right)_m x M\left(-m, \frac{3}{2}; x^2\right);$$

$$He_{2m}(x) = (-1)^m 2^m \left(\frac{1}{2}\right)_m M\left(-m, \frac{1}{2}; \frac{1}{2}x^2\right);$$

$$He_{2m+1}(x) = (-1)^m 2^{m+\frac{1}{2}} \left(\frac{3}{2}\right)_m \frac{x}{\sqrt{2}} M\left(-m, \frac{3}{2}; \frac{1}{2}x^2\right).$$

Error function, incomplete gamma function, complementary incomplete gamma function:

$$\text{erf } x \equiv \frac{2}{\sqrt{\pi}} \int_0^x e^{-t^2}\, dt = \frac{2x}{\sqrt{\pi}} M\left(\frac{1}{2}, \frac{3}{2}; -x^2\right);$$

$$\gamma(a, x) \equiv \int_0^x e^{-t} t^{a-1}\, dt = \frac{x^a}{a} M(a, a+1; -x);$$

$$\Gamma(a, x) \equiv \int_x^\infty e^{-t} t^{a-1}\, dt = x^a e^{-x} U(1, a+1; x).$$

Fresnel integrals:

$$C(x) \equiv \int_0^x \cos\left(\frac{1}{2}t^2\pi\right) dt = \frac{x}{2}\left[M\left(\frac{1}{2}, \frac{3}{2}; \frac{1}{2}ix^2\pi\right) + M\left(\frac{1}{2}, \frac{3}{2}; -\frac{1}{2}ix^2\pi\right)\right];$$

$$S(x) \equiv \int_0^x \sin\left(\frac{1}{2}t^2\pi\right) dt = \frac{x}{2i}\left[M\left(\frac{1}{2}, \frac{3}{2}; \frac{1}{2}ix^2\pi\right) - M\left(\frac{1}{2}, \frac{3}{2}; -\frac{1}{2}ix^2\pi\right)\right].$$

Exponential integral, cosine integral, and sine integral functions:

$$\text{Ei}(-z) \equiv \int_\infty^z e^{-t} \frac{dt}{t} = -e^{-z} U(1, 1; z), \quad z \notin (-\infty, 0];$$

$$\text{Ci}(x) \equiv \int_\infty^x \cos t \, \frac{dt}{t} = -\frac{1}{2}\left[e^{-ix} U(1, 1; ix) + e^{ix} U(1, 1; -ix)\right];$$

$$\text{Si}(x) \equiv \int_0^x \sin t \, \frac{dt}{t} = \frac{1}{2i}\left[e^{-ix} U(1, 1; ix) - e^{ix} U(1, 1; -ix)\right] + \frac{\pi}{2}.$$

6.9.5 Contiguous functions

Let

$$M = M(a, c; x); \quad M(a\pm) = M(a \pm 1, c; x); \quad M(c\pm) = M(a, c \pm 1; x).$$

The relations

$$xM' = a\left[M(a+) - M\right] = (c - 1)\left[M(c-) - M\right];$$

$$x M = x M' + (c - a) M - (c - a) M(a-);$$

$$cx M' = cx M - (c - a)x M(c+)$$

imply the six basic relations between contiguous functions:

$$(a - c + 1) M = a M(a+) - (c - 1) M(c-);$$

$$(2a - c + x) M = a M(a+) - (c - a) M(a-);$$

$$(a - 1 + x) M = (c - 1) M(c-) - (c - a) M(a-);$$

$$c(a + x) M = ac M(a+) + (c - a)x M(c+);$$

$$c(c - 1 + x) M = c(c - 1) M(c-) + (c - a)x M(c+);$$

$$c M = c M(a-) + x M(c+).$$

Set

$$U = U(a, c; x), \quad U(a\pm) = U(a \pm 1, c; x), \quad U(c\pm) = U(a, c \pm 1; x).$$

The identities

$$U'(a, c; x) = -a U(a + 1, c + 1; x);$$

$$U'(a, c; x) = U(a, c; x) - U(a, c + 1; x)$$

imply

$$U = a U(a+) + U(c-);$$

$$(c - a) U = x U(c+) - U(a-);$$

$$(x + a) U = x U(c+) + a(a + 1 - c) U(a+);$$

$$(x + c - 1) U = x U(c+) + (c - a - 1) U(c-);$$

$$(x + 2a - c) U = U(a-) + a(1 + a - c) U(a+);$$

$$(a + x - 1) U = U(a-) + (c - a - 1) U(c-).$$

6.9.6 Parabolic cylinder functions

A parabolic cylinder function is a solution of

$$u''(x) - \frac{x^2}{4} u(x) + \left(v + \frac{1}{2} \right) u(x) = 0.$$

Even and odd solutions, respectively, are multiples of

$$Y_{\nu 1}(x) = e^{-\frac{1}{4}x^2} M\left(-\frac{1}{2}\nu, \frac{1}{2}; \frac{1}{2}x^2\right),$$

$$Y_{\nu 2}(x) = e^{-\frac{1}{4}x^2} \frac{x}{\sqrt{2}} M\left(-\frac{1}{2}\nu + \frac{1}{2}, \frac{3}{2}; \frac{1}{2}x^2\right).$$

Solution with decay as $|x| \to \infty$:

$$D_\nu(x) = 2^{\frac{1}{2}\nu} e^{-\frac{1}{4}x^2} U\left(-\frac{1}{2}\nu, \frac{1}{2}; \frac{1}{2}x^2\right)$$

$$= 2^{\frac{1}{2}\nu} \left\{ \frac{\Gamma\left(\frac{1}{2}\right)}{\Gamma\left(\frac{1}{2} - \frac{1}{2}\nu\right)} Y_{\nu 1}(x) + \frac{\Gamma\left(-\frac{1}{2}\right)}{\Gamma\left(-\frac{1}{2}\nu\right)} Y_{\nu 2}(x) \right\}.$$

Integral representations:

$$D_\nu(x) = \frac{2^{\frac{1}{2}\nu} e^{-\frac{1}{4}x^2}}{\Gamma\left(-\frac{1}{2}\nu\right)} \int_0^\infty e^{-\frac{1}{2}tx^2} t^{-\frac{1}{2}\nu - 1} (1+t)^{\frac{1}{2}\nu - \frac{1}{2}} \, dt$$

$$= \frac{2^{\frac{1}{2}\nu} e^{-\frac{1}{4}x^2} \Gamma\left(\frac{1}{2}\right)}{\Gamma\left(\frac{1}{2} - \frac{1}{2}\nu\right) \Gamma\left(-\frac{1}{2}\nu\right)} \int_0^\infty e^{-t - \sqrt{2t}\, x} t^{-\frac{1}{2}\nu - 1} \, dt, \quad \text{Re } \nu < 0.$$

In particular,

$$D_n(x) = e^{-\frac{1}{4}x^2} He_n(x), \quad n = 0, 1, 2, \ldots$$

Behavior as $x \to 0$:

$$D_\nu(x) = 2^{\frac{1}{2}\nu} \sqrt{\pi} \left\{ \frac{1}{\Gamma\left(\frac{1}{2} - \frac{1}{2}\nu\right)} - \frac{2^{\frac{1}{2}} x}{\Gamma\left(-\frac{1}{2}\nu\right)} \right\} + O(x^2).$$

This allows determination of the relations among the four cylinder functions

$$D_\nu(x), \quad D_\nu(-x), \quad D_{-\nu-1}(ix), \quad D_{-\nu-1}(-ix).$$

In particular,

$$D_\nu(x) = \frac{\Gamma(\nu+1)}{\sqrt{2\pi}} \left\{ e^{\frac{1}{2}\nu\pi i} D_{-\nu-1}(ix) + e^{-\frac{1}{2}\nu\pi i} D_{-\nu-1}(-ix) \right\};$$

$$D_{-\nu-1}(ix) = \frac{\Gamma(-\nu)}{\sqrt{2\pi}} \left\{ i\, e^{\frac{1}{2}\nu\pi i} D_\nu(x) - i\, e^{-\frac{1}{2}\nu\pi i} D_\nu(-x) \right\}.$$

Wronskians: in particular

$$W(D_\nu(x), D_\nu(-x)) = \frac{\sqrt{2\pi}}{\Gamma(-\nu)};$$

$$W(D_\nu(x), D_{-\nu-1}(ix)) = -i\, e^{-\frac{1}{2}\pi\nu i} = e^{-\frac{1}{2}\pi(\nu+1)i};$$

$$W(D_\nu(x), D_{-\nu-1}(-ix)) = i\, e^{\frac{1}{2}\nu\pi i} = e^{\frac{1}{2}\pi(\nu+1)i}.$$

Identities for U imply

$$D_\nu'(x) + \frac{x}{2} D_\nu(x) - \nu\, D_{\nu-1}(x) = 0;$$

$$D_\nu'(x) - \frac{x}{2} D_\nu(x) + D_{\nu+1}(x) = 0;$$

$$D_{\nu+1}(x) - x\, D_\nu(x) + \nu\, D_{\nu-1}(x) = 0.$$

Asymptotics as $\nu \to +\infty$:

$$D_\nu(x) \sim \frac{2^{\frac{1}{2}\nu}}{\pi^{\frac{1}{2}}} \Gamma\left(\frac{1}{2}\nu + \frac{1}{2}\right) \cos\left(\sqrt{\nu + \frac{1}{2}}\, x - \frac{1}{2}\pi\nu\right).$$

6.9.7 Whittaker functions

Eliminating the first-order term from the confluent hypergeometric equation by a gauge transformation leads to the Whittaker equation [314]

$$u''(x) + \left[-\frac{1}{4} + \frac{\kappa}{x} + \frac{1-4\mu^2}{4x^2}\right] u(x) = 0, \quad \kappa = \frac{c}{2} - a, \quad \mu = \frac{c-1}{2}.$$

Two solutions:

$$M_{\kappa,\mu}(x) = e^{-\frac{1}{2}x} x^{\mu+\frac{1}{2}} M\left(\mu - \kappa + \frac{1}{2}, 1 + 2\mu; x\right);$$

$$M_{\kappa,-\mu}(x) = e^{-\frac{1}{2}x} x^{-\mu+\frac{1}{2}} M\left(-\mu - \kappa + \frac{1}{2}, 1 - 2\mu; x\right),$$

where

$$a = \mu - \kappa + \frac{1}{2}, \quad c = 1 + 2\mu.$$

Wronskian:

$$W(M_{\kappa,\mu}, M_{\kappa,-\mu}) = -2\mu.$$

Solution exponentially decreasing as $x \to +\infty$:

$$W_{\kappa,\mu}(x) = e^{-\frac{1}{2}x} x^{\mu+\frac{1}{2}} U\left(\mu - \kappa + \frac{1}{2}, 1 + 2\mu; x\right)$$

$$= \frac{\Gamma(-2\mu)}{\Gamma\left(-\mu - \kappa + \frac{1}{2}\right)} M_{\kappa,\mu}(x) + \frac{\Gamma(2\mu)}{\Gamma\left(\mu - \kappa + \frac{1}{2}\right)} M_{\kappa,-\mu}(x)$$

$$= W_{\kappa,-\mu}(x),$$

provided 2μ is not an integer. Wronskian:

$$W(M_{\kappa,\mu}, W_{\kappa,\mu}) = -\frac{\Gamma(1 + 2\mu)}{\Gamma\left(\mu - \kappa + \frac{1}{2}\right)}.$$

Solution exponentially decreasing at $-\infty$:

$$W_{-\kappa,\mu}(x) = \frac{\Gamma(-2\mu)}{\Gamma\left(-\mu + \kappa + \frac{1}{2}\right)} M_{-\kappa,\mu}(-x) + \frac{\Gamma(2\mu)}{\Gamma\left(\mu + \kappa + \frac{1}{2}\right)} M_{-\kappa,-\mu}(-x).$$

Asymptotics as $\kappa \to +\infty$:

$$W_{\kappa,\mu}(x) \sim \frac{\Gamma\left(\kappa + \frac{1}{4}\right) x^{\frac{1}{4}}}{\sqrt{\pi}} \cos\left(2\sqrt{\kappa x} - \kappa\pi + \frac{1}{4}\pi\right).$$

The Coulomb wave equation: normalized solution regular at the origin is

$$F_l(\eta, \rho) = \frac{C_l(\eta)}{(\pm 2i)^{l+1}} M_{\pm i\eta, l+\frac{1}{2}}(\pm 2i\rho)$$

$$= C_l(\eta) \rho^{l+1} e^{\mp i\rho} M(l + 1 \mp i\eta, 2l + 2; \pm 2i\rho),$$

where

$$C_l(\eta) = 2^l e^{-\pi\eta/2} \frac{|\Gamma(l + 1 + i\eta)|}{\Gamma(2l + 2)}.$$

This is the imaginary part of solutions

$$H_l^\pm(\eta, \rho) = (\pm i)^l e^{\pm i\sigma_l(\eta)} e^{\frac{1}{2}\pi\eta} W_{\mp i\eta, l+\frac{1}{2}}(\mp 2i\rho)$$

$$= e^{\pm i\theta_l(\eta, \rho)} (\mp 2i\rho)^{l+1\pm i\eta} U(l + 1 \pm i\eta, 2l + 2; \mp 2i\rho),$$

where

$$\sigma_l(\eta) = \arg \Gamma(l + 1 + i\eta);$$

$$\theta_l(\eta, \rho) = \rho - \eta \log(2\rho) - \frac{1}{2}l\pi + \sigma_l(\eta).$$

The real part is the singular solution

$$G_l(\eta, \rho) = \frac{1}{2} \left[H_l^+(\eta, \rho) + H_l^-(\eta, \rho) \right]. \tag{6.9.1}$$

6.10 Remarks

The Kummer functions were introduced by Kummer [168] in 1836, although the series (6.1.2) in the case $c = 2m$ had been investigated by Lagrange [170] in 1762–5. Weber introduced the parabolic cylinder functions in 1869 [308]. Whittaker [314] introduced the Whittaker functions in 1903 and showed that many known functions, including the parabolic cylinder functions and the functions in Section 6.4, can be expressed in terms of the $W_{\kappa,\mu}$. Coulomb wave functions were studied in 1928 by Gordon [116] and by Mott [208]. The current normalization and notation are due to Yost, Wheeler, and Breit [320]; see also Seaton [256]. For an application of Kummer and Whittaker functions to the study of singular and degenerate hyperbolic equations, see Beals and Kannai [23].

Three monographs devoted to confluent hypergeometric functions are Buchholz [37], Slater [260], and Tricomi [287]. Buchholz puts particular emphasis on the Whittaker functions and on applications, with many references to the applied literature, while Tricomi emphasizes the Kummer functions.

As noted above, there are several standard notations for the Kummer functions. We have chosen to use the notation M and U found in the handbooks of Abramowitz and Stegun [3], Jahnke and Emde [144], and Olver *et al.* [223, 224]. Tricomi [287] uses Φ and Ψ.

7

Cylinder functions

A cylinder function of order ν is a solution of Bessel's equation

$$x^2 u''(x) + x u'(x) + \left(x^2 - \nu^2\right) u(x) = 0. \tag{7.0.1}$$

As before we write $D = x(d/dx)$, so that Bessel's equation takes the form

$$\left(D^2 - \nu^2\right) u(x) + x^2 u(x) = 0.$$

For $x \sim 0$ this can be viewed as a perturbation of the equation $(D^2 - \nu^2)u = 0$, which has solutions $u(x) = x^{\pm \nu}$, so we expect to find solutions that have the form

$$x^{\pm \nu} f_\nu(x) \tag{7.0.2}$$

with f holomorphic near $x = 0$. Suitably normalized, solutions that have this form are the Bessel functions of the first kind, or simply "Bessel functions." For ν not an integer, one obtains two independent solutions of this form. For ν an integer, there is one solution of this form, and a second solution normalized at $x = 0$ known as a Bessel function of the second kind.

For x large, equation (7.0.1) can be viewed as a perturbation of

$$u''(x) + \frac{1}{x} u'(x) + u(x) = 0.$$

The gauge transformation $u(x) = x^{-\frac{1}{2}} v(x)$ converts this to

$$v''(x) + \left(1 + \frac{1}{4x^2}\right) v(x) = 0,$$

which can be viewed as a perturbation of $v'' + v = 0$. Therefore we expect to find solutions of (7.0.1) that have the asymptotic form

$$u(x) \sim x^{-\frac{1}{2}} e^{\pm ix} g_\nu(x), \quad x \to +\infty,$$

221

where g_ν has algebraic growth or decay. Bessel functions of the third kind, or Hankel functions, are a basis for such solutions.

We may remove the first-order term from (7.0.1) by the gauge transformation $u(x) = x^{-\frac{1}{2}} v(x)$. The equation for v is then

$$x^2 v''(x) + \left(x^2 - \nu^2 + \frac{1}{4}\right) v(x) = 0.$$

Therefore when $\nu = \pm\frac{1}{2}$, the solutions of (7.0.1) are linear combinations of

$$\frac{\cos x}{\sqrt{x}}, \quad \frac{\sin x}{\sqrt{x}}.$$

Because of this and the recurrence relations for Bessel functions, Bessel functions are elementary functions whenever ν is a half-integer.

Replacing x by ix in (7.0.1) gives the equation

$$x^2 u''(x) + x u'(x) - \left(x^2 + \nu^2\right) u(x) = 0. \tag{7.0.3}$$

Solutions are known as modified Bessel functions. The normalized solutions are real and have specified asymptotics as $x \to +\infty$.

The Airy equation

$$u''(x) - x\, u(x) = 0$$

is related to the equation (7.0.3) with $\nu^2 = \frac{1}{9}$ by a simple transformation, so its solutions can be obtained from the modified Bessel functions.

In this chapter we establish various representations of these functions in order to determine the recurrence and derivative formulas, and to determine the relations among the various normalized solutions.

7.1 Bessel functions

The simplest way to obtain solutions of Bessel's equation (7.0.1) that have the form (7.0.2) is to use the gauge transformation $u(x) = x^\nu v(x)$, so that the equation becomes

$$D^2 v(x) + 2\nu D v(x) + x^2 v(x) = 0$$

and determine the power series expansion $v(x) = \sum_{n=0}^{\infty} a_n x^n$. The coefficients must satisfy

$$n(n + 2\nu) a_n + a_{n-2} = 0.$$

It follows that $a_1 = 0$ and thus all odd terms vanish. We normalize by setting $a_0 = 1$ and obtain for the even terms

$$a_{2m} = (-1)^m \frac{1}{2m(2m-2)\cdots 2(2m+2\nu)(2m+2\nu-2)\cdots(2+2\nu)}$$

$$= (-1)^m \frac{1}{4^m m! \, (\nu+1)_m}, \quad \nu+1 \neq 0, -1, -2, \ldots, 1-m.$$

The corresponding solution of Bessel's equation is defined for ν not a negative integer:

$$x^\nu \sum_{m=0}^\infty \frac{(-1)^m}{(\nu+1)_m \, m!} \left(\frac{x}{2}\right)^{2m}. \tag{7.1.1}$$

As a function of ν for given $x > 0$, the function (7.1.1) has a simple pole at each negative integer. Division by $\Gamma(\nu+1)$ removes the pole. A further slight renormalization gives the *Bessel function of the first kind*

$$J_\nu(x) = \sum_{m=0}^\infty \frac{(-1)^m}{\Gamma(\nu+1+m)\, m!} \left(\frac{x}{2}\right)^{\nu+2m}. \tag{7.1.2}$$

The series is convergent for all complex x. Taking the principal branch of x^ν gives a function that is holomorphic on the complement of the real interval $(-\infty, 0]$.

It follows from the expansion (7.1.2) that

$$[x^\nu J_\nu]' = x^\nu J_{\nu-1}, \quad [x^{-\nu} J_\nu]' = -x^{-\nu} J_{\nu+1}, \tag{7.1.3}$$

which implies that

$$\left(\frac{1}{x}\frac{d}{dx}\right)^n \left[x^\nu J_\nu(x)\right] = x^{\nu-n} J_{\nu-n}(x); \tag{7.1.4}$$

$$\left(\frac{1}{x}\frac{d}{dx}\right)^n \left[x^{-\nu} J_\nu(x)\right] = (-1)^n x^{-\nu-n} J_{\nu+n}(x).$$

Now

$$x^{1-\nu}[x^\nu J_\nu]'(x) = x\, J_\nu'(x) + \nu J_\nu(x),$$

$$x^{1+\nu}[x^{-\nu} J_\nu]'(x) = x\, J_\nu'(x) - \nu J_\nu(x).$$

Eliminating $J'_\nu(x)$ and $J_\nu(x)$, respectively, from the resulting pairs of equations gives the recurrence and derivative relations

$$J_{\nu-1}(x) + J_{\nu+1}(x) = \frac{2\nu}{x} J_\nu(x); \qquad (7.1.5)$$

$$J_{\nu-1}(x) - J_{\nu+1}(x) = 2 J'_\nu(x). \qquad (7.1.6)$$

As remarked in the introduction to this chapter, the Bessel functions $J_{\pm\frac{1}{2}}$ must be linear combinations of $x^{-\frac{1}{2}} \cos x$ and $x^{-\frac{1}{2}} \sin x$. In general, as $x \to 0+$,

$$J_{\pm\nu}(x) \sim \frac{1}{\Gamma(\pm\nu+1)} \left(\frac{x}{2}\right)^{\pm\nu}, \quad J'_{\pm\nu}(x) \sim \pm\frac{\nu}{2} \frac{1}{\Gamma(\pm\nu+1)} \left(\frac{x}{2}\right)^{\pm\nu-1}. \tag{7.1.7}$$

It follows that

$$J_{\frac{1}{2}}(x) = \frac{\sqrt{2}\sin x}{\sqrt{\pi x}}; \quad J_{-\frac{1}{2}}(x) = \frac{\sqrt{2}\cos x}{\sqrt{\pi x}}. \tag{7.1.8}$$

These can also be derived from the series expansion (7.1.2) by using the duplication formula (2.3.1). It follows from (7.1.8) and (7.1.5) that $J_\nu(x)$ is expressible in terms of trigonometric functions and powers of x whenever $\nu + \frac{1}{2}$ is an integer.

For ν not an integer, the two solutions J_ν and $J_{-\nu}$ behave differently as $x \to 0$, so they are clearly independent. To compute the Wronskian, we note first that for any two solutions $u_1(x), u_2(x)$ of (7.0.1) the Wronskian $W(x) = W(u_1, u_2)(x)$ must satisfy $x^2 W'(x) = -x W(x)$. Therefore $W = c/x$ for some constant c. The constant is easily determined, using the identities (7.1.7) and (2.2.7), which give

$$\Gamma(\nu+1)\,\Gamma(-\nu+1) = \nu\,\Gamma(\nu)\,\Gamma(1-\nu) = \frac{\nu\pi}{\sin \nu\pi}.$$

It follows that

$$W(J_\nu, J_{-\nu})(x) = \frac{\sin \nu\pi}{\nu\pi} \begin{vmatrix} x^\nu & x^{-\nu} \\ \nu x^{\nu-1} & -\nu x^{-\nu-1} \end{vmatrix} = -\frac{2\sin \nu\pi}{\pi x}, \tag{7.1.9}$$

confirming that these solutions are independent for ν not an integer. If $\nu = -n$ is a negative integer, examination of the expansion (7.1.2) shows that the first non-vanishing term is the term with $m = n$. Setting $m = n + k$,

$$J_{-n}(x) = (-1)^n \sum_{k=0}^{\infty} \frac{(-1)^k}{k!(n+k)!} \left(\frac{x}{2}\right)^{n+2k} = (-1)^n J_n(x)$$

$$= \cos n\pi \, J_n(x). \tag{7.1.10}$$

These considerations lead to the choice of the *Bessel function of the second kind*

$$Y_\nu(x) = \frac{\cos \nu\pi \, J_\nu(x) - J_{-\nu}(x)}{\sin \nu\pi}. \tag{7.1.11}$$

In particular,

$$Y_{\frac{1}{2}}(x) = -\frac{\sqrt{2}\cos x}{\sqrt{\pi x}}; \quad Y_{-\frac{1}{2}}(x) = \frac{\sqrt{2}\sin x}{\sqrt{\pi x}}. \tag{7.1.12}$$

This solution of Bessel's equation (7.0.1) is first defined for ν not an integer. In view of (7.1.10), both the numerator and the denominator have simple zeros at integer ν, so the singularity is removable and Y_ν can be considered as a solution for all ν. The Wronskian is

$$W(J_\nu, Y_\nu)(x) = -\frac{W(J_\nu, J_{-\nu})(x)}{\sin \nu\pi} = \frac{2}{\pi x},$$

so $J_\nu(x)$ and $Y_\nu(x)$ are independent solutions for all ν. It follows from (7.1.10) and (7.1.11) that

$$Y_{-n}(x) = (-1)^n Y_n(x), \quad n = 0, 1, 2, \ldots \tag{7.1.13}$$

The identities (7.1.5) and (7.1.6) yield the corresponding identities for $Y_\nu(x)$:

$$Y_{\nu-1}(x) + Y_{\nu+1}(x) = \frac{2\nu}{x} Y_\nu(x); \tag{7.1.14}$$

$$Y_{\nu-1}(x) - Y_{\nu+1}(x) = 2 Y_\nu'(x). \tag{7.1.15}$$

Figures 7.1 and 7.2 show the graphs of these functions for some non-negative values of the argument x and the parameter ν.

The series expansion (7.1.2) can be used to find the asymptotic behavior of $J_\nu(x)$ for large values of ν so long as $y^2 = x^2/\nu$ is bounded. A formal calculation gives

$$\frac{\left(\sqrt{\nu}\, y\right)^\nu}{2^\nu \Gamma(\nu+1)} \sum_{m=0}^{\infty} \frac{(-1)^m}{m!} \left(\frac{y}{2}\right)^{2m}$$

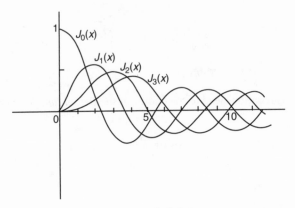

Figure 7.1 Bessel function $J_\nu(x)$, $\nu = 0, 1, 2, 3$.

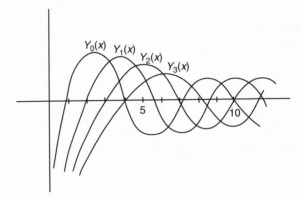

Figure 7.2 Bessel function $Y_\nu(x)$, $\nu = 0, 1, 2, 3$.

as the limiting value of the series. It is not difficult to show that

$$J_\nu\left(\sqrt{\nu}\, y\right) \sim \frac{\left(\sqrt{\nu}\, y\right)^\nu}{2^\nu \Gamma(\nu + 1)} e^{-\frac{1}{4}y^2} \tag{7.1.16}$$

as $\nu \to +\infty$, uniformly on bounded intervals. For asymptotic results when x is allowed to be comparable to ν, see Watson [306], chapter 8.

7.2 Zeros of real cylinder functions

Any real nonzero cylinder function with index ν has the form

$$u(x) = A\, J_\nu(x) + B\, Y_\nu(x), \tag{7.2.1}$$

where A and B are real constants. As noted in the introduction to this chapter, the gauge transformation $u(x) = x^{-\frac{1}{2}} v(x)$ converts Bessel's equation to a perturbation of the equation $w'' + w = 0$. Therefore we might expect that any such function u is oscillatory: that is, it has an infinite number of zeros in the half-line $(0, \infty)$, tending to infinity. This was proved for J_ν by Lommel [189]. Moreover, we might expect that the spacing between zeros is asymptotic to π.

Theorem 7.2.1 *A real nonzero cylinder function $u(x)$ has a countable number of positive zeros*

$$0 < x_1 < x_2 < \ldots < x_n < \ldots$$

The spacing $x_{n+1} - x_n$ is $\geq \pi$ if $|\nu| \geq \frac{1}{2}$, and $\leq \pi$ if $|\nu| \leq \frac{1}{2}$. As $n \to \infty$,

$$x_{n+1} - x_n = \pi + O(n^{-2}).$$

Proof As noted earlier, the gauge transformation $u(x) = x^{-\frac{1}{2}} v(x)$ converts Bessel's equation to

$$v''(x) + \left(1 - \frac{\nu^2 - \frac{1}{4}}{x^2} \right) v(x) = 0. \tag{7.2.2}$$

The cylinder function u itself is a linear combination of solutions with behavior x^ν and $x^{-\nu}$ as $x \to 0$, so v has no zeros in some interval $(0, \varepsilon]$. We make use of Sturm's comparison theorem, Theorem 3.3.3, by choosing comparison functions of the form

$$w(x) = \cos(ax + b), \quad a > 0. \tag{7.2.3}$$

Note that $w''(x) + a^2 w(x) = 0$ and the gap between zeros of $w(x)$ is π/a. If a is chosen so that a^2 is a lower bound for the zero-order coefficient

$$1 - \frac{\nu^2 - \frac{1}{4}}{x^2}, \quad \varepsilon \leq x < \infty, \tag{7.2.4}$$

then the comparison theorem implies that for $x \geq \varepsilon$ there is a zero of v between each pair of zeros of w. This proves that v, hence u, has countably many zeros. Moreover, by choosing b in (7.2.3) so that w vanishes at a given zero of v, we may conclude that the distance to the next zero of v is at most π/a.

Similarly, by choosing a so that a^2 is an upper bound for (7.2.4) and choosing b so that w vanishes at a zero of v, we conclude that the distance to the next zero of v is at least π/a. In particular, $a^2 = 1$ is a lower bound when $\nu^2 \leq \frac{1}{4}$ and an upper bound when $\nu^2 \geq \frac{1}{4}$, so π is an upper bound for the gaps in the first case and a lower bound in the second case.

This argument shows that the nth zero x_n has magnitude comparable to n. On the interval $[x_n, \infty)$, the best lower and upper bounds of (7.2.4) differ from 1 by an amount that is $O(n^{-2})$, so the gaps differ from π by a corresponding amount. $\square$

Between any two positive zeros of a real cylinder function u there is a zero of u'. It follows from equation (7.0.1) that such a zero x is simple if $x \neq v$, so $u(x)$ is a local extremum for u.

Theorem 7.2.2 *Suppose that u is a nonzero real cylinder function and suppose that the zeros of u' in the interval (v, ∞) are*

$$y_1 < y_2 < \ldots < y_n < \ldots$$

Then

$$|u(y_1)| > |u(y_2)| > \ldots > |u(y_n)| > \ldots, \tag{7.2.5}$$

and

$$\left(y_1^2 - v^2\right)^{\frac{1}{4}} |u(y_1)| < \left(y_2^2 - v^2\right)^{\frac{1}{4}} |u(y_2)| < \ldots < \left(y_n^2 - v^2\right)^{\frac{1}{4}} |u(y_n)| < \ldots \tag{7.2.6}$$

Proof Bessel's equation may be written in the form

$$[x\, u']'(x) + \left[x - \frac{v^2}{x}\right] u(x) = 0.$$

Therefore the inequalities (7.2.5) are a consequence of Proposition 3.5.2. The inequalities (7.2.6) are left as an exercise. $\square$

The inequalities (7.2.5) are a special case of results of Sturm [276]. The inequalities (7.2.6) are due to Watson [305].

The asymptotic results (7.4.8) and (7.4.9), together with the representation (7.2.1), can be used to show that the sequence in (7.2.6) has a limit and to evaluate the limit; see the exercises.

Suppose $v > 0$. Since $J_v(x) \sim (x/2)^v / \Gamma(v + 1)$ as $x \to 0+$, it follows that $J_v J_v' > 0$ for small x. By Theorem 3.3.6, J_v is positive throughout the interval $(0, v]$. Therefore the previous theorem can be sharpened for J_v itself.

Corollary 7.2.3 *Suppose $v > 0$. The local extrema of $J_v(x)$ decrease in absolute value as x increases, $x > 0$.*

A standard notation for the positive zeros of J_v is

$$0 < j_{v,1} < j_{v,2} < \ldots < j_{v,n} < \ldots \tag{7.2.7}$$

Theorem 7.2.4 *The zeros of J_ν and $J_{\nu+1}$ interlace:*

$$0 < j_{\nu,1} < j_{\nu+1,1} < j_{\nu,2} < j_{\nu+1,2} < j_{\nu,3} < \cdots$$

Proof It follows from (7.1.3) that

$$x^{-\nu} J_{\nu+1}(x) = -\frac{d}{dx}\left[x^{-\nu} J_\nu(x)\right], \quad x^{\nu+1} J_\nu(x) = \frac{d}{dx}\left[x^{\nu+1} J_{\nu+1}(x)\right].$$

$$(7.2.8)$$

The consecutive zeros $j_{\nu,k}$ and $j_{\nu,k+1}$ of $J_\nu(x)$ are also zeros of $x^{-\nu} J_\nu(x)$. Therefore there is at least one zero of the derivative of $x^{-\nu} J_\nu$, and hence of $J_{\nu+1}(x)$, between $j_{\nu,k}$ and $j_{\nu,k+1}$. Conversely, suppose that $\mu_1 < \mu_2$ are zeros of $J_{\nu+1}(x)$. Then they are also zeros of $x^{\nu+1} J_{\nu+1}(x)$, and by (7.2.8) there is at least one zero of $J_\nu(x)$ between μ_1 and μ_2. This shows that the zeros interlace. It follows from Theorem 3.3.4 that the first zero of J_ν is less than the first zero of $J_{\nu+1}$, and it follows from the remark preceding Corollary 7.2.3 that $j_{\nu,1} > \nu$. □

We conclude this section with a result that was proved by Fourier [100] for J_0 and by Lommel [189] for general real ν.

Theorem 7.2.5 *All zeros of $J_\nu(z)$ are real when $\nu > -1$. All zeros of $J'_\nu(z)$ are real when $\nu \geq 0$. When $-1 < \nu < 0$, $J'_\nu(z)$ has two imaginary zeros and all other zeros are real.*

Proof From the power series representations of $J_\nu(z)$ and $J'_\nu(z)$, it is readily seen that these functions do not have purely imaginary zeros. Bessel's equation implies that

$$(\alpha^2 - \beta^2) \int_0^z t\, J_\nu(\alpha t) J_\nu(\beta t)\, dt = z\left[\beta\, J_\nu(\alpha z) J'_\nu(\beta z) - \alpha\, J_\nu(\beta z) J'_\nu(\alpha z)\right]$$

$$(7.2.9)$$

for $\nu > -1$. Indeed, $J_\nu(\alpha z)$ satisfies the differential equation

$$\frac{1}{z}\left[z\, w'(z)\right]'(z) + \left(\alpha^2 - \frac{\nu^2}{z^2}\right) w(z) = 0.$$

Multiply this equation by $J_\nu(\beta z)$, and multiply the corresponding equation for $J_\nu(\beta z)$ by $J_\nu(\alpha z)$. Subtracting the two gives

$$J_\nu(\beta z)\left[\alpha\, z\, J'_\nu(\alpha z)\right]' - J_\nu(\alpha z)\left[\beta\, z\, J'_\nu(\beta z)\right]' = (\beta^2 - \alpha^2)z\, J_\nu(\alpha z) J_\nu(\beta z)$$

or, equivalently,

$$\left[\alpha\, z\, J_\nu(\beta z) J'_\nu(\alpha z) - \beta\, z\, J_\nu(\alpha z) J'_\nu(\beta z)\right]' = (\beta^2 - \alpha^2)z\, J_\nu(\alpha z) J_\nu(\beta z).$$

Formula (7.2.9) follows from an integration of the last equation. Now, let α be a non-real zero and $\alpha \notin i\mathbb{R}$. Then $z = \overline{\alpha}$ is also a zero. Put $z = 1$ and $\beta = \overline{\alpha}$ in (7.2.9), so that the equation becomes

$$(\alpha^2 - \overline{\alpha}^2) \int_0^1 t J_\nu(\alpha t) J_\nu(\overline{\alpha} t) dt = \left[\overline{\alpha} J_\nu(\alpha) J_\nu'(\overline{\alpha}) - \alpha J_\nu(\overline{\alpha}) J_\nu'(\alpha) \right]. \quad (7.2.10)$$

Since $J_\nu(\alpha) = J_\nu(\overline{\alpha}) = 0$, the right-hand side of (7.2.10) vanishes. Also, since $\operatorname{Re}\alpha \neq 0$ and $\operatorname{Im}\alpha \neq 0$, we deduce from (7.2.10) that

$$\int_0^1 t J_\nu(\alpha t) J_\nu(\overline{\alpha} t) dt = 0,$$

which is impossible since the integrand is positive.

If $J_\nu'(\alpha) = J_\nu'(\overline{\alpha}) = 0$, then the right-hand side of (7.2.10) again vanishes. Therefore the same argument can be used for $J_\nu'(z)$, except that now there is a pair of imaginary zeros when $-1 < \nu < 0$. This can be seen from the power series

$$\left(\frac{it}{2} \right)^{1-\nu} J_\nu'(it) = \frac{\frac{1}{2}\nu}{\Gamma(\nu + 1)} + \sum_{s=1}^{\infty} \frac{\left(s + \frac{1}{2}\nu \right)}{s! \Gamma(s + \nu + 1)} \left(\frac{t}{2} \right)^{2s},$$

where t is real. The function defined by the right-hand side of this equation is an even function, negative at $t = 0$, and monotonically increases to infinity as $t \to +\infty$. Therefore this function vanishes for two real values of t, completing the proof. $\qquad\Box$

7.3 Integral representations

An integral representation of J_ν can be obtained by using the gauge transformation $u(x) = x^\nu e^{ix} v(x)$ introduced in Section 3.7 to reduce (7.0.1) to the canonical form (3.7.3):

$$x\, v'' + (2\nu + 1 + 2ix)\, v' + (2\nu + 1)i\, v = 0. \qquad (7.3.1)$$

The change of variables $y = -2ix$ gives

$$y\, w'' + (2\nu + 1 - y)\, w' - \left(\nu + \frac{1}{2} \right) w = 0.$$

For ν not an integer, any solution of this last equation that is regular at the origin is a multiple of the Kummer function with indices $\nu + \frac{1}{2}$, $2\nu + 1$. It follows that J_ν is a multiple of the solution

$$x^\nu e^{ix} M\left(\nu + \frac{1}{2}, 2\nu + 1; -2ix \right).$$

Comparing behavior as $x \to 0$ gives the identity

$$J_\nu(x) = \frac{1}{\Gamma(\nu+1)} \left(\frac{x}{2}\right)^\nu e^{ix} M\left(\nu + \frac{1}{2}, 2\nu + 1; -2ix\right). \tag{7.3.2}$$

In view of the integral representation (6.1.3) of the Kummer function when $\mathrm{Re}\left(\nu + \frac{1}{2}\right) > 0$, we may deduce an integral representation for J_ν. Making use of the duplication identity (2.3.1) in the form

$$\frac{\Gamma(2\nu+1)}{\Gamma\left(\nu+\frac{1}{2}\right)} = \frac{2^{2\nu}}{\sqrt{\pi}} \Gamma(\nu+1),$$

we obtain the identity

$$J_\nu(x) = \frac{(2x)^\nu}{\sqrt{\pi}\,\Gamma\left(\nu+\frac{1}{2}\right)} e^{ix} \int_0^1 e^{-2ixs} s^{\nu-\frac{1}{2}} (1-s)^{\nu-\frac{1}{2}} \, ds \tag{7.3.3}$$

when $\mathrm{Re}\left(\nu + \frac{1}{2}\right) > 0$.

The change of variables $t = 1 - 2s$ in (7.3.3) leads to the *Poisson representation* [231]

$$J_\nu(x) = \frac{1}{\sqrt{\pi}\,\Gamma\left(\nu+\frac{1}{2}\right)} \left(\frac{x}{2}\right)^\nu \int_{-1}^1 \cos xt \, (1-t^2)^{\nu-\frac{1}{2}} \, dt. \tag{7.3.4}$$

The expansion (7.1.2) can be recovered from (7.3.4) by using the series expansion of $\cos xt$:

$$\int_0^1 \cos xt \, (1-t^2)^{\nu-\frac{1}{2}} \, dt = \sum_{m=0}^\infty (-1)^m \frac{\Gamma\left(m+\frac{1}{2}\right) \Gamma\left(\nu+\frac{1}{2}\right) x^{2m}}{\Gamma(m+\nu+1) \, (2m)!},$$

where we used the identity

$$\int_{-1}^1 t^{2m} (1-t^2)^{\nu-\frac{1}{2}} \, dt = \int_0^1 s^{m-\frac{1}{2}} (1-s)^{\nu-\frac{1}{2}} \, ds = \mathrm{B}\left(m + \frac{1}{2}, \nu + \frac{1}{2}\right).$$

Since $(2m)! = 2^{2m} m! \left(\frac{1}{2}\right)_m$, we obtain (7.1.2).

A second approach to the Bessel functions J_ν is closely associated with a second integral representation. We noted in Section 3.6 that functions of the form $\exp(i\mathbf{k} \cdot \mathbf{x})$ are solutions of the Helmholtz equation $\Delta u + |\mathbf{k}|^2 u = 0$. In particular, $\exp i x_2$ is a solution of $\Delta u + u = 0$. In cylindrical coordinates $x_2 = r \sin \theta$ and the Helmholtz equation is

$$\left\{ \frac{\partial^2}{\partial r^2} + \frac{1}{r} \frac{\partial}{\partial r} + \frac{1}{r^2} \frac{\partial^2}{\partial \theta^2} + 1 \right\} \left[e^{ir \sin \theta} \right] = 0. \tag{7.3.5}$$

Consider the Fourier expansion

$$e^{ir\sin\theta} = \sum_{n=-\infty}^{\infty} j_n(r)\, e^{in\theta}, \tag{7.3.6}$$

where the coefficients are given by

$$j_n(r) = \frac{1}{2\pi} \int_0^{2\pi} e^{ir\sin\theta} e^{-in\theta}\, d\theta. \tag{7.3.7}$$

Since

$$e^{-in\theta} = \frac{i}{n}\frac{d}{d\theta}\big[e^{-in\theta}\big],$$

repeated integrations by parts show that for every integer $k \geq 0$

$$|j_n(r)| \leq C_k \frac{(1+r)^k}{|n|^k}, \quad n \neq 0,$$

and similar estimates hold for derivatives of j_n. It follows that the expansion (7.3.6) may be differentiated term by term. By the uniqueness of Fourier coefficients, it follows that (7.3.5) implies Bessel's equation

$$r^2 j_n''(r) + r\, j_n'(r) + (r^2 - n^2)\, j_n(r) = 0, \quad n = 0, \pm 1, \pm 2, \ldots$$

We shall show that the functions j_n are precisely the Bessel functions J_n, by computing the series expansion. The first step is to take $t = e^{i\theta}$ in the integral representation (7.3.7), which becomes

$$j_n(r) = \frac{1}{2\pi i} \int_C \exp\left(\frac{1}{2}r[t - 1/t]\right) t^{-n-1}\, dt, \tag{7.3.8}$$

where C is the unit circle $\{|t| = 1\}$. Since the integrand is holomorphic in the plane minus the origin, we may replace the circle with a more convenient contour, one that begins and ends at $-\infty$ and encircles the origin in the positive (counterclockwise) direction. With this choice of contour we may define j_ν for all ν:

$$j_\nu(r) = \frac{1}{2\pi i} \int_C \exp\left(\frac{1}{2}r[t - 1/t]\right) t^{-\nu-1}\, dt, \tag{7.3.9}$$

where we take the curve to lie in the complement of the ray $(-\infty, 0]$ and take the argument of t in the interval $(-\pi, \pi)$. Taking $s = rt/2$ as the new variable of integration, the curve can be taken to be the same as before and (7.3.9) is

$$j_\nu(r) = \frac{1}{2\pi i} \left(\frac{r}{2}\right)^\nu \int_C \exp(s - r^2/4s)\, s^{-\nu-1}\, ds$$

$$= \frac{1}{2\pi i} \left(\frac{r}{2}\right)^\nu \sum_{m=0}^\infty \frac{(-1)^m}{m\,!} \frac{r^{2m}}{2^{2m}} \int_C e^s\, s^{-m-\nu-1}\, ds. \qquad (7.3.10)$$

According to Hankel's integral formula (2.2.8), the contour integral in the sum in (7.3.10) is $2\pi i / \Gamma(\nu + m + 1)$. Comparison with (7.1.2) shows that $j_\nu = J_\nu$. The integral representation (7.3.9) is due to Schlömilch [252].

Summarizing, we have an integral formula due to Bessel [30]

$$J_n(x) = \frac{1}{2\pi} \int_0^{2\pi} e^{ix \sin\theta - in\theta}\, d\theta \qquad (7.3.11)$$

for integer n, and

$$J_\nu(x) = \frac{1}{2\pi i} \int_C \exp\left(\frac{1}{2}x[t - 1/t]\right) t^{-\nu-1}\, dt$$

for arbitrary real ν. This last integral can also be put in the form known as the *Sommerfeld representation* [264]:

$$J_\nu(x) = \frac{1}{2\pi} \int_C e^{ix \sin\theta - i\nu\theta}\, d\theta \qquad (7.3.12)$$

by taking $t = e^{i\theta}$ and taking the path of integration in (7.3.12) to be the boundary of the strip defined by the inequalities

$$-\pi < \mathrm{Re}\, z < \pi, \quad \mathrm{Im}\, z > 0.$$

7.4 Hankel functions

Our starting point here is equation (7.3.1), which is in the canonical form discussed in Section 4.9:

$$pv'' + qv' + \lambda_\mu v = w^{-1}(pwv')' + \lambda_\mu v = 0,$$

where

$$\lambda_\mu = -q'\mu - \frac{1}{2}\mu(\mu - 1)p''.$$

As noted in that section, under certain conditions on the curve C, the integral

$$\int_C \frac{w(t)}{w(x)} \frac{p^{\mu}(t)\,dt}{(x-t)^{\mu+1}}$$

is a solution of the equation. In (7.3.1), $p(x) = x$, $q(x) = 2v + 1 + 2ix$, and $\lambda_{\mu} = (2v + 1)i$, so $\mu = -v - \frac{1}{2}$. Since $(pw)' = qw$, we take $w(x) = x^{2v}e^{2ix}$. Thus the proposed solutions $u(x) = x^{v}e^{ix}v(x)$ of the original equation (7.0.1) have the form

$$x^{-v}e^{-ix} \int_C e^{2it}t^{v-\frac{1}{2}}(x-t)^{v-\frac{1}{2}}\,dt. \qquad (7.4.1)$$

If we take C to be the interval $[0, x]$, then the conditions of Theorem 4.9.2 are satisfied at a finite endpoint 0 or x so long as $\mathrm{Re}\,v > \frac{3}{2}$. Moreover, the integral converges so long as $\mathrm{Re}\left(v + \frac{1}{2}\right) > 0$, and by analytic continuation it continues to define a solution of (7.0.1).

Up to the multiplicative constant

$$c_v = \frac{2^v}{\sqrt{\pi}\,\Gamma\left(v + \frac{1}{2}\right)}, \qquad (7.4.2)$$

(7.4.1) is (7.3.3) after the change of variables $t = x - sx$ in the integral. There are two natural alternative choices for a path of integration: the positive imaginary axis $\{is; \ s > 0\}$, and the ray $\{x + is; \ s \geq 0\}$. This leads to the *Bessel functions of the third kind*, or *Hankel functions*, defined for $\mathrm{Re}\left(v + \frac{1}{2}\right) > 0$ by the Poisson representations

$$H_v^{(1)}(x) = -2ic_v x^{-v}e^{-ix} \int_0^{\infty} e^{2i(x+is)}[(x+is)(-is)]^{v-\frac{1}{2}}\,ds$$

$$= 2c_v x^{-v}e^{i\left(x-\frac{1}{4}\pi-\frac{1}{2}v\pi\right)} \int_0^{\infty} e^{-2s}[s(x+is)]^{v-\frac{1}{2}}\,ds; \qquad (7.4.3)$$

$$H_v^{(2)}(x) = 2ic_v x^{-v}e^{-ix} \int_0^{\infty} e^{-2s}[is(x-is)]^{v-\frac{1}{2}}\,ds$$

$$= 2c_v x^{-v}e^{-i\left(x-\frac{1}{4}\pi-\frac{1}{2}v\pi\right)} \int_0^{\infty} e^{-2s}[s(x-is)]^{v-\frac{1}{2}}\,ds. \qquad (7.4.4)$$

The reason for the choice of constants is that it leads to the identity

$$J_v(x) = \frac{1}{2}\left[H_v^{(1)}(x) + H_v^{(2)}(x)\right], \quad \mathrm{Re}\left(v + \frac{1}{2}\right) > 0. \qquad (7.4.5)$$

Indeed, $2J_v(x) - H_v^{(1)}(x) - H_v^{(2)}(x)$ is an integral around a contour from $i\infty$ to $x + i\infty$ of a function that is holomorphic in the half-strip $\{0 < \mathrm{Re}\,z < x$,

Im $z > 0$} and vanishes exponentially as Im $s \to +\infty$, so by Cauchy's theorem the result is zero.

It is easy to determine the asymptotics of the Hankel functions as $x \to +\infty$, and this can be used to determine the asymptotics of J_ν. As $x \to +\infty$,

$$\int_0^\infty e^{-2s}[s(x \pm is)]^{\nu-\frac{1}{2}} \, ds \sim \int_0^\infty e^{-2s}(sx)^{\nu-\frac{1}{2}} \, ds$$

$$= \frac{x^{\nu-\frac{1}{2}}}{2^{\nu+\frac{1}{2}}} \int_0^\infty e^{-t} t^{\nu-\frac{1}{2}} \, dt = \frac{x^{\nu-\frac{1}{2}} \Gamma\left(\nu + \frac{1}{2}\right)}{2^{\nu+\frac{1}{2}}}.$$

Therefore as $|x| \to \infty$,

$$H_\nu^{(1)}(x) \sim \frac{\sqrt{2}}{\sqrt{\pi x}} e^{i\left(x - \frac{1}{4}\pi - \frac{1}{2}\nu\pi\right)}; \tag{7.4.6}$$

$$H_\nu^{(2)}(x) \sim \frac{\sqrt{2}}{\sqrt{\pi x}} e^{-i\left(x - \frac{1}{4}\pi - \frac{1}{2}\nu\pi\right)}.$$

Writing

$$(x - t)^{\nu-\frac{1}{2}} = x^{\nu-\frac{1}{2}}(1 - t/x)^{\nu-\frac{1}{2}}$$

$$\sim x^{\nu-\frac{1}{2}}\left[1 - \left(\nu - \frac{1}{2}\right)\frac{t}{x} + \cdots\right],$$

we may extend the asymptotics to full asymptotic series:

$$H_\nu^{(1)}(x) \sim \frac{\sqrt{2}}{\sqrt{\pi x}} e^{i\left(x - \frac{1}{4}\pi - \frac{1}{2}\nu\pi\right)}$$

$$\times \sum_{m=0}^\infty \frac{(-i)^m \left(-\nu + \frac{1}{2}\right)_m \left(\nu + \frac{1}{2}\right)_m}{2^m m!} x^{-m}; \tag{7.4.7}$$

$$H_\nu^{(2)}(x) \sim \frac{\sqrt{2}}{\sqrt{\pi x}} e^{-i\left(x - \frac{1}{4}\pi - \frac{1}{2}\nu\pi\right)}$$

$$\times \sum_{m=0}^\infty \frac{i^m \left(-\nu + \frac{1}{2}\right)_m \left(\nu + \frac{1}{2}\right)_m}{2^m m!} x^{-m}.$$

The verification is left as an exercise.

It follows from (7.4.6) that the Hankel functions are independent and that

$$J_\nu(x) \sim \frac{\sqrt{2}}{\sqrt{\pi x}} \cos\left(x - \frac{1}{4}\pi - \frac{1}{2}\nu\pi\right), \quad \text{Re}\,(2\nu + 1) > 0. \tag{7.4.8}$$

This result is due to Poisson [231] ($\nu = 0$), Hansen [124] ($\nu = 1$), Jacobi [141] (ν an integer), and Hankel [123]. The recurrence identity (7.1.5) implies that $-J_{\nu-2}$ has the same principal asymptotic behavior as J_ν, so (7.4.8) extends to all complex ν. This allows us to compute the asymptotics of Y_ν, using (7.1.11):

$$Y_\nu(x) \sim \frac{\sqrt{2}}{\sqrt{\pi x} \sin \nu\pi}$$

$$\times \left[\cos \nu\pi \, \cos \left(x - \frac{1}{4}\pi - \frac{1}{2}\nu\pi \right) - \cos \left(x - \frac{1}{4}\pi + \frac{1}{2}\nu\pi \right) \right].$$

The trigonometric identity

$$\cos 2b \, \cos(a - b) - \cos(a + b) = \sin 2b \, \sin(a - b)$$

with $a = x - \frac{1}{4}\pi$ and $b = \frac{1}{2}\nu\pi$ gives

$$Y_\nu(x) \sim \frac{\sqrt{2}}{\sqrt{\pi x}} \sin \left(x - \frac{1}{4}\pi - \frac{1}{2}\nu\pi \right). \tag{7.4.9}$$

For $\mathrm{Re}\left(\nu + \frac{1}{2}\right) > 0$, Y_ν is a linear combination of the Hankel functions. It follows from (7.4.6) and (7.4.9) that

$$Y_\nu(x) = \frac{1}{2i} \left[H_\nu^{(1)}(x) - H_\nu^{(2)}(x) \right]. \tag{7.4.10}$$

Conversely, (7.4.5) and (7.4.10) imply

$$H_\nu^{(1)}(x) = J_\nu(x) + iY_\nu(x); \tag{7.4.11}$$

$$H_\nu^{(2)}(x) = J_\nu(x) - iY_\nu(x),$$

for $\mathrm{Re}\left(\nu + \frac{1}{2}\right) > 0$. We may use (7.4.11) to *define* the Hankel functions for all values of ν. Then the identities (7.4.5) and (7.4.10) are valid for all complex ν, as are the asymptotics (7.4.6) and (by analytic continuation) the asymptotic series (7.4.7). In particular,

$$H_{\frac{1}{2}}^{(1)}(x) = -i\frac{\sqrt{2}}{\sqrt{\pi x}} e^{ix}; \quad H_{-\frac{1}{2}}^{(1)}(x) = \frac{\sqrt{2}}{\sqrt{\pi x}} e^{ix}; \tag{7.4.12}$$

$$H_{\frac{1}{2}}^{(2)}(x) = i\frac{\sqrt{2}}{\sqrt{\pi x}} e^{-ix}; \quad H_{-\frac{1}{2}}^{(2)}(x) = \frac{\sqrt{2}}{\sqrt{\pi x}} e^{-ix}.$$

It follows from (7.4.11) that the Wronskian is

$$W(H_\nu^{(1)}, H_\nu^{(2)})(x) = -2i\,W(J_\nu, Y_\nu)(x) = \frac{4}{\pi i x}. \tag{7.4.13}$$

It also follows from (7.4.11) that the Hankel functions satisfy the analogues of (7.1.5), (7.1.6), (7.1.14), and (7.1.15):

$$H_{\nu-1}^{(1)}(x) + H_{\nu+1}^{(1)}(x) = \frac{2\nu}{x}\, H_\nu^{(1)}(x); \tag{7.4.14}$$

$$H_{\nu-1}^{(2)}(x) - H_{\nu+1}^{(2)}(x) = 2\left[H_\nu^{(2)}\right]'(x). \tag{7.4.15}$$

For a given index ν, the functions J_ν, $J_{-\nu}$, Y_ν, $Y_{-\nu}$, $H_\nu^{(1)}$, $H_\nu^{(2)}$, $H_{-\nu}^{(1)}$, and $H_{-\nu}^{(2)}$ are all solutions of (7.0.1), so any choice of three of these functions satisfies a linear relation. The relations not already given above can easily be obtained from the asymptotics (7.4.6), (7.4.8), and (7.4.9). In particular,

$$H_{-\nu}^{(1)}(x) = e^{i\pi\nu}\, H_\nu^{(1)}(x); \tag{7.4.16}$$

$$H_{-\nu}^{(2)}(x) = e^{-i\pi\nu} H_\nu^{(2)}(x).$$

7.5 Modified Bessel functions

Replacing x by ix in Bessel's equation (7.0.1) yields

$$x^2 u''(x) + x\, u'(x) - (x^2 + \nu^2)\, u(x) = 0. \tag{7.5.1}$$

Solutions of this equation are known as *modified Bessel functions*. The most obvious way to obtain solutions is to evaluate the Bessel and Hankel functions on the positive imaginary axis. It is natural to choose one solution by modifying $J_\nu(ix)$ so that it takes real values. The result is

$$I_\nu(x) = \sum_{m=0}^{\infty} \frac{1}{\Gamma(\nu+1+m)\, m!} \left(\frac{x}{2}\right)^{\nu+2m} = e^{-\frac{1}{2}i\nu\pi} J_\nu(ix). \tag{7.5.2}$$

It follows from (7.4.6) that $H_\nu^{(1)}(ix)$ decays exponentially as $x \to +\infty$, while $H_\nu^{(2)}(ix)$ grows exponentially. Therefore it is natural to obtain a second solution by modifying $H_\nu^{(1)}(ix)$:

$$K_\nu(x) = \frac{\pi}{2}\, e^{\frac{1}{2}i(\nu+1)\pi} H_\nu^{(1)}(ix) = \frac{\pi}{2}\, \frac{I_{-\nu}(x) - I_\nu(x)}{\sin\pi\nu}. \tag{7.5.3}$$

The Poisson integral representations (7.3.4) and (7.4.3) lead to

$$I_\nu(x) = \frac{1}{\sqrt{\pi}\,\Gamma\left(\nu + \frac{1}{2}\right)} \left(\frac{x}{2}\right)^\nu \int_{-1}^{1} \cosh(xt)(1 - t^2)^{\nu - \frac{1}{2}}\, dt; \qquad (7.5.4)$$

$$K_\nu(x) = \frac{\sqrt{\pi}}{\sqrt{2x}\,\Gamma\left(\nu + \frac{1}{2}\right)} e^{-x} \int_{0}^{\infty} e^{-t} \left(t + \frac{t^2}{2x}\right)^{\nu - \frac{1}{2}}\, dt. \qquad (7.5.5)$$

A consequence is that I_ν and K_ν are positive, $0 < x < \infty$.
The derivative formula

$$\frac{d}{dx}\left[x^\nu I_\nu(x)\right] = x^\nu I_{\nu-1}(x)$$

follows from (7.1.3) or directly from the expansion (7.5.2) and leads to the relations

$$I_{\nu-1}(x) - I_{\nu+1}(x) = \frac{2\nu}{x} I_\nu(x);$$

$$I_{\nu-1}(x) + I_{\nu+1}(x) = 2\, I_\nu'(x).$$

These imply the corresponding relations for K_ν:

$$K_{\nu-1}(x) - K_{\nu+1}(x) = -\frac{2\nu}{x} K_\nu(x);$$

$$K_{\nu-1}(x) + K_{\nu+1}(x) = -2\, K_\nu'(x).$$

The asymptotic relation (7.4.8) implies

$$I_\nu(x) \sim \frac{e^x}{\sqrt{2\pi x}}; \qquad (7.5.6)$$

$$K_\nu(x) \sim \frac{\sqrt{\pi}\, e^{-x}}{\sqrt{2x}}.$$

Full asymptotic expansions may be obtained from (7.4.7). The principal terms in the derivatives come from differentiating (7.5.6), so the Wronskian is

$$W(K_\nu, I_\nu)(x) = \frac{1}{x}.$$

7.6 Addition theorems

In Section 7.3 we established the identity

$$e^{ix \sin \theta} = \sum_{n=-\infty}^{\infty} J_n(x) e^{in\theta}. \tag{7.6.1}$$

Taking $t = e^{i\theta}$, we may write this in the form of a generating function for the Bessel functions of integral order:

$$G(x, t) = \sum_{n=-\infty}^{\infty} J_n(x) t^n = e^{\frac{1}{2} x(t - 1/t)}, \quad |t| = 1. \tag{7.6.2}$$

Moreover,

$$\sum_{n=-\infty}^{\infty} J_n(x + y) e^{in\theta} = e^{i(x+y) \sin \theta} = \sum_{n=-\infty}^{\infty} J_n(x) e^{in\theta} \sum_{n=-\infty}^{\infty} J_n(y) e^{in\theta}.$$

Equating coefficients of $e^{in\theta}$ gives the *addition formula*

$$J_n(x + y) = \sum_{m=-\infty}^{\infty} J_m(x) J_{n-m}(y).$$

This is a special case of a more general addition formula. Consider a plane triangle whose vertices, in polar coordinates, are the origin and the points (r_1, θ_1) and (r_2, θ_2). Let r be the length of the third side. To fix ideas, suppose that $0 < \theta_1 < \theta_2 < \frac{1}{2}\pi$. Then the triangle lies in the first quadrant and the angle θ opposite the side of length r is $\theta_2 - \theta_1$. Projecting onto the vertical axis gives the identity

$$r \sin(\theta_2 + \varphi) = r_2 \sin \theta_2 - r_1 \sin \theta_1 = r_2 \sin \theta_2 - r_1 \sin(\theta_2 - \theta),$$

where $\theta = \theta_2 - \theta_1$ is the angle opposite the side with length r and φ is the angle opposite the side with length r_1. By analytic continuation, this identity carries over to general values of θ_2, with

$$r^2 = r_1^2 + r_2^2 - 2r_1 r_2 \cos \theta, \quad \theta = \theta_2 - \theta_1.$$

According to (7.3.12),

$$
\begin{aligned}
J_\nu(r)\, e^{i\nu\varphi} &= \frac{1}{2\pi} \int_C e^{ir\sin(\theta_2+\varphi)-i\nu\theta_2}\, d\theta_2 \\
&= \frac{1}{2\pi} \int_C e^{ir_2\sin\theta_2-ir_1\sin(\theta_2-\theta)-i\nu\theta_2}\, d\theta_2 \\
&= \frac{1}{2\pi} \int_C e^{ir_1\sin(\theta-\theta_2)}\, e^{ir_2\sin\theta_2-i\nu\theta_2}\, d\theta_2.
\end{aligned}
$$

By (7.6.1),

$$
e^{ir_1\sin(\theta-\theta_2)} = \sum_{n=-\infty}^{\infty} J_n(r_1)\, e^{in(\theta-\theta_2)}.
$$

Inserting this into the preceding integral and interchanging integration and summation gives

$$
J_\nu(r)\, e^{i\nu\varphi} = \sum_{n=-\infty}^{\infty} J_n(r_1) e^{in\theta} \cdot \frac{1}{2\pi} \int_C e^{ir_2\sin\theta_2-i(\nu+n)\theta_2}\, d\theta_2.
$$

Using (7.3.12) again, we have *Graf's addition formula*

$$
J_\nu(r)\, e^{i\nu\varphi} = \sum_{n=-\infty}^{\infty} J_n(r_1)\, J_{\nu+n}(r_2)\, e^{in\theta}. \tag{7.6.3}
$$

A deeper result is *Gegenbauer's addition formula* [110]:

$$
\frac{J_\nu(r)}{r^\nu} = 2^\nu\, \Gamma(\nu) \sum_{n=0}^{\infty} (\nu+n) \frac{J_{\nu+n}(r_1) J_{\nu+n}(r_2)}{r_1^\nu\, r_2^\nu}\, C_n^\nu(\cos\theta), \tag{7.6.4}
$$

where C_n^ν are the Gegenbauer polynomials, expressed in terms of Jacobi polynomials as

$$
C_n^\nu(x) = \frac{(2\nu)_n}{\left(\nu+\frac{1}{2}\right)_n}\, P_n^{\left(\nu-\frac{1}{2},\nu-\frac{1}{2}\right)}(x).
$$

For a simple derivation of (7.6.4) from Graf's formula when ν is an integer, see [7]. For a proof in the general case, see [219].

7.7 Fourier transform and Hankel transform

If $f(x) = f(x_1, x_2)$ is absolutely integrable, i.e.

$$\int_{-\infty}^{\infty} \int_{-\infty}^{\infty} |f(x_1, x_2)| \, dx_1 \, dx_2 < \infty,$$

then the *Fourier transform* of f is the function

$$\widehat{f}(\xi_1, \xi_2) = \frac{1}{2\pi} \int_{-\infty}^{\infty} \int_{-\infty}^{\infty} e^{-i(x_1\xi_1 + x_2\xi_2)} \, f(x_1, x_2) \, dx_1 \, dx_2.$$

If f is continuous and $\widehat{f}$ is also absolutely integrable, then

$$f(x_1, x_2) = \frac{1}{2\pi} \int_{-\infty}^{\infty} \int_{-\infty}^{\infty} e^{i(x_1\xi_1 + x_2\xi_2)} \, \widehat{f}(\xi_1, \xi_2) \, d\xi_1 \, d\xi_2;$$

see Appendix B.

If f is an absolutely integrable function on the half-line $(0, \infty)$, its nth *Hankel transform* is

$$g(y) = \int_0^{\infty} J_n(xy) \, f(x) \, x \, dx. \tag{7.7.1}$$

This Fourier inversion formula above can be used to show that the Hankel transform is its own inverse: if f is an absolutely integrable function on the half-line $[0, \infty)$ and its nth Hankel transform g is also integrable, then f is the nth Hankel transform of g:

$$f(x) = \int_0^{\infty} J_n(xy) \, g(y) \, y \, dy. \tag{7.7.2}$$

To prove (7.7.2) given (7.7.1), we first write the two-variable Fourier transform in polar coordinates (x, θ) and (y, φ):

$$\widehat{F}(y, \varphi) = \frac{1}{2\pi} \int_0^{\infty} \int_0^{2\pi} e^{-ixy\cos(\theta - \varphi)} F(x, \theta) \, d\theta \, x \, dx; \tag{7.7.3}$$

$$F(x, \theta) = \frac{1}{2\pi} \int_0^{\infty} \int_0^{2\pi} e^{ixy\cos(\theta - \varphi)} \widehat{F}(y, \varphi) \, d\varphi \, y \, dy. \tag{7.7.4}$$

Take $F(x, \theta) = f(x) \, e^{-in\theta}$. Then the integration with respect to θ in (7.7.3) gives

$$\int_0^{2\pi} e^{-ixy\cos(\theta - \varphi) - in\theta} \, d\theta = e^{-in\varphi} \int_0^{2\pi} e^{-ixy\cos\theta - in\theta} \, d\theta, \tag{7.7.5}$$

where we used periodicity to keep the limits of integration unchanged when changing the variable of integration. Since $-\cos\theta = \sin\left(\theta - \frac{1}{2}\pi\right)$, we may change variables once again and conclude that (7.7.5) is

$$(-i)^n e^{-in\varphi} \int_0^{2\pi} e^{ixy\sin\theta - in\theta} \, d\theta = (-i)^n e^{-in\varphi} 2\pi J_n(xy), \qquad (7.7.6)$$

by (7.3.11). Therefore the Fourier transform of $f(x)e^{-in\theta}$ is

$$(-i)^n e^{-in\varphi} g(y), \quad g(y) = \int_0^\infty J_n(xy) f(x) x \, dx.$$

In (7.7.4), therefore, the integral with respect to φ is

$$\int_0^{2\pi} e^{ixy\cos(\theta - \varphi) - in\varphi} \, d\varphi = e^{-in\theta} \int_0^{2\pi} e^{ixy\cos\varphi - in\varphi} \, d\varphi. \qquad (7.7.7)$$

Proceeding as above and using the identity $\cos\varphi = \sin\left(\varphi + \frac{1}{2}\pi\right)$, we find that (7.7.7) is

$$i^n e^{-in\theta} 2\pi J_n(xy).$$

Therefore the right-hand side of (7.7.4) is

$$e^{-in\theta} \int_0^\infty J_n(xy) \, g(y) \, y \, dy.$$

This proves (7.7.2), given (7.7.1).

7.8 Integrals of Bessel functions

We have shown that $J_\nu(x) \sim c_\nu x^\nu$ as $x \to 0+$ and $J_\nu(x) = O\left(x^{-\frac{1}{2}}\right)$ as $x \to \infty$. It follows that if f is any continuous function such that

$$\int_0^1 |f(x)| \, x^\nu \, dx + \int_1^\infty |f(x)| \frac{dx}{\sqrt{x}} < \infty, \qquad (7.8.1)$$

then the product $f \, J_\nu$ is absolutely integrable on $(0, \infty)$. For $\nu > -1$ it is not difficult to show that the integral can be obtained as a series by using the power series expansion (7.1.2) and integrating term by term:

$$\int_0^\infty f(x) J_\nu(x) \, dx = \sum_{m=0}^\infty \frac{(-1)^m}{\Gamma(\nu + m + 1) \, m!} \int_0^\infty f(x) \left(\frac{x}{2}\right)^{\nu + 2m} dx, \qquad (7.8.2)$$

so long as the last series is absolutely convergent. As an example, let

$$f(x) = x^{a-1} e^{-sx^2}, \quad \mathrm{Re}\,(a + \nu) > 0, \quad \mathrm{Re}\,s > 0.$$

Then

$$\int_0^\infty f(x) \left(\frac{x}{2}\right)^{\nu+2m} dx = 2^{-\nu-2m-1} \int_0^\infty e^{-sy} y^{\frac{1}{2}(\nu+a)+m} \frac{dy}{y}$$

$$= \frac{\Gamma\left(\frac{1}{2}\nu + \frac{1}{2}a + m\right)}{2^{\nu+2m+1} s^{\frac{1}{2}(\nu+a)+m}},$$

so

$$\int_0^\infty x^{a-1} e^{-sx^2} J_\nu(x) dx = \frac{1}{2^{\nu+1} s^{\frac{1}{2}(\nu+a)}} \sum_{m=0}^\infty \frac{\Gamma\left(\frac{1}{2}\nu + \frac{1}{2}a + m\right)}{\Gamma(\nu+m+1) m!} \left(-\frac{1}{4s}\right)^m$$

$$= \frac{\Gamma\left(\frac{1}{2}\nu + \frac{1}{2}a\right)}{\Gamma(\nu+1) 2^{\nu+1} s^{\frac{1}{2}(\nu+a)}} \sum_{m=0}^\infty \frac{\left(\frac{1}{2}\nu + \frac{1}{2}a\right)_m}{(\nu+1)_m m!} \left(-\frac{1}{4s}\right)^m.$$

The last sum is the Kummer function M with indices $\frac{1}{2}(\nu+a)$, $\nu+1$ evaluated at $-1/4s$, so for $\mathrm{Re}(\nu+a) > 0$ and $\mathrm{Re}\, s > 0$,

$$\int_0^\infty x^{a-1} e^{-sx^2} J_\nu(x) \, dx$$

$$= \frac{\Gamma\left(\frac{1}{2}\nu + \frac{1}{2}a\right)}{\Gamma(\nu+1) 2^{\nu+1} s^{\frac{1}{2}(\nu+a)}} M\left(\frac{1}{2}\nu + \frac{1}{2}a, \nu+1; -\frac{1}{4s}\right). \qquad (7.8.3)$$

In particular, if $a = \nu+2$ so that $\frac{1}{2}(\nu+a) = \nu+1$, this simplifies to

$$\int_0^\infty x^{\nu+1} e^{-sx^2} J_\nu(x) \, dx = \frac{e^{-1/4s}}{(2s)^{\nu+1}}.$$

As a second example, let

$$f(x) = x^{a-1} e^{-sx}, \quad a+\nu > 0, \quad s > 0,$$

to compute the Laplace transform of $x^{a-1} J_\nu(x)$. Then

$$\int_0^\infty f(x) \left(\frac{x}{2}\right)^{\nu+2m} dx = \frac{1}{2^{\nu+2m}} \int_0^\infty e^{-sx} x^{\nu+a+2m-1} dx$$

$$= \frac{\Gamma(\nu+a+2m)}{2^{\nu+2m} s^{\nu+a+2m}}.$$

Now

$$\Gamma(\nu+a+2m) = \Gamma(\nu+a)(\nu+a)_{2m}$$

$$= \Gamma(\nu+a) 2^{2m} \left(\frac{1}{2}\nu + \frac{1}{2}a\right)_m \left(\frac{1}{2}\nu + \frac{1}{2}a + \frac{1}{2}\right)_m.$$

Therefore

$$\int_0^\infty x^{a-1} e^{-sx} J_\nu(x)\, dx$$

$$= \frac{\Gamma(\nu+a)}{\Gamma(\nu+1)\, 2^\nu s^{\nu+a}} \sum_{m=0}^\infty \frac{\left(\frac{1}{2}\nu + \frac{1}{2}a\right)_m \left(\frac{1}{2}\nu + \frac{1}{2}a + \frac{1}{2}\right)_m}{(\nu+1)_m\, m!} \left(-\frac{1}{s^2}\right)^m.$$

The last sum converges for $s > 1$ to the hypergeometric function with indices $\frac{1}{2}(\nu+a)$, $\frac{1}{2}(\nu+a+1)$, $\nu+1$, evaluated at $-1/s^2$. By analytic continuation, the following identity holds for all s, ν, and a with $\mathrm{Re}\, s > 0$, $\mathrm{Re}\,(\nu+a) > 0$:

$$\int_0^\infty x^{a-1} e^{-sx} J_\nu(x)\, dx$$

$$= \frac{\Gamma(\nu+a)}{\Gamma(\nu+1)\, 2^\nu s^{\nu+a}}\, F\left(\frac{1}{2}\nu + \frac{1}{2}a, \frac{1}{2}\nu + \frac{1}{2}a + \frac{1}{2}, \nu+1; -\frac{1}{s^2}\right).$$

Corresponding to various cases (8.7.2), (8.7.5), and (8.7.6) in Section 8.7, we obtain

$$\int_0^\infty e^{-xs}\, x^\nu\, J_\nu(x)\, dx = \frac{\Gamma(2\nu+1)}{\Gamma(\nu+1)\, 2^\nu} \cdot \frac{1}{(1+s^2)^{\nu+\frac{1}{2}}}; \qquad (7.8.4)$$

$$\int_0^\infty e^{-xs}\, x^{\nu+1}\, J_\nu(x)\, dx = \frac{\Gamma(2\nu+2)}{\Gamma(\nu+1)\, 2^\nu} \cdot \frac{s}{(1+s^2)^{\nu+\frac{3}{2}}}; \qquad (7.8.5)$$

$$\int_0^\infty e^{-xs}\, x^{-1}\, J_\nu(x)\, dx = \frac{1}{\nu \left(s + \sqrt{1+s^2}\right)^\nu}; \qquad (7.8.6)$$

$$\int_0^\infty e^{-xs}\, J_\nu(x)\, dx = \frac{1}{\sqrt{1+s^2}\left(s + \sqrt{1+s^2}\right)^\nu}. \qquad (7.8.7)$$

7.9 Airy functions

If v is a solution of Bessel's equation (7.0.1) and u is defined by

$$u(x) = x^a\, v(bx^c)$$

where a, b and c are constants, then u is a solution of *Lommel's equation*

$$x^2 u''(x) + (1-2a)x\, u'(x) + \left[b^2 c^2 x^{2c} + a^2 - c^2 \nu^2\right] u(x) = 0. \qquad (7.9.1)$$

If v is, instead, a solution of the modified Bessel equation (7.5.1), then u is a solution of the corresponding modified Lommel equation

$$x^2 u''(x) + (1 - 2a)x\, u'(x) - \left[b^2 c^2 x^{2c} - a^2 + c^2 v^2 \right] u(x) = 0. \qquad (7.9.2)$$

The particular case of (7.9.2) with $v^2 = \frac{1}{9}$, $a = \frac{1}{2}$, $b = \frac{2}{3}$, $c = \frac{3}{2}$ gives the *Airy equation*

$$u''(x) - x\, u(x) = 0. \qquad (7.9.3)$$

The calculation is reversible: any solution of (7.9.3) for $x > 0$ has the form

$$u(x) = x^{\frac{1}{2}}\, v\left(\frac{2}{3} x^{\frac{3}{2}} \right),$$

where v is a solution of the modified Bessel equation. The standard choices are the *Airy functions*

$$\begin{aligned} \mathrm{Ai}\,(x) &= \frac{\sqrt{x}}{\pi \sqrt{3}}\, K_{\frac{1}{3}}\left(\frac{2}{3} x^{\frac{3}{2}} \right) \\ &= \frac{\sqrt{x}}{3}\left[I_{-\frac{1}{3}}\left(\frac{2}{3} x^{\frac{3}{2}} \right) - I_{\frac{1}{3}}\left(\frac{2}{3} x^{\frac{3}{2}} \right) \right], \end{aligned} \qquad (7.9.4)$$

and

$$\mathrm{Bi}\,(x) = \frac{\sqrt{x}}{\sqrt{3}}\left[I_{-\frac{1}{3}}\left(\frac{2}{3} x^{\frac{3}{2}} \right) + I_{\frac{1}{3}}\left(\frac{2}{3} x^{\frac{3}{2}} \right) \right]. \qquad (7.9.5)$$

It follows from (7.5.2) that the series expansions of the Airy functions are

$$\mathrm{Ai}\,(x) = \sum_{n=0}^{\infty} \left[\frac{x^{3n}}{3^{2n+\frac{2}{3}} \Gamma\left(n + \frac{2}{3}\right) n!} - \frac{x^{3n+1}}{3^{2n+\frac{4}{3}} \Gamma\left(n + \frac{4}{3}\right) n!} \right]; \qquad (7.9.6)$$

$$\mathrm{Bi}\,(x) = \sqrt{3} \sum_{n=0}^{\infty} \left[\frac{x^{3n}}{3^{2n+\frac{2}{3}} \Gamma\left(n + \frac{2}{3}\right) n!} + \frac{x^{3n+1}}{3^{2n+\frac{4}{3}} \Gamma\left(n + \frac{4}{3}\right) n!} \right],$$

which show that these are entire functions of x. It follows from these expansions that the initial conditions are

$$\mathrm{Ai}\,(0) = \frac{1}{3^{\frac{2}{3}} \Gamma\left(\frac{2}{3}\right)}, \qquad \mathrm{Ai}\,'(0) = -\frac{1}{3^{\frac{4}{3}} \Gamma\left(\frac{4}{3}\right)};$$

$$\mathrm{Bi}\,(0) = \frac{1}{3^{\frac{1}{6}} \Gamma\left(\frac{2}{3}\right)}, \qquad \mathrm{Bi}\,'(0) = \frac{1}{3^{\frac{5}{6}} \Gamma\left(\frac{4}{3}\right)}.$$

The Wronskian is constant, so the constant is

$$W(\text{Ai}, \text{Bi})(x) = \frac{2}{3^{\frac{3}{2}} \Gamma\left(\frac{2}{3}\right) \Gamma\left(\frac{4}{3}\right)}.$$

The asymptotics of the Airy functions as $x \to +\infty$ follow from the asymptotics of the modified Bessel functions. The leading terms are

$$\text{Ai}(x) \sim \frac{e^{-\frac{2}{3}x^{\frac{3}{2}}}}{2\sqrt{\pi}\, x^{\frac{1}{4}}}; \quad \text{Bi}(x) \sim \frac{e^{\frac{2}{3}x^{\frac{3}{2}}}}{\sqrt{\pi}\, x^{\frac{1}{4}}}. \tag{7.9.7}$$

The principal terms in the asymptotics of the derivatives are obtained by differentiating (7.9.7). This gives a second determination of the Wronskian:

$$W(\text{Ai}, \text{Bi})(x) = \frac{1}{\pi}.$$

The asymptotics of the Airy functions for $x \to -\infty$ can be obtained from the asymptotics of the Bessel functions J_ν. Replacing x by $-x$ in the series expansions shows that

$$\text{Ai}(-x) = \frac{\sqrt{x}}{3}\left[J_{-\frac{1}{3}}\left(\frac{2}{3}x^{\frac{3}{2}}\right) + J_{\frac{1}{3}}\left(\frac{2}{3}x^{\frac{3}{2}}\right)\right];$$

$$\text{Bi}(-x) = \frac{\sqrt{x}}{\sqrt{3}}\left[J_{-\frac{1}{3}}\left(\frac{2}{3}x^{\frac{3}{2}}\right) - J_{\frac{1}{3}}\left(\frac{2}{3}x^{\frac{3}{2}}\right)\right].$$

It follows from (7.4.8) that as $x \to +\infty$,

$$\text{Ai}(-x) \sim \frac{\cos\left(\frac{2}{3}x^{\frac{3}{2}} - \frac{1}{4}\pi\right)}{\sqrt{\pi}\, x^{\frac{1}{4}}}; \quad \text{Bi}(-x) \sim -\frac{\sin\left(\frac{2}{3}x^{\frac{3}{2}} - \frac{1}{4}\pi\right)}{\sqrt{\pi}\, x^{\frac{1}{4}}}.$$

The original function that arose in Airy's research on optics was defined by the integral

$$\frac{1}{\pi}\int_0^\infty \cos\left(\frac{1}{3}t^3 + xt\right) dt. \tag{7.9.8}$$

To see that this integral is equal to the function (7.9.6), we first make a change of variable $t = \tau/i$ so that the integral becomes

$$\frac{1}{\pi i}\int_0^{i\infty} \cosh\left(\frac{1}{3}\tau^3 - x\tau\right) d\tau = \frac{1}{2\pi i}\int_{-i\infty}^{i\infty} \exp\left(\frac{1}{3}\tau^3 - x\tau\right) d\tau. \tag{7.9.9}$$

By Cauchy's theorem, the vertical line of integration can be deformed into a contour C which begins at infinity in the sector $-\pi/2 < \arg \tau < -\pi/6$ and ends at infinity in the sector $\pi/6 < \arg\tau < \pi/2$. So far we have restricted x to be real, but the function

$$\mathrm{Ai}(z) = \frac{1}{2\pi i} \int_C \exp\left(\frac{1}{3}\tau^3 - z\tau\right) d\tau \qquad (7.9.10)$$

is an entire function in z, since the integrand vanishes rapidly at the endpoints of C. Furthermore, by differentiation under the integral sign, one can readily see that this integral satisfies the Airy equation (7.9.3). From (7.9.10), it can be verified that the integral in (7.9.8) has the Maclaurin expansion given in (7.9.6).

Let $\omega = \exp(2\pi i/3)$. In addition to $\mathrm{Ai}(z)$, the functions $\mathrm{Ai}(\omega z)$ and $\mathrm{Ai}(\omega^2 z)$ are also solutions of equation (7.9.3). With the aid of Cauchy's theorem, one can use (7.9.10) to show that these three solutions are connected by the relation

$$\mathrm{Ai}(z) + \omega\mathrm{Ai}(wz) + \omega^2\mathrm{Ai}(\omega^2 z) = 0. \qquad (7.9.11)$$

Returning to (7.9.9), we restrict x to be positive:

$$\mathrm{Ai}(x) = \frac{1}{2\pi i} \int_{-i\infty}^{i\infty} \exp\left(\frac{1}{3}\tau^3 - x\tau\right) d\tau.$$

By deforming the imaginary axis into the parallel vertical line $\mathrm{Re}\,\tau = \sqrt{x}$, one can show that

$$\mathrm{Ai}(x) = \frac{1}{2\pi} e^{-\frac{2}{3}x^{3/2}} \int_{-\infty}^{\infty} e^{-\sqrt{x}\rho^2 - \frac{1}{3}i\rho^3} \, d\rho$$

$$= \frac{1}{\pi} e^{-\frac{2}{3}x^{3/2}} \int_0^{\infty} e^{-\sqrt{x}\rho^2} \cos\left(\frac{1}{3}\rho^3\right) d\rho. \qquad (7.9.12)$$

By analytic continuation, this holds for $|\arg x| < \pi$. Let us now replace x by z, and make the change of variable $\rho^2 = u$ so that (7.9.12) becomes

$$\mathrm{Ai}(z) = \frac{1}{2\pi} e^{-\frac{2}{3}z^{3/2}} \int_0^{\infty} e^{-\sqrt{z}u} \cos\left(\frac{1}{3}u^{\frac{3}{2}}\right) \frac{du}{\sqrt{u}}, \qquad (7.9.13)$$

valid for $|\arg z| < \pi$. By expanding the cosine function into a Maclaurin series and integrating term by term, we obtain a sharper version of (7.9.7): the asymptotic expansion

$$\mathrm{Ai}(z) \sim \frac{1}{2\pi z^{\frac{1}{4}}} e^{-\frac{2}{3}z^{\frac{3}{2}}} \sum_{n=0}^{\infty} \frac{\Gamma\left(3n + \frac{1}{2}\right)}{3^{2n}(2n)!} \frac{(-1)^n}{z^{\frac{3}{2}n}}.$$

as $z \to \infty$ in $|\arg z| < \pi$.

7.10 Exercises

7.1 Show that

$$\lim_{a \to +\infty} M\left(a, \nu + 1; -\frac{x}{a}\right) = \Gamma(\nu + 1) x^{-\frac{1}{2}\nu} J_\nu\left(2\sqrt{x}\right).$$

7.2 Show that the Fourier transform of the restriction to the interval $-1 < x < 1$ of the Legendre polynomial P_n,

$$\frac{1}{\sqrt{2\pi}} \int_{-1}^{1} e^{-ix\xi} P_n(x)\, dx,$$

is

$$\frac{(-i)^n}{\sqrt{\xi}} J_{n+\frac{1}{2}}(\xi).$$

Hint: use the Rodrigues formula.

7.3 Use Exercise 7.2 and the Fourier inversion formula to show that

$$\frac{1}{\sqrt{2\pi}} \int_{-\infty}^{\infty} e^{ix\xi} J_{n+\frac{1}{2}}(\xi) \frac{d\xi}{\sqrt{\xi}} = \begin{cases} i^n P_n(x), & |x| < 1 \\ 0, & |x| > 1. \end{cases}$$

7.4 Use Exercise 7.2 to show that the expansion of the plane wave $e^{i\kappa x}$ as a sum of Legendre polynomials is

$$e^{i\kappa x} = \sum_{n=0}^{\infty} i^n \left(n + \frac{1}{2}\right) \frac{\sqrt{2\pi}}{\sqrt{\kappa}} J_{n+\frac{1}{2}}(\kappa) P_n(x).$$

7.5 Use Exercise 7.4 and the orthogonality properties of the Legendre polynomials to prove the integral formula

$$J_{n+\frac{1}{2}}(\kappa) = (-i)^n \frac{\sqrt{\kappa}}{\sqrt{2\pi}} \int_{-1}^{1} e^{i\kappa x} P_n(x)\, dx.$$

7.6 Show that the expansion of the plane wave $e^{i\kappa x}$ in terms of Gegenbauer polynomials is

$$e^{i\kappa x} = \Gamma(\lambda) \sum_{n=0}^{\infty} i^n (n + \lambda) \left(\frac{\kappa}{2}\right)^{-\lambda} J_{\lambda+n}(\kappa) C_n^\lambda(x).$$

7.7 Use the orthogonality property of the Gegenbauer polynomials to derive an integral formula for $J_{\lambda+n}(\kappa)$ involving $C_n^\lambda(x)$.

7.8 Prove (7.1.16).

7.9 Verify the following relations:

$$J_{\nu+1}(x)J_{-\nu}(x) + J_\nu(x)J_{-(\nu+1)}(x) = -\frac{2\sin\nu\pi}{\pi x};$$

$$J_{\nu+1}(x)Y_\nu(x) - J_\nu(x)Y_{\nu+1}(x) = \frac{2}{\pi x};$$

$$I_\nu(x)K_{\nu+1}(x) + I_{\nu+1}(x)K_\nu(x) = \frac{1}{x}.$$

7.10 Prove that the positive zeros of any two linearly independent real cylinder functions of the same order are interlaced.

7.11 Let $y_{\nu,n}$ denote the nth positive zero of $Y_\nu(x)$. Prove that when $\nu > -\frac{1}{2}$,

$$y_{\nu,1} < j_{\nu,1} < y_{\nu,2} < j_{\nu,2} < \ldots$$

7.12 Assume that for fixed n, we know that $j_{\nu,n}$ is a differentiable function of ν in $(-1, \infty)$.

(a) Differentiate the equation $J_\nu(j_{\nu,n}) = 0$ to get

$$J_\nu'(j_{\nu,n})\frac{dj_{\nu,n}}{d\nu} + \left[\frac{\partial J_\nu(x)}{\partial \nu}\right]_{x=j_{\nu,n}} = 0.$$

(b) Verify by differentiation

$$\int_0^x \frac{J_\mu(y)J_\nu(y)}{y}\,dy = \frac{x\{J_\mu'(x)J_\nu(x) - J_\mu(x)J_\nu'(x)\}}{\mu^2 - \nu^2}, \quad \mu^2 \ne \nu^2.$$

(c) Letting $\mu \to \nu$, show that for $\nu > 0$,

$$\int_0^{j_{\nu n}} \frac{J_\nu^2(x)}{x}\,dx = -\frac{j_{\nu,n}}{2\nu}J_\nu'(j_{\nu,n})\left[\frac{\partial J_\nu(x)}{\partial \nu}\right]_{x=j_{\nu,n}}.$$

(d) Establish the representation

$$\frac{dj_{\nu,n}}{d\nu} = \frac{2\nu}{j_{\nu,n}\{J_\nu'(j_{\nu,n})\}^2}\int_0^{j_{\nu,n}} \frac{J_\nu^2(x)}{x}\,dx, \quad \nu > 0,$$

which shows that when $\nu > 0$, $j_{\nu,n}$ is an increasing function of ν.

7.13 Prove the first statement of Theorem 7.2.2 by adapting the method of Proposition 3.5.1.

7.14 Prove the second statement of Theorem 7.2.2.

7.15 Show that the sequence (7.2.6) has a limit and determine it, assuming that u has the form (7.2.1).

7.16 Given an index $\nu \ge -\frac{1}{2}$ and a constant $\lambda > 0$, define $f_\lambda(x) = J_\nu(\lambda x)$, $x > 0$.

(a) Show that $x^2 f_\lambda'' + x f_\lambda' + (\lambda^2 x^2 - v^2) f_\lambda = 0$.

(b) Let $W(f_\lambda, f_\mu)$ be the Wronskian. Show that

$$\left[x \, W(x) \right]' = \left(\lambda^2 - \mu^2 \right) x \, f_\lambda(x) f_\mu(x),$$

and deduce that if $J_v(\lambda) = 0 = J_v(\mu)$ and $\lambda \neq \mu$, then

$$\int_0^1 x \, J_v(\lambda x) \, J_v(\mu x) \, dx = 0.$$

(c) Suppose that $J_v(\lambda) = 0$. Show that

$$\int_0^1 x \, J_v(\lambda x)^2 \, dx = \lim_{\mu \to \lambda} \int_0^1 x \, J_v(\lambda x) \, J_v(\mu x) \, dx = \frac{1}{2} J_v'(\lambda)^2.$$

(d) Use (7.1.3) to show that $J_v(\lambda) = 0$ implies $J_v'(\lambda) = -J_{v+1}(\lambda)$.

(e) Suppose that

$$f(x) = \sum_{k=0}^\infty a_k \, J_v(\lambda_k x), \quad 0 < x < 1, \tag{7.10.1}$$

where the $\{\lambda_k\}$ are the positive zeros of J_v numbered in increasing order. Assume that the series converges uniformly. Show that

$$a_k = \frac{2}{J_{v+1}(\lambda_k)^2} \int_0^1 x \, f(x) \, J_v(\lambda_k x) \, dx. \tag{7.10.2}$$

The expansion (7.10.1), (7.10.2) is called the *Fourier–Bessel expansion* of the function f. In particular, the Fourier–Bessel expansion of a function f converges to $f(x), 0 < x < 1$, if f is differentiable for $0 < x < 1$ and

$$\int_0^1 x^{\frac{1}{2}} |f(x)| \, dx < \infty;$$

see Watson [306], chapter 18.

7.17 Show that

$$|J_n(x)| \leq 1, \quad x \geq 0, \quad n = 0, 1, 2, \ldots$$

7.18 Show that

$$W\big(H_v^{(1)}, H_v^{(2)}\big)(x) = -\frac{4i}{\pi x}; \quad W\big(I_v, I_{-v}\big)(x) = -\frac{2 \sin v\pi}{\pi x}.$$

7.19 Deduce from the generating function for the Bessel functions of integer order that

$$\sin x = 2 \sum_{n=0}^{\infty} (-1)^n J_{2n+1}(x);$$

$$\cos x = J_0(z) + 2 \sum_{n=1}^{\infty} (-1)^n J_{2n}(x);$$

$$x \cos x = 2 \sum_{n=1}^{\infty} (-1)^{n+1} (2n-1)^2 J_{2n-1}(x).$$

7.20 Deduce from (7.5.5) that

$$K_\nu(x) = \frac{\sqrt{\pi} \left(\frac{1}{2}x\right)^\nu}{\Gamma\left(\nu + \frac{1}{2}\right)} \int_1^{\infty} e^{-xt} (t^2 - 1)^{\nu - \frac{1}{2}} dt, \quad \text{Re } \nu > -\frac{1}{2}.$$

Use this formula and the beta function integral to show that

$$\int_0^{\infty} t^{\mu-1} K_\nu(t) dt = 2^{\mu-2} \Gamma\left(\frac{\mu+\nu}{2}\right) \Gamma\left(\frac{\mu-\nu}{2}\right), \quad \text{Re } \mu > |\text{Re } \nu|.$$

7.21 Show that

$$e^{x \cos t} = \sum_{n=-\infty}^{\infty} (\cos nt) I_n(x) = I_0(x) + 2 \sum_{n=1}^{\infty} (\cos nt) I_n(x).$$

7.22 Show that

$$\int_0^x \cos(x-t) J_0(t) dt = x J_0(x).$$

7.23 Deduce from (7.1.6) that

$$2^m \frac{d^m}{dx^m} J_n(x) = \sum_{k=0}^{m} (-1)^{m-k} \binom{m}{k} J_{n+m-2k}(x).$$

7.24 Show that

$$\int_0^{\infty} e^{-ax} J_0(bx) \, dx = \frac{1}{\sqrt{a^2 + b^2}}, \quad a > 0, \, b > 0.$$

7.25 Show that

$$\int_0^{\infty} e^{-a^2 x^2} J_\nu(bx) x^{\nu+1} dx = \frac{b^\nu}{(2a^2)^{\nu+1}} e^{-b^2/4a^2},$$

$$a > 0, \, b > 0, \quad \text{Re } \nu > -1.$$

7.26 Using Exercise 7.22, show that for $x > 0$ and $|\text{Re } \nu| < \frac{1}{2}$,

$$J_\nu(x) = \frac{2\left(\frac{1}{2}x\right)^{-\nu}}{\sqrt{\pi}\,\Gamma\left(\frac{1}{2} - \nu\right)} \int_1^\infty \frac{\sin xt}{(t^2 - 1)^{\nu + \frac{1}{2}}} dt;$$

$$Y_\nu(x) = -\frac{2\left(\frac{1}{2}x\right)^{-\nu}}{\sqrt{\pi}\,\Gamma\left(\frac{1}{2} - \nu\right)} \int_1^\infty \frac{\cos xt}{(t^2 - 1)^{\nu + \frac{1}{2}}} dt.$$

7.27 Derive from (7.1.3) the recurrence relation

$$\int_0^x t^\mu J_\nu(t)dt = x^\mu J_{\nu+1}(x) - (\mu - \nu - 1) \int_0^x t^{\mu-1} J_{\nu+1}(t)dt,$$

$$\text{Re } (\mu + \nu) > -1.$$

7.28 Show that

$$\int_0^x J_\nu(t)dt = 2\sum_{n=0}^\infty J_{\nu+2n+1}(x), \quad \text{Re } \nu > -1.$$

7.29 Verify *Sonine's first finite integral formula* [265]: for $\text{Re } \mu > -1$ and $\text{Re } \nu > -1$,

$$\int_0^{\frac{\pi}{2}} J_\mu(x \sin \theta) \sin^{\mu+1} \theta \cos^{2\nu+1} \theta d\theta = \frac{2^\nu \Gamma(\nu + 1)}{x^{\nu+1}} J_{\mu+\nu+1}(x).$$

7.30 Verify the identities (7.8.4).

7.31 Show that $w(x) = [\text{Ai}\,(x)]^2$ satisfies the third-order equation

$$w^{(3)} - 4xw' - 2w = 0.$$

7.32 Show that the solutions of the differential equation

$$x^4 w^{(4)} + 2x^3 w^{(3)} - (1 + 2\nu^2)(x^2 w'' - xw') + (\nu^4 - 4\nu^2 + x^4)w = 0$$

are the *Kelvin functions* $\text{ber}_\nu(x)$, $\text{bei}_\nu(x)$, $\text{ber}_{-\nu}(x)$ and $\text{bei}_{-\nu}(x)$, defined by

$$\text{ber}_\nu(x) \pm i\,\text{bei}_\nu(x) = J_\nu\left(xe^{\pm 3\pi i/4}\right) = e^{\pm \nu \pi i/2} I_\nu\left(xe^{\pm \pi i/4}\right).$$

(These functions were introduced by Kelvin [153] for $\nu = 0$ and by Russell [243] and Whitehead [313] for general ν and other types. [Not *the* Russell and Whitehead, however.])

7.33 Show that

$$\int_0^\infty \text{Ai}(t)\, t^{\alpha-1}\, dt = \frac{\Gamma(\alpha)}{3^{(\alpha+2)/3}\Gamma\left(\frac{1}{3}\alpha + \frac{2}{3}\right)}.$$

7.34 Prove that for $0 < t < \infty$,

$$0 \leq \mathrm{Ai}(t) \leq \frac{1}{2\sqrt{\pi}}\, t^{-1/4} \exp\left(-\frac{2}{3}t^{3/2}\right).$$

7.35 Use Gauss's formula (2.3.6) to show that the two determinations of the Wronskian of the Airy functions Ai, Bi are the same.

7.11 Summary

7.11.1 Bessel functions

A cylinder function is a solution of Bessel's equation

$$x^2 u''(x) + x u'(x) + (x^2 - v^2) u(x) = 0.$$

One solution is the Bessel function

$$J_v(x) = \sum_{m=0}^{\infty} \frac{(-1)^m}{\Gamma(v+1+m)\, m!} \left(\frac{x}{2}\right)^{v+2m},$$

holomorphic on the complement of $(-\infty, 0]$. The series expansion implies

$$[x^v J_v]' = x^v J_{v-1}, \quad [x^{-v} J_v]' = -x^{-v} J_{v+1},$$

so

$$\left(\frac{1}{x}\frac{d}{dx}\right)^n \left[x^v J_v(x)\right] = x^{v-n} J_{v-n}(x);$$

$$\left(\frac{1}{x}\frac{d}{dx}\right)^n \left[x^{-v} J_v(x)\right] = (-1)^n x^{-v-n} J_{v+n}(x),$$

and

$$J_{v-1}(x) + J_{v+1}(x) = \frac{2v}{x} J_v(x);$$

$$J_{v-1}(x) - J_{v+1}(x) = 2\, J_v'(x).$$

For $v = \pm\frac{1}{2}$,

$$J_{\frac{1}{2}}(x) = \frac{\sqrt{2}\sin x}{\sqrt{\pi x}}; \quad J_{-\frac{1}{2}}(x) = \frac{\sqrt{2}\cos x}{\sqrt{\pi x}}.$$

The derivative relation implies that $J_v(x)$ is expressible in terms of trigonometric functions and powers of x whenever $v + \frac{1}{2}$ is an integer.

Wronskian:

$$W(J_\nu, J_{-\nu})(x) = -\frac{2 \sin \nu\pi}{\pi x}.$$

For integer values of the parameter

$$J_{-n}(x) = (-1)^n J_n(x).$$

Bessel function of the second kind:

$$Y_\nu(x) = \frac{\cos \nu\pi \, J_\nu(x) - J_{-\nu}(x)}{\sin \nu\pi}.$$

In particular,

$$Y_{\frac{1}{2}}(x) = -\frac{\sqrt{2}\cos x}{\sqrt{\pi x}}; \quad Y_{-\frac{1}{2}}(x) = \frac{\sqrt{2}\sin x}{\sqrt{\pi x}}.$$

Wronskian:

$$W(J_\nu, Y_\nu)(x) = -\frac{W(J_\nu, J_{-\nu})}{\sin \nu\pi} = \frac{2}{\pi x}.$$

For integer values

$$Y_{-n}(x) = (-1)^n Y_n(x), \quad n = 0, 1, 2, \ldots$$

The derivative and recurrence identities for the J_ν imply

$$Y_{\nu-1}(x) + Y_{\nu+1}(x) = \frac{2\nu}{x} Y_\nu(x);$$

$$Y_{\nu-1}(x) - Y_{\nu+1}(x) = 2 Y_\nu'(x).$$

The series expansion leads to

$$J_\nu\left(\sqrt{\nu}\, y\right) \sim \frac{\left(\sqrt{\nu}\, y\right)^\nu}{2^\nu \Gamma(\nu + 1)} e^{-\frac{1}{4}y^2}$$

as $\nu \to +\infty$, uniformly on bounded intervals.

7.11.2 Zeros of real cylinder functions

A real nonzero cylinder function $u(x)$ has a countable number of positive zeros

$$0 < x_1 < x_2 < \ldots < x_n < \ldots$$

The distance $x_{n+1} - x_n$ is $\geq \pi$ if $|\nu| \geq \frac{1}{2}$, and $\leq \pi$ if $|\nu| \leq \frac{1}{2}$. As $n \to \infty$,

$$x_{n+1} - x_n = \pi + O(n^{-2}).$$

If u is a nonzero real cylinder function and the zeros of the derivative u' in (ν, ∞) are

$$y_1 < y_2 < \ldots < y_n < \ldots,$$

then

$$|u(y_1)| > |u(y_2)| > \ldots > |u(y_n)| > \ldots,$$

and

$$\left(y_1^2 - \nu^2\right)^{\frac{1}{4}} |u(y_1)| < \left(y_2^2 - \nu^2\right)^{\frac{1}{4}} |u(y_2)| < \ldots < \left(y_n^2 - \nu^2\right)^{\frac{1}{4}} |u(y_n)| < \ldots$$

If $\nu > 0$, there is no zero of J_ν' in the interval $(0, \nu]$, so the previous inequalities apply to all local extrema of J_ν.

Denote the positive zeros of J_ν by

$$0 < j_{\nu,1} < j_{\nu,2} < \ldots < j_{\nu,n} < \ldots$$

The zeros of J_ν and $J_{\nu+1}$ interlace:

$$0 < j_{\nu,1} < j_{\nu+1,1} < j_{\nu,2} < j_{\nu+1,2} < j_{\nu,3} < \ldots$$

All zeros of $J_\nu(z)$ are real when $\nu > -1$, and all zeros of $J_\nu'(z)$ are real when $\nu \geq 0$.

7.11.3 Integral representations

The gauge transformation $u(x) = x^\nu e^{ix} v(x)$ and the change of variables $y = -2ix$ convert Bessel's equation to

$$y\, w'' + (2\nu + 1 - y)\, w' - \left(\nu + \tfrac{1}{2}\right) w = 0.$$

This leads to the identity

$$J_\nu(x) = \frac{1}{\Gamma(\nu + 1)} \left(\frac{x}{2}\right)^\nu e^{ix} M\left(\nu + \frac{1}{2}, 2\nu + 1; -2ix\right)$$

and to the integral representations

$$J_\nu(x) = \frac{(2x)^\nu}{\sqrt{\pi}\, \Gamma\left(\nu + \frac{1}{2}\right)} e^{ix} \int_0^1 e^{-2ixs} s^{\nu - \frac{1}{2}} (1 - s)^{\nu - \frac{1}{2}}\, ds$$

$$= \frac{1}{\sqrt{\pi}\, \Gamma\left(\nu + \frac{1}{2}\right)} \left(\frac{x}{2}\right)^\nu \int_{-1}^1 \cos xt\, (1 - t^2)^{\nu - \frac{1}{2}}\, dt, \quad \mathrm{Re}\left(\nu + \frac{1}{2}\right) > 0.$$

Analysis of the Helmholtz equation in two variables leads to Bessel's representation

$$J_n(x) = \frac{1}{2\pi} \int_0^{2\pi} e^{ix\sin\theta - in\theta} \, d\theta, \quad n = 0, \pm 1, \pm 2, \ldots,$$

the identity

$$e^{ix\sin\theta} = \sum_{n=-\infty}^{\infty} J_n(x) e^{in\theta},$$

and the representations

$$J_\nu(x) = \frac{1}{2\pi i} \int_{C_1} \exp\left(\tfrac{1}{2}x[t - 1/t]\right) t^{-\nu-1} \, dt$$

$$= \frac{1}{2\pi} \int_{C_2} e^{ix\sin\theta - i\nu\theta} \, d\theta,$$

for arbitrary real ν, where C_1 is a curve beginning and ending at $-\infty$ and enclosing the origin and C_2 encloses the strip $\{-\pi < \mathrm{Re}\, z < \pi, \, \mathrm{Im}\, z > 0\}$.

7.11.4 Hankel functions

The considerations in Section 4.9 motivate the Hankel functions, which are the solutions of the Bessel equation given by the Poisson integral representations

$$H_\nu^{(1)}(x) = 2c_\nu x^{-\nu} e^{i\left(x - \frac{1}{4}\pi - \frac{1}{2}\nu\pi\right)} \int_0^\infty e^{-2s} [s(x+is)]^{\nu-\frac{1}{2}} \, ds;$$

$$H_\nu^{(2)}(x) = 2c_\nu x^{-\nu} e^{-i\left(x - \frac{1}{4}\pi - \frac{1}{2}\nu\pi\right)} \int_0^\infty e^{-2s} [s(x-is)]^{\nu-\frac{1}{2}} \, ds,$$

for $\mathrm{Re}\,(2\nu + 1) > 0$, where

$$c_\nu = \frac{2^\nu}{\sqrt{\pi}\, \Gamma\left(\nu + \frac{1}{2}\right)}.$$

These are related to the Bessel functions of the first and second kind by

$$J_\nu(x) = \frac{1}{2}\left[H_\nu^{(1)}(x) + H_\nu^{(2)}(x)\right];$$

$$Y_\nu(x) = \frac{1}{2i}\left[H_\nu^{(1)}(x) - H_\nu^{(2)}(x)\right];$$

$$H_\nu^{(1)}(x) = J_\nu(x) + iY_\nu(x);$$

$$H_\nu^{(2)}(x) = J_\nu(x) - iY_\nu(x).$$

It follows that the Wronskian is

$$W\big(H_\nu^{(1)}, H_\nu^{(2)}\big)(x) = \frac{4}{\pi i x}.$$

The Poisson integral representation leads to asymptotic expansions

$$H_\nu^{(1)}(x) \sim \frac{\sqrt{2}}{\sqrt{\pi x}} e^{i\left(x - \frac{1}{4}\pi - \frac{1}{2}\nu\pi\right)}$$

$$\times \sum_{m=0}^{\infty} \frac{(-i)^m \left(-\nu + \frac{1}{2}\right)_m \left(\nu + \frac{1}{2}\right)_m}{2^m m!} x^{-m};$$

$$H_\nu^{(2)}(x) \sim \frac{\sqrt{2}}{\sqrt{\pi x}} e^{-i\left(x - \frac{1}{4}\pi - \frac{1}{2}\nu\pi\right)}$$

$$\times \sum_{m=0}^{\infty} \frac{i^m \left(-\nu + \frac{1}{2}\right)_m \left(\nu + \frac{1}{2}\right)_m}{2^m m!} x^{-m}.$$

There are corresponding expansions for J_ν and Y_ν, with leading terms

$$J_\nu(x) \sim \frac{\sqrt{2}}{\sqrt{\pi x}} \cos\left(x - \frac{1}{4}\pi - \frac{1}{2}\nu\pi\right);$$

$$Y_\nu(x) \sim \frac{\sqrt{2}}{\sqrt{\pi x}} \sin\left(x - \frac{1}{4}\pi - \frac{1}{2}\nu\pi\right).$$

Particular values:

$$H_{\frac{1}{2}}^{(1)}(x) = -i\frac{\sqrt{2}}{\sqrt{\pi x}} e^{ix}; \quad H_{-\frac{1}{2}}^{(1)}(x) = \frac{\sqrt{2}}{\sqrt{\pi x}} e^{ix};$$

$$H_{\frac{1}{2}}^{(2)}(x) = i\frac{\sqrt{2}}{\sqrt{\pi x}} e^{-ix}; \quad H_{-\frac{1}{2}}^{(2)}(x) = \frac{\sqrt{2}}{\sqrt{\pi x}} e^{-ix}.$$

The derivative and recurrence relations for J_ν and Y_ν imply that

$$H_{\nu-1}^{(1)}(x) + H_{\nu+1}^{(1)}(x) = \frac{2\nu}{x} H_\nu^{(1)}(x);$$

$$H_{\nu-1}^{(2)}(x) - H_{\nu+1}^{(2)}(x) = 2\big[H_\nu^{(2)}\big]'(x).$$

Moreover,

$$H_{-\nu}^{(1)}(x) = e^{i\pi\nu} H_\nu^{(1)}(x);$$

$$H_{-\nu}^{(2)}(x) = e^{-i\pi\nu} H_\nu^{(2)}(x).$$

7.11.5 Modified Bessel functions

Solutions of the modified Bessel equation

$$x^2 u''(x) + x u'(x) - (x^2 + v^2) u(x) = 0$$

can be obtained by evaluating solutions of Bessel's equation on the positive imaginary axis:

$$I_v(x) = \sum_{m=0}^{\infty} \frac{1}{\Gamma(v+1+m)\, m!} \left(\frac{x}{2}\right)^{v+2m} = e^{-\frac{1}{2}iv\pi} J_v(ix)$$

$$= \frac{1}{\sqrt{\pi}\, \Gamma\left(v+\frac{1}{2}\right)} \left(\frac{x}{2}\right)^v \int_{-1}^{1} \cosh xt\, (1-t^2)^{v-\frac{1}{2}}\, dt;$$

$$K_v(x) = \frac{\pi}{2} e^{\frac{1}{2}i(v+1)\pi} H_v^{(1)}(ix) = \frac{\pi}{2} \frac{I_{-v}(x) - I_v(x)}{\sin \pi v}$$

$$= \frac{\sqrt{\pi}}{\sqrt{2x}\, \Gamma\left(v+\frac{1}{2}\right)} e^{-x} \int_0^{\infty} e^{-t} \left(t + \frac{t^2}{2x}\right)^{v-\frac{1}{2}} dt.$$

The derivative formula

$$\frac{d}{dx} \left[x^v I_v(x)\right] = x^v I_{v-1}(x)$$

leads to the relations

$$I_{v-1}(x) - I_{v+1}(x) = \frac{2v}{x} I_v(x);$$

$$I_{v-1}(x) + I_{v+1}(x) = 2 I_v'(x);$$

$$K_{v-1}(x) - K_{v+1}(x) = -\frac{2v}{x} K_v(x);$$

$$K_{v-1}(x) + K_{v+1}(x) = -2 K_v'(x).$$

Asymptotics:

$$I_v(x) \sim \frac{e^x}{\sqrt{2\pi x}}; \quad K_v(x) \sim \frac{\sqrt{\pi}\, e^{-x}}{\sqrt{2x}}, \quad x \to +\infty.$$

Full asymptotic expansions may be obtained from the expansion of the Hankel functions.

Wronskian:

$$W(K_v, I_v)(x) = \frac{1}{x}.$$

7.11.6 Addition theorems

The identity

$$e^{ix \sin\theta} = \sum_{n=-\infty}^{\infty} J_n(x) \, e^{in\theta}$$

can be written as a generating function:

$$G(x,t) = \sum_{n=-\infty}^{\infty} J_n(x) \, t^n = e^{\frac{1}{2}x(t-1/t)}, \quad |t| = 1.$$

Addition formula:

$$J_n(x+y) = \sum_{m=-\infty}^{\infty} J_m(x) \, J_{n-m}(y).$$

Graf's formula:

$$J_\nu(r) \, e^{i\nu\varphi} = \sum_{n=-\infty}^{\infty} J_n(r_1) \, J_{\nu+n}(r_2) \, e^{in\theta},$$

where r, r_1 and r_2 are three sides of a triangle, φ is the angle opposite r_1, and θ is the angle opposite r. With the same notation, Gegenbauer's formula is

$$\frac{J_\nu(r)}{r^\nu} = 2^\nu \, \Gamma(\nu) \sum_{n=0}^{\infty} (\nu+n) \frac{J_{\nu+n}(r_1) J_{\nu+n}(r_2)}{r_1^\nu \, r_2^\nu} \, C_n^\nu(\cos\theta),$$

where C_n^ν are the Gegenbauer polynomials

$$C_n^\nu(x) = \frac{(2\nu)_n}{\left(\nu+\frac{1}{2}\right)_n} P_n^{\left(\nu-\frac{1}{2}, \nu-\frac{1}{2}\right)}(x).$$

7.11.7 Fourier transform and Hankel transform

Suppose

$$\int_0^\infty |f(x)| \, x^{\frac{1}{2}} \, dx < \infty.$$

The nth Hankel transform of f is

$$g(y) = \int_0^\infty J_n(xy) \, f(x) \, x \, dx.$$

If f is continuous and g satisfies the same integrability condition as f, then f is itself the Hankel transform of g:

$$f(x) = \int_0^\infty J_n(xy)\, g(y)\, y\, dy.$$

7.11.8 Integrals of Bessel functions

Under certain conditions

$$\int_0^\infty f(x)\, J_\nu(x)\, dx = \sum_{m=0}^\infty \frac{(-1)^m}{\Gamma(\nu+m+1)\,m!} \int_0^\infty f(x)\, \left(\frac{x}{2}\right)^{\nu+2m} dx.$$

Examples: for $\operatorname{Re} s > 0$, $\operatorname{Re}(\nu+a) > 0$,

$$\int_0^\infty x^{a-1} e^{-sx^2} J_\nu(x)\, dx = \frac{\Gamma\left(\frac{1}{2}\nu + \frac{1}{2}a\right)}{\Gamma(\nu+1)\, 2^{\nu+1} s^{\frac{1}{2}(\nu+a)}}$$

$$\times M\left(\frac{1}{2}\nu + \frac{1}{2}a,\ \nu+1;\ -\frac{1}{4s}\right);$$

$$\int_0^\infty x^{\nu+1} e^{-sx^2} J_\nu(x)\, dx = \frac{e^{-\frac{1}{4}s}}{(2s)^{\nu+1}};$$

$$\int_0^\infty x^{a-1} e^{-sx} J_\nu(x)\, dx = \frac{\Gamma(\nu+a)}{\Gamma(\nu+1)\, 2^\nu s^{\nu+a}}$$

$$\times F\left(\frac{1}{2}\nu + \frac{1}{2}a,\ \frac{1}{2}\nu + \frac{1}{2}a + \frac{1}{2},\ \nu+1;\ -\frac{1}{s^2}\right);$$

$$\int_0^\infty e^{-xs} x^\nu J_\nu(x)\, dx = \frac{\Gamma(2\nu+1)}{\Gamma(\nu+1)\, 2^\nu} \cdot \frac{1}{(1+s^2)^{\nu+\frac{1}{2}}};$$

$$\int_0^\infty e^{-xs} x^{\nu+1} J_\nu(x)\, dx = \frac{\Gamma(2\nu+2)}{\Gamma(\nu+1)\, 2^\nu} \cdot \frac{s}{(1+s^2)^{\nu+\frac{3}{2}}};$$

$$\int_0^\infty e^{-xs} x^{-1} J_\nu(x)\, dx = \frac{1}{\nu\left(s+\sqrt{1+s^2}\right)^\nu};$$

$$\int_0^\infty e^{-xs} J_\nu(x)\, dx = \frac{1}{\sqrt{1+s^2}\left(s+\sqrt{1+s^2}\right)^\nu}.$$

7.11.9 Airy functions

If

$$u(x) = x^{\frac{1}{2}} \, v\left(\frac{2}{3}x^{\frac{3}{2}}\right)$$

and v is a solution of the modified Bessel equation with index $\nu = \frac{1}{3}$, then u is a solution of the Airy equation

$$u''(x) - x\,u(x) = 0.$$

Airy functions:

$$\mathrm{Ai}\,(x) = \frac{\sqrt{x}}{\pi\sqrt{3}}\,K_{\frac{1}{3}}\left(\frac{2}{3}x^{\frac{3}{2}}\right) = \frac{\sqrt{x}}{3}\left[I_{-\frac{1}{3}}\left(\frac{2}{3}x^{\frac{3}{2}}\right) - I_{\frac{1}{3}}\left(\frac{2}{3}x^{\frac{3}{2}}\right)\right];$$

$$\mathrm{Bi}\,(x) = \frac{\sqrt{x}}{\sqrt{3}}\left[I_{-\frac{1}{3}}\left(\frac{2}{3}x^{\frac{3}{2}}\right) + I_{\frac{1}{3}}\left(\frac{2}{3}x^{\frac{3}{2}}\right)\right].$$

Series expansions:

$$\mathrm{Ai}\,(x) = \sum_{n=0}^{\infty}\left[\frac{x^{3n}}{3^{2n+\frac{2}{3}}\Gamma\left(n+\frac{2}{3}\right)n\,!} - \frac{x^{3n+1}}{3^{2n+\frac{4}{3}}\Gamma\left(n+\frac{4}{3}\right)n\,!}\right];$$

$$\mathrm{Bi}\,(x) = \sqrt{3}\sum_{n=0}^{\infty}\left[\frac{x^{3n}}{3^{2n+\frac{2}{3}}\Gamma\left(n+\frac{2}{3}\right)n\,!} + \frac{x^{3n+1}}{3^{2n+\frac{4}{3}}\Gamma\left(n+\frac{4}{3}\right)n\,!}\right].$$

Wronskian:

$$W\big(\mathrm{Ai}\,(x), \mathrm{Bi}\,(x)\big) = \frac{1}{\pi}.$$

Asymptotics:

$$\mathrm{Ai}\,(x) \sim \frac{e^{-\frac{2}{3}x^{\frac{3}{2}}}}{2\sqrt{\pi}\,x^{\frac{1}{4}}}; \quad \mathrm{Bi}\,(x) \sim \frac{e^{\frac{2}{3}x^{\frac{3}{2}}}}{\sqrt{\pi}\,x^{\frac{1}{4}}}, \quad x \to +\infty;$$

$$\mathrm{Ai}\,(-x) \sim \frac{\cos\left(\frac{2}{3}x^{\frac{3}{2}} - \frac{1}{4}\pi\right)}{\sqrt{\pi}\,x^{\frac{1}{4}}};$$

$$\mathrm{Bi}\,(-x) \sim -\frac{\sin\left(\frac{2}{3}x^{\frac{3}{2}} - \frac{1}{4}\pi\right)}{\sqrt{\pi}\,x^{\frac{1}{4}}}, \quad x \to +\infty.$$

Airy's integral form:

$$\text{Ai}(z) = \frac{1}{2\pi i} \int_{-i\infty}^{+i\infty} \exp\left(\frac{1}{3}\tau^3 - z\tau\right) d\tau.$$

Let $\omega = \exp(2\pi i/3)$. Functions $\text{Ai}(\omega^j z)$, $j = 0, 1, 2$, are solutions of (7.9.3) and

$$\text{Ai}(z) + \omega\text{Ai}(wz) + \omega^2\text{Ai}(\omega^2 z) = 0.$$

The integral form is equivalent to

$$\text{Ai}(z) = \frac{1}{2\pi} e^{-\frac{2}{3}z^{3/2}} \int_0^\infty e^{-\sqrt{z}u} \cos\left(\frac{1}{3}u^{\frac{3}{2}}\right) \frac{du}{\sqrt{u}},$$

for $|\arg z| < \pi$, which gives the full asymptotic expansion

$$\text{Ai}(z) \sim \frac{1}{2\pi z^{\frac{1}{4}}} e^{-\frac{2}{3}z^{\frac{3}{2}}} \sum_{n=0}^\infty \frac{\Gamma\left(3n + \frac{1}{2}\right)}{3^{2n}(2n)!} \frac{(-1)^n}{z^{\frac{3}{2}n}}, \quad z \to \infty, \quad |\arg z| < \pi.$$

7.12 Remarks

Bessel's equation is closely related to Riccati's equation [238], a case of which was investigated by Johann and Daniel Bernoulli starting in 1694 [28]; see Exercises 3.20 and 3.21 in Chapter 3. It also arose in investigations of the oscillations of a heavy chain (D. Bernoulli, 1734 [25]), vibrations of a circular membrane (Euler, 1759 [86]), and heat conduction in a cylinder (Fourier, 1822 [100]) or sphere (Poisson, 1823 [231]). Daniel Bernoulli gave a power series solution which is J_0. The functions J_n for integer n occur in Euler [86]; he found the series expansions and looked for a second solution, finding Y_0 but not Y_n. The J_n also appear as coefficients in studies of planetary motion (Lagrange [171], Laplace [175]). The early history is discussed in some detail in Dutka [77].

Bessel's 1824 investigation of the J_n [30] and Schlömilch's memoir [252] left Bessel's name attached both to the functions of integer order and to the functions J_ν for arbitrary ν which were introduced by Lommel [189] in 1868.

Up to factors, the Bessel function of the second kind Y_n was introduced by Hankel [123], Weber [309], and Schläfli [249]. Neumann [212] introduced a different version. The functions $H_\nu^{(i)}$ were introduced by Nielsen [216] and named in honor of Hankel. Up to factors, the functions I_ν and K_ν were introduced by Bassett [21]. The function K_ν also appears in [192] and is sometimes called *Macdonald's function*.

Airy's integral was introduced and studied by Airy in 1838 [4]. The current notation Ai (x), Bi (x) is due to Jeffreys [146].

The theory, the history, and the extensive literature on cylinder functions through the early 20th century, are surveyed in Watson's classic treatise [306]. Other references, with an emphasis on applications, are Korenev [162] and Petiau [227].

8

Hypergeometric functions

Hypergeometric functions were introduced briefly in Chapters 1 and 3. The series representations of these functions, like the series representations of Kummer functions, are examples of a more general concept of hypergeometric series.

After a brief discussion of general hypergeometric series, we discuss solutions of the hypergeometric equation

$$x(1 - x) u''(x) + \left[c - (a + b + 1)x \right] u'(x) - ab\, u(x) = 0 \qquad (8.0.1)$$

and the two classic transformations (Pfaff, Euler) from one solution to another.

There are three natural pairs of solutions of (8.0.1), normalized at the singular points $x = 0$, $x = 1$, and $x = \infty$ respectively. Any three solutions must satisfy a linear relation. In particular, for most values of the parameters (a, b, c), each solution of one normalized pair is a linear combination of the two solutions in each of the other two pairs. We find a fundamental set of such relations.

When the parameter c is an integer, the standard solutions coincide ($c = 1$), or one of them is not defined. A second solution is found by a limiting process.

Three hypergeometric functions whose respective parameters (a, b, c) differ by integers satisfy a linear relation. A basis for such relations, due to Gauss, is derived.

When the three parameters (a, b, c) satisfy certain relations, a quadratic transformation of the independent variable converts a hypergeometric function to the product of a hypergeometric function and a power of a rational function. The basic such quadratic transformations are found.

Hypergeometric functions can be transformed into other hypergeometric functions by certain combinations of multiplication, differentiation, and integration. As consequences we obtain some useful evaluations in closed form and some useful integral representations. Jacobi polynomials, rescaled to the interval $[0, 1]$, are multiples of hypergeometric functions. This leads to some additional explicit evaluations, recalled from Chapter 4.

8.1 Hypergeometric series

A *hypergeometric series* is a power series of the form

$$\sum_{n=0}^{\infty} \frac{(a_1)_n (a_2)_n \cdots (a_p)_n}{(c_1)_n (c_2)_n \cdots (c_q)_n \, n!} \, x^n, \tag{8.1.1}$$

where again the extended factorials are defined by

$$(a)_n = a(a+1)(a+2) \cdots (a+n-1) = \frac{\Gamma(a+n)}{\Gamma(a)}$$

and so on. It is assumed in (8.1.1) that no c_j is a non-positive integer. If some a_j is a non-positive integer then (8.1.1) is a polynomial; we exclude this case in the following general remarks.

The ratio test shows that the radius of convergence of the series (8.1.1) is zero if $p > q + 1$, while the radius of convergence is 1 if $p = q + 1$ and infinite if $p \le q$. Therefore it is assumed that $p \le q + 1$. The function defined by (8.1.1) for $|x| < 1$ is denoted by

$$_pF_q(a_1, a_2, \ldots, a_p; c_1, c_2, \ldots, c_q; x). \tag{8.1.2}$$

This function can be characterized as the solution of the generalized hypergeometric equation

$$L_{(a),(c)} F(x) = 0, \quad F(0) = 1, \tag{8.1.3}$$

where $L_{(a),(c)}$ denotes the differential operator

$$L_{(a),(c)} = x^{-1} D \prod_{j=1}^{q} (D + c_j - 1) - \prod_{k=1}^{p} (D + a_k), \quad D = D_x = x \frac{d}{dx}. \tag{8.1.4}$$

Indeed $D[x^n] = nx^n$, so if $F(x) = \sum_{n=0}^{\infty} b_n x^n$ is a solution of (8.1.3), the coefficients satisfy

$$n \prod_{j=1}^{q} (c_j + n - 1) \cdot b_n = \prod_{k=1}^{p} (a_k + n - 1) \cdot b_{n-1}.$$

Therefore $b_0 = F(0) = 1$ implies that the solution F is given by the series (8.1.1).

8.1.1 Examples of generalized hypergeometric functions

An empty product (no factors) is always taken to be 1, so

$$_0F_0(x) = \sum_{n=0}^{\infty} \frac{x^n}{n!} = e^x;$$

the corresponding operator is $d/dx - 1$. The binomial expansion gives

$$_1F_0(a; x) = \sum_{n=0}^{\infty} \frac{(a)_n}{n!} x^n = \frac{1}{(1-x)^a};$$

the corresponding operator is $(1 - x) d/dx - a$.

Since

$$\frac{1}{(2n)!} = \frac{1}{4^n \left(\frac{1}{2}\right)_n n!}, \qquad \frac{1}{(2n+1)!} = \frac{1}{4^n \left(\frac{3}{2}\right)_n n!},$$

it follows that

$$\cos x = {}_0F_1\left(\frac{1}{2}; -\frac{1}{4}x^2\right); \tag{8.1.5}$$

$$\cosh x = {}_0F_1\left(\frac{1}{2}; \frac{1}{4}x^2\right);$$

$$\frac{\sin x}{x} = {}_0F_1\left(\frac{3}{2}; -\frac{1}{4}x^2\right);$$

$$\frac{\sinh x}{x} = {}_0F_1\left(\frac{3}{2}; \frac{1}{4}x^2\right).$$

It is clear from the series representation (7.1.2) that the Bessel function J_ν is

$$J_\nu(x) = \frac{(x/2)^\nu}{\Gamma(\nu+1)} \cdot {}_0F_1\left(\nu+1; -\frac{1}{4}x^2\right).$$

Integrating the series representations of $(1+t)^{-1}$, $(1+t^2)^{-1}$, $(1-t^2)^{-\frac{1}{2}}$, $(1-t^2)^{-1}$, and $(1+t^2)^{-\frac{1}{2}}$ from $t = 0$ to $t = \pm x$ or $t = \pm x^2$ gives the identities

$$\frac{\log(1+x)}{x} = {}_2F_1(1, 1; 2; -x); \tag{8.1.6}$$

$$\frac{\tan^{-1} x}{x} = {}_2F_1\left(\frac{1}{2}, 1; \frac{3}{2}; -x^2\right);$$

$$\frac{\sin^{-1} x}{x} = {}_2F_1\left(\frac{1}{2}, \frac{1}{2}; \frac{3}{2}; x^2\right);$$

$$\frac{\tanh^{-1} x}{x} = {}_2F_1\left(\frac{1}{2}, 1; \frac{3}{2}; x^2\right);$$

$$\frac{\sinh^{-1} x}{x} = {}_2F_1\left(\frac{1}{2}, \frac{1}{2}; \frac{3}{2}; -x^2\right).$$

Manipulation of the coefficients in the series expansions leads to the identities

$$\frac{1}{2}\left[(1+x)^{-a} + (1-x)^{-a}\right] = {}_2F_1\left(\frac{1}{2}a, \frac{1}{2}a + \frac{1}{2}, \frac{1}{2}; x^2\right); \qquad (8.1.7)$$

$$\frac{1}{2a}\left[(1+x)^{-a} - (1-x)^{-a}\right] = -x\,{}_2F_1\left(\frac{1}{2}a + \frac{1}{2}, \frac{1}{2}a + 1, \frac{3}{2}; x^2\right).$$

The order of the differential operator (8.1.4) is $q + 1$. It is not surprising that for most applications, the cases of interest are those when the operator has order two, i.e. $q = 1$. The term *hypergeometric function* or *Gauss hypergeometric function* is usually reserved for the case $q = 1$, $p = 2$, and the subscripts are usually dropped:

$$
{}_2F_1(a, b; c; x) = F(a, b, c; x) = \sum_{n=0}^{\infty} \frac{(a)_n (b)_n}{(c)_n n!} x^n, \quad c \neq 0, -1, -2, \ldots
$$
$$(8.1.8)$$

The case $q = 1$, $p = 1$ is the "confluent hypergeometric" case of Chapter 6. The terminology comes from the idea of replacing x by x/b in (8.1.8), so that the singularities of the equation are at 0, b, and ∞, and then letting $b \to +\infty$ so that the latter two singularities flow together. The formal limit of the series in (8.1.8) is the series for ${}_1F_1(a, c; x) = M(a, c; x)$.

8.2 Solutions of the hypergeometric equation

The operator associated with the series (8.1.8) is the hypergeometric operator

$$L_{abc} = x(1-x)\frac{d^2}{dx^2} + \left[c - (a+b+1)x\right]\frac{d}{dx} - ab. \qquad (8.2.1)$$

Any solution of $L_{abc} F = 0$ in a region in the complex plane extends analytically to any simply connected plane region that does not contain the singular points $x = 0$, $x = 1$. The series (8.1.8),

$$F(a, b, c; x) = \sum_{n=0}^{\infty} \frac{(a)_n (b)_n}{(c)_n \, n!} x^n \qquad (8.2.2)$$

is the solution that is regular at the origin and satisfies $F(0) = 1$. It has a single-valued analytic continuation to the complement of the real ray $\{x \geq 1\}$. In the various formulas that follow, we choose principal branches on the complement of this ray. Note the identity

$$\frac{d}{dx}\big[F(a, b, c; x)\big] = \frac{ab}{c} F(a+1, b+1, c+1; x). \qquad (8.2.3)$$

In the notation of (8.1.4), with $D = D_x = x(d/dx)$, the hypergeometric operator (8.2.1) is

$$L_{abc} = x^{-1} D(D + c - 1) - (D + a)(D + b). \qquad (8.2.4)$$

Recall (6.1.5): for any constant b,

$$x^{-b} D\{x^b u(x)\} = (D + b)u(x). \qquad (8.2.5)$$

It follows that conjugating by x^{1-c} converts the operator (8.2.1) to the operator

$$x^{c-1} L_{abc} x^{1-c}$$

$$= x^{-1} D(D + 1 - c) - (D + a + 1 - c)(D + b + 1 - c)$$

$$= L_{a+1-c, \, b+1-c, \, 2-c}.$$

Therefore a second solution of the equation $L_{abc} F = 0$ is provided through the gauge transformation $u(x) = x^{1-c} v(x)$:

$$x^{1-c} F(a + 1 - c, b + 1 - c, 2 - c; x), \qquad (8.2.6)$$

provided that c is not a positive integer. (This is one of many provisos that exclude certain integer values of combinations of the indices a, b, c. The exceptional cases will be discussed separately.)

The hypergeometric operators L_{abc} can be characterized as the linear second-order differential operators that have exactly three singular points on the Riemann sphere, $0, 1, \infty$, each of them regular. (For the concepts of regular and irregular singular points, see Coddington and Levinson [55], Hille [129], or Ince [135]; for a look at irregular singular points, see the exercises for Chapter 10.) We mention this because it explains an important invariance property of the set of hypergeometric operators $\{L_{abc}\}$: this set is invariant under changes of coordinates on the Riemann sphere $\mathbf{C} \cup \{\infty\}$ by linear fractional transformations (Möbius transformations) that map the set of singular points $\{0, 1, \infty\}$ to itself. This provides a way of producing solutions that have specified behavior at $x = 1$ or at $x = \infty$. These transformations are generated by the transformation $y = 1 - x$ that interchanges 0 and 1 and fixes ∞ and the transformation $y = 1/x$ that interchanges 0 and ∞ and fixes 1.

Consider the first of these transformations. Let $u(x) = v(1 - x)$. Then equation (8.0.1) is equivalent to

$$y(1 - y) v''(y) + \left[c' - (a + b + 1)y\right] v'(x) - ab\, v(y) = 0,$$
$$c' = a + b + 1 - c.$$

Therefore there is a solution regular at $y = 0$ ($x = 1$) and another that behaves like $y^{1-c'} = (1 - x)^{1-c'}$ at $y = 0$. These solutions are multiples of the two solutions

$$F(a, b, a + b + 1 - c; 1 - x), \tag{8.2.7}$$

$$(1 - x)^{c-a-b} F(c - a, c - b, 1 + c - a - b; 1 - x),$$

respectively, provided that $c - a - b$ is not an integer.

The inversion $y = 1/x$ takes D_x to $-D_y$, so

$$L_{abc} = yD_y(D_y + 1 - c) - (D_y - a)(D_y - b);$$

$$(-x)^a L_{abc}(-x)^{-a} = -\left[D_y(D_y - b + a) - y(D_y + a)(D_y + 1 - c + a)\right],$$

where we have made use of (8.2.5). It follows from this and from interchanging the roles of a and b that there are solutions that behave like $(-x)^{-a}$ and $(-x)^{-b}$ at ∞. They are multiples of the two solutions

$$(-x)^{-a} F\left(a, 1 - c + a, a - b + 1; \frac{1}{x}\right), \tag{8.2.8}$$

$$(-x)^{-b} F\left(1 - c + b, b, b - a + 1; \frac{1}{x}\right),$$

respectively, provided that $a - b$ is not an integer.

These results can be used to generate identities for hypergeometric functions. For example, composing $x \to 1 - x$ with inversion and composing the result with $x \to 1 - x$ gives the map $y = x(x - 1)^{-1}$ that fixes the origin and interchanges 1 and ∞. This leads to a different expression for the solution that is regular at the origin, given as a function of $x(1 - x)^{-1}$; this is *Pfaff's identity* [229]:

$$F(a, b, c; x) = (1 - x)^{-b} F\left(c - a, b, c; \frac{x}{x - 1}\right). \tag{8.2.9}$$

(The special case of this with $b = 1$ was known to Stirling [274].) There is, of course, a companion identity with a and b interchanged:

$$F(a, b, c; x) = (1 - x)^{-a} F\left(a, c - b, c; \frac{x}{x - 1}\right). \tag{8.2.10}$$

If we iterate (8.2.9), fixing the index $c - a$ on the right, we obtain an identity due to Euler [90]:

$$F(a, b, c; x) = (1 - x)^{c-a-b} F(c - a, c - b, c; x). \tag{8.2.11}$$

Another derivation of Pfaff's identity is given in the next section.

Kummer [168] listed 24 solutions of the hypergeometric equation. Four of them occur in (8.2.9), (8.2.10), and (8.2.11). The remaining 20 are generated in the same way, starting with the five solutions (8.2.6), (8.2.7), and (8.2.8).

If we replace a by $a + v$ and b by $-v$, the operator (8.2.1) has the form

$$x(1 - x)\frac{d^2}{dx^2} + \left[c - (a + 1)x\right]\frac{d}{dx} + v(a + v),$$

so that $\lambda(v) = v(a + v)$ appears as a natural parameter associated with the fixed operator

$$x(1 - x)\frac{d^2}{dx^2} + \left[c - (a + 1)x\right]\frac{d}{dx}.$$

The following asymptotic result due to Darboux [62] will be proved in Chapter 10:

$$F\left(a + v, -v, c; \sin^2\left(\frac{1}{2}\theta\right)\right)$$

$$= \frac{\Gamma(c) \cos\left(v\theta + \frac{1}{2}a\theta - \frac{1}{2}c\pi + \frac{1}{4}\pi\right) + O(v^{-1})}{\sqrt{\pi}\left(v \sin \frac{1}{2}\theta\right)^{c-\frac{1}{2}}\left(\cos \frac{1}{2}\theta\right)^{\frac{1}{2}+(a-c)}} \tag{8.2.12}$$

as $v \to +\infty$, for $0 < \theta < \pi$.

8.3 Linear relations of solutions

In the previous section we identified six solutions of the hypergeometric equation $L_{abc} F = 0$ that have specified behavior at the singular points $\{0, 1, \infty\}$:

$$F(a, b, c; x) \sim 1, \quad x \to 0; \tag{8.3.1}$$

$$x^{1-c} F(a + 1 - c, b + 1 - c, 2 - c; x) \sim x^{1-c}, \quad x \to 0;$$

$$F(a, b, a + b + 1 - c; 1 - x) \sim 1, \quad x \to 1;$$

$$(1 - x)^{c-a-b} F(c - a, c - b, 1 + c - a - b; 1 - x) \sim (1 - x)^{c-a-b}, \quad x \to 1;$$

$$(-x)^{-a} F(a, 1 + a - c, 1 + a - b; 1/x) \sim (-x)^{-a}, \quad x \to -\infty;$$

$$(-x)^{-b} F(b, 1 + b - c, 1 + b - a; 1/x) \sim (-x)^{-b}, \quad x \to -\infty.$$

The hypergeometric operator has degree 2, so any three solutions must be linearly related. Our principal tool for computing coefficients in these relations is an integral representation due to Euler [88], which is obtained in the same way as the integral representation (6.1.3) for Kummer's confluent hypergeometric function.

Proposition 8.3.1 (Euler's integral representation) *Suppose* $\operatorname{Re} c > \operatorname{Re} a > 0$. *Then*

$$F(a, b, c; x) = \frac{1}{\mathrm{B}(a, c - a)} \int_0^1 s^{a-1} (1 - s)^{c-a-1} (1 - sx)^{-b} ds. \quad (8.3.2)$$

Proof Since

$$\frac{(a)_n}{(c)_n} = \frac{\Gamma(a + n) \Gamma(c)}{\Gamma(a) \Gamma(c + n)} = \frac{\mathrm{B}(a + n, c - a)}{\mathrm{B}(a, c - a)},$$

the integral representation (2.1.7) for the beta function implies

$$F(a, b, c; x) = \frac{1}{\mathrm{B}(a, c - a)} \int_0^1 s^{a-1} (1 - s)^{c-a-1} \left[\sum_{n=0}^{\infty} \frac{(b)_n}{n!} (sx)^n \right] ds.$$

Summing the series in brackets gives (8.3.2). $\qquad \square$

The identity (8.3.2) provides an explicit analytic continuation for x in the complement of $[1, \infty)$ when $\operatorname{Re} c > \operatorname{Re} a > 0$.

The change of variables $t = 1 - s$ in (8.3.2) converts the integral to

$$(1 - x)^{-b} \int_0^1 t^{c-a-1} (1 - t)^{a-1} \left(1 - \frac{tx}{x - 1} \right)^{-b} dt.$$

In view of (8.3.2) and the values at $x = 0$, we obtain Pfaff's identity (8.2.9) in the case $\operatorname{Re} c > \operatorname{Re} a > 0$. Analytic continuation in the parameters gives (8.2.9) for all values.

Let us return to the six solutions (8.3.1). In principle, the first is a linear combination of the third and fourth:

$$F(a, b, c; x) = C_1(a, b, c) F(a, b, a + b + 1 - c; 1 - x)$$

$$+ C_2(a, b, c) (1 - x)^{c-a-b}$$

$$\times F(c - a, c - b, 1 + c - a - b; 1 - x). \quad (8.3.3)$$

The coefficients C_1 and C_2 are analytic functions of the parameters, so it is enough to compute these coefficients under special assumptions. Assuming that $c - a - b > 0$, the fourth solution vanishes at $x = 1$, so the coefficient of the third solution is $F(a, b, c; 1)$. This value can be obtained immediately from (8.3.2) if we assume also that $c > a > 0$ or $c > b > 0$:

$$C_1(a, b, c) = F(a, b, c; 1) = \frac{\Gamma(c)\,\Gamma(c - a - b)}{\Gamma(c - a)\,\Gamma(c - b)}.$$

This extends by analytic continuation to the full case, giving *Gauss's summation formula* [103]:

$$\sum_{n=0}^{\infty} \frac{(a)_n\,(b)_n}{(c)_n\,n!} = \frac{\Gamma(c)\,\Gamma(c - a - b)}{\Gamma(c - a)\,\Gamma(c - b)}, \quad \mathrm{Re}\,(c - a - b) > 0. \qquad (8.3.4)$$

If b is a negative integer, the sum is finite, and (8.3.4) reduces to a combinatorial identity usually attributed to Vandermonde in 1772 [295], but known to Chu in 1303 [54]; see Lecture 7 of [14]:

$$F(a, -n, c; 1) = \sum_{k=0}^{n} (-1)^k \binom{n}{k} \frac{(a)_k}{(c)_k} = \frac{(c - a)_n}{(c)_n}, \qquad (8.3.5)$$

valid for $c \neq 0, -1, -2, \ldots$

We may use Euler's identity (8.2.11) to rewrite (8.3.3) as

$$F(c - a, c - b, c; x)$$
$$= C_1(a, b, c)\,(1 - x)^{a+b-c}\,F(a, b, a + b + 1 - c; 1 - x)$$
$$+ C_2(a, b, c)\,F(c - a, c - b, 1 + c - a - b; 1 - x).$$

Assuming that $a + b > c$, we evaluate C_2 by computing $F(c - a, c - b, c; 1)$. By (8.3.4) (with a change of indices) the result is

$$C_2(a, b, c) = F(c - a, c - b, c; 1) = \frac{\Gamma(c)\,\Gamma(a + b - c)}{\Gamma(a)\,\Gamma(b)}.$$

Therefore

$$F(a, b, c; x) = \frac{\Gamma(c)\,\Gamma(c - a - b)}{\Gamma(c - a)\,\Gamma(c - b)}\,F(a, b, a + b + 1 - c; 1 - x)$$
$$+ \frac{\Gamma(c)\,\Gamma(a + b - c)}{\Gamma(a)\,\Gamma(b)}\,(1 - x)^{c-a-b}$$
$$\times F(c - a, c - b, 1 + c - a - b; 1 - x). \qquad (8.3.6)$$

This identity can be inverted by replacing x by $1 - x$ and c by $a + b + 1 - c$ in order to express the first of the solutions normalized at $x = 1$ as a linear combination of the solutions normalized at $x = 0$:

$$F(a, b, a + b + 1 - c; 1 - x) = \frac{\Gamma(a + b + 1 - c)\,\Gamma(1 - c)}{\Gamma(a + 1 - c)\,\Gamma(b + 1 - c)}\, F(a, b, c; x)$$

$$+ \frac{\Gamma(a + b + 1 - c)\,\Gamma(c - 1)}{\Gamma(a)\,\Gamma(b)}\, x^{1-c}\, F(a + 1 - c, b + 1 - c, 2 - c; x).$$

$$(8.3.7)$$

This identity and a change of indices allows one to obtain the second of the solutions normalized at $x = 1$ as a linear combination of the solutions normalized at $x = 0$.

Similarly, the first solution in (8.3.1) is a linear combination of the last two, and it is enough to obtain one coefficient, under the assumption that $c > a > b$. Under this assumption, take $x \to -\infty$ in (8.3.2) to obtain

$$F(a, b, c; x) \sim \frac{(-x)^{-b}}{B(a, c - a)} \int_0^1 s^{a-b}(1 - s)^{c-a} \frac{ds}{s(1 - s)}$$

$$= \frac{(-x)^{-b}}{B(a, c - a)}\, B(a - b, c - a) = (-x)^{-b}\, \frac{\Gamma(c)\,\Gamma(a - b)}{\Gamma(c - b)\,\Gamma(a)}.$$

By symmetry and analytic continuation, we obtain

$$F(a, b, c; x) = \frac{\Gamma(c)\,\Gamma(b - a)}{\Gamma(c - a)\,\Gamma(b)}\, (-x)^{-a}\, F\left(a, 1 + a - c, 1 + a - b; \frac{1}{x}\right)$$

$$+ \frac{\Gamma(c)\,\Gamma(a - b)}{\Gamma(c - b)\,\Gamma(a)}\, (-x)^{-b}\, F\left(b, 1 + b - c, 1 + b - a; \frac{1}{x}\right).$$

$$(8.3.8)$$

This can be inverted to give

$$(-x)^{-a}\, F\left(a, 1 + a - c, 1 + a - b; \frac{1}{x}\right)$$

$$= \frac{\Gamma(1 + a - b)\,\Gamma(1 - c)}{\Gamma(a + 1 - c)\,\Gamma(1 - b)}\, F(a, b, c; x)$$

$$+ \frac{\Gamma(1 + a - b)\,\Gamma(c - 1)}{\Gamma(c - b)\,\Gamma(a)}\, x^{1-c}\, F(a + 1 - c, b + 1 - c, 2 - c; x).$$

$$(8.3.9)$$

The identities (8.3.6), (8.3.7), (8.3.8), and (8.3.9) are valid when all coefficients are well-defined, no third index is a non-positive integer, and all arguments x and $\varphi(x)$ are in the complement of the ray $[1, \infty)$. Additional identities can be obtained by applying the Pfaff transformation or the Euler transformation to some or all of the terms.

8.4 Solutions when c is an integer

As is the case for the confluent hypergeometric functions, when c is an integer $\neq 1$, one of the two solutions (8.1.8) and (8.2.6) of the hypergeometric equation (8.0.1) is not defined, while if $c = 1$ these two solutions coincide. We can find a second solution by adapting the procedure used in Section 6.3.

Assume first that neither a nor b is an integer. Assuming that $c \neq 0$, $-1, -2, \ldots$, let

$$N(a, b, c; x) \equiv \frac{\Gamma(a)\,\Gamma(b)}{\Gamma(c)}\, F(a, b, c; x) = \sum_{n=0}^{\infty} \frac{\Gamma(a+n)\,\Gamma(b+n)}{\Gamma(c+n)\,n!}\, x^n.$$

The series expansion is well-defined for all values of c. Note that if $c = -k$ is a non-positive integer, then the first $k+1$ terms of the series vanish. In particular, if $c = m$ is a positive integer,

$$N(a + 1 - m, b + 1 - m, 2 - m; x)$$

$$= \sum_{n=m-1}^{\infty} \frac{\Gamma(a+1-m+n)\,\Gamma(b+1-m+n)}{\Gamma(2-m+n)\,n!}\, x^n$$

$$= x^{m-1} \sum_{k=0}^{\infty} \frac{\Gamma(a+k)\,\Gamma(b+k)}{\Gamma(m+k)\,k!}\, x^k = x^{m-1}\, N(a, b, m; x). \qquad (8.4.1)$$

We define a solution of (8.0.1) by analogy with the Kummer function of the second kind:

$$U(a, b, c; x) = \frac{\Gamma(1-c)}{\Gamma(a+1-c)\,\Gamma(b+1-c)}\, F(a, b, c; x)$$

$$+ \frac{\Gamma(c-1)}{\Gamma(a)\,\Gamma(b)}\, x^{1-c}\, F(a+1-c, b+1-c, 2-c; x)$$

$$(8.4.2)$$

$$= \frac{\pi}{\sin \pi c\, \Gamma(a)\,\Gamma(a+1-c)\,\Gamma(b)\,\Gamma(b+1-c)}$$

$$\times \left[N(a, b, c; x) - x^{1-c}\, N(a+1-c, b+1-c, 2-c; x) \right].$$

$$(8.4.3)$$

In view of (8.4.1), the difference in brackets has limit zero as $c \to m$, m a positive integer. Therefore, by l'Hôpital's rule,

$$U(a, b, m; x) = \frac{(-1)^m}{\Gamma(a)\,\Gamma(a+1-m)\,\Gamma(b)\,\Gamma(b+1-m)}$$

$$\times \frac{\partial}{\partial c} \big[N(a, b, c; x) - x^{1-c}$$

$$\times N(a+1-c, b+1-c, 2-c; x) \big]\big|_{c=m}.$$

For non-integer values of a and b and positive integer values of m, calculating the derivative shows that

$$U(a, b, m; x) = \frac{(-1)^m}{\Gamma(a+1-m)\,\Gamma(b+1-m)\,(m-1)!}$$

$$\times \left\{ \log x \, F(a, b, m; x) + \sum_{n=0}^{\infty} \frac{(a)_n (b)_n}{(m)_n \, n!} \big[\psi(a+n) + \psi(b+n) \right.$$

$$\left. - \psi(n+1) - \psi(m+n) \big] x^n \right\}$$

$$+ \frac{(m-2)!}{\Gamma(a)\,\Gamma(b)} x^{1-m} \sum_{n=0}^{m-2} \frac{(a+1-m)_n (b+1-m)_n}{(2-m)_n \, n!} x^n,$$

$$(8.4.4)$$

where $\psi(b) = \Gamma'(b)/\Gamma(b)$ and the last sum is taken to be zero if $m = 1$.

The function in (8.4.4) is well-defined for all values of a and b. By a continuity argument, it is a solution of (8.0.1) for all values of a, b, c and all values of $x \notin (-\infty, 0]$. If neither a nor b is an integer less than m, then $U(a, b, m; x)$ has a logarithmic singularity at $x = 0$ and is therefore independent of the solution $F(a, b, c; x)$. If neither a nor b is a non-positive integer and one or both is an integer less than m, then the coefficient of the term in brackets vanishes and $U(a, b, c; x)$ is the finite sum, which is a rational function that is again independent of $F(a, b, c; x)$.

If a and/or b is a non-positive integer, then $U(a, b, m; x) \equiv 0$. To obtain a solution in this case we start with non-integer a and b and multiply (8.4.4) by $\Gamma(a)$ and/or $\Gamma(b)$. The limiting value of the resulting function as a and/or b approaches a non-positive integer is a well-defined solution of (8.0.1) that has a logarithmic singularity at $x = 0$.

We have found a second solution of (8.0.1) when c is a positive integer. When c is a negative integer, we may take advantage of the identity

$$U(a, b, c; x) = x^{1-c} U(a+1-c, b+1-c, 2-c; x). \qquad (8.4.5)$$

The various identities of Section 8.3 can be extended to exceptional cases by using the solution U. As an example, to obtain the analogue of (8.3.6) when $1 + a + b - c$ is a non-positive integer, we may start with the general case and express the right-hand side of (8.3.6) using the solutions

$$(1 - x)^{c-a-b} F(c - a, c - b, 1 + c - a - b; 1 - x),$$

$$U(a, b, 1 + a + b - c; 1 - x).$$

The result in this case is

$$F(a, b, c; x) = \Gamma(c) U(a, b, 1 + a + b - c; 1 - x).$$

8.5 Contiguous functions

As in the case of confluent hypergeometric functions, two hypergeometric functions are said to be *contiguous* if two of the indices of one function are the same as those of the other, and the third indices differ by 1. Gauss [103] showed that there is a linear relationship between a hypergeometric function and any two of its contiguous functions, with coefficients that are linear functions of the indices a, b, c and the variable x. By iteration, it follows that if the respective indices of three hypergeometric functions differ by integers, then they satisfy a linear relationship, with coefficients that are rational functions of the indices a, b, c and the variable x.

It is convenient to use again a shorthand notation: fixing indices a, b, c, let F be the function $F(a, b, c; x)$ and denote the six contiguous functions by $F(a\pm)$, $F(b\pm)$, $F(c\pm)$, where

$$F(a\pm) = F(a \pm 1, b, c; x)$$

and so on. Since there are 15 pairs of these six functions, there are 15 contiguous relations. Because of the symmetry between a and b, however, there are nine distinct relations: we do not need the five that involve b but not a, and the relation that involves $F(a-)$ and $F(b+)$ follows from the one that involves $F(a+)$ and $F(b-)$.

These relations can be derived in a way similar to that used for Kummer functions in Section 6.5. The coefficient of x^n in the expansion of F is

$$\varepsilon_n = \frac{(a)_n (b)_n}{(c)_n n!}.$$

The coefficients of x^n in the expansions of $F(a+)$ and $F(c-)$ are

$$\frac{(a + 1)_n}{(a)_n} \varepsilon_n = \frac{a + n}{a} \varepsilon_n, \qquad \frac{(c)_n}{(c - 1)_n} \varepsilon_n = \frac{c - 1 + n}{c - 1} \varepsilon_n$$

respectively. Since $D[x^n] = nx^n$, the corresponding coefficient for DF is $n\varepsilon_n$. It follows that

$$DF = a\left[F(a+) - F\right] = b\left[F(b+) - F\right] = (c - 1)\left[F(c-) - F\right].$$

These identities give

$$(a - b) F = a F(a+) - b F(b+); \tag{8.5.1}$$

$$(a - c + 1) F = a F(a+) - (c - 1) F(c-). \tag{8.5.2}$$

The coefficient of x^n in the expansion of F' is

$$\frac{(n + 1)(a)_{n+1}(b)_{n+1}}{(c)_{n+1}(n + 1)!} = \frac{(a + n)(b + n)}{(c + n)}\,\varepsilon_n. \tag{8.5.3}$$

Now

$$\frac{(a + n)(b + n)}{c + n} = n + (a + b - c) + \frac{(c - a)(c - b)}{c + n},$$

while the coefficient of x^n in the expansion of $F(c+)$ is

$$\frac{(c)_n}{(c + 1)_n}\,\varepsilon_n = \frac{c}{c + n}\,\varepsilon_n.$$

Therefore (8.5.3) implies that

$$F' = DF + (a + b - c) F + \frac{(c - a)(c - b)}{c} F(c+).$$

Multiplying by x gives

$$(1 - x) DF = x\left[(a + b - c)F + \frac{(c - a)(c - b)}{c} F(c+)\right].$$

Since

$$(1 - x) DF = (1 - x) a\left[F(a+) - F\right],$$

it follows that

$$[a + (b - c)x] F = a(1 - x) F(a+) - \frac{(c - a)(c - b)x}{c} F(c+). \tag{8.5.4}$$

The procedure that led to (8.5.4) can be applied to $F(a-)$ to yield another such relation. In fact, the coefficient of x^n in $F'(a-)$ is

$$\frac{(a - 1)(b + n)}{c + n}\,\varepsilon_n = \left[(a - 1) - \frac{(a - 1)(c - b)}{c + n}\right]\varepsilon_n.$$

Multiplying by x gives

$$DF(a-) = (a - 1)x\, F - \frac{(a - 1)(c - b)\, x}{c} F(c+).$$

Replacing a by $a - 1$ in a previous identity gives

$$DF(a-) = (a - 1)\big[F - F(a-)\big].$$

Therefore

$$(1 - x)\, F = F(a-) - \frac{(c - b)\, x}{c} F(c+). \tag{8.5.5}$$

Identities (8.5.1)–(8.5.5) can be used to generate the remaining five identities. Eliminating $F(c+)$ from (8.5.4) and (8.5.5) gives

$$[2a - c + (b - a)x]\, F = a(1 - x)\, F(a+) - (c - a)\, F(a-). \tag{8.5.6}$$

Eliminating $F(a+)$ from (8.5.2) and (8.5.4) gives

$$[(c - 1) + (a + b + 1 - 2c)x]\, F$$
$$= (c - 1)(1 - x)\, F(c-) - \frac{(c - a)(c - b)x}{c} F(c+). \tag{8.5.7}$$

Eliminating $F(c+)$ from (8.5.5) and (8.5.7) gives

$$[1 - a + (c - b - 1)x]\, F = (c - a)\, F(a-) - (c - 1)(1 - x)\, F(c-). \tag{8.5.8}$$

Eliminating $F(c+)$ from (8.5.4) and (8.5.5), with a replaced by b in (8.5.5), gives

$$(a + b - c)\, F = a(1 - x)\, F(a+) - (c - b)\, F(b-). \tag{8.5.9}$$

Eliminating $F(a+)$ from (8.5.6) and (8.5.9) gives

$$(b - a)(1 - x)\, F = (c - a)\, F(a-) - (c - b)\, F(b-). \tag{8.5.10}$$

8.6 Quadratic transformations

Suppose that φ is a quadratic transformation of the Riemann sphere, i.e. a two-to-one rational map. Under what circumstances is the function $F(a, b, c; \varphi(x))$ a hypergeometric function:

$$F\big(a, b, c; \varphi(x)\big) = F(a', b', c'; x)? \tag{8.6.1}$$

Assume first that φ is a polynomial of degree 2. An equation of the form (8.6.1) implies that φ takes the singular points $\{0, 1\}$ of the equation satisfied by the right-hand side of (8.6.1) to the singular points $\{0, 1\}$ of the equation satisfied by the left-hand side. Furthermore, by comparison of these two equations, it is readily seen that the unique double point of φ, the zero of φ', must go to 0 or 1. The right side is holomorphic at $x = 0$, so necessarily $\varphi(0) = 0$. Finally, the origin must be a simple zero of φ. The unique such polynomial φ is $4x(1 - x)$. Considering asymptotics at infinity for the two sides of (8.6.1), we must have $\{a', b'\} = \{2a, 2b\}$. Considering behavior at 0 and at 1, we must have $1 - c = 1 - c'$ and $1 - c = c' - a' - b'$ so, up to interchanging a' and b',

$$a' = 2a, \quad b' = 2b, \quad c = c' = a + b + \frac{1}{2}.$$

A comparison of the two differential equations and of the behavior at $x = 0$ shows that these necessary conditions are also sufficient:

$$F\left(a, b, a + b + \frac{1}{2}; 4x(1 - x)\right) = F\left(2a, 2b, a + b + \frac{1}{2}; x\right). \quad (8.6.2)$$

This can also be written in the inverse form

$$F\left(a, b, a + b + \frac{1}{2}; x\right) = F\left(2a, 2b, a + b + \frac{1}{2}; \frac{1}{2} - \frac{1}{2}\sqrt{1 - x}\right). \quad (8.6.3)$$

The general quadratic two-to-one rational map of the sphere has the form $\varphi(x) = p(x)/q(x)$, where p and q are polynomials of degree ≤ 2 with no common factor, and at least one has degree 2. The requirement for an identity of the form (8.6.1) is that φ map the set $\{0, 1, \infty\}$, together with any double points, into the set $\{0, 1, \infty\}$ and that the origin be a simple zero. This last requirement can be dropped if we look for a more general form

$$F\big(a, b, c; \varphi(x)\big) = (1 - \alpha x)^{\beta} F(a', b', c'; x).$$

Indeed, if $\alpha\beta = a'b'/c'$ then the right-hand side will have a double zero at the origin, i.e. the derivative vanishes at $x = 0$. A candidate for φ here is $x^2/(2 - x)^2$, which takes both 1 and ∞ to 1. In this case we would expect to take $\alpha = \frac{1}{2}$ to compensate for the singularity of the left-hand side at $x = 2$. Since

$$1 - \frac{x^2}{(2 - x)^2} = \frac{4(1 - x)}{(2 - x)^2} \sim 4(1 - x), \quad x \to 1;$$

$$\sim -\frac{4}{x}, \quad x \to \infty,$$

comparison of the behavior of the two sides as $x \to 1$ gives the condition $c - a - b = c' - a' - b'$, while comparison of the two sides as $x \to \infty$ gives $\{0, c - a - b\} = \{\beta - a', \beta - b'\}$. Up to interchanging a' and b', these conditions together with $\frac{1}{2}\beta = a'b'/c'$ imply that $\beta = b'$, $c' = 2a'$. Comparison of behavior as $x \to 2$ shows that, up to interchanging a and b, we should have $2a = \beta$, $2b = \beta + 1$. Then $c = a' + \frac{1}{2}$. Our proposed identity is therefore

$$F\left(a, a + \frac{1}{2}, c; \frac{x^2}{(2-x)^2}\right) = \left(1 - \frac{x}{2}\right)^{2a} F\left(c - \frac{1}{2}, 2a, 2c - 1; x\right).$$

(8.6.4)

It can be shown that both sides satisfy the same modification of the hypergeometric equation. The inverse form is

$$F\left(a, a + \frac{1}{2}, c; x\right) = \left(1 + \sqrt{x}\right)^{-2a} F\left(c - \frac{1}{2}, 2a, 2c - 1; \frac{2\sqrt{x}}{1 + \sqrt{x}}\right).$$

(8.6.5)

Starting with (8.6.2), inverting it, applying Pfaff's identity (8.2.9) to the right-hand side, and repeating this process yields the following sequence of identities. (At each step we reset the indices a, b, c.)

$$F\left(a, b, \frac{1}{2}(a + b + 1); x\right)$$

$$= F\left(\frac{1}{2}a, \frac{1}{2}b, \frac{1}{2}(a + b + 1); 4x(1 - x)\right);$$

(8.6.6)

$$F\left(a, b, a + b + \frac{1}{2}; x\right)$$

$$= F\left(2a, 2b, a + b + \frac{1}{2}; \frac{1}{2} - \frac{1}{2}\sqrt{1 - x}\right)$$

(8.6.7)

$$= \left(\frac{1}{2} + \frac{1}{2}\sqrt{1 - x}\right)^{-2a} F\left(2a, a - b + \frac{1}{2}, a + b + \frac{1}{2}; \frac{\sqrt{1 - x} - 1}{\sqrt{1 - x} + 1}\right);$$

(8.6.8)

$$F(a, b, a - b + 1; x)$$

$$= (1 - x)^{-a} F\left(\frac{1}{2}a, \frac{1}{2}a - b + \frac{1}{2}, a - b + 1; -\frac{4x}{(1 - x)^2}\right)$$

(8.6.9)

$$= (1 + x)^{-a} F\left(\frac{1}{2}a, \frac{1}{2}a + \frac{1}{2}, a - b + 1; \frac{4x}{(1 + x)^2}\right);$$

(8.6.10)

$$F\left(a, a + \frac{1}{2}, c; x\right)$$

$$= \left(\frac{1}{2} + \frac{1}{2}\sqrt{1 - x}\right)^{-2a} F\left(2a, 2a - c + 1, c; \frac{1 - \sqrt{1 - x}}{1 + \sqrt{1 - x}}\right)$$

(8.6.11)

$$= (1 - x)^{-a} F \left(2a, 2c - 2a - 1, c; \frac{\sqrt{1-x} - 1}{2\sqrt{1-x}} \right); \tag{8.6.12}$$

$$F \left(a, b, \frac{1}{2}(a + b + 1), x \right)$$

$$= (1 - 2x)^{-2a} F \left(\frac{1}{2}a, \frac{1}{2}a + \frac{1}{2}, \frac{1}{2}(a + b + 1); -\frac{4x(1-x)}{(1-2x)^2} \right). \tag{8.6.13}$$

Applying (8.2.9) to the right-hand side of (8.6.13) returns us to (8.6.6).

Similarly, starting with (8.6.4), inverting it, applying (8.2.9) to the right-hand side, and repeating this process yields the following sequence of identities. (Again we reset the indices a, b, c.)

$$F(a, b, 2b; x)$$

$$= \left(1 - \frac{1}{2}x \right)^{-a} F \left(\frac{1}{2}a, \frac{1}{2}a + \frac{1}{2}, b + \frac{1}{2}; \frac{x^2}{(2-x)^2} \right); \tag{8.6.14}$$

$$F \left(a, a + \frac{1}{2}, c; x \right)$$

$$= \left(1 + \sqrt{x} \right)^{-2a} F \left(2a, c - \frac{1}{2}, 2c - 1; \frac{2\sqrt{x}}{1 + \sqrt{x}} \right) \tag{8.6.15}$$

$$= \left(1 - \sqrt{x} \right)^{-2a} F \left(2a, c - \frac{1}{2}, 2c - 1; -\frac{2\sqrt{x}}{1 - \sqrt{x}} \right); \tag{8.6.16}$$

$$F(a, b, 2b; x)$$

$$= (1 - x)^{-\frac{1}{2}a} F \left(\frac{1}{2}a, b - \frac{1}{2}a, b + \frac{1}{2}; \frac{x^2}{4x - 4} \right). \tag{8.6.17}$$

Applying (8.2.9) to the right-hand side of (8.6.17) returns us to (8.6.14).

One more collection of identities can be generated by starting with (8.6.17) and following it with (8.6.8), then proceeding to invert and to apply (8.2.9) on the right:

$$F(a, b, 2b; x) = \left(\frac{1}{2} + \frac{1}{2}\sqrt{1-x} \right)^{-2a}$$

$$\times F \left(a, a - b + \frac{1}{2}, b + \frac{1}{2}; \left[\frac{1 - \sqrt{1-x}}{1 + \sqrt{1-x}} \right]^2 \right); \tag{8.6.18}$$

$$F(a, b, a - b + 1; x) = \left(1 + \sqrt{x}\right)^{-2a}$$

$$\times F\left(a, a - b + \frac{1}{2}, 2a - 2b + 1; \frac{4\sqrt{x}}{\left(1 + \sqrt{x}\right)^2}\right) \qquad (8.6.19)$$

$$= \left(1 - \sqrt{x}\right)^{-2a}$$

$$\times F\left(a, a - b + \frac{1}{2}, 2a - 2b + 1; -\frac{4\sqrt{x}}{\left(1 - \sqrt{x}\right)^2}\right); \qquad (8.6.20)$$

$$F(a, b, 2b, x) = (1 - x)^{-\frac{1}{2}a}$$

$$\times F\left(a, 2b - a, b + \frac{1}{2}; -\frac{\left(1 - \sqrt{1 - x}\right)^2}{4\sqrt{1 - x}}\right). \qquad (8.6.21)$$

Applying (8.2.9) to the right-hand side of (8.6.21) returns us to (8.6.18).

Additional identities can be generated from (8.6.6)–(8.6.21) by applying (8.2.9) (in the first or second index) and (8.2.11) to one or both sides, or by a change of variables. We mention in particular the identity obtained by applying (8.2.11) to the left-hand side of (8.6.2):

$$(1 - 2x) F\left(a + \frac{1}{2}, b + \frac{1}{2}, a + b + \frac{1}{2}; 4x(1 - x)\right)$$

$$= F\left(2a, 2b, a + b + \frac{1}{2}; x\right) \qquad (8.6.22)$$

and an identity obtained by a change of variable in (8.6.18):

$$F\left(a, b, 2b; \frac{4x}{(1 + x)^2}\right) = (1 + x)^{2a} F\left(a, a - b + \frac{1}{2}, b + \frac{1}{2}; x^2\right). \qquad (8.6.23)$$

Each of the identities (15.3.15)–(15.3.32) in [3] can be obtained in this way. The basic identities are due to Kummer [168]; a complete list is found in Goursat [117].

8.7 Transformations and special values

A given hypergeometric function may be transformed into another by operations that involve multiplication and differentiation or integration. Two examples are

$$F(a, b, c; x) = x^{1+k-a} \frac{d^k}{dx^k} \left[\frac{x^{a-1}}{(a-k)_k} F(a-k, b, c; x) \right]$$

$$= x^{1-c} \frac{d^k}{dx^k} \left[\frac{x^{c+k-1}}{(c)_k} F(a, b, c+k; x) \right], \quad k = 0, 1, 2, \ldots$$

$$(8.7.1)$$

The proofs are left as exercises.

Formulas like this are primarily of interest when the hypergeometric function on the right-hand side can be expressed in closed form, such as

$$F(a, b, a; x) = (1-x)^{-b}. \qquad (8.7.2)$$

This example, together with (8.7.1), shows that $F(a, b, c; x)$ can be written in closed form whenever $c - a$ or $c - b$ is a non-negative integer.

The identities (8.7.1) allow us to decrease the indices a, b by integers or to increase c by an integer. The following integral transform allows us, under certain conditions, to increase an upper index or to decrease the lower index by integer or fractional amounts.

Given complex constants α and β with positive real parts, we define an integral transform $E_{\alpha,\beta}$ that acts on functions that are defined on an interval containing the origin:

$$E_{\alpha,\beta} f(x) = \frac{\Gamma(\alpha+\beta)}{\Gamma(\alpha)\Gamma(\beta)} \int_0^1 s^{\alpha-1}(1-s)^{\beta-1} f(sx) \, ds. \qquad (8.7.3)$$

Then

$$E_{\alpha,\beta}[x^n] = \frac{\Gamma(\alpha+\beta)}{\Gamma(\alpha)\Gamma(\beta)} B(\alpha+n, \beta) x^n = \frac{(\alpha)_n}{(\alpha+\beta)_n} x^n.$$

Taking $\alpha = c$, or $\alpha + \beta = a$, respectively, we obtain

$$E_{c,\beta} F(a, b, c; \cdot) = F(a, b, c+\beta; \cdot)$$

and

$$E_{\alpha,a-\alpha} F(a, b, c; \cdot) = F(\alpha, b, c; \cdot).$$

Reversing the point of view, we may write a given hypergeometric function as an integral. For $\operatorname{Re} a > 0$ and $\operatorname{Re} \varepsilon > 0$ or for $\operatorname{Re} c > \operatorname{Re} \varepsilon > 0$, respectively,

$$F(a, b, c; x) = E_{a,\varepsilon}[F(a + \varepsilon, b, c; \cdot)](x)$$

$$= \frac{\Gamma(a + \varepsilon)}{\Gamma(a)\,\Gamma(\varepsilon)} \int_0^1 s^{a-1}(1 - s)^{\varepsilon-1} F(a + \varepsilon, b, c; sx)\,ds, \quad (8.7.4)$$

$$F(a, b, c; x) = E_{c-\varepsilon,\varepsilon}[F(a, b, c - \varepsilon; \cdot)](x)$$

$$= \frac{\Gamma(c)}{\Gamma(c - \varepsilon)\,\Gamma(\varepsilon)} \int_0^1 s^{c-\varepsilon-1}(1 - s)^{\varepsilon-1} F(a, b, c - \varepsilon; sx)\,ds.$$

Therefore if $\operatorname{Re} c > \operatorname{Re} a > 0$ we may take $\varepsilon = c - a$ and use (8.7.2) and (8.7.4) to recover the integral form (8.3.2)

$$F(a, b, c; x) = E_{a,c-a}\big[(1 - x)^{-b}\big].$$

The collection of well-understood hypergeometric functions can be enlarged by combining (8.7.2) with the various quadratic transformations in the previous section, choosing values of the indices so that one of the first two is equal to the third and (8.7.2) is applicable. In many cases the result is a simple algebraic identity; for example, taking $c = a + \frac{1}{2}$ in (8.6.5) yields

$$(1 - x)^{-a} = \big(1 + \sqrt{x}\big)^{-2a} \left(\frac{1 - \sqrt{x}}{1 + \sqrt{x}}\right)^{-a}.$$

In addition, however, one can obtain some less obvious identities.

Taking $b = a + \frac{1}{2}$ in (8.6.3) gives

$$F\left(a, a + \frac{1}{2}, 2a + 1; x\right) = \left(\frac{1 + \sqrt{1 - x}}{2}\right)^{-2a}. \quad (8.7.5)$$

Taking $c = 2a$ in (8.6.11) gives

$$F\left(a, a + \frac{1}{2}, 2a; x\right) = \frac{1}{\sqrt{1 - x}} \left(\frac{1 + \sqrt{1 - x}}{2}\right)^{1-2a}. \quad (8.7.6)$$

Another category of special values occurs when one of the first two indices (a, b) is a non-positive integer, say $a = -n$, so that $F(a, b, c; x)$ is a polynomial of degree n. As we noted in Chapter 4, this polynomial is a rescaling of a Jacobi polynomial, provided $c > 0$ and $b - c - n + 1 > 0$. The identity (4.6.12) is equivalent to

$$F(a + n, -n, c; x) = \frac{n!}{(c)_n} P_n^{(c-1, a-c)}(1 - 2x). \quad (8.7.7)$$

The arguments that led to the formulas (4.7.14) and (4.7.19) for the Chebyshev polynomials, as well as to (4.7.24) and (4.7.25), are valid for general values of n. It follows from this fact and from (8.7.7) (carried over to general values of n) that

$$F\left(\nu, -\nu, \frac{1}{2}; \frac{1}{2}(1 - \cos\theta)\right) = \cos\nu\theta;$$

$$F\left(\nu + 2, -\nu, \frac{3}{2}; \frac{1}{2}(1 - \cos\theta)\right) = \frac{\sin(\nu + 1)\theta}{(\nu + 1)\sin\theta};$$

$$F\left(\nu + 1, -\nu, \frac{1}{2}; \frac{1}{2}(1 - \cos\theta)\right) = \frac{\cos\left(\nu + \frac{1}{2}\right)\theta}{\cos\frac{1}{2}\theta};$$

$$F\left(\nu + 1, -\nu, \frac{3}{2}; \frac{1}{2}(1 - \cos\theta)\right) = \frac{\sin\left(\nu + \frac{1}{2}\right)\theta}{(2\nu + 1)\sin\frac{1}{2}\theta}.$$

These can be rewritten, setting $x = \frac{1}{2}(1 - \cos\theta)$, to obtain

$$F\left(\nu, -\nu, \frac{1}{2}; x\right) = \mathrm{Re}\left\{\left[1 - 2x + i\sqrt{4x(1 - x)}\,\right]^\nu\right\}; \tag{8.7.8}$$

$$F\left(\nu + 2, -\nu, \frac{3}{2}; x\right) = \frac{\mathrm{Im}\left\{\left[1 - 2x + i\sqrt{4x(1 - x)}\,\right]^{\nu+1}\right\}}{(\nu + 1)\sqrt{4x(1 - x)}}; \tag{8.7.9}$$

$$F\left(\nu + 1, -\nu, \frac{1}{2}; x\right) = \frac{\mathrm{Re}\left\{\left[1 - 2x + i\sqrt{4x(1 - x)}\,\right]^{\nu+\frac{1}{2}}\right\}}{\sqrt{1 - x}}; \tag{8.7.10}$$

$$F\left(\nu + 1, -\nu, \frac{3}{2}; x\right) = \frac{\mathrm{Im}\left\{\left[1 - 2x + i\sqrt{4x(1 - x)}\,\right]^{\nu+\frac{1}{2}}\right\}}{(2\nu + 1)\sqrt{x}}. \tag{8.7.11}$$

In addition to these identities, we note that the identities (8.1.6) and (8.1.7) involve $F = {}_2F_1$.

An integral transform that is more specialized in its application is

$$E_{\alpha,\beta}^{(2)} f(x) = \frac{\Gamma(2\alpha + 2\beta)}{\Gamma(2\alpha)\,\Gamma(2\beta)} \int_0^1 s^{2\alpha-1}(1 - s)^{2\beta-1} f(s^2 x)\, ds. \tag{8.7.12}$$

Then

$$E_{\alpha,\beta}^{(2)}\left[x^n\right] = \frac{(2\alpha)_{2n}}{(2\alpha + 2\beta)_{2n}}\, x^n = \frac{(\alpha)_n\left(\alpha + \frac{1}{2}\right)_n}{(\alpha + \beta)_n\left(\alpha + \beta + \frac{1}{2}\right)_n}\, x^n.$$

It follows that

$$F\left(a, a + \frac{1}{2}, c; x\right) = E^{(2)}_{a,\frac{1}{4}}\left[F\left(a + \frac{1}{4}, a + \frac{3}{4}, c; x\right)\right]. \qquad (8.7.13)$$

For use in Chapter 9 we give two examples of these considerations. The first uses (8.7.3) and (8.7.10):

$$F(v + 1, -v, 1; x) = E_{\frac{1}{2},\frac{1}{2}}\left[F\left(v + 1, -v; \frac{1}{2}; x\right)\right]$$

$$= \frac{1}{\pi} \int_0^1 \frac{\mathrm{Re}\left\{\left[1 - 2sx + i\sqrt{4sx(1-sx)}\right]^{v+\frac{1}{2}}\right\}}{\sqrt{1-sx}} \frac{ds}{\sqrt{s(1-s)}}. \qquad (8.7.14)$$

The second uses (8.7.13) and (8.7.6): for $\mathrm{Re}\, v > -1$,

$$F\left(\frac{1}{2}v + \frac{1}{2}, \frac{1}{2}v + 1; v + \frac{3}{2}; x\right)$$

$$= E^{(2)}_{\frac{1}{2}(v+1),\frac{1}{4}}\left[F\left(\frac{1}{2}v + \frac{3}{4}, \frac{1}{2}v + \frac{5}{4}; v + \frac{3}{2}; x\right)\right]$$

$$= \frac{\Gamma\left(v + \frac{3}{2}\right)}{\sqrt{\pi}\,\Gamma(v+1)} \int_0^1 \left(\frac{1 + \sqrt{1 - s^2x}}{2}\right)^{-v-\frac{1}{2}} \frac{s^v\, ds}{\sqrt{1 - s^2x}\,\sqrt{1 - s}}. \qquad (8.7.15)$$

8.8 Exercises

8.1 Show that

$$\lim_{b \to +\infty} F\left(a, b, c; \frac{x}{b}\right) = {}_1F_1(a, c; x), \quad |x| < 1.$$

8.2 Verify the identity (8.2.3).
8.3 Verify the identities (8.1.6).
8.4 Verify the identities (8.1.7).
8.5 Show that

$$\log \frac{1 + x}{1 - x} = 2x\, F\left(\frac{1}{2}, 1, \frac{3}{2}; x^2\right).$$

8.6 Show that for $|t| < 1$ and $|(1 - x)t| < 1$,

$$\sum_{n=0}^{\infty} \frac{(c)_n}{n!} F(a, -n, c; x) t^n = (1 - t)^{a-c}[1 - (1 - x)t]^{-a}.$$

8.7 Show that for $\operatorname{Re} a > 0$ and $x \notin [1, \infty)$,

$$\int_0^x t^{a-1}(1 - t)^{b-1} dt = \frac{x^a}{a} F(a, 1 - b, a + 1; x).$$

The integral is called the *incomplete beta function*, denoted $B_x(a, b)$. The identity is due to Gauss [103].

8.8 Use the reflection formula (2.2.7) and the integral formula (8.3.2) to verify the integral formula for $\operatorname{Re} c > \operatorname{Re} a > 0$:

$$F(a, b, c; x) = e^{-i\pi a} \frac{\Gamma(1 - a) \Gamma(c)}{\Gamma(c - a)} \frac{1}{2\pi i} \int_C \frac{t^{a-1}(1 + t)^{b-c}}{(1 + t - xt)^b} dt,$$

where the curve C runs from $+\infty$ to 0 along the upper edge of the cut on $[0, \infty)$ and returns to $+\infty$ along the lower edge.

8.9 Verify the integral formula, for $\operatorname{Re} a > 0$ and arbitrary complex b, c:

$$F(a, b, c; x)$$
$$= \frac{\Gamma(c) \Gamma(1 + a - c)}{\Gamma(a)} \frac{1}{2\pi i} \int_C s^{a-1}(s - 1)^{c-a-1}(1 - xs)^{-b} ds,$$

where C is a counterclockwise loop that passes through the origin and encloses the point $s = 1$. Hint: assume first that $\operatorname{Re} c > \operatorname{Re} a > 0$ and change the contour to run along the interval $[0, 1]$ and back.

8.10 Derive Kummer's identity (6.1.10) from Pfaff's transformation (8.2.9).

8.11 Show that

$$\int_0^{\pi/2} \frac{d\varphi}{\sqrt{1 - k^2 \sin^2 \varphi}} = \frac{\pi}{2} F\left(\frac{1}{2}, \frac{1}{2}, 1; k^2\right), \quad |k| < 1;$$

$$\int_0^{\pi/2} \sqrt{1 - k^2 \sin^2 \varphi} \, d\varphi = \frac{\pi}{2} F\left(\frac{1}{2}, -\frac{1}{2}, 1; k^2\right), \quad |k| < 1.$$

The functions $K = K(k)$ and $E = E(k)$ that are defined by these integrals are called the complete elliptic integral of the first kind and second kind, respectively; see Chapter 11. Hint: evaluate

$$\int_0^{\pi/2} \sin^{2n} \varphi \, d\varphi.$$

8.12 Let u_1 and u_2 be two solutions of the hypergeometric equation (8.0.1). Show that the Wronskian has the form

$$W(u_1, u_2)(x) = A\, x^{-c}(1-x)^{c-a-b-1}$$

for some constant A.

8.13 Denote the following six solutions of the hypergeometric equation by

$$F_1(x) = F(a, b, c; x);$$

$$F_2(x) = x^{1-c}\, F(a+1-c, b+1-c, 2-c; x);$$

$$F_3(x) = F(a, b, a+b+1-c; 1-x);$$

$$F_4(x) = (1-x)^{c-a-b} F(c-a, c-b, 1+c-a-b; 1-x);$$

$$F_5(x) = (-x)^{-a} F(a, 1-c+a, a-b+1; 1/x);$$

$$F_6(x) = (-x)^{-b} F(1-c+b, b, b-a+1; 1/x).$$

(a) Compute the Wronskians

$$W(F_1, F_2)(x), \quad W(F_3, F_4)(x), \quad W(F_5, F_6)(x).$$

(b) Compute the Wronskians

$$W(F_1, F_3)(x), \quad W(F_1, F_4)(x), \quad W(F_1, F_5)(x), \quad W(F_1, F_6)(x).$$

Hint: use (a) and the relations in Section 8.3.

8.14 Exercise 8.13 lists six of Kummer's 24 solutions of the hypergeometric equation. Use (8.2.9)–(8.2.11) to find the remaining 18 solutions.

8.15 Verify the contiguous relations (8.5.6)–(8.5.10).

8.16 Suppose that $Q(x)$ is a quadratic polynomial and
$u(x) = F(a', b', c'; Q(x))$ satisfies the hypergeometric equation (8.0.1) with indices a, b, c.

(a) Let $y = Q(x)$. Show that $x(1-x)[Q'(x)]^2 = Ay(1-y)$ for some constant A.

(b) Show that $A = 4$ and $Q(x) = 4x(1-x)$.

(c) Show that $c = \frac{1}{2}(a+b+1)$ and $a' = \frac{1}{2}a$, $b' = \frac{1}{2}b$, $c' = c$.

8.17 Show that

$$F(a, 1-a, c; x)$$

$$= (1-x)^{c-1} F\left(\frac{1}{2}[c-a], \frac{1}{2}[c+a-1], c; 4x[1-x]\right).$$

8.18 Show that

$$F\left(a, 1-a, c, \frac{1}{2}\right) = 2^{1-c}\,\frac{\Gamma(c)\,\Gamma\left(\frac{1}{2}\right)}{\Gamma\left(\frac{1}{2}[c+a]\right)\Gamma\left(\frac{1}{2}[c-a+1]\right)}.$$

8.19 Use (8.6.23) (with a and b interchanged) to derive Kummer's quadratic transformation (6.1.11).

8.20 Show that

$$F\left(a, b, \frac{1}{2}[a+b+1]; \frac{1}{2}\right) = \frac{\Gamma\left(\frac{1}{2}[a+b+1]\right)\Gamma\left(\frac{1}{2}\right)}{\Gamma\left(\frac{1}{2}[a+1]\right)\Gamma\left(\frac{1}{2}[b+1]\right)}.$$

8.21 Verify the identity (8.4.5).

8.22 Suppose $c > 0$, $c' > 0$. Let $w(x) = x^{c-1}(1-x)^{c'-1}, 0 < x < 1$.

(a) Show that the hypergeometric operator

$$L = x(1-x)\frac{d^2}{dx^2} + \left[c - (c+c')x\right]\frac{d}{dx}$$

is symmetric in the Hilbert space $L^2\left([0, 1],\, w(x)\,dx\right)$.

(b) Given $\lambda > 0$ and $f \in L_w^2$, the equation $Lu + \lambda u = f$ has a unique solution $u \in L_w^2$, expressible in the form

$$u(x) = \int_0^1 G_\lambda(x, y)\, f(y)\, dy.$$

(Note that if we set $a = c + c' - 1$ and $v > \max\{a, 0\}$ is chosen so that $\lambda = v(v-a)$, then $L + \lambda$ is the hypergeometric operator with indices $(a - v, v, c)$.) Compute the Green's function G_λ. Hint: see Section 3.3. The appropriate boundary conditions here are regularity at $x = 0$ and at $x = 1$.

8.23 Let $c, c', w(x), L$, and a be as in Exercise 8.22. Let $F_v(x) = F(a + v, -v, c; x)$, so that $F_0, F_1, F_2, \ldots$ are orthogonal polynomials for the weight w. What is wrong with the following argument? Suppose that $(a + v)v \neq (a + n)n, n = 0, 1, 2, \ldots$ Then

$$-\lambda_v(F_v, F_n)_w = (LF_v, F_n)_w = (F_v, LF_n)_w = -\lambda_n(F_v, F_n)_w.$$

Therefore $(F_v, F_n)_w = 0$ for all $n = 0, 1, 2, \ldots$ By Theorem 4.1.5, the orthogonalized polynomials

$$P_n(x) = \|F_n\|_w^{-1} F_n(x)$$

are complete in L_w^2. Therefore $F_v = 0$.

8.24 Verify the identities (8.7.1). Show that similar identities hold for the Kummer functions.

8.25 Change a to $2a$ and b to $a + 1$ in (8.6.10) and show that the resulting identity is a special case of (8.7.1).

8.26 Change a to $2a + 1$ and b to a in (8.6.14) and show that the resulting identity is a special case of (8.7.1).

8.27 Show that for $c > 2$ and $x > 0$,

$$F(1, 1, c; -x) = \frac{\Gamma(c)}{\Gamma(c - 2)} \int_0^1 (1 - s)^{c-3} \frac{\log(1 + sx)}{x}\, ds.$$

8.28 Show that

$$\cos ax = F\left(\frac{1}{2}a, -\frac{1}{2}a, \frac{1}{2}; \sin^2 x\right);$$

$$\sin ax = a \sin x\, F\left(\frac{1}{2} + \frac{1}{2}a, \frac{1}{2} - \frac{1}{2}a, \frac{3}{2}; \sin^2 x\right).$$

8.9 Summary

8.9.1 Hypergeometric series

Hypergeometric series have the form

$$\sum_{n=0}^{\infty} \frac{(a_1)_n (a_2)_n \cdots (a_p)_n}{(c_1)_n (c_2)_n \cdots (c_q)_n\, n!} x^n,$$

$c \neq 0, -1, -2, \ldots$ and $p \leq q + 1$. The corresponding function

$$_pF_q(a_1, a_2, \ldots, a_p; c_1, c_2, \ldots, c_q; x), \quad |x| < 1$$

is the solution of

$$L_{(a),(c)} F = 0, \quad F(0) = 1,$$

$$L_{(a),(c)} = x^{-1} D \prod_{j=1}^{q} (D + c_j - 1) - \prod_{k=1}^{p} (D + a_k), \quad D = D_x = x\frac{d}{dx}.$$

Examples:

$$_0F_0(x) = \sum_{n=0}^{\infty} \frac{x^n}{n!} = e^x;$$

$$_1F_0(a; x) = \sum_{n=0}^{\infty} \frac{(a)_n}{n!} x^n = \frac{1}{(1 - x)^a};$$

$$_0F_1\left(\frac{1}{2}; -\frac{1}{4}x^2\right) = \cos x;$$

$$_0F_1\left(\frac{3}{2}; -\frac{1}{4}x^2\right) = \frac{\sin x}{x};$$

$$_0F_1\left(\nu + 1; -\frac{1}{4}x^2\right) = \Gamma(\nu + 1)\left(\frac{2}{x}\right)^\nu J_\nu(x).$$

More examples, involving $F = {}_2F_1$, are listed below, where F is the Gauss hypergeometric function

$$_2F_1(a, b; c; x) = F(a, b, c; x) = \sum_{n=0}^{\infty} \frac{(a)_n (b)_n}{(c)_n n!} x^n, \quad c \neq 0, -1, -2, \ldots$$

The case $q = 1$, $p = 1$ is the "confluent hypergeometric" case of Chapter 6: $_1F_1 = M$.

8.9.2 Solutions of the hypergeometric equation

Two solutions of (8.0.1) are

$$F(a, b, c; x), \quad x^{1-c} F(a + 1 - c, b + 1 - c, 2 - c; x), \quad c \neq 0, \pm 1, \pm 2, \ldots$$

Other solutions can be produced by using linear fractional transformations that permute the singular points $\{0, 1, \infty\}$:

$$F(a, b, a + b + 1 - c; 1 - x);$$

$$(1 - x)^{c-a-b} F(c - a, c - b, 1 + c - a - b; 1 - x);$$

$$(-x)^{-a} F(a, 1 - c + a, a - b + 1; 1/x);$$

$$(-x)^{-b} F(1 - c + b, b, b - a + 1; 1/x),$$

with certain restrictions on the indices.

Pfaff's identity:

$$F(a, b, c; x) = (1 - x)^{-b} F\left(c - a, b, c; \frac{x}{x - 1}\right).$$

Euler's identity:

$$F(a, b, c; x) = (1 - x)^{c-a-b} F(c - a, c - b, c; x).$$

As $\nu \to +\infty$, for $0 < \theta < \pi$,

$$F\left(a + \nu, -\nu, c; \sin^2\left(\frac{1}{2}\theta\right)\right) = \frac{\Gamma(c) \cos\left(\nu\theta + \frac{1}{2}a\theta - \frac{1}{2}c\pi + \frac{1}{4}\pi\right) + O(\nu^{-1})}{\sqrt{\pi}\left(\nu \sin \frac{1}{2}\theta\right)^{c-\frac{1}{2}}\left(\cos \frac{1}{2}\theta\right)^{\frac{1}{2}+(a-c)}}.$$

$$\text{(8.9.1)}$$

8.9.3 Linear relations of solutions

The solutions listed above satisfy linear relations. When all coefficients are well-defined, no third index is a non-positive integer, and all arguments x and $\varphi(x)$ are in the complement of the ray $[1, \infty)$:

$$F(a, b, c; x) = \frac{\Gamma(c)\,\Gamma(c - a - b)}{\Gamma(c - a)\,\Gamma(c - b)} F(a, b, a + b + 1 - c; 1 - x)$$

$$+ \frac{\Gamma(c)\,\Gamma(a + b - c)}{\Gamma(a)\,\Gamma(b)}(1 - x)^{c-a-b} F(c - a, c - b, 1 + c - a - b; 1 - x);$$

$$F(a, b, a + b + 1 - c, 1 - x) = \frac{\Gamma(a + b + 1 - c)\,\Gamma(1 - c)}{\Gamma(a + 1 - c)\,\Gamma(b + 1 - c)} F(a, b, c; x)$$

$$+ \frac{\Gamma(a + b + 1 - c)\,\Gamma(c - 1)}{\Gamma(a)\,\Gamma(b)} x^{1-c} F(a + 1 - c, b + 1 - c, 2 - c; x);$$

$$F(a, b, c; x) = \frac{\Gamma(c)\,\Gamma(b - a)}{\Gamma(c - a)\,\Gamma(b)}(-x)^{-a} F\left(a, 1 + a - c, 1 + a - b; \frac{1}{x}\right)$$

$$+ \frac{\Gamma(c)\,\Gamma(a - b)}{\Gamma(c - b)\,\Gamma(a)}(-x)^{-b} F\left(b, 1 + b - c, 1 + b - a; \frac{1}{x}\right);$$

$$(-x)^{-a} F\left(a, 1 + a - c, 1 + a - b; \frac{1}{x}\right) = \frac{\Gamma(1 + a - b)\,\Gamma(1 - c)}{\Gamma(a + 1 - c)\,\Gamma(1 - b)} F(a, b, c; x)$$

$$+ \frac{\Gamma(1 + a - b)\,\Gamma(c - 1)}{\Gamma(c - b)\,\Gamma(a)} x^{1-c} F(a + 1 - c, b + 1 - c, 2 - c; x).$$

Additional identities can be obtained by using the Pfaff and Euler transforms. Particular cases give Gauss's summation formula

$$\sum_{n=0}^{\infty} \frac{(a)_n\,(b_n)}{(c)_n\,n!} = \frac{\Gamma(c)\,\Gamma(c - a - b)}{\Gamma(c - a)\,\Gamma(c - b)}, \quad \operatorname{Re}(c - a - b) > 0$$

and the Chu–Vandermonde identity

$$\sum_{k=0}^{n}(-1)^k\binom{n}{k}\frac{(a)_k}{(c)_k} = \frac{(c - a)_n}{(c)_n}, \quad c \neq 0, -1, -2, \ldots$$

8.9.4 Solutions when c is an integer

A second solution of (8.0.1) when c is a positive integer m is obtained as the value of

$$U(a, b, c; x) = \frac{\Gamma(1-c)}{\Gamma(a+1-c)\,\Gamma(b+1-c)}\, F(a, b, c; x)$$

$$+ \frac{\Gamma(c-1)}{\Gamma(a)\,\Gamma(b)}\, x^{1-c}\, F(a+1-c, b+1-c, 2-c; x)$$

at $c = m$, which is

$$U(a, b, m; x) = \frac{(-1)^m}{\Gamma(a+1-m)\,\Gamma(b+1-m)\,(m-1)!}$$

$$\times \left\{ \log x\, F(a, b, m; x) + \sum_{n=0}^{\infty} \frac{(a)_n (b)_n}{(m)_n\, n!} \left[\psi(a+n) + \psi(b+n) \right. \right.$$

$$\left. - \psi(n+1) - \psi(m+n) \right] x^n \Big\}$$

$$+ \frac{(m-2)!}{\Gamma(a)\,\Gamma(b)}\, x^{1-m} \sum_{n=0}^{m-2} \frac{(a+1-m)_n (b+1-m)_n}{(2-m)_n\, n!}\, x^n,$$

where $\psi(b) = \Gamma'(b)/\Gamma(b)$ and the last sum is taken to be zero if $m = 1$.
For c a non-positive integer we may take advantage of the identity

$$U(a, b, c; x) = x^{1-c}\, U(a+1-c, b+1-c, 2-c; x).$$

This solution, or a transformed version, may be substituted in the various linear relations among solutions in the exceptional cases.

8.9.5 Contiguous functions

Two hypergeometric functions are said to be contiguous if two indices are the same and the third indices differ by 1. Let

$$F(a\pm) = F(a \pm 1, b, c; x), \quad \text{etc.}$$

The basic identities follow from

$$x\, F' = a\big[F(a+) - F\big] = (c-1)\big[F(c-) - F\big].$$

They are

$$(a - b) F = a F(a+) - b F(b+);$$

$$(a - c + 1) F = a F(a+) - (c - 1) F(c-);$$

$$[a + (b - c)x] F = a(1 - x) F(a+) - \frac{(c - a)(c - b)x}{c} F(c+);$$

$$(1 - x) F = F(a-) - \frac{(c - b) x}{c} F(c+);$$

$$[2a - c + (b - a)x] F = a(1 - x) F(a+) - (c - a) F(a-);$$

$$[(c - 1) + (a + b + 1 - 2c)x] F = (c - 1)(1 - x) F(c-)$$
$$- \frac{(c - a)(c - b)x}{c} F(c+);$$

$$[1 - a + (c - b - 1)x] F = (c - a) F(a-) - (c - 1)(1 - x) F(c-);$$

$$(a + b - c) F = a(1 - x) F(a+) - (c - b) F(b-);$$

$$(b - a)(1 - x) F = (c - a) F(a-) - (c - b) F(b-).$$

8.9.6 Quadratic transformations

Quadratic transformations of the plane lead to the identities

$$F\left(a, b, \frac{1}{2}(a + b + 1); x\right) = F\left(\frac{1}{2}a, \frac{1}{2}b, \frac{1}{2}(a + b + 1); 4x(1 - x)\right);$$

$$F\left(a, b, a + b + \frac{1}{2}; x\right) = F\left(2a, 2b, a + b + \frac{1}{2}; \frac{1}{2} - \frac{1}{2}\sqrt{1 - x}\right)$$

$$= \left(\frac{1}{2} + \frac{1}{2}\sqrt{1 - x}\right)^{-2a} F\left(2a, a - b + \frac{1}{2}, a + b + \frac{1}{2}; \frac{\sqrt{1 - x} - 1}{\sqrt{1 - x} + 1}\right);$$

$$F(a, b, a - b + 1; x)$$

$$= (1 - x)^{-a} F\left(\frac{1}{2}a, \frac{1}{2}a - b + \frac{1}{2}, a - b + 1; -\frac{4x}{(1 - x)^2}\right)$$

$$= (1 + x)^{-a} F\left(\frac{1}{2}a, \frac{1}{2}a + \frac{1}{2}, a - b + 1; \frac{4x}{(1 + x)^2}\right)$$

$$= \left(1 + \sqrt{x}\right)^{-2a} F\left(a, a - b + \frac{1}{2}, 2a - 2b + 1; \frac{4\sqrt{x}}{(1 + \sqrt{x})^2}\right)$$

$$= \left(1 - \sqrt{x}\right)^{-2a} F\left(a, a - b + \frac{1}{2}, 2a - 2b + 1; -\frac{4\sqrt{x}}{(1 - \sqrt{x})^2}\right);$$

$$F\left(a, a + \frac{1}{2}, c; x\right)$$

$$= \left(\frac{1}{2} + \frac{1}{2}\sqrt{1 - x}\right)^{-2a} F\left(2a, 2a - c + 1, c; \frac{1 - \sqrt{1 - x}}{1 + \sqrt{1 - x}}\right)$$

$$= (1 - x)^{-a} F\left(2a, 2c - 2a - 1, c; \frac{\sqrt{1 - x} - 1}{2\sqrt{1 - x}}\right);$$

$$F\left(a, b, \frac{1}{2}(a + b + 1), x\right)$$

$$= (1 - 2x)^{-2a} F\left(\frac{1}{2}a, \frac{1}{2}a + \frac{1}{2}, \frac{1}{2}(a + b + 1); -\frac{4x(1 - x)}{(1 - 2x)^2}\right);$$

$$F(a, b, 2b; x) = \left(1 - \frac{1}{2}x\right)^{-a} F\left(\frac{1}{2}a, \frac{1}{2}a + \frac{1}{2}, b + \frac{1}{2}; \frac{x^2}{(2 - x)^2}\right)$$

$$= (1 - x)^{-\frac{1}{2}a} F\left(\frac{1}{2}a, b - \frac{1}{2}a, b + \frac{1}{2}; \frac{x^2}{4x - 4}\right)$$

$$= \left(\frac{1}{2} + \frac{1}{2}\sqrt{1 - x}\right)^{-2a} F\left(a, a - b + \frac{1}{2}, b + \frac{1}{2}; \left[\frac{1 - \sqrt{1 - x}}{1 + \sqrt{1 - x}}\right]^2\right)$$

$$= (1 - x)^{-\frac{1}{2}a} F\left(a, 2b - a, b + \frac{1}{2}; -\frac{(1 - \sqrt{1 - x})^2}{4\sqrt{1 - x}}\right);$$

$$F\left(a, a + \frac{1}{2}, c; x\right)$$

$$= \left(1 + \sqrt{x}\right)^{-2a} F\left(2a, c - \frac{1}{2}, 2c - 1; \frac{2\sqrt{x}}{1 + \sqrt{x}}\right)$$

$$= \left(1 - \sqrt{x}\right)^{-2a} F\left(2a, c - \frac{1}{2}, 2c - 1; -\frac{2\sqrt{x}}{1 - \sqrt{x}}\right).$$

8.9.7 Transformations and special values

Apart from some exceptional values of the parameters,

$$F(a, b, c; x) = x^{1+k-a} \frac{d^k}{dx^k} \left[\frac{x^{a-1}}{(a-k)_k} F(a-k, b, c; x) \right]$$

$$= x^{1-c} \frac{d^k}{dx^k} \left[\frac{x^{c+k-1}}{(c)_k} F(a, b, c+k; x) \right], \quad k = 0, 1, 2, \ldots$$

For $\operatorname{Re} a > 0$, $\operatorname{Re} b > 0$, define the integral transform

$$E_{\alpha,\beta} f(x) = \frac{\Gamma(\alpha+\beta)}{\Gamma(\alpha)\,\Gamma(\beta)} \int_0^1 s^{\alpha-1}(1-s)^{\beta-1} f(sx)\, ds.$$

Then for $\operatorname{Re} a > 0$ and $\operatorname{Re} \varepsilon > 0$ or for $\operatorname{Re} c > \operatorname{Re} \varepsilon > 0$ respectively,

$$F(a, b, c; x) = E_{a,\varepsilon}[F(a+\varepsilon, b, c; \cdot)](x)$$

$$= \frac{\Gamma(a+\varepsilon)}{\Gamma(a)\,\Gamma(\varepsilon)} \int_0^1 s^{a-1}(1-s)^{\varepsilon-1} F(a+\varepsilon, b, c; sx)\, ds,$$

$$F(a, b, c; x) = E_{c-\varepsilon,\varepsilon}[F(a, b, c-\varepsilon; \cdot)](x)$$

$$= \frac{\Gamma(c)}{\Gamma(c-\varepsilon)\,\Gamma(\varepsilon)} \int_0^1 s^{c-\varepsilon-1}(1-s)^{\varepsilon-1} F(a, b, c+\varepsilon; sx)\, ds.$$

A special case is Euler's integral form

$$F(a, b, c; x) = E_{a,c-a}[(1-x)^{-b}], \quad \operatorname{Re} c > \operatorname{Re} a > 0,$$

which uses the identity

$$F(a, b, a; x) = (1-x)^{-b}.$$

Other useful special cases are

$$F\left(a, a+\frac{1}{2}, 2a+1; x\right) = \left(\frac{1+\sqrt{1-x}}{2}\right)^{-2a};$$

$$F\left(a, a+\frac{1}{2}, 2a; x\right) = \frac{1}{\sqrt{1-x}} \left(\frac{1+\sqrt{1-x}}{2}\right)^{1-2a};$$

$$F\left(\frac{1}{2}a, \frac{1}{2}a+\frac{1}{2}, \frac{1}{2}; x^2\right) = \frac{1}{2}\left[(1+x)^{-a} + (1-x)^{-a}\right];$$

$$F\left(\frac{1}{2}a + \frac{1}{2}, \frac{1}{2}a + 1, \frac{3}{2}; x^2\right) = -\frac{1}{2ax}\left[(1+x)^{-a} - (1-x)^{-a}\right];$$

$$F(a+n, -n, c; x) = \frac{n!}{(c)_n} P_n^{(c-1, a-c)}(1 - 2x);$$

$$F\left(\nu, -\nu, \frac{1}{2}; \frac{1}{2}(1 - \cos\theta)\right) = \cos\nu\theta;$$

$$F\left(\nu + 2, -\nu, \frac{3}{2}; \frac{1}{2}(1 - \cos\theta)\right) = \frac{\sin(\nu + 1)\theta}{(\nu + 1)\sin\theta};$$

$$F\left(\nu + 1, -\nu, \frac{1}{2}; \frac{1}{2}(1 - \cos\theta)\right) = \frac{\cos\left(\nu + \frac{1}{2}\right)\theta}{\cos\frac{1}{2}\theta};$$

$$F\left(\nu + 1, -\nu, \frac{3}{2}; \frac{1}{2}(1 - \cos\theta)\right) = \frac{\sin\left(\nu + \frac{1}{2}\right)\theta}{(2\nu + 1)\sin\frac{1}{2}\theta};$$

$$F(1, 1, 2; -x) = \frac{\log(1 + x)}{x};$$

$$F\left(\frac{1}{2}, 1, \frac{3}{2}; -x^2\right) = \frac{\tan^{-1} x}{x};$$

$$F\left(\frac{1}{2}, \frac{1}{2}, \frac{3}{2}; x^2\right) = \frac{\sin^{-1} x}{x};$$

$$F\left(\frac{1}{2}, 1, \frac{3}{2}; x^2\right) = \frac{\tanh^{-1} x}{x};$$

$$F\left(\frac{1}{2}, \frac{1}{2}, \frac{3}{2}; -x^2\right) = \frac{\sinh^{-1} x}{x}.$$

A second transform is

$$E_{\alpha, \beta}^{(2)} f(x) = \frac{\Gamma(2\alpha + 2\beta)}{\Gamma(2\alpha)\,\Gamma(2\beta)} \int_0^1 s^{2\alpha - 1}(1 - s)^{2\beta - 1} f(s^2 x)\, ds.$$

Then

$$F\left(a, a + \frac{1}{2}, c; x\right) = E_{a, \frac{1}{4}}^{(2)}\left[F\left(a + \frac{1}{4}, a + \frac{3}{4}, c; x\right)\right].$$

Two examples used in Chapter 9 are

$$F(\nu + 1, -\nu, 1; x) = E_{\frac{1}{2}, \frac{1}{2}} \left[F\left(\nu + 1, -\nu; \frac{1}{2}; x \right) \right]$$

$$= \frac{1}{\pi} \int_0^1 \frac{\mathrm{Re}\left\{ \left[1 - 2sx + i\sqrt{4sx(1 - sx)}\right]^{\nu + \frac{1}{2}} \right\}}{\sqrt{1 - sx}} \frac{ds}{\sqrt{s(1 - s)}};$$

$$F\left(\frac{1}{2}\nu + \frac{1}{2}, \frac{1}{2}\nu + 1; \nu + \frac{3}{2}; x \right)$$

$$= E^{(2)}_{\frac{1}{2}(\nu+1), \frac{1}{4}} \left[F\left(\frac{1}{2}\nu + \frac{3}{4}, \frac{1}{2}\nu + \frac{5}{4}; \nu + \frac{3}{2}; x \right) \right]$$

$$= \frac{\Gamma\left(\nu + \frac{3}{2}\right)}{\Gamma(\nu + 1)\sqrt{\pi}} \int_0^1 \left(\frac{1 + \sqrt{1 - s^2 x}}{2} \right)^{-\nu - \frac{1}{2}} \frac{s^\nu \, ds}{\sqrt{1 - s^2 x}\sqrt{1 - s}}.$$

8.10 Remarks

The hypergeometric equation was studied by Euler and Gauss, among others. According to Askey in the addendum to [280], both the hypergeometric equation and its series solution were probably first written down by Euler in a manuscript dated 1778 and published in 1794 [90]. The six solutions in Section 8.2 and the relations among them were obtained by Kummer in 1836 [168]; see the discussion by Prosser [234]. Another method for obtaining such relations was introduced by Barnes [20]. The contiguous relations were obtained by Gauss in his 1812 paper on hypergeometric series [103]. They include as special cases many of the recurrence relations given elsewhere.

Riemann [239] studied hypergeometric functions from a function-theoretic point of view. Riemann's study of the conformal mappings defined by the quotient of two solutions of the same equation was carried out systematically by Schwarz [254]. The theory and history, pre-Riemann and post-Riemann, are treated in Klein's classic work [156], which has an extensive bibliography. The history, and its roots in work of Wallis, Newton, and Stirling, is also discussed in Dutka [75].

Hypergeometric functions are discussed in almost every text on special functions; see the relatively recent books by Andrews, Askey, and Roy [7] and by Seaborn [255]. In particular, [7] contains a different treatment of quadratic transformations and more information about general hypergeometric series. Various algorithms have been developed to establish identities involving hypergeometric series; see Koepf [159], Petkovšek, Wilf, and Zeilberger [228], Wilf and Zeilberger [317]. Algebraic solutions of the hypergeometric equation are treated in Matsuda [199].

Separation of variables in various singular or degenerate partial differential equations leads to solutions that are expressible in terms of hypergeometric functions. These include the Euler–Poisson–Darboux equation, see Darboux [63], Copson [57]; the Tricomi equation, see Delache and Leray [68]; certain subelliptic operators, see Beals, Gaveau, and Greiner [22]; and certain singular or degenerate hyperbolic operators, see Beals and Kannai [23].

A modified version of hypergeometric series is called "basic hypergeo-metric series." The basic series that corresponds to the series for the Gauss hypergeometric function $F = {}_2F_1$ is

$$2\Phi_1(a, b, c; q; x) = \sum_{n=0}^{\infty} \frac{(a; q)_n \, (b; q)_n}{(c; q)_n \, (q; q)_n} x^n,$$

in which the extended factorials $(a)_n$ and so on are replaced by the products

$$(a; q)_n = (1 - a)(1 - aq) \cdots (1 - aq^{n-1})$$

and so on. Identities involving these expressions also go back to Euler and Gauss. This "q-analysis" has applications in number theory and in combinato-rial analysis; see Andrew, Askey, and Roy [7], Bailey [18], Fine [97], Gasper and Rahman [102], and Slater [261].

There are a number of multidimensional generalizations of hypergeometric series and integrals; see Appell and Kampé de Fériet [10], Dwork [78], Mathai and Saxena [197], and Varchenko [297].

9

Spherical functions

Spherical functions are solutions of the equation

$$(1 - x^2) u''(x) - 2x u'(x) + \left[v(v + 1) - \frac{m^2}{1 - x^2} \right] u(x) = 0, \qquad (9.0.1)$$

that arises from separating variables in Laplace's equation $\Delta u = 0$ in spherical coordinates. *Surface harmonics* are the restrictions to the unit sphere of *harmonic functions* (solutions of Laplace's equation) in three variables. For surface harmonics, m and v are non-negative integers. The case $m = 0$ is Legendre's equation

$$(1 - x^2) u''(x) - 2x u'(x) + v(v + 1) u(x) = 0, \qquad (9.0.2)$$

with v a non-negative integer. The solutions to equation (9.0.1) that satisfy the associated boundary conditions are Legendre polynomials and certain multiples of their derivatives. These functions are the building blocks for all surface harmonics. They satisfy a number of important identities.

For general values of the parameter v, Legendre's equation (9.0.2) has linearly independent solutions $P_v(z)$, holomorphic for z in the complement of $(-\infty, -1]$, and $Q_v(z)$, holomorphic in the complement of $(-\infty, 1]$. These Legendre functions satisfy a number of identities and have several representations as integrals.

For most values of the parameter v, the four functions

$$P_v(z), \quad P_v(-z), \quad Q_v(z), \quad Q_{-v-1}(z)$$

are distinct solutions of (9.0.2), so there are linear relations connecting any three. Integer and half-integer values of v are exceptional cases.

The solutions of the spherical harmonic equation (9.0.1) with $m = 1, 2, \ldots$ are known as *associated Legendre functions*. They are closely related to derivatives $P_v^{(m)}$ and $Q_v^{(m)}$ of the Legendre functions. The associated Legendre

300

functions also satisfy a number of important identities and have representations as integrals.

9.1 Harmonic polynomials; surface harmonics

A polynomial $P(x, y, z)$ in three variables is said to be *homogeneous* of degree n if it is a linear combination of monomials of degree n:

$$P(x, y, z) = \sum_{j+k+l=n} a_{jkl} x^j y^k z^l.$$

By definition, the zero polynomial is homogeneous of all degrees. A homogeneous polynomial P of degree n is said to be *harmonic* if it satisfies Laplace's equation

$$\Delta P = P_{xx} + P_{yy} + P_{zz} = 0.$$

The homogeneous polynomials of degree n and the harmonic polynomials of degree n are vector spaces over the real or complex numbers. (Note that our convention here is that a harmonic polynomial is, by definition, homogeneous of some degree.)

Proposition 9.1.1 *The space of homogeneous polynomials of degree n has dimension $(n + 2)(n + 1)/2$. The space of harmonic polynomials of degree n has dimension $2n + 1$.*

Proof The monomials of degree n are a basis for the homogeneous polynomials of degree n. Such a monomial has the form

$$x^{n-j-k} y^j z^k,$$

and is uniquely determined by the pair $(j, j + k + 1)$, which corresponds to a choice of two distinct elements from the set $\{0, 1, \ldots, n + 1\}$. This proves the first statement. To prove the second, we note that Δ maps homogeneous polynomials of degree n to homogeneous polynomials of degree $n - 2$. It is enough to show that this map is surjective (onto), since then its kernel – the space of harmonic polynomials of degree n – has dimension

$$\frac{(n + 2)(n + 1)}{2} - \frac{n(n - 1)}{2} = 2n + 1.$$

To prove surjectivity we note that $\Delta x^n = n(n - 1)x^{n-2}$, and in general

$$\Delta\left(x^{n-j-k} y^j z^k\right) = (n - j - k)(n - j - k - 1)x^{n-2-j-k} y^j z^k + R_{njk}$$

where R_{njk} has total degree $j + k - 2$ in y and z. It follows by recursion on j that monomials $x^{n-2-j} y^j$ are in the range of Δ, and then by recursion on k that monomials $x^{n-2-j-k} y^j z^k$ are in the range of Δ. $\square$

In spherical coordinates (3.6.7), monomials take the form

$$x^j y^k z^l = r^{j+k+l} \cos^j \varphi \, \sin^k \varphi \, \sin^{j+k} \theta \, \cos^l \theta.$$

In particular, a harmonic polynomial of degree n has the form

$$P(r, \theta, \varphi) = r^n Y(\theta, \varphi),$$

where Y is a trigonometric polynomial in θ and φ, of degree at most n in each, and

$$\frac{1}{\sin^2 \theta} Y_{\varphi\varphi} + \frac{1}{\sin \theta} [\sin \theta \, Y_\theta]_\theta + n(n + 1) Y = 0; \qquad (9.1.1)$$

see (3.6.10). The function Y can be regarded as the restriction of the harmonic polynomial P to the unit sphere $\{x^2 + y^2 + z^2 = 1\}$; it is called a surface harmonic of degree n.

As in Section 3.6 we may seek to solve (9.1.1) by separating variables: if $Y(\theta, \varphi) = \Theta(\theta)\Phi(\varphi)$, then wherever the product $\Theta\Phi \neq 0$ we must have

$$\frac{\Phi''}{\Phi} + \left\{ \sin \theta \, \frac{[\sin \theta \, \Theta']'}{\Theta} + n(n + 1) \sin^2 \theta \right\} = 0.$$

The first summand is a function of φ alone and the second is a function of θ alone, so each summand is constant. Since Y is a trigonometric polynomial in φ of degree at most n, it follows that $\Phi'' = -m^2 \Phi$ for some integer $m = -n, -n + 1, \ldots, n - 1, n$. This gives us $2n + 1$ linearly independent choices for Φ:

$$\cos m\varphi, \quad m = 0, 1, \ldots, n; \quad \sin m\varphi, \quad m = 1, 2, \ldots, n,$$

or the complex version

$$e^{im\varphi}, \quad m = 0, \pm 1, \ldots, \pm n.$$

Let us take $\Phi(x) = e^{imx}$. Then Θ is a solution of the equation

$$\frac{1}{\sin \theta} \frac{d}{d\theta} \left[\sin \theta \, \frac{d\Theta}{d\theta} \right] + \left[n(n + 1) - \frac{m^2}{\sin^2 \theta} \right] \Theta = 0.$$

As noted in Section 3.6, the change of variables $x = \cos\theta$ converts the preceding equation to the spherical harmonic equation

$$(1 - x^2)u''(x) - 2x\,u'(x) + \left[n(n+1) - \frac{m^2}{1-x^2}\right]u(x) = 0, \quad 0 < x < 1.$$
(9.1.2)

Suppose for the moment that $m \geq 0$. As noted in Section 3.7, the gauge transformation $u(x) = (1 - x^2)^{\frac{1}{2}m} v(x)$ reduces (9.1.2) to the equation

$$(1 - x^2)\,v'' - 2(m+1)x\,v' + (n-m)(n+m+1)\,v = 0. \quad (9.1.3)$$

This has as a solution the Jacobi polynomial $P_{n-m}^{(m,m)}$, so (9.1.2) has a solution

$$(1 - x^2)^{\frac{1}{2}m} P_{n-m}^{(m,m)}(x).$$

In view of (4.3.11), $P_{n-m}^{(m,m)}$ is a multiple of the mth derivative $P_n^{(m)}$ of the Legendre polynomial of degree n:

$$P_{n-m}^{(m,m)} = \frac{2^m n!}{(n+m)!} P_n^{(m)}. \quad (9.1.4)$$

Therefore Θ_{nm} is a multiple of the associated Legendre function

$$P_n^m(x) = (1 - x^2)^{\frac{1}{2}m} P_n^{(m)}(x). \quad (9.1.5)$$

Formula (9.1.5) may be used to obtain an integral formula for P_n^m. The Rodrigues formula (4.2.12) for P_n and the Cauchy integral representation for the derivative give

$$P_n^m(x) = (-1)^n \frac{1}{2^n n!} (1 - x^2)^{\frac{1}{2}m} \frac{(n+m)!}{2\pi i} \int_C \frac{(1-s^2)^n\,ds}{(s-x)^{n+m+1}},$$

where C is a contour enclosing x. Let us take C to be the circle centered at x with radius $\rho = \sqrt{1 - x^2}$:

$$s(\alpha) = x + \rho e^{i\alpha} = x + \sqrt{1 - x^2}\,e^{i\alpha}.$$

Then

$$1 - s(\alpha)^2 = -2\rho e^{i\alpha}(x + i\rho\sin\alpha)$$

so for $m \geq 0$,

$$P_n^m(\cos\theta) = \frac{(m+n)!}{n!} \frac{1}{2\pi} \int_0^{2\pi} e^{-im\alpha}(\cos\theta + i\sin\theta\sin\alpha)^n\,d\alpha. \quad (9.1.6)$$

In defining Θ_{nm} and Y_{nm}, we normalize so that the solution has L^2 norm 1 on the interval $[-1, 1]$ or, equivalently, the corresponding multiple of $P_{n-m}^{(m,m)}$

has L^2 norm 1 with respect to the weight $(1 - x^2)^m$. Using (4.2.15) and the Rodrigues formula we obtain the solution

$\Theta_{nm}(x)$

$$= \frac{1}{2^m n!} \left[\left(n + \frac{1}{2} \right) (n - m)!(n + m)! \right]^{\frac{1}{2}} (1 - x^2)^{\frac{1}{2}m} P_{n-m}^{(m,m)}(x)$$

$$= (-1)^{n-m} \frac{1}{2^n n!} \left[\frac{2n + 1}{2} \frac{(n + m)!}{(n - m)!} \right]^{\frac{1}{2}} (1 - x^2)^{-\frac{1}{2}m} \frac{d^{n-m}}{dx^{n-m}} \left[(1 - x^2)^n \right].$$

$$(9.1.7)$$

Combining (9.1.7), (9.1.4), and the Rodrigues formula for P_n, we obtain

$\Theta_{nm}(x)$

$$= \left[\frac{2n + 1}{2} \frac{(n - m)!}{(n + m)!} \right]^{\frac{1}{2}} (1 - x^2)^{\frac{1}{2}m} P_n^{(m)}(x)$$

$$= (-1)^n \frac{1}{2^n n!} \left[\frac{2n + 1}{2} \frac{(n - m)!}{(n + m)!} \right]^{\frac{1}{2}} (1 - x^2)^{\frac{1}{2}m} \frac{d^{n+m}}{dx^{n+m}} \left[(1 - x^2)^n \right].$$

$$(9.1.8)$$

The expressions (9.1.7) and (9.1.8) were obtained under the assumption that $m \geq 0$. Comparing the last part of each, it is natural to define Θ_{nm} for $m = -1, -2, \ldots$ by

$$\Theta_{n,-m}(x) = (-1)^m \Theta_{nm}(x), \quad m = 1, 2, \ldots, n. \tag{9.1.9}$$

Then the last parts of (9.1.7) and (9.1.8) are valid for all integers $m = -n$, $-n + 1, \ldots, n$.

The previous considerations lead to surface harmonics

$$Y_{nm}(\theta, \varphi) = \frac{1}{\sqrt{2\pi}} e^{im\varphi} \Theta_{nm}(\cos \theta), \quad -n \leq m \leq n. \tag{9.1.10}$$

The factor $1/\sqrt{2\pi}$ is introduced so that the $\{Y_{nm}\}$ are orthonormal with respect to the normalized measure $\sin \theta \, d\theta \, d\varphi$ on the sphere.

Combining (9.1.6) with (9.1.10) and (9.1.8) gives Laplace's integral [175]:

$$Y_{nm}(\theta, \varphi) = \frac{A_{nm}}{2\pi} \int_0^{2\pi} e^{im(\varphi - \alpha)} (\cos \theta + i \sin \theta \sin \alpha)^n \, d\alpha, \tag{9.1.11}$$

$$A_{nm} = \frac{1}{n!} \left[\frac{2n + 1}{4\pi} (n - m)!(n + m)! \right]^{\frac{1}{2}}. \tag{9.1.12}$$

In view of (9.1.8), this is valid for all $m = -n, \ldots, n$. This leads to an integral formula for the corresponding harmonic polynomial:

$$r^n Y_{nm}(\theta, \varphi) = \frac{A_{nm}}{2\pi} \int_0^{2\pi} e^{-im\alpha} [r \cos\theta + ir \sin\theta \sin(\alpha + \varphi)]^n \, d\alpha.$$

We have proved the following.

Theorem 9.1.2 *The functions Y_{nm} of (9.1.10) are a basis for the surface harmonics of degree n. They are orthonormal with respect to normalized surface measure $\sin\theta \, d\theta \, d\varphi$. The corresponding harmonic polynomials of degree n have the form*

$$r^n Y_{nm}(\theta, \varphi) = \frac{A_{nm}}{2\pi} \int_0^{2\pi} e^{-im\alpha} (z + ix \sin\alpha + iy \cos\alpha)^n \, d\alpha \qquad (9.1.13)$$

where A_{nm} is given by (9.1.12).

Remarks. 1. If we allow both indices n, m to vary, the functions Y_{nm} are still orthonormal. This is clear when the second indices differ, since the factors involving φ are orthogonal. When the second indices are the same, but the first indices differ, the second factors involve polynomials $P_k^{(m,m)}$ with different indices k, and again are orthogonal.

2. A second solution of the equation $r^2 R'' + 2r R' - n(n+1)R = 0$ is $R(r) = r^{-n-1}$. Thus to each surface harmonic Y_{nm} there corresponds a function

$$r^{-n-1} Y_{nm}(\theta, \varphi)$$

which is harmonic away from the origin, and, in particular, in the exterior of the sphere.

Suppose that O is a rotation about the origin in $\mathbf{R}^3$ (an orthogonal transformation with determinant 1). The Laplacian Δ is invariant with respect to O:

$$\Delta[f(Op)] = [\Delta f](Op), \quad p \in \mathbf{R}^3.$$

If $P(p)$ is a homogeneous polynomial of degree n, then so is $Q(p) = P(Op)$. Therefore the space of harmonic polynomials of degree n is invariant under rotations. The surface measure $\sin\theta \, d\theta \, d\varphi$ is also invariant under rotations. It follows that any rotation takes the orthonormal basis $\{Y_{nm}\}$ to an orthonormal basis; thus it induces a unitary transformation in the (complex) space of spherical harmonics of degree n. This implies that the sum

$$F_n(\theta, \varphi; \theta', \varphi') = \sum_{m=-n}^{n} Y_{nm}(\theta, \varphi) \overline{Y}_{nm}(\theta', \varphi'),$$

where $\overline{Y}_{nm}$ is the complex conjugate, is left unchanged if both points (θ, φ) and (θ', φ') are subjected to the same rotation O. It follows that this sum is a function of the inner product between the points with spherical coordinates $(1, \theta, \varphi)$ and $(1, \theta', \varphi')$:

$$(\cos\varphi \sin\theta, \sin\varphi \sin\theta, \cos\theta) \cdot (\cos\varphi' \sin\theta', \sin\varphi' \sin\theta', \cos\theta')$$

$$= \cos(\varphi - \varphi') \sin\theta \sin\theta' + \cos\theta \cos\theta'.$$

If we take $(\theta', \varphi') = (0, 0)$, which corresponds to the point with Cartesian coordinates $(0, 0, 1)$, then the inner product is $\cos\theta$. With this choice, the function is independent of rotations around the z-axis, which implies that it is a multiple of Y_{n0}:

$$c_n Y_{n0}(\theta, \varphi) = \sum_{m=-n}^{n} Y_{nm}(\theta, \varphi) \overline{Y}_{nm}(0, 0). \qquad (9.1.14)$$

The constant c_n may be determined by multiplying both sides by $\overline{Y}_{n0}(\theta, \varphi)$ and integrating over the sphere. This computation, together with (9.1.10), (9.1.7), and (4.6.4), yields

$$c_n = \overline{Y}_{n0}(0, 0) = \frac{1}{\sqrt{2\pi}} \Theta_{n0}(1)$$

$$= \frac{\sqrt{2n+1}}{\sqrt{4\pi}}.$$

The identities (9.1.8) and (9.1.9), together with (9.1.10), imply that

$$\frac{\sqrt{2n+1}}{\sqrt{4\pi}} Y_{n0}(\theta, \varphi) = \frac{2n+1}{4\pi} P_n(\cos\theta);$$

$$Y_{n0}(\theta, \varphi) \overline{Y}_{n0}(\theta', \varphi') = \frac{2n+1}{4\pi} P_n(\cos\theta) P_n(\cos\theta');$$

$$Y_{n,\pm m}(\theta, \varphi) \overline{Y}_{n,\pm m}(\theta', \varphi')$$

$$= \frac{2n+1}{4\pi} \frac{(n-m)!}{(n+m)!} e^{\pm im(\varphi-\varphi')} P_n^m(\cos\theta) P_n^m(\cos\theta'), \quad m = 1, 2, \ldots, n.$$

This shows that the left-hand side, and each summand on the right-hand side, of (9.1.14) is a function of the difference $\varphi - \varphi'$. We take $\varphi' = 0$ and obtain

Legendre's *addition formula* [181]:

$$P_n(\cos \varphi \sin \theta \sin \theta' + \cos \theta \cos \theta')$$

$$= P_n(\cos \theta) P_n(\cos \theta') + 2 \sum_{m=1}^{n} \frac{(n-m)!}{(n+m)!} \cos(m\varphi) P_n^m(\cos \theta) P_n^m(\cos \theta').$$

$$(9.1.15)$$

We have chosen the surface harmonics Y_{nm} to be orthonormal in $L^2(\Sigma)$, the space of functions on the sphere whose squares are integrable with respect to the measure $\sin \theta \, d\theta \, d\varphi$.

Theorem 9.1.3 *The surface harmonics $\{Y_{nm}\}$ are a complete orthonormal set in $L^2(\Sigma)$.*

Proof The restrictions to the sphere Σ of polynomials are dense in the space $L^2(\Sigma)$. Therefore it is sufficient to show that any such restriction can be written as a sum of harmonic polynomials. It is enough to consider homogeneous polynomials. For this, see Exercise 9.2. □

9.2 Legendre functions

A Legendre function is a solution of the Legendre differential equation

$$(1 - z^2) u''(z) - 2z \, u'(z) + \nu(\nu + 1) u(z) = 0, \qquad (9.2.1)$$

where the parameter ν is not necessarily a non-negative integer. When ν is a non-negative integer, one solution is the Legendre polynomial P_ν. We know from Section 4.3 that this solution has the integral representation

$$P_\nu(z) = \frac{1}{2^{\nu+1}\pi i} \int_C \frac{(t^2 - 1)^\nu}{(t - z)^{\nu+1}} \, dt, \qquad (9.2.2)$$

where C is a contour that encloses z. As noted in Section 4.9, the function P_ν defined by (9.2.2) for general values of ν is still a solution of (4.7.2) if C is a closed curve lying in a region where $[(t^2 - 1)/(t - z)]^\nu$ is holomorphic. We use (9.2.2) to define a solution that is holomorphic in the complement of the interval $(-\infty, -1]$ by choosing C in this complement in such a way as to enclose both $t = z$ and $t = 1$. The resulting solution satisfies $P_\nu(1) = 1$, and is called the *Legendre function of the first kind*.

As noted in Sections 3.4, 4.6, and 9.1, the change of variables $y = \frac{1}{2}(1 - z)$ converts (9.2.1) to the hypergeometric equation

$$y(1 - y) v''(y) + (1 - 2y) v'(y) + v(v + 1) v(y) = 0.$$

It follows that

$$P_v(z) = F\left(v + 1, -v, 1; \frac{1}{2}(1 - z)\right). \tag{9.2.3}$$

Since the hypergeometric function is unchanged if the first two indices are interchanged, it follows that

$$P_{-v-1}(z) = P_v(z). \tag{9.2.4}$$

Suppose that $z > 1$. Then we may take the curve C in (9.2.2) to be the circle of radius $R = \sqrt{z^2 - 1}$ centered at z. Let

$$t = t(\varphi) = z + \sqrt{z^2 - 1}\, e^{i\varphi}, \quad -\pi \le \varphi \le \pi.$$

Then (9.2.2) becomes the general form of Laplace's integral:

$$P_v(z) = \frac{1}{\pi} \int_0^\pi \left[z + \sqrt{z^2 - 1}\, \cos\varphi\right]^v d\varphi. \tag{9.2.5}$$

Combining this with (9.2.4) gives

$$P_v(z) = \frac{1}{\pi} \int_0^\pi \frac{d\varphi}{\left[z + \sqrt{z^2 - 1}\, \cos\varphi\right]^{v+1}}. \tag{9.2.6}$$

Each of these formulas extends analytically to z in the complement of $(-\infty, 1]$.
The change of variables

$$e^\alpha = z + \sqrt{z^2 - 1}\, \cos\varphi$$

leads to the Dirichlet–Mehler integral representation [71, 201]:

$$P_v(\cosh\theta) = \frac{1}{\pi} \int_{-\theta}^\theta \frac{e^{-\left(v+\frac{1}{2}\right)\alpha}\, d\alpha}{\sqrt{2\cosh\theta - 2\cosh\alpha}}. \tag{9.2.7}$$

This representation shows that

$$P_v(x) > 0, \quad \text{for} \quad 1 < x < \infty, \quad v \text{ real}. \tag{9.2.8}$$

Note that if u is a solution of (9.2.1), then so is $v(x) = u(-x)$. The third index in the hypergeometric function in (9.2.3) is the sum of the first two, so this is one of the exceptional cases discussed in Section 8.4; for non-integer v the hypergeometric function in (9.2.3) has a logarithmic singularity at $z = 0$. For non-integer v, $P_v(x)$ and $P_v(-x)$ are independent solutions of (9.2.1).

However for integer v, $P_v = P_{-v-1}$ is a Legendre polynomial and $P_v(-x) = (-1)^v P_v(x)$.

To find a second solution of (9.2.1) we proceed as suggested in Section 4.9 by adapting the formula (9.2.2) to a different contour. The *Legendre function of the second kind* is defined to be

$$Q_v(z) = \frac{1}{2^{v+1}} \int_{-1}^1 \frac{(1-s^2)^v\, ds}{(z-s)^{v+1}}, \quad v \neq -1, -2, \dots \tag{9.2.9}$$

Here we take the principal branch of the power w^{-v} on the complement of the negative real axis, so that Q_v is holomorphic in the complement of the interval $(-\infty, 1]$.

The function Q_v is a multiple of a hypergeometric function, but with argument z^{-2}. To verify this, suppose that $|z| > 1$ and $\mathrm{Re}\,(v+1) > 0$. Then

$$Q_v(z) = \frac{1}{(2z)^{v+1}} \int_{-1}^1 \frac{(1-s^2)^v\, ds}{(1-s/z)^{v+1}}$$

$$= \frac{1}{(2z)^{v+1}} \sum_{k=0}^\infty \left[\frac{(v+1)_k}{k!} \int_{-1}^1 (1-s^2)^v s^k\, ds \right] z^{-k}.$$

The integral vanishes for odd k. For $k = 2n$,

$$\int_{-1}^1 (1-s^2)^v s^{2n}\, ds = \int_0^1 (1-t)^v t^{n-\frac{1}{2}}\, dt$$

$$= \mathrm{B}\left(v+1, n+\frac{1}{2}\right) = \frac{\Gamma(v+1)\,\Gamma\left(n+\frac{1}{2}\right)}{\Gamma\left(v+n+\frac{3}{2}\right)}$$

$$= \frac{\Gamma(v+1)\,\Gamma\left(\frac{1}{2}\right)\left(\frac{1}{2}\right)_n}{\Gamma\left(v+\frac{3}{2}\right)\left(v+\frac{3}{2}\right)_n}.$$

Also

$$\frac{(v+1)_{2n}}{(2n)!} = \frac{2^{2n}\left(\frac{1}{2}v+\frac{1}{2}\right)_n \left(\frac{1}{2}v+1\right)_n}{2^{2n}\left(\frac{1}{2}\right)_n n!}.$$

Therefore, for $|z| > 1$ and $\mathrm{Re}\,(v+1) > 0$,

$$Q_v(z) = \frac{\Gamma(v+1)\,\Gamma\left(\frac{1}{2}\right)}{\Gamma\left(v+\frac{3}{2}\right)(2z)^{v+1}} \sum_{n=0}^\infty \frac{\left(\frac{1}{2}v+\frac{1}{2}\right)_n \left(\frac{1}{2}v+1\right)_n}{\left(v+\frac{3}{2}\right)_n n!} z^{-2n}$$

$$= \frac{\Gamma(v+1)\,\sqrt{\pi}}{\Gamma\left(v+\frac{3}{2}\right)(2z)^{v+1}} F\left(\frac{1}{2}v+\frac{1}{2}, \frac{1}{2}v+1, v+\frac{3}{2}; \frac{1}{z^2}\right). \tag{9.2.10}$$

This formula allows us to define Q_ν for all values of ν except $\nu = -1, -2, \ldots$ The apparent difficulty for $\nu + \frac{3}{2}$ a non-positive integer is overcome by noting that the coefficient of $z^{-\nu-1-2n}$ has $\Gamma\left(\nu + \frac{3}{2} + n\right)$ rather than $\left(\nu + \frac{3}{2}\right)_n$ in the denominator. A consequence is that when $\nu = -m - \frac{1}{2}$ for m a positive integer, these coefficients vanish for $n < m$.

We take the principal branch of the power z^ν. It follows that

$$Q_\nu(-z) = -e^{\nu \pi i}\, Q_\nu(z), \quad \operatorname{Im} z > 0. \tag{9.2.11}$$

Suppose $z > 1$. The change of variables

$$s = z - \sqrt{z^2 - 1}\, e^\theta$$

in the integral (9.2.9) gives the integral representation

$$Q_\nu(z) = \int_0^\alpha \left[z - \sqrt{z^2 - 1}\, \cosh\theta\right]^\nu d\theta, \tag{9.2.12}$$

where $\coth\alpha = z$. This extends by analytic continuation to all z in the complement of $(-\infty, 1]$.

A further change of variables

$$\left[z - \sqrt{z^2 - 1}\, \cosh\theta\right]^{-1} = z + \sqrt{z^2 - 1}\, \cosh\varphi$$

leads to a form similar to (9.2.6), due to Heine [125]:

$$Q_\nu(z) = \int_0^\infty \frac{d\varphi}{\left[z + \sqrt{z^2 - 1}\, \cosh\varphi\right]^{\nu+1}}, \quad \operatorname{Re}\nu > 0. \tag{9.2.13}$$

Let $z = \cosh\theta$. The change of variables

$$e^\alpha = \cosh\theta + \sinh\theta\,\cosh\varphi$$

leads to a form similar to (9.2.7):

$$Q_\nu(\cosh\theta) = \int_\theta^\infty \frac{e^{-\left(\nu+\frac{1}{2}\right)\alpha}\, d\alpha}{\sqrt{2\cosh\alpha - 2\cosh\theta}}. \tag{9.2.14}$$

This representation shows that

$$Q_\nu(x) > 0, \quad \text{for} \quad 1 < x < \infty, \quad \nu \text{ real}, \quad \nu \neq -1, -2, \ldots \tag{9.2.15}$$

9.3 Relations among the Legendre functions

Although $P_\nu = P_{-\nu-1}$, the same is not generally true for Q_ν. To sort out the remaining relationships among the solutions

$$P_\nu(z), \quad P_\nu(-z), \quad Q_\nu(z), \quad Q_{-\nu-1}(z)$$

of the Legendre equation (9.2.1), we start by computing the Wronskian of Q_ν and $Q_{-\nu-1}$. If u_1 and u_2 are any two solutions, the Wronskian $W = u_1 u_2' - u_2 u_1'$ satisfies

$$(1 - z^2) W'(z) = 2z W(z).$$

Therefore wherever $W(z) \neq 0$, it satisfies $W'(z)/W(z) = -(1 - z^2)'/(1 - z^2)$:

$$W(u_1, u_2)(z) = \frac{C}{1 - z^2}, \quad C = \text{constant}.$$

To compute the constant it is enough to compute asymptotics of u_j and u_j', say as $z = x \to +\infty$. It follows from (9.2.10) or directly from (9.2.9) that as $x \to +\infty$,

$$Q_\nu(x) \sim c_\nu x^{-\nu-1}, \quad Q_\nu'(x) \sim -c_\nu(\nu + 1) x^{-\nu-2};$$

$$c_\nu = \frac{\Gamma(\nu + 1)\sqrt{\pi}}{\Gamma\left(\nu + \frac{3}{2}\right) 2^{\nu+1}},$$

so long as ν is not a negative integer. Therefore the Wronskian is asymptotically

$$c_\nu c_{-\nu-1} \begin{vmatrix} x^{-\nu-1} & x^\nu \\ -(\nu+1)x^{-\nu-2} & \nu x^{\nu-1} \end{vmatrix} = \frac{(2\nu + 1) c_\nu c_{-\nu-1}}{x^2}.$$

The reflection formula (2.2.7) gives

$$c_\nu c_{-\nu-1} = \frac{\Gamma(\nu + 1)\Gamma(-\nu)\pi}{\left(\nu + \frac{1}{2}\right)\Gamma\left(\nu + \frac{1}{2}\right)\Gamma\left(-\nu + \frac{1}{2}\right) 2}$$

$$= \frac{\pi \sin\left(\nu\pi + \frac{1}{2}\pi\right)}{(2\nu + 1)\sin(-\nu\pi)} = -\frac{\pi \cos\nu\pi}{(2\nu + 1)\sin\nu\pi}.$$

It follows that

$$W(Q_\nu, Q_{-\nu-1})(z) = \frac{\pi \cot\nu\pi}{1 - z^2}. \tag{9.3.1}$$

Recall that Q_ν is not defined when ν is a negative integer, so $Q_{-\nu-1}$ is not defined when ν is a non-negative integer. Therefore the right-hand side of (9.3.1) is well-defined for all admissible values, and these two solutions of the Legendre equation are independent if and only if $\nu + \frac{1}{2}$ is not an integer.

Assuming that ν is neither an integer nor a half-integer, every solution of the Legendre equation (9.2.1) is a linear combination of Q_ν and $Q_{-\nu-1}$. The coefficients for

$$P_\nu = A_\nu Q_\nu + B_\nu Q_{-\nu-1}$$

are analytic functions of ν, so it is enough to consider the case $-\frac{1}{2} < \nu < 0$. Then the integral form (8.3.2) gives

$$P_\nu(x) = \frac{1}{\Gamma(\nu+1)\,\Gamma(-\nu)} \int_0^1 s^\nu (1-s)^{-\nu-1} \left(1 - s\,\frac{1-x}{2}\right)^\nu \, ds. \quad (9.3.2)$$

It follows that as $x \to +\infty$,

$$P_\nu(x) \sim \frac{B(2\nu+1, -\nu)}{2^\nu\,\Gamma(\nu+1)\,\Gamma(-\nu)} x^\nu$$

$$= \frac{\Gamma(2\nu+1)}{2^\nu\,\Gamma(\nu+1)^2} x^\nu = \frac{\Gamma\!\left(\nu+\frac{1}{2}\right)}{\Gamma(\nu+1)\,\sqrt{\pi}} (2x)^\nu,$$

where we have used Legendre's duplication formula (2.3.1). In the range $-\frac{1}{2} < \nu < 0$, $Q_{-\nu-1} \sim c_{-\nu-1} x^\nu$ and Q_ν decays more rapidly, so the coefficient B_ν is the ratio

$$\frac{\Gamma\!\left(\nu+\frac{1}{2}\right) 2^\nu}{\Gamma(\nu+1)\,\sqrt{\pi}} \left[\frac{\Gamma(-\nu)\,\sqrt{\pi}}{\Gamma\!\left(-\nu+\frac{1}{2}\right) 2^{-\nu}}\right]^{-1}$$

$$= \frac{\Gamma\!\left(\nu+\frac{1}{2}\right)\Gamma\!\left(-\nu+\frac{1}{2}\right)}{\Gamma(\nu+1)\,\Gamma(-\nu)\,\pi} = -\frac{\sin\nu\pi}{\sin\!\left(\nu\pi+\frac{1}{2}\pi\right)\pi} = -\frac{\tan\nu\pi}{\pi}.$$

Since $P_{-\nu-1} = P_\nu$, the coefficient $B_\nu = A_{-\nu-1} = -A_\nu$. Thus

$$P_\nu = \frac{\tan\nu\pi}{\pi}\left[Q_\nu - Q_{-\nu-1}\right]. \quad (9.3.3)$$

Equations (9.3.3) and (9.3.1) allow the computation of the Wronskian

$$W(Q_\nu, P_\nu)(z) = -\frac{\tan\nu\pi}{\pi}\,W(Q_\nu, Q_{-\nu-1})(z) = -\frac{1}{1-z^2}. \quad (9.3.4)$$

The identity (9.3.3) can also be derived by a complex variable argument. Given $0 < \theta < 2\pi$, the function $f(\alpha) = \cos\theta - \cos\alpha$ has no zeros in the strip

$0 < \operatorname{Im} \alpha < 2\pi$ and is periodic with period 2π, continuous up to the boundary except at $\alpha = \pm\theta$ and $\alpha = \pm\theta + 2\pi$. Assuming that $-1 < \nu < 0$,

$$\frac{e^{i\left(\nu+\frac{1}{2}\right)\alpha}}{\sqrt{2\cos\alpha - 2\cos\theta}}$$

is integrable over the boundary and has integral zero over the oriented boundary. Keeping track of the argument on the various portions of the boundary, the result is again (9.3.3).

Multiplying (9.3.3) by $\cos\nu\pi$ and taking the limit as ν approaches a half-integer $n - \frac{1}{2}$, we obtain

$$Q_{n-\frac{1}{2}} = Q_{-n-\frac{1}{2}}, \quad n = 0, \pm 1, \pm 2, \dots \tag{9.3.5}$$

Recall that P_ν is holomorphic for $z \notin (-\infty, -1]$, and therefore continuous at the interval $(-1, 1)$. We show next that Q_ν has finite limits $Q_\nu(x \pm i0)$ from the upper and lower half-planes for $x \in (-1, 1)$ and that these limits are linear combinations of $P_\nu(x)$ and $P_\nu(-x)$. It follows from (9.3.4) that each limit is independent of P_ν. We define Q_ν on the interval to be the average:

$$Q_\nu(x) = \frac{1}{2}\big[Q_\nu(x+i0) + Q_\nu(x-i0)\big], \quad -1 < x < 1. \tag{9.3.6}$$

To compute the average and the jump, note that (9.3.3) and (9.2.11) imply that

$$P_\nu(-z) = -\frac{\tan\nu\pi}{\pi}\big[e^{\nu\pi i}Q_\nu(z) + e^{-\nu\pi i}Q_{-\nu-1}(z)\big], \quad \operatorname{Im} z > 0.$$

Eliminating $Q_{-\nu-1}$ between this equation and (9.3.3) gives

$$Q_\nu(z) = \frac{\pi}{2\sin\nu\pi}\big[e^{-\nu\pi i}P_\nu(z) - P_\nu(-z)\big], \quad \operatorname{Im} z > 0. \tag{9.3.7}$$

Similarly

$$Q_\nu(z) = \frac{\pi}{2\sin\nu\pi}\big[e^{\nu\pi i}P_\nu(z) - P_\nu(-z)\big], \quad \operatorname{Im} z < 0. \tag{9.3.8}$$

It follows from the previous two equations that the average Q_ν is

$$Q_\nu(x) = \frac{\pi}{2}\cot\nu\pi\, P_\nu(x) - \frac{\pi}{2\sin\nu\pi}P_\nu(-x), \quad -1 < x < 1, \tag{9.3.9}$$

while the jump is

$$Q_\nu(x+i0) - Q_\nu(x-i0) = -i\pi\, P_\nu(x), \quad -1 < x < 1. \tag{9.3.10}$$

In terms of hypergeometric functions, then

$$Q_\nu(x) = \frac{\pi}{2} \cot \nu\pi \ F\left(\nu + 1, -\nu, 1; \frac{1}{2}(1 - x)\right)$$

$$- \frac{\pi}{2 \sin \nu\pi} F\left(\nu + 1, -\nu, 1; \frac{1}{2}(1 + x)\right), \quad -1 < x < 1.$$
$$(9.3.11)$$

The recurrence and derivative identities satisfied by the Legendre polynomials, (4.7.3), (4.7.4), (4.7.5) carry over to the Legendre functions of the first and second kinds. They can be derived from the integral formulas as in Section 4.9 or checked directly from the series expansions (see the next section). As an alternative, the identities

$$(\nu + 1)P_{\nu+1}(x) - (2\nu + 1)x \ P_\nu(x) + \nu P_{\nu-1}(x) = 0, \qquad (9.3.12)$$

$$(\nu + 1)Q_{\nu+1}(x) - (2\nu + 1)x \ Q_\nu(x) + \nu Q_{\nu-1}(x) = 0,$$

and

$$P'_{\nu+1}(x) - P'_{\nu-1}(x) - (2\nu + 1)P_\nu(x) = 0, \qquad (9.3.13)$$

$$Q'_{\nu+1}(x) - Q'_{\nu-1}(x) - (2\nu + 1)Q_\nu(x) = 0,$$

follow easily from the integral representations (9.2.7) and (9.2.14). To derive (9.3.12) for P_ν from (9.2.7), we write

$$-\frac{d}{d\alpha}\left\{e^{-\left(\nu+\frac{1}{2}\right)\alpha}\sqrt{2\cosh\theta - 2\cosh\alpha}\right\}$$

$$= e^{-\left(\nu+\frac{1}{2}\right)\alpha} \frac{\left(\nu + \frac{1}{2}\right)\left[2\cosh\theta - e^\alpha - e^{-\alpha}\right] + \frac{1}{2}(e^\alpha - e^{-\alpha})}{\sqrt{2\cosh\theta - 2\cosh\alpha}}$$

as a linear combination of the integrands of

$$P_{\nu+1}(\cosh\theta), \quad \cosh\theta \ P_\nu(\cosh\theta), \quad P_{\nu-1}(\cosh\theta)$$

and integrate. The proof for Q_ν is essentially the same, using (9.2.14).

To derive (9.3.13) from the integral representation, note that the integrand of

$$P_{\nu+1}(\cosh\theta) - P_{\nu-1}(\cosh\theta)$$

is $e^{-\left(\nu+\frac{1}{2}\right)\alpha}$ multiplied by the derivative with respect to α of

$$2\sqrt{2\cosh\theta - 2\cosh\alpha}.$$

Integrating by parts,

$$P_{v+1}(\cosh\theta) - P_{v-1}(\cosh\theta) =$$

$$\frac{2v+1}{\pi} \int_{-\theta}^{\theta} e^{-\left(v+\frac{1}{2}\right)\alpha} \sqrt{2\cosh\theta - 2\cosh\alpha}\, d\alpha.$$

Differentiation with respect to θ gives (9.3.13) for P_v. The proof for Q_v is essentially the same, using (9.2.14).

Differentiating (9.3.12) and combining the result with (9.3.13) gives

$$P'_{v+1}(x) - x\, P'_v(x) = (v+1)\, P_v(x); \qquad (9.3.14)$$

$$Q'_{v+1}(x) - x\, Q'_v(x) = (v+1)\, Q_v(x).$$

Subtracting (9.3.14) from (9.3.13) gives

$$x\, P'_v(x) - P'_{v-1}(x) = v\, P_v(x); \qquad (9.3.15)$$

$$x\, Q'_v(x) - Q'_{v-1}(x) = v\, Q_v(x).$$

Multiplying (9.3.15) by x and subtracting it from the version of (9.3.14) with v replaced by $v-1$ gives

$$(1 - x^2)\, P'_v(x) = -vx\, P_v(x) + v\, P_{v-1}(x); \qquad (9.3.16)$$

$$(1 - x^2)\, Q'_v(x) = -vx\, Q_v(x) + v\, Q_{v-1}(x).$$

9.4 Series expansions and asymptotics

Expansions of the Legendre functions $P_v(z)$ and $Q_v(z)$ as $z \to \infty$ in the complement of $(-\infty, 1]$ follow from the representation (9.2.10) of Q_v as a hypergeometric function, together with the representation (9.3.3) of P_v as a multiple of $Q_v - Q_{-v-1}$, for v not a half-integer.

To find expansions for P_v and Q_v on the interval $(-1, 1)$ at $x = 0$, we begin by noting that taking $y = z^2$ converts the Legendre equation (9.2.1) to the hypergeometric equation

$$y(1 - y)\, v''(y) + \left(\frac{1}{2} - \frac{3}{2}y\right) v'(y) + \frac{1}{4}(v+1)v\, v(y) = 0.$$

It follows that P_v is a linear combination of two solutions

$$P_v(x) = A_v\, F\left(\frac{1}{2}[v+1], -\frac{1}{2}v, \frac{1}{2}; x^2\right)$$

$$+ B_v\, x\, F\left(\frac{1}{2}v + 1, -\frac{1}{2}[v-1], \frac{3}{2}; x^2\right).$$

Then

$$P_\nu(0) = A_\nu, \quad P'_\nu(0) = B_\nu.$$

To determine $P_\nu(0)$ and $P'_\nu(0)$ we assume $-1 < \nu < 0$ and use (9.3.2) to find that

$$P_\nu(0) = \frac{1}{\Gamma(\nu+1)\,\Gamma(-\nu)} \int_0^1 s^\nu (1-s)^{-\nu-1} \left(1 - \frac{1}{2}s\right)^\nu ds;$$

$$P'_\nu(0) = \frac{\nu}{2\,\Gamma(\nu+1)\,\Gamma(-\nu)} \int_0^1 s^{\nu+1} (1-s)^{-\nu-1} \left(1 - \frac{1}{2}s\right)^{\nu-1} ds.$$

To evaluate the integrals we first let $s = 1 - t$ so that the first integral becomes

$$2^{-\nu} \int_0^1 (1-t^2)^\nu \, t^{-\nu-1} \, dt.$$

This suggests letting $u = t^2$, so that the integral is

$$2^{-1-\nu} \int_0^1 (1-u)^\nu u^{-\frac{1}{2}\nu-1} \, du = 2^{-1-\nu}\, \mathrm{B}\left(\nu+1, -\frac{1}{2}\nu\right).$$

Use of the reflection formula (2.2.7) and the duplication formula (2.3.1) leads to the evaluation

$$P_\nu(0) = \frac{\Gamma\left(\frac{1}{2}\nu + \frac{1}{2}\right)}{\Gamma\left(\frac{1}{2}\nu + 1\right)\sqrt{\pi}} \cos\left(\frac{1}{2}\nu\pi\right).$$

The same procedure applied to the integral in the expression for $P'_\nu(0)$ leads to

$$2^{-\nu} \int_0^1 (1-u)^{\nu-1} \left(1 - 2\sqrt{u} + u\right) u^{-\frac{1}{2}\nu-1} \, du$$

$$= 2^{-\nu}\left[\mathrm{B}\left(\nu, -\frac{1}{2}\nu\right) - 2\mathrm{B}\left(\nu, -\frac{1}{2}\nu + \frac{1}{2}\right) + \mathrm{B}\left(\nu, -\frac{1}{2}\nu + 1\right)\right].$$

The first and third summands cancel. Since $\nu\,\Gamma(\nu) = \Gamma(\nu+1)$, use of the reflection formula and the duplication formula leads to the evaluation

$$P'_\nu(0) = \frac{2\Gamma\left(\frac{1}{2}\nu + 1\right)}{\Gamma\left(\frac{1}{2}\nu + \frac{1}{2}\right)\sqrt{\pi}} \sin\left(\frac{1}{2}\nu\pi\right).$$

The result expresses P_ν as the sum of an even function and an odd function of x:

$$P_\nu(x) = \frac{\Gamma\left(\frac{1}{2}\nu + \frac{1}{2}\right)}{\Gamma\left(\frac{1}{2}\nu + 1\right)\sqrt{\pi}} \cos\left(\frac{1}{2}\nu\pi\right) F\left(\frac{1}{2}\nu + \frac{1}{2}, -\frac{1}{2}\nu, \frac{1}{2}; x^2\right)$$

$$+ \frac{2\Gamma\left(\frac{1}{2}\nu + 1\right)}{\Gamma\left(\frac{1}{2}\nu + \frac{1}{2}\right)\sqrt{\pi}} \sin\left(\frac{1}{2}\nu\pi\right) x F\left(\frac{1}{2}\nu + 1, \frac{1}{2} - \frac{1}{2}\nu, \frac{3}{2}; x^2\right).$$

$$(9.4.1)$$

Thus P_ν is an even function precisely when ν is an even integer, and an odd function when ν is an odd integer: the case of Legendre polynomials.

The identities (9.3.7) and (9.3.8), together with (9.4.1), give the corresponding expression for Q_ν in general:

$$Q_\nu(x) = e^{\mp\frac{1}{2}\nu\pi i}\left[\frac{\Gamma\left(\frac{\nu}{2} + 1\right)\sqrt{\pi}}{\Gamma\left(\frac{1}{2}\nu + \frac{1}{2}\right)} x F\left(\frac{1}{2}\nu + 1, \frac{1}{2} - \frac{1}{2}\nu, \frac{3}{2}; x^2\right)\right.$$

$$\left.\mp i\,\frac{\Gamma\left(\frac{1}{2}\nu + \frac{1}{2}\right)\sqrt{\pi}}{2\Gamma\left(\frac{1}{2}\nu + 1\right)} F\left(\frac{1}{2}\nu + \frac{1}{2}, -\frac{1}{2}\nu, \frac{1}{2}; x^2\right)\right], \quad \pm\mathrm{Im}\,x > 0.$$

$$(9.4.2)$$

This is mainly of interest on the interval $(-1, 1)$, where (9.3.6) gives a formula dual to (9.4.1) as a sum of odd and even parts:

$$Q_\nu(x) = \frac{\Gamma\left(\frac{1}{2}\nu + 1\right)\sqrt{\pi}}{\Gamma\left(\frac{1}{2}\nu + \frac{1}{2}\right)} \cos\left(\frac{1}{2}\nu\pi\right) x F\left(\frac{1}{2}\nu + 1, \frac{1}{2} - \frac{1}{2}\nu, \frac{3}{2}; x^2\right)$$

$$- \frac{\Gamma\left(\frac{1}{2}\nu + \frac{1}{2}\right)\sqrt{\pi}}{2\Gamma\left(\frac{1}{2}\nu + 1\right)} \sin\left(\frac{1}{2}\nu\pi\right) F\left(\frac{1}{2}\nu + \frac{1}{2}, -\frac{1}{2}\nu, \frac{1}{2}; x^2\right).$$

$$(9.4.3)$$

The following asymptotic results of Laplace [175], Darboux [62], and Heine [125] will be proved in Chapter 10:

$$P_\nu(\cosh\theta) = \frac{e^{\left(\nu + \frac{1}{2}\right)\theta}}{\sqrt{2\nu\pi\,\sinh\theta}}\left[1 + O\left(|\nu|^{-1}\right)\right]; \tag{9.4.4}$$

$$Q_\nu(\cosh\theta) = \frac{\sqrt{\pi}\,e^{-\left(\nu + \frac{1}{2}\right)\theta}}{\sqrt{2\nu\,\sinh\theta}}\left[1 + O\left(|\nu|^{-1}\right)\right], \quad |\arg\nu| \le \frac{1}{2}\pi - \delta,$$

as $|\nu| \to \infty$, uniformly for $0 < \delta \le \theta \le \delta^{-1}$.

Asymptotics on the interval $-1 < x < 1$ can be computed from (9.2.3) and (9.3.9) using (8.2.12):

$$P_\nu(\cos\theta) = \frac{\sqrt{2}\,\cos\left(\nu\theta + \frac{1}{2}\theta - \frac{1}{4}\pi\right) + O\left(|\nu|^{-1}\right)}{\sqrt{\nu\pi\sin\theta}}; \qquad (9.4.5)$$

$$Q_\nu(\cos\theta) = -\frac{\sqrt{\pi}\,\sin\left(\nu\theta + \frac{1}{2}\theta - \frac{1}{4}\pi\right) + O\left(|\nu|^{-1}\right)}{\sqrt{2\nu\sin\theta}}, \qquad (9.4.6)$$

as $|\nu| \to \infty$, uniformly for $0 < \delta \le \theta \le \pi - \delta$.

9.5 Associated Legendre functions

The *associated Legendre functions* are the solutions of the spherical harmonic equation (9.1.2)

$$(1 - z^2)\,u''(z) - 2z\,u'(z) + \left[\nu(\nu + 1) - \frac{m^2}{1 - z^2}\right]u(z) = 0, \quad m = 1, 2, \ldots \qquad (9.5.1)$$

As in Section 9.1, the gauge transformation $u(z) = (1 - z^2)^{\frac{1}{2}m}\,v(z)$ reduces this to

$$(1 - z^2)\,v''(z) - 2(m + 1)z\,v'(z) + (\nu - m)(\nu + m + 1)\,v(z) = 0. \qquad (9.5.2)$$

Repeated differentiation shows that the mth derivative of a solution of the Legendre equation (9.2.1) is a solution of (9.5.2). Therefore the functions

$$P_\nu^m(z) = (z^2 - 1)^{\frac{1}{2}m}\frac{d^m}{dz^m}\big[P_\nu(z)\big], \qquad (9.5.3)$$

$$Q_\nu^m(z) = (z^2 - 1)^{\frac{1}{2}m}\frac{d^m}{dz^m}\big[Q_\nu(z)\big],$$

are solutions of (9.5.1). These are conveniently normalized for $|z| > 1$. Various normalizations are used for associated Legendre functions on the interval $(-1, 1)$, including

$$P_\nu^m(x) = (-1)^m (1 - x^2)^{\frac{1}{2}m}\frac{d^m}{dx^m}\big[P_\nu(x)\big], \qquad (9.5.4)$$

$$Q_\nu^m(x) = (-1)^m (1 - x^2)^{\frac{1}{2}m}\frac{d^m}{dx^m}\big[Q_\nu(x)\big].$$

In this section, z will always denote a complex number in the complement of $(-\infty, 1]$ and x will denote a real number in the interval $(-1, 1)$. It follows from (9.2.3), (8.2.3), and the definitions (9.5.3), (9.5.4) that

$$P_\nu^m(z) = (-1)^m \frac{(\nu+1)_m (-\nu)_m}{2^m m!} (z^2 - 1)^{\frac{1}{2}m}$$

$$\times F\left(\nu + 1 + m, m - \nu, m + 1; \frac{1}{2}(1-z)\right)$$

$$= \frac{\Gamma(\nu + m + 1)}{2^m \Gamma(\nu - m + 1) m!} (z^2 - 1)^{\frac{1}{2}m}$$

$$\times F\left(\nu + 1 + m, m - \nu, m + 1; \frac{1}{2}(1-z)\right); \tag{9.5.5}$$

$$P_\nu^m(x) = (-1)^m \frac{\Gamma(\nu + m + 1)}{2^m \Gamma(\nu - m + 1) m!} (1 - x^2)^{\frac{1}{2}m}$$

$$\times F\left(\nu + 1 + m, m - \nu, m + 1; \frac{1}{2}(1-x)\right).$$

To obtain a representation of Q_ν^m as a multiple of a hypergeometric function, we differentiate the series representation (9.2.10) m times. Since

$$\frac{d^m}{dz^m}\left[z^{-\nu-1-2n}\right]$$

$$= (-1)^m (\nu + 1 + 2n)_m z^{-\nu-1-2n-m}$$

$$= (-1)^m \frac{(\nu + 1)_m (\nu + 1 + m)_{2n}}{(\nu + 1)_{2n}} z^{-\nu-1-2n-m}$$

$$= (-1)^m \frac{(\nu+1)_m \left(\frac{1}{2}\nu + \frac{1}{2}m + \frac{1}{2}\right)_n \left(\frac{1}{2}\nu + \frac{1}{2}m + 1\right)_n}{\left(\frac{1}{2}\nu + \frac{1}{2}\right)_n \left(\frac{1}{2}\nu + 1\right)_n} z^{-\nu-1-2n-m},$$

it follows from the series expansion in (9.2.10) that

$$\frac{d^m}{dz^m}[Q_\nu(z)] = (-1)^m \frac{\Gamma(\nu + 1 + m)\sqrt{\pi}}{\Gamma\left(\nu + \frac{3}{2}\right) 2^{\nu+1} z^{\nu+1+m}}$$

$$\times F\left(\frac{1}{2}\nu + \frac{1}{2}m + \frac{1}{2}, \frac{1}{2}\nu + \frac{1}{2}m + 1, \nu + \frac{3}{2}; \frac{1}{z^2}\right).$$

Therefore

$$Q_\nu^m(z) = (-1)^m (z^2 - 1)^{\frac{1}{2}m} \frac{\Gamma(\nu + 1 + m)\sqrt{\pi}}{\Gamma\left(\nu + \frac{3}{2}\right) 2^{\nu+1} z^{\nu+1+m}}$$

$$\times F\left(\frac{1}{2}\nu + \frac{1}{2}m + \frac{1}{2}, \frac{1}{2}\nu + \frac{1}{2}m + 1, \nu + \frac{3}{2}; \frac{1}{z^2}\right). \tag{9.5.6}$$

To obtain integral representations of P_ν^m and Q_ν^m, we begin with the representations (9.2.6) and (9.2.13). Set

$$A(z, \varphi) = z + \sqrt{z^2 - 1} \cos \varphi,$$

so that (9.2.6) is

$$P_\nu(z) = \frac{1}{\pi} \int_0^\pi \frac{d\varphi}{A(z, \varphi)^{\nu+1}}.$$

The identity

$$\frac{\partial}{\partial z} \left[\frac{\sin^{2k} \varphi}{A(z, \varphi)^r} \right] = \frac{r(r - 2k - 1)}{2k + 1} \frac{\sin^{2k+2} \varphi}{A(z, \varphi)^{r+1}}$$
$$- \frac{r}{2k + 1} \frac{\partial}{\partial \varphi} \left[\frac{\sin^{2k+1} \varphi}{A(z, \varphi)^r \sqrt{z^2 - 1}} \right] \qquad (9.5.7)$$

implies that

$$\frac{d}{dz} \int_0^\pi \frac{\sin^{2k} \varphi \, d\varphi}{A(z, \varphi)^r} = \frac{r(r - 2k - 1)}{2k + 1} \int_0^\pi \frac{\sin^{2k+2} \varphi \, d\varphi}{A(z, \varphi)^{r+1}}.$$

Applying this identity m times starting with $k = 0$, $r = \nu + 1$ shows that (9.2.6) and (9.5.3) imply

$$P_\nu^m(z) = (z^2 - 1)^{\frac{1}{2}m} \frac{(-m + \nu + 1)_{2m}}{\pi 2^m \left(\frac{1}{2}\right)_m}$$
$$\times \int_0^\pi \frac{\sin^{2m} \varphi \, d\varphi}{\left[z + \sqrt{z^2 - 1} \cos \varphi\right]^{\nu+1+m}}. \qquad (9.5.8)$$

Similarly, let

$$B(z, \varphi) = z + \sqrt{z^2 - 1} \cosh \varphi.$$

Then

$$\frac{\partial}{\partial z} \left[\frac{\sinh^{2k} \varphi}{B(z, \varphi)^r} \right] = -\frac{r(r - 2k - 1)}{2k + 1} \frac{\sinh^{2k+2} \varphi}{B(z, \varphi)^{r+1}}$$
$$- \frac{r}{2k + 1} \frac{\partial}{\partial \varphi} \left[\frac{\sinh^{2k+1} \varphi}{B(z, \varphi)^r \sqrt{z^2 - 1}} \right]. \qquad (9.5.9)$$

Together with (9.2.13) and (9.5.3), this implies that

$$Q_\nu^m(z) = (-1)^m (z^2 - 1)^{\frac{1}{2}m} \frac{(-m + \nu + 1)_{2m}}{2^m \left(\frac{1}{2}\right)_m}$$

$$\times \int_0^\infty \frac{\sinh^{2m} \varphi \, d\varphi}{\left[z + \sqrt{z^2 - 1} \, \cosh \varphi\right]^{\nu+1+m}}. \tag{9.5.10}$$

These integral representations will be used in Chapter 10 to obtain the asymptotics on the interval $-1 < x < \infty$:

$$P_\nu^m(\cosh \theta) = \frac{e^{\left(\nu + \frac{1}{2}\right)\theta}}{\sqrt{2\pi \sinh \theta}} (m + 1 + \nu)^{m - \frac{1}{2}} \left[1 + O\left(\{m + 1 + \nu\}^{-\frac{1}{2}}\right)\right]; \tag{9.5.11}$$

$$Q_\nu^m(\cosh \theta) = (-1)^m \frac{e^{-\left(\nu + \frac{1}{2}\right)\theta} \sqrt{\pi}}{\sqrt{2 \sinh \theta}} (m + 1 + \nu)^{m - \frac{1}{2}} \left[1 + O\left(\{m + 1 + \nu\}^{-\frac{1}{2}}\right)\right]$$

as $\nu \to \infty$, uniformly on intervals $0 < \delta \le \theta \le \delta^{-1}$.

9.6 Relations among associated functions

As noted above, the mth derivatives $P_\nu^{(m)}$ and $Q_\nu^{(m)}$ are solutions of (9.5.2). Putting these derivatives into the equation and using (9.5.3), we obtain the recurrence relations

$$P_\nu^{m+2}(z) + \frac{2(m + 1)z}{\sqrt{z^2 - 1}} P_\nu^{m+1}(z) - (\nu - m)(\nu + m + 1) P_\nu^m(z) = 0; \tag{9.6.1}$$

$$Q_\nu^{m+2}(z) + \frac{2(m + 1)z}{\sqrt{z^2 - 1}} Q_\nu^{m+1}(z) - (\nu - m)(\nu + m + 1) Q_\nu^m(z) = 0,$$

for $z \notin (-\infty, 1]$. Similarly, on the interval $(-1, 1)$:

$$P_\nu^{m+2}(x) + \frac{2(m + 1)x}{\sqrt{1 - x^2}} P_\nu^{m+1}(x) + (\nu - m)(\nu + m + 1) P_\nu^m(x) = 0; \tag{9.6.2}$$

$$Q_\nu^{m+2}(x) + \frac{2(m + 1)x}{\sqrt{1 - x^2}} Q_\nu^{m+1}(x) + (\nu - m)(\nu + m + 1) Q_\nu^m(x) = 0,$$

for $-1 < x < 1$. Now $P_\nu^0 = P_\nu$, and it follows from (9.3.16) and (9.5.3) that

$$P_\nu^1(z) = \frac{\nu z}{\sqrt{z^2 - 1}} P_\nu(z) - \frac{\nu}{\sqrt{z^2 - 1}} P_{\nu-1}(z),$$

so P_ν^m and Q_ν^m can be computed recursively from the case $m = 0$.

Other relations can be obtained from general results for Jacobi polynomials. As an alternative we may use the results above for the case $m = 0$. Differentiating (9.3.12) and (9.3.13) m times and multipling by $(z^2 - 1)^{\frac{1}{2}m}$ gives analogous identities involving the P_ν^m, Q_ν^m. Proceeding from these identities as in the derivation of (9.3.14)–(9.3.16) gives identities that generalize (9.3.14)–(9.3.16).

In addition, a pure recurrence relation can be obtained. Differentiate (9.3.13) $(m - 1)$ times, differentiate (9.3.12) m times, use the latter to eliminate the term $P_\nu^{(m-1)}$ from the former, and multiply the result by $(z^2 - 1)^{\frac{1}{2}m}$ to obtain the recurrence relations

$$(\nu + 1 - m) P_{\nu+1}^m(z) - (2\nu + 1) z P_\nu^m(z) + (\nu + m) P_{\nu-1}^m(z) = 0; \quad (9.6.3)$$

$$(\nu + 1 - m) Q_{\nu+1}^m(z) - (2\nu + 1) z Q_\nu^m(z) + (\nu + m) Q_{\nu-1}^m(z) = 0.$$

Differentiating the relations (9.2.4), (9.2.11), (9.3.3), (9.3.5), (9.3.7), and (9.3.8), and multiplying by $(z^2 - 1)^{\frac{1}{2}m}$, gives the corresponding relations for the various solutions of (9.1.2):

$$P_\nu^m(z) = P_{-\nu-1}^m(z);$$

$$Q_{n-\frac{1}{2}}^m(z) = Q_{-n-\frac{1}{2}}^m(z);$$

$$Q_\nu^m(-z) = -e^{\pm\nu\pi i} Q_\nu^m(z), \quad \pm\text{Im } z > 0;$$

$$P_\nu^m(z) = \frac{\tan \nu\pi}{\pi} [Q_\nu^m(z) - Q_{-\nu-1}^m(z)];$$

$$Q_\nu^m(z) = \frac{\pi}{2 \sin \nu\pi} [e^{\mp\nu\pi i} P_\nu^m(z) - P_\nu^m(-z)], \quad \pm\text{Im } z > 0.$$

In view of these relations, to compute Wronskians we only need to compute $W(P_\nu^m, Q_\nu^m)$. Differentiating (9.5.3) gives

$$[P_\nu^m]'(z) = (z^2 - 1)^{-\frac{1}{2}} P_\nu^{m+1}(z) + mz(z^2 - 1)^{-1} P_\nu^m(z);$$

$$[Q_\nu^m]'(z) = (z^2 - 1)^{-\frac{1}{2}} Q_\nu^{m+1}(z) + mz(z^2 - 1)^{-1} Q_\nu^m(z).$$

It follows that

$$W(P_\nu^m, Q_\nu^m)(z) = (z^2 - 1)^{-\frac{1}{2}} \left[P_\nu^m(z) Q_\nu^{m+1}(z) - P_\nu^{m+1}(z) Q_\nu^m(z) \right].$$

The recurrence relation (9.6.1) implies that

$$P_\nu^m Q_\nu^{m+1} - P_\nu^{m+1} Q_\nu^m = (\nu + m)(m - \nu - 1) \left[P_\nu^{m-1} Q_\nu^m - P_\nu^m Q_\nu^{m-1} \right],$$

so that the computation leads to

$$P_\nu(z) Q_\nu^1(z) - P_\nu^1(z) Q_\nu(z) = (z^2 - 1)^{\frac{1}{2}} \left[P_\nu(z) Q_\nu'(z) - P_\nu'(z) Q_\nu(z) \right]$$

$$= (z^2 - 1)^{\frac{1}{2}} W(P_\nu, Q_\nu)(z) = \frac{(z^2 - 1)^{\frac{1}{2}}}{1 - z^2}.$$

Combining these results,

$$W(P_\nu^m, Q_\nu^m)(z) = \frac{\Gamma(\nu + 1 + m) \Gamma(-\nu + m)}{\Gamma(\nu + 1) \Gamma(-\nu)} \frac{1}{1 - z^2} = \frac{(\nu + 1)_m (-\nu)_m}{1 - z^2}.$$

9.7 Exercises

9.1 Suppose that $f(\mathbf{x})$ and $g(\mathbf{x})$ are smooth functions in $\mathbf{R}^n$. Show that

$$\Delta(fg) = (\Delta f) g + 2 \sum_{j=1}^n \frac{\partial f}{\partial x_j} \frac{\partial g}{\partial x_j} + f \Delta g.$$

Show that if $f(\lambda \mathbf{x}) = \lambda^m f(\mathbf{x})$ for all $\lambda > 0$, then

$$\sum_{j=1}^n x_j \frac{\partial f}{\partial x_j} = m f.$$

9.2 Suppose that $p(x, y, z)$ is a homogeneous polynomial of degree n. Use Exercise 9.1 to show that there are harmonic polynomials p_{n-2j} of degree $n - 2j$, $0 \le j \le n/2$, such that

$$p = p_n + r^2 p_{n-2} + r^4 p_{n-4} + \dots$$

9.3 Suppose that p is a harmonic polynomial of degree m in three variables and $r^2 = x^2 + y^2 + z^2$. Show that

$$\Delta(r^k p) = k(k + 2m + 1) r^{k-2} p.$$

Use this fact to show that the harmonic polynomials in Exercise 9.2 can be identified in sequence, starting with p_0 if n is even, or p_1 if n is odd. This procedure is due to Gauss [106].

9.4 Use the Taylor expansion of $f(s) = 1/r(x, y, z - s)$ to show that

$$P_n\left(\frac{z}{r}\right) = \frac{(-1)^n r^{n+1}}{n!} \frac{\partial^n}{\partial z^n}\left[\frac{1}{r}\right].$$

9.5 Verify that the change of variables after (9.2.4) converts (9.2.2) to (9.2.5).

9.6 Show that the change of variables after (9.2.6) gives the identity

$$(z^2 - 1) \sin^2 \varphi = e^{\alpha}(2z - 2 \cosh \alpha)$$

and show that (9.2.6) becomes (9.2.7).

9.7 Use (9.2.2) to prove that for $|z| < 1$,

$$P_\nu(z) = \frac{1}{2\pi} \int_0^{2\pi} \left(z + i\sqrt{1 - z^2} \sin \varphi\right)^\nu d\varphi.$$

9.8 Show that for $|z| < 1$,

$$P_\nu(z) = \frac{1}{2\pi} \int_0^{2\pi} \frac{1}{\left(z + i\sqrt{1 - z^2} \sin \varphi\right)^{\nu+1}} d\varphi.$$

9.9 Use the identity

$$\frac{1}{\alpha^{\nu+1}} = \frac{1}{\Gamma(\nu + 1)} \int_0^\infty e^{-\alpha s} s^\nu \, ds, \quad \operatorname{Re} \alpha > 0,$$

to show that

$$P_\nu(\cos \theta) = \frac{1}{\Gamma(\nu + 1)} \int_0^\infty e^{-\cos \theta s} J_0(\sin \theta s) s^\nu \, ds,$$

where J_0 is the Bessel function of order 0.

9.10 Show that

$$P_\nu(\cosh \theta) = \frac{1}{\Gamma(\nu + 1)} \int_0^\infty e^{-\cosh \theta s} I_0(\sinh \theta s) s^\nu \, ds,$$

where I_0 is one of the modified Bessel functions.

9.11 Prove the Rodrigues-type identity

$$Q_n(z) = \frac{(-1)^n}{2^{n+1} n!} \frac{d^n}{dz^n} \int_{-1}^1 \frac{(1 - s^2)^n}{z - s} \, ds, \quad z \neq (-\infty, 1].$$

9.12 Prove the generating function identity

$$\sum_{n=0}^\infty Q_n(z)t^n = \frac{1}{2R} \log \frac{z - t + R}{z - t - R}, \quad R = \sqrt{1 + t^2 - 2tz}.$$

9.13 Show that

$$Q_n\left(\frac{z}{r}\right) = \frac{(-1)^n r^{n+1}}{n!} \frac{\partial^n}{\partial z^n}\left[\frac{1}{2r} \log\frac{r+z}{r-z}\right],$$

where $r^2 = x^2 + y^2 + z^2$.

9.14 Show that the change of variables after (9.2.11) converts (9.2.9) to (9.2.12).

9.15 Show that the change of variables after (9.2.12) leads to the identities

$$R^2 \sinh^2\varphi = s^2 - 2sz + 1 = s^2 R^2 \sinh^2\theta,$$

where

$$R = \sqrt{z^2 - 1}, \quad s = z + R\cosh\varphi = [z - R\cosh\theta]^{-1}.$$

9.16 Use Exercise 9.15 to show that (9.2.12) implies (9.2.13).

9.17 Show that the change of variables after (9.2.13) leads to the identity

$$e^\alpha(2\cosh\alpha - 2\cosh\theta) = \sinh^2\theta \sinh^2\varphi$$

and show that (9.2.13) implies (9.2.14).

9.18 Show that for $x < -1$,

$$P_\nu(x+i0) - P_\nu(x-i0) = 2i \sin\nu\pi \, P_\nu(-x);$$

$$Q_\nu(x+i0) - Q_\nu(x-i0) = 2i \sin\nu\pi \, Q_\nu(-x).$$

9.19 Use the method of proof of (9.4.1) to prove

$$P_\nu(z) = F\left(\frac{1}{2}\nu + \frac{1}{2}, -\frac{1}{2}\nu, 1; 1 - z^2\right), \quad |1 - z^2| < 1.$$

9.20 Prove (9.4.6). Hint: $2\cos A \cos B = \cos(A+B) + \cos(A-B)$, and $\cos(A+B) - \cos(A-B) = -2\sin A \sin B$.

9.21 Use (8.3.8) and (9.4.6) to give a different proof of (9.3.3).

9.22 Prove by induction the Jacobi lemma

$$\frac{d^{m-1}}{d\mu^{m-1}}\left[\sin^{2m-1}\varphi\right] = (-1)^{m-1}\frac{(2m)!}{m2^m m!}\sin(m\varphi)$$

where $\mu = \cos\varphi$ and deduce that

$$\cos(m\varphi) = (-1)^{m-1}\frac{2^m m!}{(2m)!}\frac{d^m}{d\mu^m}\left[\sin^{2m-1}\varphi\right]\frac{d\mu}{d\varphi}.$$

9.23 Use Exercise 9.22 to give another derivation of the integral representation (9.5.8).

9.24 Use (9.2.2) and (9.5.3) to derive the identities

$$P_\nu^m(z) = (z^2 - 1)^{\frac{1}{2}m} \frac{(\nu + 1)_m}{2^{\nu+1}\pi i} \int_C \frac{(t^2 - 1)^\nu}{(t - z)^{\nu+1+m}} \, dt$$

and

$$P_\nu^m(z) = \frac{(\nu + 1)_m}{\pi} \int_0^\pi \left[z + \sqrt{z^2 - 1} \cos\varphi\right]^\nu \cos(m\varphi) \, d\varphi$$

$$= \frac{(-\nu)_m}{\pi} \int_0^\pi \frac{\cos(m\varphi) \, d\varphi}{\left[z + \sqrt{z^2 - 1} \cos\varphi\right]^{\nu+1}}.$$

9.25 Use (9.2.9) and (9.5.3) to derive the identity

$$Q_\nu^m(z) = (-1)^m (z^2 - 1)^{\frac{1}{2}m} \frac{(\nu + 1)_m}{2^{\nu+1}} \int_{-1}^1 \frac{(1 - t^2)^\nu}{(z - t)^{\nu+1+m}} \, dt$$

$$= (-1)^m (\nu + 1)_m$$

$$\times \int_0^\alpha \left[z - \sqrt{z^2 - 1} \cosh\varphi\right]^\nu \cosh(m\varphi) \, d\varphi, \quad \coth\alpha = z.$$

9.26 Verify (9.5.7) and (9.5.9).

9.8 Summary

9.8.1 Harmonic polynomials; surface harmonics

Harmonic polynomials of degree n are homogeneous polynomials of degree n that are solutions of Laplace's equation

$$\Delta P = P_{xx} + P_{yy} + P_{zz} = 0.$$

They are a vector space of dimension $2n + 1$. In spherical coordinates they have the form

$$P(r, \theta, \varphi) = r^n Y(\theta, \varphi),$$

where Y is a trigonometric polynomial in θ and φ and

$$\frac{1}{\sin^2\theta} Y_{\varphi\varphi} + \frac{1}{\sin\theta} \left[\sin\theta Y_\theta\right]_\theta + n(n + 1) Y = 0.$$

The function Y is called a surface harmonic of degree n. Basis of solutions:

$$Y_{nm}(\theta, \varphi) = \frac{1}{\sqrt{2\pi}} e^{im\varphi} \, \Theta_{nm}(\cos \theta), \quad -n \le m \le n,$$

$$\Theta_{nm}(x) = (-1)^n \frac{1}{2^n n!} \left[\frac{2n+1}{2} \frac{(n-m)!}{(n+m)!} \right]^{\frac{1}{2}}$$

$$\times (1 - x^2)^{\frac{1}{2}m} \frac{d^{n+m}}{dx^{n+m}} \left[(1 - x^2)^n \right].$$

Another characterization is in terms of derivatives of the Legendre polynomial P_n:

$$\Theta_{nm}(x) = \left[\frac{2n+1}{2} \frac{(n-m)!}{(n+m)!} \right]^{\frac{1}{2}} (1 - x^2)^{\frac{1}{2}m} P_n^{(m)}(x)$$

for $m \ge 0$, with

$$\Theta_{n,-m} = (-1)^m \, \Theta_{nm}, \quad m = 1, 2, \ldots, n.$$

Since

$$P_n^m(\cos \theta) = \frac{(m+n)!}{n!} \frac{1}{2\pi} \int_0^{2\pi} e^{-im\alpha} (\cos \theta + i \sin \theta \sin \alpha)^n \, d\alpha,$$

the corresponding surface harmonic and harmonic polynomial have the form

$$Y_{nm}(\theta, \varphi) = \frac{A_{nm}}{2\pi} \int_0^{2\pi} e^{im(\varphi - \alpha)} (\cos \theta + i \sin \theta \sin \alpha)^n \, d\alpha;$$

$$r^n Y_{nm}(\theta, \varphi) = \frac{A_{nm}}{2\pi} \int_0^{2\pi} e^{-im\alpha} (z + ix \sin \alpha + iy \cos \alpha)^n \, d\alpha,$$

$$A_{nm} = \frac{1}{n!} \left[\frac{2n+1}{4\pi} (n-m)!(n+m)! \right]^{\frac{1}{2}}.$$

The set $\{Y_{nm}\}$, $n = 0, 1, 2, \ldots$, $m = -n, \ldots, n$ is an orthonormal basis for functions on the sphere, with respect to normalized surface measure.

The functions

$$r^{-n-1} Y_{nm}(\theta, \varphi)$$

are harmonic in the complement of the origin in $\mathbf{R}^3$.

These functions satisfy the addition formula

$$P_n(\cos\varphi\sin\theta\sin\theta' + \cos\theta\cos\theta') = P_n(\cos\theta)\,P_n(\cos\theta')$$

$$+ 2\sum_{m=1}^{n} \frac{(n-m)!}{(n+m)!}\cos(m\varphi)\,P_n^m(\cos\theta)\,P_n^m(\cos\theta').$$

9.8.2 Legendre functions

Legendre functions are solutions of the Legendre equation

$$(1-z^2)\,u''(z) - 2z\,u'(z) + \nu(\nu+1)\,u(z) = 0.$$

Legendre function of the first kind:

$$P_\nu(z) = \frac{1}{2^{\nu+1}\pi i}\int_C \frac{(t^2-1)^\nu\,dt}{(t-z)^{\nu+1}} = F\left(\nu+1, -\nu, 1; \frac{1}{2}(1-z)\right)$$

$$= P_{-\nu-1}(z),$$

holomorphic in the complement of $(-\infty, -1]$. For $\nu = n = 0, 1, 2, \ldots, P_n$ is the Legendre polynomial of degree n. Moreover,

$$P_\nu(x) > 0, \quad 1 < x < \infty, \quad \nu \text{ real.}$$

Legendre function of the second kind:

$$Q_\nu(z) = \frac{1}{2^{\nu+1}}\int_{-1}^{1}\frac{(1-s^2)^\nu\,ds}{(z-s)^{\nu+1}}$$

$$= \frac{\Gamma(\nu+1)\sqrt{\pi}}{\Gamma\left(\nu+\frac{3}{2}\right)(2z)^{\nu+1}}\,F\left(\frac{1}{2}\nu+\frac{1}{2}, \frac{1}{2}\nu+1, \nu+\frac{3}{2}; \frac{1}{z^2}\right),$$

$$\nu \neq -1, -2, \ldots.$$

Q_ν is holomorphic for $z \notin (-\infty, 1]$, and

$$Q_\nu(-z) = -e^{\nu\pi i}\,Q_\nu(z), \quad \text{Im}\,z > 0;$$

$$Q_\nu(x) > 0, \quad 1 < x < \infty, \quad \nu \text{ real}, \nu \neq -1, -2, \ldots$$

Integral representations:

$$P_\nu(\cosh\theta) = \frac{1}{\pi} \int_0^\pi \left[\cosh\theta + \sinh\theta\,\cos\varphi\right]^\nu d\varphi$$

$$= \frac{1}{\pi} \int_0^\pi \frac{d\varphi}{\left[\cosh\theta + \sinh\theta\,\cos\varphi\right]^{\nu+1}}$$

$$= \frac{1}{\pi} \int_{-\theta}^\theta \frac{e^{-\left(\nu+\frac{1}{2}\right)\alpha}\,d\alpha}{\sqrt{2\cosh\theta - 2\cosh\alpha}};$$

$$Q_\nu(\cosh\theta) = \int_0^\alpha \left[\cosh\theta - \sinh\theta\,\cosh\varphi\right]^\nu d\varphi, \quad \coth\alpha = \cosh\theta$$

$$= \int_0^\infty \frac{d\varphi}{\left[\cosh\theta + \sinh\theta\,\cosh\varphi\right]^{\nu+1}}$$

$$= \int_\theta^\infty \frac{e^{-\left(\nu+\frac{1}{2}\right)\alpha}\,d\alpha}{\sqrt{2\cosh\alpha - 2\cosh\theta}}.$$

9.8.3 Relations among Legendre functions

The functions

$$P_\nu(z), \quad P_\nu(-z), \quad Q_\nu(z), \quad Q_{-\nu-1}(z)$$

are solutions of the Legendre equation. For $\nu = n - \frac{1}{2}$ a half-integer,

$$Q_{n-\frac{1}{2}} = Q_{-n-\frac{1}{2}}, \quad n = 0, \pm 1, \pm 2, \ldots.$$

Otherwise Q_ν and $Q_{-\nu-1}$ are independent with Wronskian

$$W(Q_\nu, Q_{-\nu-1})(z) = \frac{\pi\,\cot\nu\pi}{1 - z^2}.$$

Relations among solutions:

$$P_\nu = \frac{\tan\nu\pi}{\pi}\left[Q_\nu - Q_{-\nu-1}\right];$$

$$Q_\nu(z) = \frac{\pi}{2\sin\nu\pi}\left[e^{-\nu\pi i}P_\nu(z) - P_\nu(-z)\right], \quad \operatorname{Im} z > 0;$$

$$Q_\nu(z) = \frac{\pi}{2\sin\nu\pi}\left[e^{\nu\pi i}P_\nu(z) - P_\nu(-z)\right], \quad \operatorname{Im} z < 0.$$

For $-1 < x < 1$, by definition:

$$Q_\nu(x) = \frac{1}{2}\left[Q_\nu(x + i0) + Q_\nu(x - i0)\right]$$

$$= \frac{\pi}{2} \cot \nu\pi \, P_\nu(x) - \frac{\pi}{2 \sin \nu\pi} \, P_\nu(-x)$$

$$= \frac{\pi}{2} \cot \nu\pi \, F\left(\nu + 1, -\nu, 1; \frac{1}{2}(1 - x)\right)$$

$$- \frac{\pi}{2 \sin \nu\pi} \, F\left(\nu + 1, -\nu, 1; \frac{1}{2}(1 + x)\right), \quad -1 < x < 1.$$

Jump across the interval:

$$Q_\nu(x + i0) - Q_\nu(x - i0) = -i\pi \, P_\nu(x), \quad -1 < x < 1.$$

Recurrence and derivative relations:

$$(\nu + 1)P_{\nu+1}(x) - (2\nu + 1)x \, P_\nu(x) + \nu P_{\nu-1}(x) = 0;$$

$$P'_{\nu+1}(x) - P'_{\nu-1}(x) - (2\nu + 1)P_\nu(x) = 0;$$

$$(1 - x^2)P'_\nu(x) + \nu x P_\nu(x) - \nu P_{\nu-1}(x) = 0;$$

$$P'_\nu(x) - 2x \, P'_{\nu-1}(x) + P'_{\nu-2}(x) - P_{\nu-1}(x) = 0;$$

$$(\nu + 1)Q_{\nu+1}(x) - (2\nu + 1)x \, Q_\nu(x) + \nu Q_{\nu-1}(x) = 0;$$

$$Q'_{\nu+1}(x) - Q'_{\nu-1}(x) - (2\nu + 1)Q_\nu(x) = 0;$$

$$(1 - x^2)Q'_\nu(x) + \nu x Q_\nu(x) - \nu Q_{\nu-1}(x) = 0;$$

$$Q'_\nu(x) - 2x \, Q'_{\nu-1}(x) + Q'_{\nu-2}(x) - Q_{\nu-1}(x) = 0.$$

9.8.4 Series expansions and asymptotics

Expansions of $P_\nu(z)$ and $Q_\nu(z)$ as $z \to \infty$ follow from the representation of Q_ν as a hypergeometric function, together with the representation (9.3.3) of P_ν as a multiple of $Q_\nu - Q_{-\nu-1}$, for ν not a half-integer.

Expansions around $x = 0$ are obtained from

$$P_\nu(x) = \frac{\Gamma\left(\frac{1}{2}\nu + \frac{1}{2}\right)}{\Gamma\left(\frac{1}{2}\nu + 1\right)\sqrt{\pi}} \cos\left(\frac{1}{2}\nu\pi\right) F\left(\frac{1}{2}\nu + \frac{1}{2}, -\frac{1}{2}\nu, \frac{1}{2}; x^2\right)$$

$$+ \frac{2\Gamma\left(\frac{1}{2}\nu + 1\right)}{\Gamma\left(\frac{1}{2}\nu + \frac{1}{2}\right)\sqrt{\pi}} \sin\left(\frac{1}{2}\nu\pi\right) x \, F\left(\frac{1}{2}\nu + 1, \frac{1}{2} - \frac{1}{2}\nu, \frac{3}{2}; x^2\right);$$

$$Q_\nu(x) = \frac{\Gamma\left(\frac{1}{2}\nu+1\right)\sqrt{\pi}}{\Gamma\left(\frac{1}{2}\nu+\frac{1}{2}\right)} \cos\left(\frac{1}{2}\nu\pi\right) x\, F\left(\frac{1}{2}\nu+1, \frac{1}{2}-\frac{1}{2}\nu, \frac{3}{2}; x^2\right)$$

$$-\frac{\Gamma\left(\frac{1}{2}\nu+\frac{1}{2}\right)\sqrt{\pi}}{2\,\Gamma\left(\frac{1}{2}\nu+1\right)} \sin\left(\frac{1}{2}\nu\pi\right) F\left(\frac{1}{2}\nu+\frac{1}{2}, -\frac{1}{2}\nu, \frac{1}{2}; x^2\right),$$

$$-1 < x < 1.$$

Asymptotics as $|\nu| \to \infty$:

$$P_\nu(\cosh\theta) = \frac{e^{\left(\nu+\frac{1}{2}\right)\theta}}{\sqrt{2\nu\pi\,\sinh\theta}}\left[1 + O(|\nu|^{-1})\right];$$

$$Q_\nu(\cosh\theta) = \frac{\sqrt{\pi}\,e^{-\left(\nu+\frac{1}{2}\right)\theta}}{\sqrt{2\nu\,\sinh\theta}}\left[1 + O(|\nu|^{-1})\right] \quad |\arg\nu| \le \frac{1}{2}\pi - \delta$$

as $|\nu| \to \infty$, uniformly for $\delta \le \theta \le \delta^{-1}$. Also

$$P_\nu(\cos\theta) = \frac{\sqrt{2}\,\cos\left(\nu\theta + \frac{1}{2}\theta - \frac{1}{4}\pi\right) + O(|\nu|^{-1})}{\sqrt{\nu\pi\,\sin\theta}};$$

$$Q_\nu(\cos\theta) = -\frac{\sqrt{\pi}\,\sin\left(\nu\theta + \frac{1}{2}\theta - \frac{1}{4}\pi\right) + O(|\nu|^{-1})}{\sqrt{2\nu\,\sin\theta}}$$

as $|\nu| \to \infty$, uniformly for $\delta \le \theta \le \pi - \delta$.

9.8.5 Associated Legendre functions

These are the solutions of

$$\left[(1-z^2)u'\right]'(z) + \left[\nu(\nu+1) - \frac{m^2}{1-z^2}\right]u(z) = 0, \quad m = 1, 2, \ldots$$

Solutions include the functions

$$P_\nu^m(z) = (z^2-1)^{m/2}\frac{d^m}{dz^m}\left[P_\nu(z)\right];$$

$$Q_\nu^m(z) = (z^2-1)^{m/2}\frac{d^m}{dz^m}\left[Q_\nu(z)\right]$$

for $z \notin (-\infty, 1]$, and

$$P_\nu^m(x) = (-1)^m(1-x^2)^{m/2}\frac{d^m}{dx^m}\left[P_\nu(x)\right];$$

$$Q_\nu^m(x) = (-1)^m(1-x^2)^{m/2}\frac{d^m}{dx^m}\left[Q_\nu(x)\right], \quad -1 < x < 1.$$

These functions are multiples of hypergeometric functions:

$$P_\nu^m(z) = \frac{\Gamma(\nu + m + 1)}{2^m \, \Gamma(\nu - m + 1) \, m!} \, (z^2 - 1)^{m/2}$$

$$\times F\left(\nu + 1 + m, m - \nu, m + 1; \frac{1}{2}(1 - z)\right);$$

$$P_\nu^m(x) = (-1)^m \frac{\Gamma(\nu + m + 1)}{2^m \, \Gamma(\nu - m + 1) \, m!} \, (1 - x^2)^{m/2}$$

$$\times F\left(\nu + 1 + m, m - \nu, m + 1; \frac{1}{2}(1 - x)\right), \quad -1 < x < 1;$$

$$Q_\nu^m(z) = \frac{(-1)^m \Gamma(\nu + m + 1) \sqrt{\pi}}{\Gamma\left(\nu + \frac{3}{2}\right) 2^{\nu+1} z^{\nu+1+m}} \, (z^2 - 1)^{m/2}$$

$$\times F\left(\frac{1}{2}\nu + \frac{1}{2}m + \frac{1}{2}, \frac{1}{2}\nu + \frac{1}{2}m + 1, \nu + \frac{3}{2}; \frac{1}{z^2}\right).$$

For $z = \cosh\theta \notin (-\infty, 1]$ they have integral representations

$$P_\nu^m(z) = (z^2 - 1)^{m/2} \frac{(\nu + 1)_m}{2^{\nu+1}\pi i} \int_C \frac{(t^2 - 1)^\nu}{(t - z)^{\nu+1+m}} \, dt$$

$$= \frac{(\nu + 1)_m}{\pi} \int_0^\pi \left[z + \sqrt{z^2 - 1} \, \cos\varphi\right]^\nu \cos(m\varphi) \, d\varphi$$

$$= \frac{(-\nu)_m}{\pi} \int_0^\pi \frac{\cos(m\varphi) \, d\varphi}{\left[z + \sqrt{z^2 - 1} \cos\varphi\right]^{\nu+1}}$$

$$= (z^2 - 1)^{m/2} \frac{(-m + \nu + 1)_{2m}}{\pi \, 2^m \left(\frac{1}{2}\right)_m} \int_0^\pi \frac{\sin^{2m}\varphi \, d\varphi}{\left[z + \sqrt{z^2 - 1} \cos\varphi\right]^{\nu+1+m}},$$

and

$$Q_\nu^m(z) = (-1)^m (z^2 - 1)^{m/2} \frac{(\nu + 1)_m}{2^{\nu+1}} \int_{-1}^1 \frac{(1 - t^2)^\nu}{(z - t)^{\nu+1+m}} \, dt$$

$$= (-1)^m (\nu + 1)_m \int_0^\alpha \left[z - \sqrt{z^2 - 1} \, \cosh\varphi\right]^\nu \cosh(m\varphi) \, d\varphi,$$

$$\coth\alpha = z$$

$$= (-1)^m (z^2 - 1)^{m/2} \frac{(1 + \nu - m)_{2m}}{2^m \left(\frac{1}{2}\right)_m} \int_0^\infty \frac{\sinh^{2m}\varphi \, d\varphi}{\left[z + \sqrt{z^2 - 1} \cosh\varphi\right]^{\nu+1+m}}.$$

Asymptotics:

$$P_\nu^m(\cosh\theta) = \frac{e^{\left(\nu+\frac{1}{2}\right)\theta}}{\sqrt{2\pi\,\sinh\theta}}\,(m+1+\nu)^{m-\frac{1}{2}}\left[1 + O\left(\{m+1+\nu\}^{-\frac{1}{2}}\right)\right];$$

$$Q_\nu^m(\cosh\theta) = (-1)^m \frac{e^{-\left(\nu+\frac{1}{2}\right)\theta}\sqrt{\pi}}{\sqrt{2\,\sinh\theta}}\,(m+1+\nu)^{m-\frac{1}{2}}\left[1 + O\left(\{m+1+\nu\}^{-\frac{1}{2}}\right)\right]$$

as $\nu \to \infty$, uniformly on intervals $0 < \delta \le \theta \le \delta^{-1}$.

9.8.6 Relations among associated functions

The functions P_ν^m and Q_ν^m can be computed recursively from the case $m = 0$ using

$$P_\nu^0 = P_\nu,$$

$$P_\nu^1 = \frac{\nu\,z}{\sqrt{z^2-1}}\,P_\nu - \frac{\nu}{\sqrt{z^2-1}}\,P_{\nu-1},$$

$$P_\nu^{m+2} + \frac{2(m+1)z}{\sqrt{z^2-1}}\,P_\nu^{m+1} - (\nu-m)(\nu+m+1)\,P_\nu^m = 0,$$

$$Q_\nu^{m+2} + \frac{2(m+1)z}{\sqrt{z^2-1}}\,Q_\nu^{m+1} - (\nu-m)(\nu+m+1)\,Q_\nu^m = 0,$$

for $z \notin (-\infty, 1]$.

For $-1 < x < 1$:

$$P_\nu^{m+2} + \frac{2(m+1)x}{\sqrt{1-x^2}}\,P_\nu^{m+1} + (\nu-m)(\nu+m+1)\,P_\nu^m = 0;$$

$$Q_\nu^{m+2} + \frac{2(m+1)x}{\sqrt{1-x^2}}\,Q_\nu^{m+1} + (\nu-m)(\nu+m+1)\,Q_\nu^m = 0.$$

The associated Legendre functions also satisfy

$$(\nu+1-m)\,P_{\nu+1}^m - (2\nu+1)\,z\,P_\nu^m + (\nu+m)P_{\nu-1}^m = 0;$$

$$(\nu+1-m)\,Q_{\nu+1}^m - (2\nu+1)\,z\,Q_\nu^m + (\nu+m)Q_{\nu-1}^m = 0;$$

$$P_\nu^m(z) = P_{-\nu-1}^m(z);$$

$$Q_{n-\frac{1}{2}}^m(z) = Q_{-n-\frac{1}{2}}^m(z);$$

$$Q_\nu^m(-z) = -e^{\pm\nu\pi i} Q_\nu^m(z), \quad \pm\mathrm{Im}\, z > 0;$$

$$P_\nu^m(z) = \frac{\tan \nu\pi}{\pi} \left[Q_\nu^m(z) - Q_{-\nu-1}^m(z) \right];$$

$$Q_\nu^m(z) = \frac{\pi}{2 \sin \nu\pi} \left[e^{\mp\nu\pi i} P_\nu^m(z) - P_\nu^m(-z) \right], \quad \pm\mathrm{Im}\, z > 0.$$

Wronskians can be computed from the preceding relations using

$$W(P_\nu^m, Q_\nu^m)(z) = \frac{\Gamma(\nu+1+m)\,\Gamma(-\nu+m)}{\Gamma(\nu+1)\,\Gamma(-\nu)} \frac{1}{1-z^2}.$$

9.9 Remarks

The surface harmonics Y_{n0} occur in Laplace's work on celestial mechanics [175]. In his study of potential theory for celestial bodies he introduced Laplace's equation and found solutions by separating variables in spherical coordinates. As noted earlier, Legendre polynomials were studied by Legendre in 1784 [180, 181]. See the end of Chapter 4 for remarks on the early history of Legendre polynomials. The functions P_ν were defined for general ν by Schläfli [250]. Associated Legendre functions for non-negative integer values of ν were introduced by Ferrers [96], and for general values of ν by Hobson [131].

Laplace [175] gave the principal term in the asymptotics (9.4.4) of P_n for positive integer n, and Heine [125] investigated asymptotics for both P_n and Q_n for integer n. Darboux [62] proved (9.4.4) and (9.4.5) for integer n. Hobson [131] proved the general form of (9.4.4), (9.4.5), and (9.5.11).

The term "spherical harmonics" is often used to include all the topics covered in this chapter. Two classical treatises are Ferrers [96] and Heine [125]. More recent references include Hobson [132], MacRobert [194], Robin [240], and Sternberg and Smith [269]. Müller [209] treats the subject in $\mathbf{R}^n$ and in $\mathbf{C}^n$.

10

Asymptotics

In this chapter we prove various asymptotic results for special functions and classical orthogonal polynomials that have been stated without proof in previous chapters.

The method of proof used in the first three sections is to reduce the second-order differential equation to the point where it takes one of the following two forms:

$$u''(x) + \lambda^2 u(x) = f(x) u(x),$$

a perturbation of the wave equation; or:

$$v''(\lambda x) + \frac{1}{\lambda x} v'(\lambda x) + \left(1 - \frac{v^2}{(\lambda x)^2}\right) v(\lambda x) = g(\lambda x) v(\lambda x),$$

a perturbation of Bessel's equation. In each case λ is a large parameter and one is interested in the asymptotic behavior of solutions as $\lambda \to +\infty$.

Taking into account initial conditions, these equations can be converted to integral equations of the form

$$u(x) = u_0(x) + \frac{1}{\lambda} \int_0^x \sin(\lambda x - \lambda y) f(y) u(y) \, dy$$

or

$$v(\lambda x) = v_0(\lambda x) + \frac{1}{\lambda} \int_0^{\lambda x} G_v(\lambda x, \lambda y) g(\lambda y) v(\lambda y) \, d(\lambda y),$$

where u_0 and v_0 are solutions of the unperturbed equations

$$u_0'' + \lambda^2 u_0 = 0; \quad x^2 v_0'' + x v_0' + (x^2 - v^2) v_0 = 0,$$

and G_v is a Green's function for Bessel's equation. The asymptotic results follow easily.

335

Asymptotics of the Legendre and associated Legendre functions P_ν^m, Q_ν^m on the interval $1 < x < \infty$ are obtained from integral representations of these functions, by concentrating on the portion of the contour of integration that contributes most strongly to the value.

Other asymptotic results can be obtained from integral formulas by means of elaborations of this method. One of these is the "method of steepest descents," another is the "method of stationary phase." These two methods are illustrated in the final section with alernative derivations of the asymptotics of the Laguerre polynomials and the asymptotics of Bessel functions of the first kind, J_ν.

10.1 Hermite and parabolic cylinder functions

The equations (6.6.1) and (6.6.2),

$$u''(x) \mp \frac{x^2}{4} u(x) + \left(\nu + \frac{1}{2} \right) u(x) = 0, \qquad (10.1.1)$$

can be viewed as perturbations of the equation

$$u''(x) + \left(\nu + \frac{1}{2} \right) u(x) = 0. \qquad (10.1.2)$$

The solutions of the corresponding inhomogeneous equation

$$v''(x) + \left(\nu + \frac{1}{2} \right) v(x) = f(x)$$

have the form

$$v(x) = u_0(x) + \int_0^x \frac{\sin \lambda (x - y)}{\lambda} f(y)\, dy, \quad \lambda = \sqrt{\nu + \frac{1}{2}},$$

where u_0 is a solution of (10.1.2). Note that $v(0) = u_0(0)$, $v'(0) = u_0'(0)$. Thus solving (10.1.1) is equivalent to solving the integral equation

$$u(x) = u_0(x) \pm \int_0^x \frac{\sin \lambda (x - y)}{4\lambda} y^2 u(y)\, dy, \qquad (10.1.3)$$

where u_0 is a solution of (10.1.2). For any given choice of u_0, the solution of (10.1.3) can be obtained by the method of successive approximations, i.e. as the limit

$$u(x) = \lim_{m \to \infty} u_m(x),$$

where $u_{-1}(x) \equiv 0$ and

$$u_m(x) = u_0(x) \pm \int_0^x \frac{\sin \lambda (x - y)}{4\lambda} y^2 u_{m-1}(y) \, dy, \quad m = 0, 1, 2, \dots$$

Since $|u_0(x)| \leq A$ for some constant A, it follows by induction that

$$|u_m(x) - u_{m-1}(x)| \leq \frac{A}{m!} \frac{x^{3m}}{(12\lambda)^m}. \tag{10.1.4}$$

Therefore the sequence $\{u_m\}$ converges uniformly on bounded intervals, and

$$u(x) = u_0(x) + O(\lambda^{-1})$$

uniformly on bounded intervals.

To apply this to the parabolic cylinder functions D_ν given by (6.6.6), we take

$$u_0(x) = D_\nu(0) \cos \lambda x + D_\nu'(0) \frac{\sin \lambda x}{\lambda}.$$

It follows from (6.6.10) and (2.2.7) that

$$D_\nu(0) = \frac{2^{\frac{1}{2}\nu} \sqrt{\pi}}{\Gamma(\frac{1}{2} - \frac{1}{2}\nu)} = \frac{2^{\frac{1}{2}\nu}}{\sqrt{\pi}} \Gamma\left(\frac{1}{2}\nu + \frac{1}{2}\right) \cos \frac{1}{2}\nu\pi;$$

$$D_\nu'(0) = -\frac{2^{\frac{1}{2}\nu + \frac{1}{2}} \sqrt{\pi}}{\Gamma(-\frac{1}{2}\nu)} = \frac{2^{\frac{1}{2}\nu + \frac{1}{2}}}{\sqrt{\pi}} \Gamma\left(\frac{1}{2}\nu + 1\right) \sin \frac{1}{2}\nu\pi.$$

Formula (2.1.9) implies that

$$\Gamma\left(\frac{1}{2}\nu + 1\right) \sim \left(\frac{1}{2}\nu\right)^{\frac{1}{2}} \Gamma\left(\frac{1}{2}\nu + \frac{1}{2}\right)$$

as $\nu \to +\infty$, so we obtain

$$u_0 \sim \frac{2^{\frac{1}{2}\nu}}{\sqrt{\pi}} \Gamma\left(\frac{1}{2}\nu + \frac{1}{2}\right) \left[\cos \frac{1}{2}\nu\pi \cos \lambda x + \sin \frac{1}{2}\nu\pi \sin \lambda x\right]$$

$$= \frac{2^{\frac{1}{2}\nu}}{\sqrt{\pi}} \Gamma\left(\frac{1}{2}\nu + \frac{1}{2}\right) \cos\left(\sqrt{\nu + \frac{1}{2}} \, x - \frac{1}{2}\nu\pi\right),$$

which gives (6.6.19).

The asymptotics of the Hermite polynomials can be obtained in a similar way. Recall that H_n is the solution of

$$H_n''(x) - 2x \, H_n'(x) + 2n H_n(x) = 0,$$

with

$$H_{2m}(0) = (-1)^m \frac{(2m)!}{m!}, \quad H'_{2m}(0) = 0$$

and

$$H_{2m+1}(0) = 0, \quad H'_{2m+1}(0) = (-1)^m \frac{2(2m+1)!}{m!}.$$

The gauge transformation $H_n(x) = e^{\frac{1}{2}x^2} h_n(x)$ gives the equation

$$h_n''(x) - x^2 h_n(x) + (2n+1) h_n(x) = 0,$$

with conditions

$$h_n(0) = H_n(0), \quad h_n'(0) = H_n'(0).$$

Stirling's approximation (2.5.1) implies that

$$\frac{(2m)!}{m!} \sim 2^{m+\frac{1}{2}} \left(\frac{2m}{e}\right)^m \sim 2^m \frac{[(2m)!]^{\frac{1}{2}}}{(m\pi)^{\frac{1}{4}}};$$

$$\frac{2(2m+1)!}{\sqrt{4m+3}\,m!} \sim 2^{m+1} \left(\frac{2m+1}{e}\right)^{m+\frac{1}{2}} \sim 2^{m+\frac{1}{2}} \frac{[(2m+1)!]^{\frac{1}{2}}}{[(m+\frac{1}{2})\pi]^{\frac{1}{4}}}$$

as $n \to \infty$. It follows that for even $n = 2m$,

$$H_n(x) = (-1)^m 2^m \frac{[(2m)!]^{\frac{1}{2}}}{(m\pi)^{\frac{1}{4}}} e^{\frac{1}{2}x^2} \left[\cos(\sqrt{2n+1}\,x) + O\left(n^{-\frac{1}{2}}\right)\right]$$

as $n \to \infty$, while for odd $n = 2m + 1$,

$$H_n(x) = (-1)^m 2^{m+\frac{1}{2}} \frac{[(2m+1)!]^{\frac{1}{2}}}{[(m+\frac{1}{2})\pi]^{\frac{1}{4}}} e^{\frac{1}{2}x^2} \left[\sin(\sqrt{2n+1}\,x) + O\left(n^{-\frac{1}{2}}\right)\right]$$

as $n \to \infty$. These two results can be combined as

$$H_n(x) = 2^{\frac{1}{2}n} \frac{2^{\frac{1}{4}}(n!)^{\frac{1}{2}}}{(n\pi)^{\frac{1}{4}}} e^{\frac{1}{2}x^2} \left[\cos\left(\sqrt{2n+1}\,x - \frac{1}{2}n\pi\right) + O\left(n^{-\frac{1}{2}}\right)\right],$$

which is (4.4.20).

10.2 Confluent hypergeometric functions

Starting with the confluent hypergeometric equation

$$x\,u''(x) + (c - x)\,u'(x) - a\,u(x) = 0, \qquad (10.2.1)$$

we use the Liouville transformation. This begins with the change of variables $y = 2\sqrt{x}$ that leads to

$$v''(y) + \left[\frac{2c - 1}{y} - \frac{y}{2} \right] v'(y) - a\,v(y) = 0,$$

followed by the gauge transformation

$$v(y) = y^{\frac{1}{2} - c} e^{\frac{1}{8} y^2}\, w(y)$$

to remove the first-order term. This leads to the equation

$$w''(y) - \left[\frac{(c - 1)^2 - \frac{1}{4}}{y^2} + \frac{y^2}{16} - \frac{c}{2} + a \right] w(y) = 0. \qquad (10.2.2)$$

We are considering this equation on the half-line $y > 0$. If we could identify $u(y_0)$ and $u'(y_0)$ at some point we could use the method of the previous section: solve (10.2.2) as a perturbation of $w'' - a\,w = 0$. The singularity at $y = 0$ prevents the use of $y_0 = 0$ and it is difficult to determine values at other points, such as $y_0 = 1$. Instead, we shall introduce a first-order term in such a way that (10.2.2) can be converted to a perturbation of Bessel's equation. The gauge transformation

$$w(y) = y^{\frac{1}{2}}\, W(y)$$

converts (10.2.2) to

$$W''(y) + \frac{1}{y} W'(y) - \left[\frac{(c - 1)^2}{y^2} - \frac{c}{2} + a + \frac{y^2}{16} \right] W(y) = 0.$$

Let us assume that $c - 2a$ is positive. The final step in getting to a perturbation of Bessel's equation is to take $y = \left(\frac{1}{2}c - a \right)^{-\frac{1}{2}} z$, so that the last equation becomes

$$V''(z) + \frac{1}{z} V'(z) + \left[1 - \frac{(c - 1)^2}{z^2} \right] V(z) = \frac{z^2}{(2c - 4a)^2}\, V(z). \qquad (10.2.3)$$

This is indeed a perturbation of Bessel's equation with index $\nu = c - 1$. Tracing this argument back, a solution of the confluent hypergeometric equation (10.2.1) has the form

$$y^{1-c} e^{\frac{1}{8} y^2} V(\beta y), \quad \beta = \sqrt{\frac{1}{2} c - a},$$

where V is a solution of (10.2.3). Since $y = 2\sqrt{x}$, this has the form

$$x^{\frac{1}{2}(1-c)} e^{\frac{1}{2} x} V\left(\sqrt{(2c - 4a)x}\right). \tag{10.2.4}$$

The Kummer function $u(x) = M(a, c; x)$ is the solution of (10.2.1) with $u(0) = 1$. The corresponding function $V(z)$ must look like a multiple of $z^{c-1} = z^{\nu}$ as $z \to 0$. Let us assume for the moment that $c \geq 1$. To obtain a solution of (10.2.3) with this behavior we begin with the Bessel function J_{ν} and convert (10.2.3) to the integral equation

$$V(z) = J_{\nu}(z) + \gamma \int_0^z G_{\nu}(z, \zeta) \zeta^2 V(\zeta) \, d\zeta, \quad \gamma = \frac{1}{(2c - 4a)^2}, \tag{10.2.5}$$

where $G_{\nu}(z, \zeta)$ is the Green's function

$$G_{\nu}(z, \zeta) = \frac{Y_{\nu}(z) J_{\nu}(\zeta) - J_{\nu}(z) Y_{\nu}(\zeta)}{W(J_{\nu}(\zeta), Y_{\nu}(\zeta))}.$$

The solution of (10.2.5) can be obtained by the method of successive approximations:

$$V(z) = \lim_{m \to \infty} V_m(z),$$

where $V_{-1}(z) \equiv 0$ and

$$V_m(z) = J_{\nu}(z) + \gamma \int_0^z G_{\nu}(z, \zeta) \zeta^2 V_{m-1}(\zeta) \, d\zeta, \quad m = 0, 1, 2, \ldots$$

It follows from results in Sections 7.1 and 7.4 that there are constants $A = A_{\nu}$ and $B = B_{\nu}$ such that

$$|J_{\nu}(z)| \leq A z^{\nu}, \quad |G_{\nu}(z, \zeta)| \leq B \left(\frac{z}{\zeta}\right)^{\nu}.$$

It follows from these inequalities and induction that

$$|V_m(z) - V_{m-1}(z)| \leq A z^{\nu} \frac{\left(\gamma B z^3\right)^m}{3^m \, m!}. \tag{10.2.6}$$

Therefore the sequence $\{V_m\}$ converges uniformly on bounded intervals, and

$$V(z) = J_{\nu}(z) + O\left([c - 2a]^{-2}\right)$$

uniformly on bounded intervals. Up to a constant factor, the Kummer function $M(a, c; x)$ is given by (10.2.4). As $x \to 0$, the function defined by (10.2.4) has limiting value

$$\left[\frac{\sqrt{2c-4a}}{2}\right]^{c-1}\frac{1}{\Gamma(\nu+1)}.$$

Therefore, for $c \geq 1$ and $c - 2a > 0$,

$$M(a,c;x) = \Gamma(\nu+1)\left[\frac{1}{2}cx - ax\right]^{\frac{1}{2}(1-c)} e^{\frac{1}{2}x} V\left(\sqrt{2cx-4ax}\right)$$

$$= \Gamma(c)\left[\frac{1}{2}cx - ax\right]^{\frac{1}{2}-\frac{1}{2}c} e^{\frac{1}{2}x}$$

$$\times \left[J_{c-1}\left(\sqrt{2cx-4ax}\right) + O\left([c-2a]^{-2}\right)\right]. \quad (10.2.7)$$

According to (7.4.8),

$$J_{c-1}(z) = \frac{\sqrt{2}}{\sqrt{\pi z}}\left[\cos\left(z - \frac{1}{2}c\pi + \frac{1}{4}\pi\right) + O(z^{-1})\right]$$

as $z \to +\infty$. Combining this with (10.2.7) gives

$$M(a,c;x) = \frac{\Gamma(c)}{\sqrt{\pi}}\left(\frac{1}{2}cx - ax\right)^{\frac{1}{4}-\frac{1}{2}c} e^{\frac{1}{2}x}$$

$$\times \left[\cos\left(\sqrt{2cx-4ax} - \frac{1}{2}c\pi + \frac{1}{4}\pi\right) + O\left(|a|^{-\frac{1}{2}}\right)\right], \quad (10.2.8)$$

as $a \to -\infty$, which is the Erdélyi–Schmidt–Fejér result (6.1.12). The convergence is uniform for x in any interval $0 < \delta \leq x \leq \delta^{-1}$.

We have proved (10.2.8) under the assumption $c \geq 1$. Suppose that $0 < c < 1$. The contiguous relation (6.5.6) with c replaced by $c + 1$ is

$$M(a,c;x) = \frac{c+x}{c} M(a,c+1;x) + \frac{a-c-1}{c(c+1)} x M(a,c+2;x).$$

The asymptotics of the two terms on the right are given by (10.2.8). They imply that

$$M(a,c;x) \sim \frac{ax}{c(c+1)} M(a,c+2;x)$$

$$\sim \frac{\Gamma(c)}{\sqrt{\pi}} \frac{-ax}{\left(\frac{1}{2}cx - ax + x\right)^{\frac{3}{4}+\frac{1}{2}c}} e^{\frac{1}{2}x}$$

$$\times \cos\left(\sqrt{(2c-4a+4)x} - \frac{1}{2}c\pi + \frac{1}{4}\pi\right)$$

as $a \to -\infty$. Since $-ax$, $\frac{1}{2}cx - ax + x$, and $\frac{1}{2}cx - ax$ agree up to $O\big([-a]^{-1}\big)$ for $x > 0$, it follows that (10.2.8) is true for $c > 0$. This argument can be iterated to show that (10.2.8) is valid for all real indices $c \neq 0, -1, \ldots$.

The asymptotics of the function $U(a, c; x)$ can be obtained from (10.2.8) and the identity (6.2.6). We use the reflection formula (2.2.7) to rewrite (6.2.6):

$$U(a, c; x) = \frac{\Gamma(1 - c)\,\Gamma(c - a)\,\sin(c - a)\pi}{\pi}\,M(a, c; x)$$
$$+ \frac{\Gamma(c - 1)\,\Gamma(1 - a)\,\sin a\pi}{\pi}\,x^{1-c}\,M(a + 1 - c, 2 - c; x).$$

Therefore the asymptotics are

$$U(a, c; x)$$

$$\sim \frac{\Gamma(1 - c)\,\Gamma(c)\,\Gamma(c - a)}{\pi^{\frac{3}{2}}}\,(-ax)^{\frac{1}{4} - \frac{1}{2}c}\,e^{\frac{1}{2}x}$$

$$\times \sin(c\pi - a\pi)\,\cos\!\left(y - \frac{1}{2}c\pi + \frac{1}{4}\pi\right)$$

$$+ \frac{\Gamma(c - 1)\,\Gamma(2 - c)\,\Gamma(1 - a)}{\pi^{\frac{3}{2}}}\,e^{\frac{1}{2}x}\,x^{1-c}(-ax)^{\frac{1}{2}c - \frac{3}{4}}$$

$$\times \sin a\pi\,\cos\!\left(y + \frac{1}{2}c\pi - \frac{3}{4}\pi\right)$$

$$= \frac{e^{\frac{1}{2}x}\,x^{\frac{1}{4} - \frac{1}{2}c}}{\sqrt{\pi}\,\sin c\pi}\left[(-a)^{\frac{1}{4} - \frac{1}{2}c}\,\Gamma(c - a)\,\sin(c\pi - a\pi)\,\cos\!\left(y - \frac{1}{2}c\pi + \frac{1}{4}\pi\right)\right.$$

$$\left. -(-a)^{\frac{1}{2}c - \frac{3}{4}}\,\Gamma(1 - a)\,\sin a\pi\,\cos\!\left(y + \frac{1}{2}c\pi - \frac{3}{4}\pi\right)\right],$$

where $y = \sqrt{(2c - 4a)x}$; we used (2.2.7) again. It follows from (2.1.9) that

$$(-a)^{\frac{1}{4} - \frac{1}{2}c}\,\Gamma(c - a) \sim \Gamma\!\left(\frac{1}{2}c - a + \frac{1}{4}\right) \sim (-a)^{\frac{1}{2}c - \frac{3}{4}}\,\Gamma(1 - a).$$

Let $z = y - \frac{1}{2}c\pi + \frac{1}{4}\pi$. Then

$$\sin(c\pi - a\pi)\,\cos\!\left(y - \frac{1}{2}c\pi + \frac{1}{4}\pi\right) - \sin a\pi\,\cos\!\left(y + \frac{1}{2}c\pi - \frac{3}{4}\pi\right)$$

$$= \sin(c\pi - a\pi)\,\cos z + \sin a\pi\,\cos(z + c\pi)$$

$$= \sin c\pi\,\cos(z + a\pi) = \sin c\pi\,\cos\!\left(y - \frac{1}{2}c\pi + a\pi + \frac{1}{4}\pi\right).$$

Therefore

$$U(a, c; x) \sim \frac{\Gamma\left(\frac{1}{2}c - a + \frac{1}{4}\right)}{\sqrt{\pi}} x^{\frac{1}{4} - \frac{1}{2}c} e^{\frac{1}{2}x}$$

$$\times \cos\left(\sqrt{2cx - 4ax} - \frac{1}{2}c\pi + a\pi + \frac{1}{4}\pi\right),$$

which gives (6.2.11).

As noted in (4.5.11), the Laguerre polynomial

$$L_n^{(\alpha)}(x) = \frac{(\alpha + 1)_n}{n!} M(-n, \alpha + 1; x).$$

It follows from this, (2.1.10), and (10.2.8) that as $n \to \infty$,

$$L_n^{(\alpha)}(x) = \frac{e^{\frac{1}{2}x} n^{\frac{1}{2}\alpha - \frac{1}{4}}}{\sqrt{\pi} \, x^{\frac{1}{2}\alpha + \frac{1}{4}}} \left[\cos\left(2\sqrt{nx} - \frac{1}{2}\alpha\pi - \frac{1}{4}\pi\right) + O\left(n^{-\frac{1}{2}}\right)\right],$$

which is Fejér's result (4.5.12).

10.3 Hypergeometric functions, Jacobi polynomials

Consider the eigenvalue problem

$$x(1 - x) u''(x) + [c - (a + 1)x] u'(x) + \lambda u(x) = 0. \tag{10.3.1}$$

As we have noted several times, the change of variables $y = 1 - 2x$ converts the hypergeometric equation (10.3.1) to the equation

$$(1 - y^2) v''(y) + [a + 1 - 2c - (a + 1)y] v'(y) + \lambda v(y) = 0.$$

Setting $\alpha = c - 1$, $\beta = a - c$, this equation is

$$(1 - y^2) v''(y) + [\beta - \alpha - (\alpha + \beta + 2)y] v'(y) + \lambda v(y) = 0. \tag{10.3.2}$$

We noted in Section 4.6 that the change of variables $\theta = \cos^{-1} y$ followed by the gauge transformation

$$v(y) = \left(\sin \frac{1}{2}\theta\right)^{-\alpha - \frac{1}{2}} \left(\cos \frac{1}{2}\theta\right)^{-\beta - \frac{1}{2}} w(\theta)$$

converts (10.3.2) to

$$w''(\theta) + \frac{1 - 4\alpha^2}{16 \sin^2 \frac{1}{2}\theta} w(\theta) + \left[\frac{1 - 4\beta^2}{16 \cos^2 \frac{1}{2}\theta} + \lambda_1\right] w(\theta) = 0, \tag{10.3.3}$$

where $\lambda_1 = \lambda + \frac{1}{4}(\alpha + \beta + 1)^2$. Now

$$\frac{1}{\sin^2 \frac{1}{2}\theta} = \frac{4}{\theta^2} + \frac{1}{3} + O(\theta^2); \qquad \frac{1}{\cos^2 \frac{1}{2}\theta} = 1 + O(\theta^2),$$

so (10.3.3) can be written as

$$w''(\theta) + \frac{1 - 4\alpha^2}{4\theta^2} w(\theta) + \left[\frac{1 - 4\alpha^2}{48} + \frac{1 - 4\beta^2}{16} + \lambda_1 \right] w(\theta) = r(\theta)\, w(\theta),$$

where $|r(\theta)| \le A\theta^2$. As in the previous section, we can convert this to a perturbation of Bessel's equation. First, taking $w(\theta) = \theta^{\frac{1}{2}} W(\theta)$ leads to

$$W''(\theta) + \frac{1}{\theta} W'(\theta) + \left[\mu^2 - \frac{\alpha^2}{\theta^2} \right] W(\theta) = r(\theta)\, W(\theta), \qquad (10.3.4)$$

where

$$\mu^2 = \frac{1 - 4\alpha^2}{48} + \frac{1 - 4\beta^2}{16} + \lambda + \frac{(\alpha + \beta + 1)^2}{4}.$$

We are interested in the limit $\lambda \to +\infty$, so suppose that μ is positive. The change of variables $z = \mu\theta$ leads to the equation

$$V''(z) + \frac{1}{z} V'(z) + \left[1 - \frac{(c - 1)^2}{z^2} \right] V(z) = \frac{R(z)\, V(z)}{\mu^4}, \qquad (10.3.5)$$

with $|R(z)| \le Az^2$. This equation is very close to (10.2.3) and can be analyzed in exactly the same way.

So far we have shown that a solution of (10.3.1) has the form

$$u(x) = v(1 - 2x) = \left(\sin \frac{1}{2}\theta \right)^{-\alpha - \frac{1}{2}} \left(\cos \frac{1}{2}\theta \right)^{-\beta - \frac{1}{2}} w(\theta)$$

$$= \theta^{\frac{1}{2}} \left(\sin \frac{1}{2}\theta \right)^{-\alpha - \frac{1}{2}} \left(\cos \frac{1}{2}\theta \right)^{-\beta - \frac{1}{2}} W(\theta)$$

$$= \theta^{\frac{1}{2}} \left(\sin \frac{1}{2}\theta \right)^{-\alpha - \frac{1}{2}} \left(\cos \frac{1}{2}\theta \right)^{-\beta - \frac{1}{2}} V(\mu\theta),$$

where V is a solution of the perturbed Bessel equation (10.3.5). Note that

$$x = \frac{1 - y}{2} = \frac{1 - \cos\theta}{2} = \sin^2 \frac{1}{2}\theta \qquad (10.3.6)$$

so that as $x \to 0+$, $\theta \sim 2\sqrt{x}$ and

$$u(x) \sim \sqrt{2}\, x^{-\frac{1}{2}\alpha} V(\mu\theta) = \sqrt{2}\, x^{\frac{1}{2}(1-c)} V(\mu\theta).$$

We suppose first that $c - 1 > 0$, and we want to choose V so that $u(x)$ is a hypergeometric function: the solution of (10.3.1) that is characterized by $u(0) = 1$. Since

$$J_{c-1}(\mu\theta) \sim \left(\frac{\mu\theta}{2}\right)^{c-1} \frac{1}{\Gamma(c)} \sim \frac{\mu^{c-1}}{\Gamma(c)} x^{\frac{1}{2}(c-1)}$$

as $x \to 0$, we obtain $V = \lim_{m\to\infty} V_m$ as in the previous section, with

$$V_0(z) = \frac{\Gamma(c)}{\sqrt{2}} \mu^{1-c} J_{c-1}(z),$$

which has asymptotics

$$V_0(z) \sim \frac{\Gamma(c)}{\sqrt{\pi z}} \mu^{1-c} \cos\left(z - \frac{1}{2}c\pi + \frac{1}{4}\pi\right)$$

as $z \to +\infty$.

Taking $\lambda = \nu(a + \nu)$, the hypergeometric solution to (10.3.1) is the function with indices $a + \nu, -\nu, c$. As $\nu \to +\infty$ we may replace $\mu = \sqrt{\lambda}$ in the asymptotics by ν, or by the more precise $\nu + \frac{1}{2}a$. Putting all this together, we have, for $0 < \theta < \pi$,

$$F\left(a + \nu, -\nu, c; \sin^2\left(\frac{1}{2}\theta\right)\right)$$

$$= \frac{\Gamma(c) \cos\left(\nu\theta + \frac{1}{2}a\theta - \frac{1}{2}c\pi + \frac{1}{4}\pi\right) + O(\nu^{-1})}{\sqrt{\pi} \left(\nu \sin\frac{1}{2}\theta\right)^{c-\frac{1}{2}} \left(\cos\frac{1}{2}\theta\right)^{\frac{1}{2}+(a-c)}}$$

as $\nu \to +\infty$. This is Darboux's result (8.2.12).

We have proved this asymptotic result under the assumption $c > 1$. The contiguous relation (8.5.7) shows that the asymptotics with index $c - 1$ can be computed from those with indices c and $c + 1$ in the range $c > 1$, and that the result extends to $c > 0$. By induction, it extends to all real values of c, $c \neq 0, -1, -2, \ldots$

As a corollary, we obtain Darboux's asymptotic result (4.6.11) for Jacobi polynomials. In fact

$$P_n^{(\alpha,\beta)}(\cos\theta) = \frac{(\alpha + 1)_n}{n!} F\left(a + n, -n, c; \sin^2\left(\frac{1}{2}\theta\right)\right),$$

with $c = \alpha + 1$, $a = \alpha + \beta + 1$. Since

$$\frac{(c)_n}{n!} = \frac{\Gamma(c + n)}{\Gamma(c)\,\Gamma(n + 1)} \sim \frac{n^{c-1}}{\Gamma(c)}$$

as $n \to \infty$, by (2.1.9), it follows that

$$P_n^{(\alpha,\beta)}(\cos\theta) = \frac{\cos\left(n\theta + \frac{1}{2}[\alpha + \beta + 1]\theta - \frac{1}{2}\alpha\pi - \frac{1}{4}\pi\right) + O(n^{-1})}{\sqrt{n\pi}\left(\sin\frac{1}{2}\theta\right)^{\alpha+\frac{1}{2}}\left(\cos\frac{1}{2}\theta\right)^{\beta+\frac{1}{2}}}.$$

10.4 Legendre functions

As noted in Section 9.4, the asymptotics in ν for the Legendre functions P_ν and Q_ν on the interval $-1 < x < 1$ follow from the asymptotics of the hypergeometric function. To determine the asymptotics on the interval $1 < x < \infty$ for the Legendre functions and associated Legendre functions, we make use of the integral representations in Section 9.5, written in the form

$$P_\nu^m(\cosh\theta) = (\sinh\theta)^m \frac{(-m + \nu + 1)_{2m}}{2^m\left(\frac{1}{2}\right)_m} \pi$$

$$\times \int_0^\pi \frac{(\sin\alpha)^{2m}\,d\alpha}{(\cosh\theta - \sinh\theta\,\cos\alpha)^{m+1+\nu}}; \qquad (10.4.1)$$

$$Q_\nu^m(\cosh\theta) = (-1)^m(\sinh\theta)^m \frac{(-m + \nu + 1)_{2m}}{2^m\left(\frac{1}{2}\right)_m}$$

$$\times \int_0^\infty \frac{(\sinh\alpha)^{2m}\,d\alpha}{(\cosh\theta + \sinh\theta\,\cosh\alpha)^{m+1+\nu}}. \qquad (10.4.2)$$

These representations, and the asymptotic formulas below, apply in particular to the Legendre functions $P_\nu = P_\nu^0$ and $Q_\nu = Q_\nu^0$.

As $\nu \to \infty$ the principal contribution to each integral comes where the denominator is smallest, which is near $\alpha = 0$. We use Laplace's method: make a change of variables so that the denominator takes a form for which the asymptotics are easily computed.

For the integral in (10.4.1) we take

$$s(\alpha) = \log(A - B\cos\alpha), \quad A = e^\theta\cosh\theta, \quad B = e^\theta\sinh\theta,$$

so

$$\cosh\theta - \sinh\theta\,\cos\alpha = e^{-\theta}e^{s(\alpha)}.$$

Then s is strictly increasing for $0 \le \alpha \le \pi$, with $s(0) = 0$, $s(\pi) = 2\theta$. For $\alpha \approx 0$,

$$A - B \cos\alpha = 1 + \frac{1}{2}B\alpha^2 + O(\alpha^4), \quad s(\alpha) = \frac{1}{2}B\alpha^2 + O(\alpha^4),$$

so

$$\frac{d\alpha}{ds} = \frac{1}{\sqrt{2Bs}}[1 + O(s)], \quad \sin\alpha = \frac{\sqrt{2s}}{\sqrt{B}}[1 + O(s)].$$

Therefore the integral in (10.4.1) is

$$\frac{2^{m-\frac{1}{2}}e^{(m+1+\nu)\theta}}{B^{m+\frac{1}{2}}} \int_0^{2\theta} e^{-(m+1+\nu)s} s^{m-\frac{1}{2}} [1 + O(s)]\, ds$$

$$= \frac{2^{m-\frac{1}{2}}e^{(m+1+\nu)\theta}}{B^{m+\frac{1}{2}}(m+1+\nu)^{m+\frac{1}{2}}}$$

$$\times \int_0^{(m+1+\nu)2\theta} e^{-t} t^{m-\frac{1}{2}} \left[1 + O\left(t[m+1+\nu]^{-1}\right)\right] ds.$$

Up to an error that is exponentially small in ν, we may replace the preceding integral by an integral over the line and write it as the sum of two parts. The first part is $\Gamma\left(m + \frac{1}{2}\right)$, and the second part is dominated by

$$\frac{\Gamma\left(m + 1 + \frac{1}{2}\right)}{m+1+\nu} = \frac{m + \frac{1}{2}}{m+1+\nu}\Gamma\left(m + \frac{1}{2}\right).$$

Therefore the integral in (10.4.1) is

$$\frac{2^{m-\frac{1}{2}}e^{(m+1+\nu)\theta}\Gamma\left(m + \frac{1}{2}\right)}{B^{m+\frac{1}{2}}(m+1+\nu)^{m+\frac{1}{2}}}\left[1 + O\left([m+1+\nu]^{-1}\right)\right] \quad (10.4.3)$$

as $\nu \to \infty$.

To complete the calculation we note that as $\nu \to \infty$, (2.1.9) implies that

$$(-m + \nu + 1)_{2m} = \frac{\Gamma(m + \nu + 1)}{\Gamma(-m + \nu + 1)}$$

$$= (m + \nu + 1)^{2m}\left[1 + O\left([m+1+\nu]^{-1}\right)\right],$$

while

$$\frac{\Gamma\left(m + \frac{1}{2}\right)}{\left(\frac{1}{2}\right)_m} = \Gamma\left(\frac{1}{2}\right) = \sqrt{\pi}.$$

Combining these with (10.4.1) and (10.4.3) gives the Hobson–Darboux–
Laplace result

$$P_\nu^m(\cosh\theta) = \frac{e^{\left(\nu+\frac{1}{2}\right)\theta}}{\sqrt{2\pi\sinh\theta}} (m+1+\nu)^{m-\frac{1}{2}} \left[1 + O\big([m+1+\nu]^{-1}\big)\right].$$

$$(10.4.4)$$

The same idea is used to compute the asymptotics of Q_ν^m: the integral in
(10.4.2) is rewritten using the variable

$$s(\alpha) = \log(A + B\cosh\alpha), \quad A = e^\theta\cosh\theta, \quad B = e^\theta\sinh\theta.$$

The calculation is essentially the same and gives Hobson's result

$$Q_\nu^m(\cosh\theta) = (-1)^m \frac{e^{-\left(\nu+\frac{1}{2}\right)\theta}\sqrt{\pi}}{\sqrt{2\sinh\theta}} (m+1+\nu)^{m-\frac{1}{2}}$$

$$\times \left[1 + O\left([m+1+\nu]^{-1}\right)\right].$$

$$(10.4.5)$$

10.5 Steepest descents and stationary phase

In this section we discuss briefly two methods of deriving asymptotics from
integral representations. Both methods address integrals that are in, or can be
put into, the form

$$I(\lambda) = \int_C e^{\lambda\varphi(t)} f(t)\,dt,$$

$$(10.5.1)$$

and one is interested in the behavior as the parameter $\lambda \to +\infty$.

Suppose that the functions φ and f in (10.5.1) are holomorphic in some
domain that contains the contour C, except perhaps the endpoints. If φ is
real on the contour, then clearly the main contribution to $I(\lambda)$ for large λ
will occur near points where φ has a maximum value; any such point t_0 that
is not an endpoint of C will be a critical value: $\varphi'(t_0) = 0$. The idea is to
deform the contour C, if possible, so that φ is real along C and attains a global
maximum. If this is not possible, we look for a contour such that Re φ attains a
global maximum and φ is real near any points where the maximum is attained.
Thus in a neighborhood of such a point the curve will follow the paths of
steepest descent: the paths along which the real part decreases most rapidly.
This accounts for the terminology: *method of steepest descents*. One can then
use a method like that used in Section 10.4 to get the asymptotic result.

As an illustration, we consider the Laguerre polynomials. The generating function is given by (4.5.7):

$$\sum_{n=0}^{\infty} L_n^{(\alpha)}(x) \, z^n = (1 - z)^{-\alpha-1} \exp\left(-\frac{xz}{1-z}\right), \quad |z| < 1.$$

By Cauchy's theorem,

$$L_n^{(\alpha)}(x) = \frac{1}{2\pi i} \int_{C_0} (1 - z)^{-\alpha-1} \exp\left(-\frac{xz}{1-z}\right) \frac{dz}{z^{n+1}}, \tag{10.5.2}$$

where the path of integration encloses $z = 0$ but not $z = 1$. The change of variable $z = \exp(t/\sqrt{n})$ converts (10.5.2) to

$$L_n^{(\alpha)}(x) = \frac{1}{2\pi\sqrt{n}i} \int_C \exp\left\{\sqrt{n}\left(\frac{x}{t} - t\right)\right\} f\left(\frac{t}{\sqrt{n}}\right) dt, \tag{10.5.3}$$

where the amplitude function

$$f(s) = (1 - e^s)^{-\alpha-1} \exp\left\{-x\left(\frac{e^s}{1-e^s} + \frac{1}{s}\right)\right\}$$

is holomorphic in the strip $\{|\operatorname{Im} s| < 2\pi\}$ cut along $[0, \infty)$. The contour C starts at $+\infty$, follows the lower edge of the cut, encircles the origin in the clockwise direction, and returns to $+\infty$ along the upper edge of the cut.

The real part of the phase function $\varphi(t) = -t + x/t$ has the same sign as $\operatorname{Re} t(x - |t|^2)$, and its critical points for $x > 0$ are $t = \pm i\sqrt{x}$. The second derivative at the point $\pm i\sqrt{x}$ is $\pm 2i/\sqrt{x}$. It follows that there is a change of variables $u = u(t)$ near $\pm i\sqrt{x}$ such that

$$t = \sqrt{x}\left(\pm i \pm e^{\pm i\pi/4} u\right) + O(u^2), \quad \frac{x}{t} - t = \sqrt{x}(\mp 2i - u^2).$$

Therefore each of these points is a *saddle point* for the real part of φ: at $t = i\sqrt{x}$ the real part decreases in the southwest and northeast directions and increases in the southeast and northwest directions, while the opposite is true at $t = -i\sqrt{x}$. The path C can be chosen so that it lies in the region where $\operatorname{Re}\varphi(t) < 0$, except at the points $\pm i\sqrt{x}$, while near these points it coincides with the paths of steepest descent.

These considerations imply that, apart from quantities that are exponentially small in $\sqrt{n}$ as $n \to \infty$, the integral in (10.5.3) can be replaced by the sum of two integrals $I_{\pm}$, along short segments through the points $t = \pm i\sqrt{x}$ coinciding with the paths of steepest descent. Moreover, up to terms of lower order in

$\sqrt{n}$ we may replace $f(t/\sqrt{n})$ by its value at $t = \pm i\sqrt{x}$. Taking $u(t)$ as above, and taking into account the orientation of C, the integral I_+ is then

$$\int_{-\delta}^{\delta} f\left(\frac{i\sqrt{x}}{\sqrt{n}}\right) \exp\left\{-\sqrt{nx}(2i + u^2)\right\} \sqrt{x}\, e^{i\pi/4}\, du$$

$$= f\left(\frac{i\sqrt{x}}{\sqrt{n}}\right) e^{i\left(-2\sqrt{nx}+\frac{1}{4}\pi\right)} \sqrt{x} \int_{-\delta}^{\delta} \exp\left(-\sqrt{nx}\, u^2\right)\, du,$$

while I_- is the negative of the complex conjugate of I_+.

Up to an error that is exponentially small in $\sqrt{n}$, we may replace the last integral over $[-\delta, \delta]$ with the integral over the line and use

$$\int_{-\infty}^{\infty} e^{-\sqrt{nx}u^2}\, du = \frac{1}{(nx)^{\frac{1}{4}}} \int_{-\infty}^{\infty} e^{-s^2}\, ds = \frac{\sqrt{\pi}}{(nx)^{\frac{1}{4}}}.$$

As $n \to \infty$,

$$f\left(\frac{i\sqrt{x}}{\sqrt{n}}\right) \approx e^{\frac{1}{2}(\alpha+1)\pi i}\, x^{-\frac{1}{2}(\alpha+1)}\, n^{\frac{1}{2}(\alpha+1)}\, e^{\frac{1}{2}x}.$$

Collecting these results, we find again Fejér's result

$$L_n^{(\alpha)}(x) = \frac{1}{\sqrt{\pi}}\, x^{-\frac{1}{2}\alpha-\frac{1}{4}}\, n^{\frac{1}{2}\alpha-\frac{1}{4}}\, e^{\frac{1}{2}x} \cos\left(2\sqrt{nx} - \frac{1}{2}\alpha\pi - \frac{1}{4}\pi\right)$$

$$+ O\left(n^{\frac{1}{2}\alpha-\frac{3}{4}}\right).$$

Suppose now that the function in the exponent of (10.5.1) is purely imaginary. We change notation and write $\exp i\lambda\varphi$. We assume that the functions f and φ have some degree of smoothness but are not necessarily analytic. The idea here is that wherever $\varphi' \neq 0$ the exponential factor oscillates rapidly as $\lambda \to \infty$ and therefore there is cancellation in the integral. If φ' does not vanish at any point of C, we may integrate by parts using the identity

$$e^{i\lambda\varphi(t)} = \frac{1}{i\lambda\varphi'(t)} \frac{d}{dt}\left[e^{i\lambda\varphi(t)}\right]$$

to introduce a factor $1/\lambda$. On the other hand, if, say, $\varphi'(t_0) = 0$ and $f(t_0)\varphi''(t_0) \neq 0$, then it turns out that the part of the integral near t_0 contributes an amount that is $O(\lambda^{-\frac{1}{2}})$. Thus once again the main contributions to the asymptotics come from points where $\varphi' = 0$, i.e. points of stationary phase, hence the terminology: *method of stationary phase*.

As an example, we consider the asymptotics of the Bessel function $J_\nu(x)$ as $x \to +\infty$. In Chapter 7 the asymptotics were obtained by expressing J_ν in terms of Hankel functions. Here we give a direct argument by the method of stationary phase applied to the Sommerfeld–Bessel integral (7.3.12):

$$J_\nu(x) = \frac{1}{2\pi} \int_C e^{ix\sin\theta - i\nu\theta} \, d\theta,$$

where the path of integration consists of the boundary of the strip $\{|\mathrm{Re}\,\theta| < \pi,\ \mathrm{Im}\,\theta > 0\}$. The critical points of the phase function $\sin\theta$ on this path are at $\theta = \pm\pi/2$. As noted above, on any straight part of the path of integration that does not contain a critical point, we may integrate by parts to get an estimate that is $O(x^{-1})$. Thus we are led to consider integration over two small intervals, each containing one of the two points $\pm\pi/2$. We may change variables $u = u(\theta)$ in these intervals in such a way that

$$\theta = \pm\frac{\pi}{2} + u + O(u^2), \quad \sin\theta = \pm\left(1 - \frac{u^2}{2}\right).$$

Up to terms of lower order, we may replace $\exp(-i\nu\theta)$ by $\exp(\mp i\nu\pi/2)$. Thus the integral over the interval that contains $\theta = \pi/2$ becomes

$$\frac{e^{ix - \frac{1}{2}i\nu\pi}}{2\pi} \int_{-\delta}^{\delta} e^{-\frac{1}{2}ixu^2} \, du,$$

while the integral over the interval that contains $\theta = -\pi/2$ becomes the complex conjugate. Up to quantities of order $1/\sqrt{x}$, we may replace the last integral by the integral over the line, interpreted as

$$\int_{-\infty}^{\infty} e^{-\frac{1}{2}ixu^2} \, du = \lim_{\varepsilon \to 0+} \int_{-\infty}^{\infty} e^{-\frac{1}{2}x(i+\varepsilon)u^2} \, du. \tag{10.5.4}$$

Recall that for $a > 0$,

$$\int_{-\infty}^{\infty} e^{-au^2} \, du = a^{-1/2} \int_{-\infty}^{\infty} e^{-t^2} \, dt = \frac{\sqrt{\pi}}{\sqrt{a}}.$$

This formula remains valid, by analytic continuation, for complex a with $\mathrm{Re}\,a > 0$, so (10.5.4) gives

$$\int_{-\infty}^{\infty} e^{-\frac{1}{2}ixu^2} \, du = \lim_{\varepsilon \downarrow 0} \frac{\sqrt{2\pi}}{\sqrt{x(i+\varepsilon)}} = \frac{\sqrt{2\pi}}{\sqrt{x}} e^{-\frac{1}{4}\pi i}.$$

Combining these results, we obtain the Jacobi–Hankel result

$$J_\nu(x) = \frac{\sqrt{2}}{\sqrt{\pi x}} \cos\left(x - \frac{1}{2}\nu\pi - \frac{1}{4}\pi\right) + O(x^{-1})$$

as $x \to +\infty$, which is (7.4.8).

10.6 Exercises

10.1 Verify (10.1.4).

10.2 Verify (10.2.6).

10.3 Consider the second-order differential equation

$$w''(z) + f(z) \, w'(z) + g(z) \, w(z) = 0,$$

where $f(z)$ and $g(z)$ have convergent power series expansions

$$f(z) = \sum_{n=0}^{\infty} \frac{f_n}{z^n}, \quad g(z) = \sum_{n=0}^{\infty} \frac{g_n}{z^n}.$$

Assume that not all coefficients f_0, g_0 and g_1 are zero, i.e. $z = \infty$ is an *irregular singular point*. Show that this equation has a formal series solution of the form

$$w(z) = e^{\lambda z} z^{\mu} \sum_{n=0}^{\infty} \frac{a_n}{z^n},$$

where

$$\lambda^2 + f_0 \lambda + g_0 = 0, \quad (f_0 + 2\lambda)\mu = -(f_1 \lambda + g_1),$$

and the coefficients a_n are constants. Derive the equation that determines the coefficients a_n recursively.

10.4 The quadratic equation of λ in Exercise 10.3 is called the *characteristic equation*, and it has two *characteristic roots* λ_1, λ_2. Solutions with the kind of expansions given in Exercise 10.3 are called *normal solutions*. Suppose that $f_0^2 = 4g_0$, so $\lambda_1 = \lambda_2$. Show that the *Fabry transformation* [92]

$$w(z) = e^{-f_0 z/2} \, W, \quad t = z^{1/2}$$

takes the differential equation in Exercise 10.3 into the new equation

$$W''(t) + F(t) \, W'(t) + G(t) \, W(t) = 0,$$

where

$$F(t) = \frac{2f_1 - 1}{t} + \frac{2f_2}{t^3} + \cdots,$$

$$G(t) = (4g_1 - 2f_0 f_1) + \frac{4g_2 - 2f_0 f_2}{t^2} + \cdots$$

If $4g_1 = 2f_0 f_1$, then infinity is a regular singular point of the new equation, so it has a convergent power series solution. If $4g_1 \neq 2f_0 f_1$, then the new equation has distinct characteristic roots. By Exercise 10.3, we can write down formal series solutions in the form given there, with z replaced by t or, equivalently, $z^{1/2}$. Series solutions of this kind are called *subnormal solutions*.

10.5 Let λ_1 and λ_2 be the two characteristic values in Exercise 10.3, and suppose that $\operatorname{Re} \lambda_1 \geq \operatorname{Re} \lambda_2$. Consider first $j = 1$ and, for convenience, drop the subscript and write

$$w(z) = L_n(z) + \varepsilon_n(z), \qquad L_n(z) = e^{\lambda z} z^\mu \sum_{m=0}^{n-1} \frac{a_m}{z^m}.$$

Use the recursive formulas in Exercise 10.3 to show that

$$L_n''(z) + f(z) L_n'(z) + g(z) L_n(z) = e^{\lambda z} z^\mu R_n(z),$$

where $R_n(z) = O(z^{-n-1})$ as $z \to \infty$. Show that the error term $\varepsilon_n(z)$ satisfies the integral equation

$$\varepsilon_n(z) = \int_z^{e^{-i\omega}\infty} K(z,t)\{e^{\lambda t} t^\mu R_n(t) + [f(t) - f_0]\varepsilon_n'(t)$$

$$+ [g(t) - g_0]\varepsilon_n(t)\} dt,$$

where

$$K(z,t) = \frac{e^{\lambda_1(z-t)} - e^{\lambda_2(z-t)}}{\lambda_1 - \lambda_2}, \qquad \omega = \arg(\lambda_2 - \lambda_1).$$

Recall that we have used λ for λ_1. Use the method of successive approximation to prove that for sufficiently large n,

$$\varepsilon_n(z) = O(e^{\lambda_1 z} z^{\mu_1 - n})$$

as $z \to \infty$ in the sector $|\arg(\lambda_2 z - \lambda_1 z)| \leq \pi$.

10.6 In Exercise 10.5, we established that for all sufficiently large n, the differential equation in Exercise 10.3 has a solution $w_{n,1}(z)$ given by

$$w_{n,1}(z) = e^{\lambda_1 z} z^{\mu_1} \left[\sum_{m=0}^{n-1} \frac{a_{m,1}}{z^m} + O\left(\frac{1}{z^n}\right) \right]$$

as $z \to \infty$, $\left|\arg(\lambda_2 z - \lambda_1 z)\right| \leq \pi$. By relabeling, we get another solution

$$w_{n,2}(z) = e^{\lambda_2 z} z^{\mu_2} \left[\sum_{m=0}^{n-1} \frac{a_{m,2}}{z^m} + O\left(\frac{1}{z^n}\right) \right]$$

as $z \to \infty$, $|\arg(\lambda_2 z - \lambda_1 z)| \le \pi$. Show that $w_{n,1}(z)$ and $w_{n,2}(z)$ are independent of n.

10.7 Show that the equation

$$\frac{d^2 w}{dz^2} = \left(1 + \frac{4}{z}\right)^{\frac{1}{2}} w$$

has two linearly independent solutions given by

$$w_1(z) \sim \frac{1}{z} e^{-z} \left(1 - \frac{2}{z} + \frac{5}{z^2} - \frac{44}{3z^3} + \cdots\right)$$

as $z \to \infty$, $|\arg z| \le \pi$, and

$$w_2(z) \sim z e^z \left(1 + \frac{1}{z} - \frac{1}{2z^2} + \frac{2}{3z^3} + \cdots\right)$$

as $z \to \infty$, $|\arg(-z)| \le \pi$.

10.8 Find a change of variable $z \to \xi$ that transforms Airy's equation $w'' - zw = 0$ into the equation

$$w''(\xi) + \frac{1}{3\xi} w'(\xi) - w(\xi) = 0.$$

Show that the new equation has two linearly independent asymptotic solutions

$$w_1(\xi) \sim e^{-\xi} \xi^{-1/6} \left(1 - \frac{5}{2^3 \cdot 3^2} \frac{1}{\xi} + \frac{5 \cdot 7 \cdot 11}{2^7 \cdot 3^4} \frac{1}{\xi^2} + \cdots\right),$$

$$w_2(\xi) \sim e^{\xi} \xi^{-1/6} \left(1 + \frac{5}{2^3 \cdot 3^2} \frac{1}{\xi} + \frac{5 \cdot 7 \cdot 11}{2^7 \cdot 3^4} \frac{1}{\xi^2} + \cdots\right);$$

the first valid in the sector $|\arg \xi| \le \pi$, and the second in the sector $|\arg(-\xi)| \le \pi$.

10.9 Show that the modified Bessel equation (7.5.1) has an irregular singular point at ∞, and that for x real, two linearly independent asymptotic solutions are

$$w_1(x) \sim x^{-1/2} e^x \left[1 - \frac{(4\nu^2 - 1^2)}{1! 8x} + \frac{(4\nu^2 - 1^2)(4\nu^2 - 3^2)}{2!(8x)^2} - \cdots\right],$$

$$w_2(x) \sim x^{-1/2} e^{-x} \left[1 + \frac{(4\nu^2 - 1^2)}{1! 8x} + \frac{(4\nu^2 - 1^2)(4\nu^2 - 3^2)}{2!(8x)^2} + \cdots\right]$$

as $x \to \infty$. The modified Bessel function $I_\nu(x)$ grows exponentially as $x \to \infty$, and its asymptotic expansion is given by $w_1(x)$ multiplied by the constant $(2\pi)^{-1/2}$. The function $K_\nu(x)$ decays exponentially as $x \to \infty$, and its asymptotic expansion is given by $w_2(x)$ multiplied by $(\pi/2)^{1/2}$; see (7.5.6).

10.10 Show that the equation

$$x\,y''(x) - (x+1)\,y(x) = 0, \quad x > 0,$$

has solutions of the forms

$$y_1(x) \sim x^{\frac{1}{2}}e^x \sum_{n=0}^{\infty} \frac{a_n}{x^n}, \quad y_2(x) \sim x^{-\frac{1}{2}}e^{-x}\sum_{n=0}^{\infty}\frac{b_n}{x^n}.$$

Furthermore, the two series involved are both divergent for all values of x.

10.11 Show that the equation

$$y''(x) - x^4\,y(x) = 0$$

has two linearly independent formal solutions given by

$$y_1(x) \sim e^{\frac{1}{3}x^3}\sum_{n=0}^{\infty}\frac{(3n)!}{18^n(n!)^2}x^{-3n-1},$$

$$y_2(x) \sim e^{-\frac{1}{3}x^3}\sum_{n=0}^{\infty}\frac{(-1)^n(3n)!}{18^n(n!)^2}x^{-3n-1}$$

as $x \to \infty$; see de Bruijn [65].

10.12 Consider the equation

$$y''(x) + \left[\lambda^2 a(x) + b(x)\right]y(x) = 0.$$

Assume that $a(x)$ is positive and twice continuously differentiable in a finite or infinite interval (a_1, a_2) and that $b(x)$ is continuous. Show that the Liouville transformation

$$\xi = \int a^{1/2}(x)dx, \quad w = a^{1/4}(x)y(x)$$

takes this equation into

$$\frac{d^2w}{d\xi^2} + \left[\lambda^2 + \psi(\xi)\right]w = 0,$$

where

$$\psi(\xi) = \frac{5}{16}\frac{a'^2(x)}{a^3(x)} - \frac{1}{4}\frac{a''(x)}{a^2(x)} + \frac{b(x)}{a(x)}.$$

Discarding ψ in the transformed equation gives two linearly independent solutions $e^{\pm i\lambda\xi}$. In terms of the original variables, one obtains formally the WKB approximation

$$y(x) \sim Aa^{-1/4}(x)\exp\left\{i\lambda \int a^{1/2}(x)dx\right\}$$

$$+ Ba^{-1/4}\exp\left\{-i\lambda \int a^{1/2}(x)dx\right\}$$

as $\lambda \to \infty$, where A and B are arbitrary constants. (See the remarks at the end of this chapter concerning the history and terminology of the WKB approximation.) This formula sometimes also holds as $x \to \infty$ with λ fixed. For a discussion of this double asymptotic feature of the WKB approximation, see chapter 6, p. 203 of Olver [222].
Use this idea to show that the equation in Exercise 10.3 has solutions

$$w(z) \sim C \exp(\lambda z + \mu \log z),$$

where C is a constant and

$$\lambda = \frac{-f_0 \pm (f_0^2 - 4g_0)^{1/2}}{2}, \quad \mu = -\frac{f_1\lambda + g_1}{f_0 + 2\lambda}.$$

This is the leading term of the formal series solution given in Exercise 10.3.

10.13 In the transformed equation in Exercise 10.12, substitute $w(\xi) = e^{i\lambda\xi}[1 + h(\xi)]$. Show that

$$h''(\xi) + i2\lambda h'(\xi) = -\psi(\xi)[1 + h(\xi)].$$

Convert this to the integral equation

$$h(\xi) = -\frac{1}{i2\lambda} \int_\alpha^\xi \left\{1 - e^{i2\lambda(v-\xi)}\right\} \psi(v)[1 + h(v)]dv,$$

where α is the value of $x = c$, $c = a_1$ or $c = a_2$. Assume that α is finite. Verify that any solution of this integral equation is also a solution of the second-order differential equation satisfied by $h(\xi)$. Use the method of successive approximation to show that

$$|h(\xi)| \le \exp\left\{\frac{1}{\lambda}\Psi(\xi)\right\} - 1,$$

where

$$\Psi(\xi) = \int_\alpha^\xi |\psi(v)|dv.$$

10.14 Introduce the *control function*

$$F(x) = \int \left\{ \frac{1}{a^{1/4}} \frac{d^2}{dx^2} \left[\frac{1}{a^{1/4}} \right] + \frac{b}{a^{1/2}} \right\} dx$$

and the notation

$$\mathcal{V}_{c,x}(F) = \int_c^x |F'(t)| dt$$

for the total variation of F over the interval (c, x). Verify that $\Psi(\xi) = \mathcal{V}_{c,x}(F)$, where Ψ is defined in Exercise 10.13. By summarizing the results in Exercises 10.12 and 10.13, show that the equation

$$y''(x) + \left\{ \lambda^2 a(x) + b(x) \right\} y(x) = 0$$

has two linearly independent solutions

$$y_1(x) = a^{-1/4}(x) \exp\left\{ i\lambda \int a^{1/2}(x) dx \right\} [1 + \varepsilon_1(\lambda, x)],$$

$$y_2(x) = a^{-1/4}(x) \exp\left\{ -i\lambda \int a^{1/2}(x) dx t \right\} [1 + \varepsilon_2(\lambda, x)],$$

where

$$|\varepsilon_j(\lambda, x)| \leq \exp\left\{ \frac{1}{\lambda} \mathcal{V}_{c,x}(F) \right\} - 1, \quad j = 1, 2.$$

For fixed x and large λ, the right-hand side of the last equation is $O(\lambda^{-1})$. Hence a general solution has the asymptotic behavior given by the WKB approximation.

10.15 By following the argument outlined in Exercises 10.12–10.14, prove that the equation

$$y''(x) - \left\{ \lambda^2 a(x) + b(x) \right\} y(x) = 0$$

has two linearly independent solutions

$$y_1(x) = a^{-1/4}(x) \exp\left\{ \lambda \int a^{1/2}(x) dx \right\} [1 + \varepsilon_1(\lambda, x)],$$

$$y_2(x) = a^{-1/4}(x) \exp\left\{ -\lambda \int a^{1/2}(x) dx \right\} [1 + \varepsilon_2(\lambda, x)],$$

where

$$|\varepsilon_j(\lambda, x)| \leq \exp\left\{ \frac{1}{2\lambda} \mathcal{V}_{c,x}(F) \right\} - 1, \quad j = 1, 2.$$

If $\mathcal{V}_{a_1,a_2}(F) < \infty$, then these inequalities imply that

$$y_j(x) = a^{-1/4}(x) \exp\left\{(-1)^{j-1}\lambda \int a^{1/2}(x)dx\right\}\left[1 + O(\lambda^{-1})\right]$$

with $j = 1, 2$, uniformly for $x \in (a_1, a_2)$.

10.16 Let $N = 2n + 1$ and $x = N^{1/2}\zeta$. From the generating function of the Hermite polynomial $H_n(x)$ given in (4.4.6), derive the integral representation

$$H_n(x) = \frac{(-1)^n n!}{N^{n/2}} \frac{1}{2\pi i} \int_C g(t)e^{-Nf(t,\zeta)}dt,$$

where

$$g(t) = t^{-1/2}, \quad f(t,\zeta) = 2\zeta t + t^2 + \frac{1}{2}\log t$$

and C is the steepest descent path passing through the two saddle points

$$t_\pm = \frac{-\zeta \pm \sqrt{\zeta^2 - 1}}{2}.$$

Note that for fixed $x \in (0, \infty)$, $\zeta \to 0^+$ as $n \to \infty$ and, hence, $t_\pm$ are two well-separated complex numbers both approaching the imaginary axis. Show that

$$\frac{(-1)^n n!}{2\pi i N^{n/2}} g(t_+) e^{-Nf(t_+,\zeta)} \left(\frac{-2\pi}{Nf''(t_+,\zeta)}\right)^{1/2}$$

$$\sim e^{x^2/2 - i\sqrt{N}x} 2^{(n-1)/2}\left(\frac{n}{e}\right)^{n/2} e^{n\pi i/2} \quad \text{as} \quad n \to \infty.$$

Use this to prove that

$$H_n(x) \sim 2^{(n+1)/2}\left(\frac{n}{e}\right)^{n/2} e^{x^2/2} \cos\left(\sqrt{2n + 1}x - \frac{1}{2}n\pi\right),$$

as $n \to \infty$; thus establishing (4.4.20).

10.17 Returning to (10.5.3), we deform the contour C into an infinite loop which consists of (i) a circle centered at the origin with radius $\sqrt{x}$ and (ii) two straight lines along the upper and lower edges of the positive real axis from $\sqrt{x}$ to $+\infty$. (a) Show that the contribution from the two straight lines is of magnitude $\varepsilon_n(x) = O\left(n^{\frac{1}{2}|\alpha|+1}e^{-\sqrt{nx}}\right)$. On the circle we introduce the parametrization $t = \sqrt{x}e^{i\theta}$, $\theta \in (0, 2\pi)$. Show that

$$L_n^{(\alpha)}(x) = \int_0^{2\pi} \psi(\theta) e^{i\lambda\phi(\theta)} d\theta + \varepsilon_n(x),$$

where $\lambda = 2\sqrt{nx}$, $\phi(\theta) = -\sin\theta$,

$$\psi(\theta) = -\frac{\sqrt{x}}{2\pi\sqrt{n}} e^{i\theta} f\left(\sqrt{\frac{x}{n}} e^{i\theta}\right),$$

$$f(s) = (1 - e^s)^{-\alpha-1} \exp\left\{-x\left(\frac{e^s}{1 - e^s} + \frac{1}{s}\right)\right\}.$$

The oscillatory integral is now exactly in the form to which the method of stationary phase applies. The stationary points, where $\phi'(\theta) = 0$, occur at $\theta = \theta_1 = \frac{\pi}{2}$ and $\theta = \theta_2 = \frac{3\pi}{2}$. The contribution from θ_1 is

$$\psi(\theta_1)\sqrt{\frac{2\pi}{\lambda\phi''(\theta_1)}} e^{i\left[\lambda\phi(\theta_1) + \frac{1}{4}\pi\right]} = \frac{e^{\frac{1}{2}x}n^{\frac{1}{2}\alpha - \frac{1}{4}}}{2\sqrt{\pi}x^{\frac{1}{2}\alpha + \frac{1}{4}}} e^{\left(\frac{1}{2}\alpha\pi + \frac{1}{4} - 2\sqrt{nx}\right)i}.$$

The contribution from θ_2 is simply the complex conjugate of the last expression, thus proving (4.5.12).

10.18 (a) Use (4.6.3) to derive the representation

$$P_n^{(\alpha,\beta)}(x)$$

$$= \binom{n+\alpha}{n}\left(\frac{x+1}{2}\right)^n \sum_{k=0}^{n} \frac{(n-k+1)_k}{(\alpha+1)_k} \binom{n+\beta}{k}\left(\frac{x-1}{x+1}\right)^k$$

$$= \binom{n+\alpha}{n}\left(\frac{x+1}{2}\right)^n F\left(-n, -n-\beta, \alpha+1; \frac{x-1}{x+1}\right).$$

(b) Show that the formula in (a) can be written as

$$P_n^{(\alpha,\beta)}(z)$$

$$= \frac{1}{2\pi i}\int_C \left(1 + \frac{x+1}{2}z\right)^{n+\alpha}\left(1 + \frac{x-1}{2}z\right)^{n+\beta} z^{-n-1}\,dz,$$

where we assume $x \neq \pm 1$, and where C is a closed curve that encircles the origin in the positive sense and excludes the points $-2(x \pm 1)^{-1}$.

(c) Prove that the formula in (a) can also be written in the form

$$P_n^{(\alpha,\beta)}(x) = \frac{1}{2\pi i}\int_{C'}\left(\frac{1}{2}\frac{t^2 - 1}{t - x}\right)^n\left(\frac{1-t}{1-x}\right)^\alpha\left(\frac{1+t}{1+x}\right)^\beta \frac{dt}{t - x},$$

where $x \neq \pm 1$ and C' is a contour encircling the point $t = x$ in the positive sense but not the points $t = \pm 1$. The functions $(1 - t/1 - x)^\alpha$ and $(1 + t/1 + x)^\beta$ are assumed to be reduced to 1 for $t = x$.

(d) Let $t - x = w$, and establish the representation

$$2^{-\frac{1}{2}(\alpha+\beta+1)}(1 + x)^\beta (1 - x)^\alpha P_n^{(\alpha,\beta)}(x)$$

$$= \frac{1}{2\pi i} \int_{C'} g(w) e^{Nf(w,x)} dw,$$

where

$$N = n + \frac{1}{2}(\alpha + \beta + 1),$$

$$f(w, x) = \log \frac{(w + x + 1)(w + x - 1)}{2w},$$

$$g(w) = \frac{(1 - w - x)^\alpha}{(w + x - 1)^{(\alpha+\beta+1)/2}} \frac{(w + x + 1)^{(-\alpha+\beta-1)/2}}{w^{(1-\alpha-\beta)/2}}.$$

The integration path is the steepest descent curve passing through the relevant saddle points. For $x \in (0, 1)$, we write $x = \cos \theta$ with $\theta \in \left(0, \frac{1}{2}\pi\right)$. The two saddle points of $f(w, x)$ are located at $w_\pm = \pm\sqrt{1 - x^2} = \pm i \sin \theta$. Show that

$$g(w_+) = 2^{(\alpha+\beta-3)/2} e^{-\alpha\pi i/2 - \pi i/2} \left(\sin \frac{1}{2}\theta\right)^{\alpha-1} \left(\cos \frac{1}{2}\theta\right)^{\beta-1} e^{-i\theta/2},$$

$$e^{Nf(w_+,x)} = e^{N\theta i}, \quad f''(w_+, x) = \frac{-ie^{-\theta i}}{2 \sin \frac{1}{2}\theta \, \cos \frac{1}{2}\theta}.$$

From these, deduce the asymptotic formula (4.6.11).

10.19 Use Exercise 8.8 of Chapter 8 to show that

$$F(a + v, -v, c; x) = e^{v\pi i} \frac{\Gamma(c)\Gamma(1 + v)}{\Gamma(c + v)} \frac{1}{2\pi i} \int_C g(t) e^{vh(t)} dt,$$

where

$$h(t) = \log \frac{1 + t}{t(1 + t - xt)}, \quad g(t) = \frac{(1 + t)^{a-c}}{t(1 + t - xt)^a}$$

and C comes from $+\infty$ to 0 along the upper edge of the cut $[0, \infty)$ and returns along the lower edge. Use the steepest descent method to derive the asymptotic formula

$$F(a + v, -v, c; x)$$
$$= \frac{\Gamma(c)\Gamma(1 + v)}{\Gamma(c + v)\sqrt{v\pi}} \operatorname{Re}\left[e^{v\pi i} i^{\frac{1}{2} + a - c} x^{\frac{1}{4} - \frac{1}{2}c} (1 - x)^{\frac{1}{2}c - \frac{1}{2}a - \frac{1}{4}} \right.$$
$$\left. \times \left(\sqrt{x} + i\sqrt{1 - x}\right)^{-a} \left(\sqrt{x} - i\sqrt{1 - x}\right)^{2v}\right]\left[1 + O(v^{-1})\right].$$

Deduce (8.2.12) from the last equation.

10.20 Use the change of variables

$$t = \sqrt{-ax + \frac{1}{2}cx}\, s$$

to show that the integral in Exercise 6.3 of Chapter 6 is equal to

$$\left(-ax + \frac{1}{2}cx\right)^{-\frac{1}{2}c + \frac{1}{2}} \int_{C'} \exp\left\{-a \log\left(1 - \frac{\sqrt{x}}{\left(-a + \frac{1}{2}c\right)^{\frac{1}{2}} s}\right)\right.$$
$$\left. + \left(-ax + t\frac{1}{2}cx\right)^{\frac{1}{2}} s\right\} s^{-c}\, ds,$$

where C' is the image of C. Show that the quantity inside the braces is

$$\left(-ax + \frac{1}{2}cx\right)^{\frac{1}{2}} \left(s - \frac{1}{s}\right) - \frac{x}{2s^2} + O\left(\left[-a + \frac{1}{2}c\right]^{-\frac{1}{2}}\right)$$

and

$$M(a, c; x) = \Gamma(c) \left(-ax + \frac{1}{2}cx\right)^{\frac{1}{2} - \frac{1}{2}c}$$
$$\times \frac{1}{2\pi i} \int_{C'} e^{Nf(s)} g(s)\, ds \left[1 + O\left(\left[-a + \frac{1}{2}c\right]^{-\frac{1}{2}}\right)\right],$$

where $N = \left(-ax + \frac{1}{2}cx\right)^{\frac{1}{2}}$, $f(s) = s - 1/s$ and $g(s) = e^{-x/2s^2} s^{-c}$.
The saddle points are at $\pm i$. Show that the contribution from the saddle point at $s = i$ is

$$\frac{\Gamma(c)}{2\sqrt{\pi}} e^{\frac{1}{2}x} \left(-ax + \frac{1}{2}cx\right)^{\frac{1}{4} - \frac{1}{2}c} \exp\left\{i\left[(2cx - 4ax)^{\frac{1}{2}} - \frac{1}{2}c\pi + \frac{1}{4}\pi\right]\right\}.$$

Deduce from this the asymptotic formula (6.1.10).

10.21 (a) Use Exercise 8.9 of Chapter 8 and (5.5.6) to prove that for $\mathrm{Re}\,(b + x) > 0$, the Meixner polynomial

$$m_n(x; b, c) = \frac{(b)_n}{c^n} \frac{\Gamma(x + 1)\Gamma(b)}{\Gamma(b + x)} \frac{1}{2\pi i}$$
$$\times \int_C \frac{t^{b+x-1}}{(t - 1)^{x+1}} [1 - (1 - c)t]^n \, dt,$$

where C is a counterclockwise loop through $t = 0$ that encloses $t = 1$. Therefore

$$m_n(n\alpha; b, c) = \frac{(b)_n}{c^n} \frac{\Gamma(x + 1)\Gamma(b)}{\Gamma(b + x)} \frac{1}{2\pi i} \int_C \frac{t^{b-1}}{t - 1} e^{-nf(t)} dt,$$

where

$$x = n\alpha, \quad f(t) = -\alpha \log t + \alpha \log(t - 1) - \log\left[1 - (1 - c)t\right].$$

(b) Show that for $0 < c < 1$, fixed x (i.e. $\alpha = O(n^{-1})$) and large n, the phase function $f(t)$ has two saddle points

$$t_+ \sim 1 - \frac{c\alpha}{1 - c}, \quad t_- \sim \frac{\alpha}{1 - c}$$

in the interval $(0, 1)$. Note that for large n (i.e. small α), the two saddle points are well separated. Deform the contour in (a) to run from $t = 0$ to $t = t_+$ through t_- and with $\arg(t - 1) = -\pi$, and from t_+ in the lower half-plane $\mathrm{Im}\,t < 0$ to $t_c = 1/1 - c$, where the integrand vanishes; from t_c the contour runs in the half-plane $\mathrm{Im}\,t > 0$ to t_+, and returns to the origin with $\arg(t - 1) = \pi$; see the figure below.

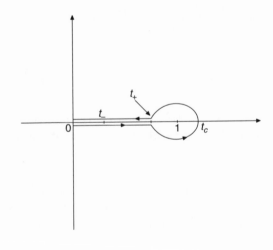

Denote the sum of the integrals along $[0, t_+]$ by I_1 and the remaining portion of the contour by I_2. Show that

$$m_n(n\alpha; b, c) = \frac{(b)_n}{c^n} \frac{\Gamma(x+1)\Gamma(b)}{\Gamma(b+x)} (I_1 + I_2),$$

where

$$I_1 = -\frac{\sin \pi x}{\pi} \int_0^{t_+} \frac{t^{b+x-1}}{(1-t)^{x+1}} \left[1 - (1-c)t\right]^n dt,$$

$$I_2 = \frac{1}{2\pi i} \int_{t_+}^{(1+)} \frac{t^{b+x-1}}{(t-1)^{x+1}} \left[1 - (1-c)t\right]^n dt.$$

(c) With $x = n\alpha$, show that

$$I_1 = -\frac{\sin \pi x}{\pi} \int_0^{t_+} \frac{t^{b-1}}{1-t} e^{-nf(t)} dt,$$

where

$$f(t) = -\alpha \log t + \alpha \log(1-t) - \log(1 - (1-c)t).$$

This function has the same saddle points as the phase function given in (a). Note that on the interval $(0, t_+)$, this function has only one saddle point which occurs at $t = t_-$. Furthermore, as $\alpha \to 0$, this point coalesces with the endpoint $t = 0$. In fact, this point disappears when $\alpha = 0$, since the function $\log(1 - (1-c)t)$ has no saddle point in the interval $(0, t_+)$. Thus, the steepest descent method does not work. Show that the integral

$$\int_0^\infty s^{n\alpha} e^{-ns} ds = \int_0^\infty e^{-n(s-\alpha \log s)} ds$$

has the same asymptotic phenomenon as I_1, namely, the phase function $\phi(s) = s - \alpha \log s$ also has a saddle point (i.e. $s = \alpha$) which coincides with the endpoint $s = 0$ as $\alpha \to 0$. This suggests the change of variable $t \to s$ defined implicitly by

$$f(t) = -\alpha \log s + s + A,$$

where A is a constant independent of t or s. Show that to make the mapping $t \to s$ one-to-one and analytic in the interval $0 \le t < t_+$, one should choose A so that $t = t_-$ corresponds to $s = \alpha$, i.e.,

$$A = f(t_-) + \alpha \log \alpha - \alpha.$$

This gives

$$I_1 = -\frac{\sin \pi x}{\pi} e^{-nA} \int_0^{s_+} h(s) s^{b+x-1} e^{-ns} ds,$$

where

$$h(s) = \left(\frac{t}{s}\right)^{b-1} \frac{1}{1-t} \frac{dt}{ds}$$

and s_+ is implicitly defined via the equation
$g(t) = -\alpha \log s + s + A$ by substituting $t = t_+$ and $s = s_+$.

(d) Show that the function $t = t(s)$ defined in (c) satisfies

$$\left.\frac{dt}{ds}\right|_{s=\alpha} = \frac{1}{\sqrt{\alpha f''(t_-)}}.$$

Furthermore, the integral I_1 in (c) has the leading-order
approximation

$$I_1 \sim -\frac{\sin \pi x}{\pi} e^{-nA} h(\alpha) \frac{\Gamma(b+n)}{n^{b+x}}$$

as $n \to \infty$, uniformly with respect to $\alpha \in [0, \alpha_0]$, $\alpha_0 > 0$. Deduce
from this the asymptotic formula (5.5.14):

$$m_n(x, b, c) \sim -\frac{\Gamma(b+n) \Gamma(x+1)}{c^n (1-c)^{x+b} n^{b+x}} \frac{\sin \pi x}{\pi}.$$

The derivation of this formula, as presented in this exercise, is due
to N. M. Temme (private communication).

10.22 By following the same argument as in Exercise 10.21, prove the
asymptotic formula for the Charlier polynomial $C_n(x; a)$ given in
(5.3.12).

10.7 Summary

10.7.1 Hermite and parabolic cylinder functions

The equations

$$u''(x) \mp \frac{x^2}{4} u(x) + \left(v + \frac{1}{2}\right) u(x) = 0$$

are equivalent to

$$u(x) = u_0(x) \pm \int_0^x \frac{\sin \lambda(x - y)}{4\lambda} y^2 u(y) \, dy,$$

with solutions

$$u(x) = \lim_{m \to \infty} u_m(x),$$

where $u_{-1}(x) \equiv 0$ and

$$u_m(x) = u_0(x) \pm \int_0^x \frac{\sin \lambda(x - y)}{4\lambda} y^2 u_{m-1}(y) \, dy, \quad m = 0, 1, 2, \ldots$$

For the parabolic cylinder functions D_ν,

$$u_0 \sim \frac{2^{\frac{1}{2}\nu}}{\sqrt{\pi}} \Gamma\left(\frac{1}{2}\nu + \frac{1}{2}\right) \left[\cos \frac{1}{2}\nu\pi \cos \lambda x + \sin \frac{1}{2}\nu\pi \sin \lambda x\right]$$

$$= \frac{2^{\frac{1}{2}\nu}}{\sqrt{\pi}} \Gamma\left(\frac{1}{2}\nu + \frac{1}{2}\right) \cos\left(\sqrt{\nu + \frac{1}{2}} x - \frac{1}{2}\nu\pi\right).$$

The gauge transformation $H_n(x) = e^{\frac{1}{2}x^2} h_n(x)$ gives the equation

$$h_n''(x) - x^2 h_n(x) + (2n + 1) h_n(x) = 0,$$

and leads to the asymptotics of the Hermite polynomials

$$H_n(x) = 2^{\frac{1}{2}n} \frac{2^{\frac{1}{4}} (n!)^{\frac{1}{2}}}{(n\pi)^{\frac{1}{4}}} e^{\frac{1}{2}x^2} \left[\cos\left(\sqrt{2n + 1}\, x - \frac{1}{2}n\pi\right) + O\left(n^{-\frac{1}{2}}\right)\right].$$

10.7.2 Confluent hypergeometric functions

The Liouville transformation, followed by a gauge transformation and a change of variables, converts the confluent hypergeometric equation

$$x u''(x) + (c - x) u'(x) - a u(x) = 0$$

to

$$V''(z) + \frac{1}{z} V'(z) + \left(1 - \frac{(c - 1)^2}{z^2}\right) V(z) = \frac{z^2}{(2c - 4a)^2} V(z),$$

a perturbation of Bessel's equation, and leads to the integral equation

$$V(z) = J_\nu(z) + \gamma \int_0^z G_\nu(z, \zeta) \zeta^2 V(\zeta) \, d\zeta, \quad \gamma = \frac{1}{(2c - 4a)^2},$$

where $G_\nu(z, \zeta)$ is the Green's function

$$G_\nu(z, \zeta) = \frac{Y_\nu(z)J_\nu(\zeta) - J_\nu(z)Y_\nu(\zeta)}{W(J_\nu(\zeta), Y_\nu(\zeta))}.$$

The solution is

$$V(z) = \lim_{m \to \infty} V_m(z),$$

where $V_{-1}(z) \equiv 0$ and

$$V_m(z) = J_\nu(z) + \gamma \int_0^z G_\nu(z, \zeta)\zeta^2 V_{m-1}(\zeta)\, d\zeta, \quad m = 0, 1, 2, \ldots$$

This leads to the asymptotic results

$$M(a, c; x) = \frac{\Gamma(c)}{\sqrt{\pi}} \left(\frac{1}{2}cx - ax\right)^{\frac{1}{4} - \frac{1}{2}c} e^{\frac{1}{2}x}$$
$$\times \left[\cos\left(\sqrt{2cx - 4ax} - \frac{1}{2}c\pi + \frac{1}{4}\pi\right) + O\left(|a|^{-\frac{1}{2}}\right)\right];$$

$$U(a, c; x) \sim \frac{\Gamma\left(\frac{1}{2}c - a + \frac{1}{4}\right)}{\sqrt{\pi}} x^{\frac{1}{4} - \frac{1}{2}c} e^{\frac{1}{2}x}$$
$$\times \cos\left(\sqrt{2cx - 4ax} - \frac{1}{2}c\pi + a\pi + \frac{1}{4}\pi\right).$$

As a corollary, the Laguerre polynomials

$$L_n^{(\alpha)}(x) = \frac{(\alpha + 1)_n}{n!} M(-n, \alpha + 1; x)$$

have asymptotics

$$L_n^{(\alpha)}(x) = \frac{e^{\frac{1}{2}x} n^{\frac{1}{2}\alpha - \frac{1}{4}}}{\sqrt{\pi}\, x^{\frac{1}{2}\alpha + \frac{1}{4}}} \left[\cos\left(2\sqrt{nx} - \frac{1}{2}\alpha\pi - \frac{1}{4}\pi\right) + O\left(n^{-\frac{1}{2}}\right)\right].$$

10.7.3 Hypergeometric functions, Jacobi polynomials

After changes of variables and a gauge transformation, the hypergeometric equation has the form

$$w''(\theta) + \frac{1 - 4\alpha^2}{4\theta^2}\, w(\theta)$$
$$+ \left[\frac{1 - 4\alpha^2}{48} + \frac{1 - 4\beta^2}{16} + \lambda + \frac{(\alpha + \beta + 1)^2}{4}\right] w(\theta) = r(\theta)\, w(\theta).$$

This can be converted to a perturbation of Bessel's equation by a change of scale and a gauge transformation, leading to

$$V''(z) + \frac{1}{z} V'(z) + \left[1 - \frac{(c-1)^2}{z^2} \right] V(z) = \frac{R(z)\, V(z)}{\mu^4}.$$

This can be solved as above with

$$V_0(z) = \frac{\Gamma(c)}{\sqrt{2}} \mu^{1-c} J_{c-1}(z),$$

giving the asymptotics

$$F\left(a + \nu, -\nu, c; \sin^2\left(\tfrac{1}{2}\theta \right) \right)$$

$$= \frac{\Gamma(c) \cos\left(\nu\theta + \tfrac{1}{2}a\theta - \tfrac{1}{2}c\pi + \tfrac{1}{4}\pi \right) + O\left(\nu^{-1} \right)}{\sqrt{\pi} \left(\nu \sin \tfrac{1}{2}\theta \right)^{c - \frac{1}{2}} \left(\cos \tfrac{1}{2}\theta \right)^{\frac{1}{2} + (a-c)}}$$

as $\nu \to +\infty$. The asymptotics of the Jacobi polynomials are a corollary: as $n \to \infty$,

$$P_n^{(\alpha, \beta)}(\cos \theta) = \frac{\cos\left(n\theta + \tfrac{1}{2}[\alpha + \beta + 1]\theta - \tfrac{1}{2}\alpha\pi - \tfrac{1}{4}\pi \right) + O\left(n^{-1} \right)}{\sqrt{n\pi} \left(\sin \tfrac{1}{2}\theta \right)^{\alpha + \frac{1}{2}} \left(\cos \tfrac{1}{2}\theta \right)^{\beta + \frac{1}{2}}}.$$

10.7.4 Legendre functions

Asymptotics in ν for the Legendre functions P_ν and Q_ν on the interval $-1 < x < 1$ follow from the asymptotics of the hypergeometric function. For the interval $1 < x < \infty$, we use

$$P_\nu^m(\cosh \theta) = (\sinh \theta)^m \frac{(-m + \nu + 1)_{2m}}{2^m \left(\tfrac{1}{2} \right)_m \pi}$$

$$\times \int_0^\pi \frac{(\sin \alpha)^{2m}\, d\alpha}{(\cosh \theta - \sinh \theta \, \cos \alpha)^{m+1+\nu}};$$

$$Q_\nu^m(\cosh \theta) = (-1)^m (\sinh \theta)^m \frac{(-m + \nu + 1)_{2m}}{2^m \left(\tfrac{1}{2} \right)_m}$$

$$\times \int_0^\infty \frac{(\sinh \alpha)^{2m}\, d\alpha}{(\cosh \theta + \sinh \theta \, \cosh \alpha)^{m+1+\nu}}.$$

The principal contributions come where the denominator is smallest. Ignoring terms of lower order, we change variables near $\alpha = 0$ and are led to

$$\frac{2^{m-\frac{1}{2}}\, e^{(m+1+v)\theta}}{B^{m+\frac{1}{2}}(m+1+v)^{m+\frac{1}{2}}} \int_0^\infty e^{-t}\, t^{m-\frac{1}{2}}\, dt, \quad B = e^\theta \sinh\theta,$$

which leads in turn to

$$P_v^m(\cosh\theta) = \frac{e^{\left(v+\frac{1}{2}\right)\theta}}{\sqrt{2\pi \sinh\theta}} (m+1+v)^{m-\frac{1}{2}} \left[1 + O\left(\{m+1+v\}^{-1}\right)\right].$$

A similar computation leads to

$$Q_v^m(\cosh\theta)$$

$$= (-1)^m \frac{e^{-\left(v+\frac{1}{2}\right)\theta}\, \sqrt{\pi}}{\sqrt{2 \sinh\theta}} (m+1+v)^{m-\frac{1}{2}} \left[1 + O\left(\{m+1+v\}^{-1}\right)\right].$$

10.7.5 Steepest descents and stationary phase

These are methods for obtaining asymptotics of integrals of the form

$$I(\lambda) = \int_C e^{\lambda \varphi(t)} f(t)\, dt.$$

If the functions are holomorphic, one seeks to deform the contour so that $\operatorname{Re}\varphi$ has one or more strict local maxima; the main contributions to the integral come at these points.

Starting from the generating function for the Laguerre polynomials and applying Cauchy's theorem and a change of variables, we obtain an integral of the form

$$L_n^{(\alpha)}(x) = \frac{1}{2\pi \sqrt{n} i} \int_C \exp\left\{\sqrt{n}\left(\frac{x}{t} - t\right)\right\} f\left(\frac{t}{\sqrt{n}}\right) dt.$$

For $x > 0$ the main contributions come near the points $s = \pm i\sqrt{x}$. Changing variables near these points leads to integrals of the form

$$\int_{-\infty}^\infty e^{-\sqrt{nx}\, u^2}\, du = (nx)^{-\frac{1}{4}} \int_{-\infty}^\infty e^{-s^2}\, ds = \frac{\sqrt{\pi}}{(nx)^{\frac{1}{4}}}$$

and to asymptotics

$$L_n^{(\alpha)}(x) = \frac{1}{\sqrt{\pi}}\, x^{-\frac{1}{2}\alpha - \frac{1}{4}}\, n^{\frac{1}{2}\alpha - \frac{1}{4}}\, e^{\frac{1}{2}x}\, \cos\left(2\sqrt{nx} - \frac{1}{2}\alpha\pi - \frac{1}{4}\pi\right)$$
$$+ O\left(n^{\frac{1}{2}\alpha - \frac{3}{4}}\right).$$

If the phase function φ is purely imaginary, one expects the principal contribution to the integral to come from points where $\varphi' = 0$; elsewhere there is cancellation due to rapid oscillation. In the integral

$$J_\nu(x) = \frac{1}{2\pi} \int_C e^{ix\sin\theta - i\nu\theta}\, d\theta,$$

changes of variables near the critical points $\theta = \pm\pi/2$ lead to integrals of the form

$$\int_{-\infty}^{\infty} e^{\pm\frac{1}{2}ixu^2}\, du$$

and to asymptotics

$$J_\nu(x) = \frac{\sqrt{2}}{\sqrt{\pi x}}\, \cos\left(x - \frac{1}{2}\nu\pi - \frac{1}{4}\pi\right) + O\left(x^{-1}\right)$$

as $x \to +\infty$.

10.8 Remarks

Some asymptotic results for special functions are treated in detail by Olver [222]. The book by Erdélyi [81] is a concise introduction to the general methods. For more detail, see Bleistein and Handelsman [32], Copson [58], van der Corput [293], and Wong [318]. Asymptotic expansions for solutions of general ordinary differential equations are treated by Wasow [304]. We have not touched on another powerful method for asymptotics, an adaptation of the steepest descent method known as the Riemann–Hilbert method; see Deift [67].

The method used in Sections 10.1–10.3 and in Exercise 10.12 goes back at least to Carlini in 1817 [40]. In the mathematical literature it is sometimes called the Liouville or Liouville–Green approximation [120, 187]. In the physics literature it is called the WKB or WKBJ approximation, referring to papers by Jeffreys [145], and by Wentzel [312], Kramers [165] and Brillouin [35], who also developed connection formulas. The method was adapted by Steklov [268] and Uspensky [291] to obtain asymptotics of classical orthogonal polynomials.

The method of steepest descents is an adaptation of the method of Laplace that we used to obtain the asymptotics of the Legendre functions. It was developed by Cauchy and Riemann, and adapted by Debye [66] to study the asymptotics of Bessel functions.

The method of stationary phase was developed by Stokes [275] and Kelvin [152] for the asymptotic evaluation of integrals that occur in the study of fluid mechanics.

11

Elliptic functions

Integrating certain functions of x that involve expressions $\sqrt{P(x)}$, P a quadratic polynomial, leads to trigonometric functions and their inverses. For example, the sine function can be defined implicitly by inverting a definite integral:

$$\theta = \int_0^{\sin\theta} \frac{ds}{\sqrt{1-s^2}}.$$

Integrating functions involving expressions $\sqrt{P(x)}$, where P is a polynomial of degree 3 or more, leads to new types of transcendental functions. When the degree is 3 or 4 the functions are called elliptic functions. For example, the Jacobi elliptic function $\operatorname{sn}(u) = \operatorname{sn}(u, k)$ is defined implicitly by

$$u = \int_0^{\operatorname{sn}u} \frac{ds}{\sqrt{(1-s^2)(1-k^2s^2)}}. \tag{11.0.1}$$

As functions of a complex variable, the trigonometric functions are periodic with a real period. Elliptic functions are doubly periodic, having two periods whose ratio is not real.

This chapter begins with the question of integrating a function $R(z, \sqrt{P(z)})$, where R is a rational function of two variables and P is a polynomial of degree 3 or 4 with no repeated roots. The general case can be reduced to

$$P(z) = (1-z^2)(1-k^2z^2).$$

We sketch Legendre's reduction of the integration of $R(z, \sqrt{P(z)})$ to three cases, called elliptic integrals of the first, second, and third kinds. The elliptic integral of the first kind is (11.0.1), leading to the Jacobi elliptic function $\operatorname{sn}u$ and then to the associated Jacobi elliptic functions $\operatorname{cn}u, \operatorname{dn}u$.

Jacobi eventually developed a second approach to elliptic functions as quotients of certain entire functions that just miss being doubly periodic, the theta functions. After covering the basics of theta functions and their relation to the Jacobi elliptic functions, we turn to the Weierstrass approach, which is based on a single function $\wp(u)$, doubly periodic with double poles.

11.1 Integration

Unlike differentiation, integration of relatively simple functions is not a more or less mechanical process.

Any rational function (quotient of polynomials) can be integrated, in principle, by factoring the denominator and using the partial fractions decomposition. The result is a sum of a rational function and a logarithmic term. Any function of the form

$$f(z) = \frac{p\left(z, \sqrt{P(z)}\right)}{q\left(z, \sqrt{P(z)}\right)}, \tag{11.1.1}$$

where p and q are polynomials in two variables and P is a quadratic polynomial, can be integrated by reducing it to a rational function. In fact, after a linear change of variable (possibly complex), we may assume that $P(z) = 1 - z^2$. Setting $z = 2u/(1 + u^2)$ converts the integrand to a rational function of u.

This process breaks down at the next degree of algebraic complexity: integrating a function of the form (11.1.1) when P is a polynomial of degree 3 or 4 with no multiple roots. The functions obtained by integrating such rational functions of z and $\sqrt{P(z)}$ are known as elliptic functions. The terminology stems from the fact that calculating the arc length of an ellipse, say as a function of the angle in polar coordinates, leads to such an integral. The same is true for another classical problem, calculating the arc length of a lemniscate (the locus of points the product of whose distances from two fixed points at distance $2d$ is equal to d^2; Jacob Bernoulli [26]).

In this section we prove Legendre's result [183], that any such integration can be reduced to one of three basic forms:

$$\int \frac{dz}{\sqrt{P(z)}}, \quad \int \frac{z^2\,dz}{\sqrt{P(z)}}, \quad \int \frac{dz}{(1 + az^2)\sqrt{P(z)}}. \tag{11.1.2}$$

The omitted steps in the proof are included as exercises.

The case of a polynomial of degree 3 can be reduced to that of a polynomial of degree 4 by a linear fractional transformation, and conversely; see the exercises for this and for subsequent statements for which no argument is supplied. Suppose that P has degree 4, with no multiple roots. Up to a linear fractional transformation, we may assume for convenience that the roots are ± 1 and $\pm 1/k$. Thus we may assume that

$$P(z) = \left(1 - z^2\right)\left(1 - k^2 z^2\right).$$

Suppose that $r(z, w)$ is a rational function in two variables. It can be written as the sum of two rational functions, one that is an even function of w and one that is an odd function of w:

$$r(z, w) = \frac{1}{2}\left[r(z, w) + r(z, -w)\right] + \frac{1}{2}\left[r(z, w) - r(z, -w)\right]$$

$$= r_1\left(z, w^2\right) + r_2\left(z, w^2\right) w.$$

Therefore, in integrating $r(z, w)$ with $w^2 = P(z)$, we may reduce to the case when the integrand has the form $r(z)\sqrt{P(z)}$, where r is a rational function of z.

At the next step we decompose r into even and odd parts, so that we are considering

$$\int r_1(z^2) \sqrt{P(z)}\, dz + \int r_2(z^2) \sqrt{P(z)}\, z\, dz.$$

Since P is a function of z^2, the substitution $s = z^2$ converts the integral on the right to the integral of a rational function of s and $\sqrt{Q(s)}$ with Q quadratic. As for the integral on the left, multiplying numerator and denominator by $\sqrt{P(z)}$ converts it to the form

$$\int \frac{R(z^2)\, dz}{\sqrt{P(z)}},$$

where R is a rational function. We use the partial fractions decomposition of R to express the integral as a linear combination of integrals with integrands

$$J_n(z) = \frac{z^{2n}}{\sqrt{P(z)}}, \quad n = 0, \pm 1, \pm 2, \ldots;$$

$$K_m(z) = K_m(a, z) = \frac{1}{(1 + az^2)^m \sqrt{P(z)}}, \quad a \neq 0,\ m = 1, 2, 3, \ldots$$

(This is a temporary notation, not to be confused with the notation for cylinder functions.)

At the next step we show that integration of J_{n+2}, $n \geq 0$, can be reduced to that of J_{n+1} and J_n. This leads recursively to the first two cases of (11.1.2). The idea is to relate these terms via a derivative:

$$
\left[z^{2n+1} \sqrt{P(z)} \right]' = (2n+1)\, z^{2n} \sqrt{P(z)} + \frac{1}{2} z^{2n+1} \frac{P'(z)}{\sqrt{P(z)}}
$$

$$
= \frac{(2n+1)\, z^{2n} P(z) + \frac{1}{2} z^{2n+1} P'(z)}{\sqrt{P(z)}}
$$

$$
= \frac{(2n+3)k^2 z^{2n+4} - (2n+2)(1+k^2)z^{2n+2} + (2n+1)z^{2n}}{\sqrt{P(z)}}
$$

$$
= (2n+3)k^2 J_{n+2} - (2n+2)(1+k^2)\, J_{n+1} + (2n+1)\, J_n.
$$

Therefore, up to a multiple of $z^{2n+1}\sqrt{P(z)}$, the integral of J_{n+2} is a linear combination of the integrals of J_{n+1} and J_n. The same calculation can be used to move indices upward if $n < 0$, leading again to J_0 and J_1.

A similar idea is used to reduce integration of K_{m+1} to integration of K_k, $k \le m$, together with J_0 if $m = 2$ and also J_1 if $m = 1$:

$$
\left[\frac{z\sqrt{P(z)}}{(1+az^2)^m} \right]' = \frac{\sqrt{P(z)}}{(1+az^2)^m} + \frac{zP'(z)}{2(1+az^2)^m \sqrt{P(z)}} - \frac{2maz^2\sqrt{P(z)}}{(1+az^2)^{m+1}}
$$

$$
= \frac{P(z)(1+az^2) + \frac{1}{2}zP'(z)(1+az^2) - 2amz^2 P(z)}{(1+az^2)^{m+1}\sqrt{P(z)}}.
$$

$$(11.1.3)$$

The numerator in the last expression is a polynomial in z^2 of degree 3. Writing it as a sum of powers of $1 + az^2$, we may rewrite the last expression as a combination of K_{m+1}, K_m, K_{m-1}, and K_{m-2} if $m \ge 3$. For $m = 2$ we have K_3, K_2, K_1, and J_0, and for $m = 1$ we have K_2, K_1, J_0, and J_1. This completes the proof of Legendre's result.

This result leads to the following classification: *the elliptic integral of the first kind*

$$
F(z) = F(k, z) = \int_0^z \frac{d\zeta}{\sqrt{(1-\zeta^2)(1-k^2\zeta^2)}};
$$

$$(11.1.4)$$

the elliptic integral of the second kind

$$
E(z) = E(k, z) = \int_0^z \sqrt{\frac{1-k^2\zeta^2}{1-\zeta^2}}\, d\zeta;
$$

$$(11.1.5)$$

the elliptic integral of the third kind

$$\Pi(a, z) = \Pi(a, k, z) = \int_0^z \frac{d\zeta}{(1 + a\zeta^2)\sqrt{(1 - \zeta^2)(1 - k^2\zeta^2)}}. \qquad (11.1.6)$$

Note that the integrand in (11.1.5) is just $J_0(\zeta) - k^2 J_1(\zeta)$.

11.2 Elliptic integrals

We begin with the properties of the elliptic integral of the first kind

$$F(z) = F(k, z) = \int_0^z \frac{d\zeta}{\sqrt{(1 - \zeta^2)(1 - k^2\zeta^2)}}. \qquad (11.2.1)$$

The parameter k is called the *modulus*. For convenience we assume that $0 < k < 1$; the various formulas to follow extend by analytic continuation to all values $k \neq \pm 1$.

Since the integrand is multiple-valued, we shall consider the integral to be taken over paths in the Riemann surface of the function $\sqrt{(1 - z^2)(1 - k^2 z^2)}$. This surface can be visualized as two copies of the complex plane with slits on the intervals $[-1/k, -1]$ and $[1, 1/k]$, the two copies being joined across the slits. In fact we adjoin the point at infinity to each copy of the plane. The result is two copies of the Riemann sphere joined across the slits. The resulting surface is a torus.

The function $\sqrt{(1 - z^2)(1 - k^2 z^2)}$ is single-valued and holomorphic on each slit sphere; we take the value at $z = 0$ to be 1 on the "upper" sphere and -1 on the "lower" sphere. The function F is multiple-valued, the value depending on the path taken in the Riemann surface. The integral converges as the upper limit goes to infinity, so we also allow curves that pass through infinity in one or both spheres.

Up to homotopy (continuous deformation on the Riemann surface), two paths between the same two points can differ by a certain number of circuits from -1 to 1 through the upper (resp. lower) sphere and back to -1 through the lower (resp. upper) sphere, or by a certain number of circuits around one or both of the slits. The value of the integral does not change under homotopy. Taking symmetries into account, this means that determinations of F can differ by integer multiples of $4K$, where

$$K = \int_0^1 \frac{dz}{\sqrt{(1 - z^2)(1 - k^2 z^2)}}, \qquad (11.2.2)$$

or by integer multiples of $2iK'$, where

$$K' = \int_1^{1/k} \frac{dz}{\sqrt{(z^2 - 1)(1 - k^2 z^2)}}. \tag{11.2.3}$$

Here the integrals are taken along line segments in the upper sphere, so that K and K' are positive. In fact, $4K$ corresponds to integrating from -1 to 1 in the upper sphere and returning to -1 in the lower sphere, while $2iK'$ corresponds to integrating from 1 to $1/k$ along the upper edge of the slit in the upper sphere and back along the lower edge of the slit.

The change of variables

$$z = \frac{1}{k}\sqrt{1 - k'^2 \zeta^2}$$

in (11.2.3) shows that the constants K' and K are related by

$$K' = \int_0^1 \frac{d\zeta}{\sqrt{(1 - \zeta^2)(1 - k'^2 \zeta^2)}} = K(k'),$$

where $k' = \sqrt{1 - k^2}$ is called the *complementary modulus*. Because of these considerations, we consider values of F to be determined only up to the addition of elements of the *period lattice*

$$\Lambda = \left\{ 4mK + 2inK', \quad m, n = 0, \pm 1, \pm 2, \ldots \right\}.$$

It follows from the definitions that

$$F(k, 1) = K, \quad F(k, 1/k) = K + iK'.$$

Integrating along line segments from 0 in the upper sphere shows that

$$F(k, -z) = -F(k, z).$$

Integrating from 0 in the upper sphere to 0 in the lower sphere and then to z in the lower sphere shows that

$$F(k, z-) = 2K - F(k, z+), \tag{11.2.4}$$

where $z+$ and $z-$ refer to representatives of z in the upper and lower spheres, respectively.

Integrating from 0 to $1/k$ in the upper sphere and then to 0 in the lower sphere and from there to z in the lower sphere shows that

$$F(k, z-) = 2K + 2iK' - F(k, z+). \tag{11.2.5}$$

Integrating along the positive imaginary axis in the upper sphere gives

$$F(k, \infty) = i \int_0^\infty \frac{ds}{\sqrt{(1 + s^2)(1 + k^2 s^2)}}. \tag{11.2.6}$$

The change of variables

$$s = \frac{\zeta}{\sqrt{1 - \zeta^2}}$$

in the integral in (11.2.6) shows that

$$F(k, \infty) = i K(k') = i K'. \tag{11.2.7}$$

Three classical transformations of F can be accomplished by changes of variables. Let $k_1 = (1 - k')/(1 + k')$. The change of variables

$$\zeta = \varphi(t) \equiv (1 + k')t \sqrt{\frac{1 - t^2}{1 - k^2 t^2}}$$

in the integrand for $F(k_1, \cdot)$ leads to the identity

$$F\big(k_1, \varphi(z)\big) = \int_0^{\varphi(z)} \frac{d\zeta}{\sqrt{(1 - \zeta^2)(1 - k_1^2 \zeta^2)}}$$

$$= (1 + k') \int_0^z \frac{dt}{\sqrt{(1 - t^2)(1 - k^2 t^2)}}.$$

This is *Landen's transformation* [173]:

$$F(k_1, z_1) = (1 + k') F(k, z); \tag{11.2.8}$$

$$k_1 = \frac{1 - k'}{1 + k'}, \quad z_1 = (1 + k')z \sqrt{\frac{1 - z^2}{1 - k^2 z^2}}.$$

Now take $k_1 = 2\sqrt{k}/(1 + k)$. Then the change of variables

$$\zeta = \varphi(t) = \frac{(1 + k)t}{1 + kt^2}$$

in the integrand for $F(k_1, \cdot)$ leads to the identity

$$F(k_1, z_1) = (1 + k) F(k, z); \quad k_1 = \frac{2\sqrt{k}}{1 + k}, \quad z_1 = \frac{(1 + k)z}{1 + kz^2}. \tag{11.2.9}$$

This is known as *Gauss's transformation* [104] or the *descending Landen transformation*.

These two transformations lend themselves to computation. To use (11.2.9), let $k_2 = 2\sqrt{k_1}/(1 + k_1)$; then

$$k_2' = (1 - k_1)/(1 + k_1) = \frac{(k_1')^2}{(1 + k_1)^2}.$$

Continuing, let $k_{n+1} = 2\sqrt{k_n}/(1 + k_n)$. So long as $0 < k_1 < 1$ it follows that the sequence $\{k_n'\}$ decreases rapidly to zero, so $k_n \to 1$. If, for example, $|z_1| < 1$, the corresponding sequence z_n will converge rapidly to a limit Z, giving

$$F(k_1, z_1) = \prod_{n=1}^{\infty} (1 + k_n)^{-1} F(1, Z),$$

where

$$F(1, z) \equiv \lim_{k \to 1} F(k, z) = \tanh^{-1} z. \tag{11.2.10}$$

Similar considerations apply to (11.2.8) and lead to an evaluation involving

$$F(0, z) \equiv \lim_{k \to 0} F(k, z) = \sin^{-1} z. \tag{11.2.11}$$

The change of variables

$$\zeta = \frac{is}{\sqrt{1 - s^2}}$$

in the integrand for $F(k, \cdot)$ leads to the identity

$$F(k, ix) = i \int_0^{x/\sqrt{1+x^2}} \frac{ds}{\sqrt{(1 - s^2)(1 - k'^2 s^2)}}.$$

This is *Jacobi's imaginary transformation*:

$$F(k, ix) = i F\left(k', \frac{x}{\sqrt{1 + x^2}}\right),$$

or equivalently

$$i F(k, y) = F\left(k', \frac{iy}{\sqrt{1 - y^2}}\right). \tag{11.2.12}$$

Our final result concerning the function F is Euler's *addition formula* [87]:

$$F(k, x) + F(k, y) = F\left(k, \frac{x\sqrt{P(y)} + y\sqrt{P(x)}}{1 - k^2 x^2 y^2}\right), \tag{11.2.13}$$

where as before $P(x) = (1 - x^2)(1 - k^2x^2)$. To prove (11.2.13) it is enough to assume that $0 < x, y < 1$ and that we have chosen a parametrized curve $\{(x(t), y(t))\}$ with $y(0) = 0$ and $x(T) = x$, $y(T) = y$, such that

$$F\big(k, x(t)\big) + F\big(k, y(t)\big) = F(k, C_1), \quad \text{constant.} \tag{11.2.14}$$

Differentiating with respect to t gives

$$\frac{x'}{\sqrt{P(x)}} + \frac{y'}{\sqrt{P(y)}} = 0.$$

(We are abusing notation and writing x and y for $x(t)$ and $y(t)$.) We follow an argument of Darboux. After reparametrizing, we may assume

$$x'(t) = \sqrt{P\big(x(t)\big)}, \quad y'(t) = -\sqrt{P\big(y(t)\big)}. \tag{11.2.15}$$

As often, it is helpful to consider the Wronskian $W = xy' - yx'$. Differentiating (11.2.15) gives x'' and y'' as functions of x and y and leads to

$$W' = xy'' - yx'' = 2k^2xy\big(y^2 - x^2\big).$$

Using (11.2.15) again,

$$W (xy)' = (xy')^2 - \big(yx'\big)^2 = \big(x^2 - y^2\big)\big(1 - k^2x^2y^2\big).$$

Thus

$$\frac{W'}{W} = -\frac{2k^2(xy)(xy)'}{1 - k^2x^2y^2} = \Big[\log(1 - k^2x^2y^2)\Big]',$$

so

$$\frac{x(t)\sqrt{P(y(t))} + y(t)\sqrt{P(x(t))}}{1 - k^2x(t)^2y(t)^2} = C_2, \quad \text{constant.} \tag{11.2.16}$$

Taking $t = 0$ in (11.2.14) and (11.2.16) shows that

$$C_1 = x(0) = C_2.$$

This proves (11.2.13). The case $x = y$ is known as *Fagnano's duplication formula* [93]. Fagnano's formula inspired Euler to find the full addition formula; see Ayoub [16], D'Antonio [61].

The definite integrals over the unit interval

$$K(k) = \int_0^1 \frac{dt}{\sqrt{(1 - t^2)(1 - k^2t^2)}}, \quad E(k) = \int_0^1 \frac{\sqrt{1 - k^2t^2}\, dt}{\sqrt{1 - t^2}},$$

are known as the *complete elliptic integrals of the first and second kind*, respectively. They can be expressed as hypergeometric functions:

$$K(k) = \frac{\pi}{2} F\left(\frac{1}{2}, \frac{1}{2}, 1; k^2\right), \quad E(k) = \frac{\pi}{2} F\left(-\frac{1}{2}, \frac{1}{2}, 1; k^2\right); \quad (11.2.17)$$

see Exercise 8.11 of Chapter 8.

11.3 Jacobi elliptic functions

The integral defining F is analogous to the simpler integral

$$\int_0^x \frac{dt}{\sqrt{1 - t^2}}.$$

This is also multiple-valued: values differ by integer multiples of 2π. One obtains a single-valued entire function by taking the inverse:

$$u = \int_0^{\sin u} \frac{dt}{\sqrt{1 - t^2}}.$$

Jacobi defined the function $\operatorname{sn} u = \operatorname{sn}(u, k)$ by

$$u = \int_0^{\operatorname{sn} u} \frac{d\zeta}{\sqrt{(1 - \zeta^2)(1 - k^2\zeta^2)}}. \quad (11.3.1)$$

It follows from the discussion in Section 11.2 that sn is *doubly periodic*, with a real period $4K$ and a period $2iK'$:

$$\operatorname{sn}(u + 4K) = \operatorname{sn}(u + 2iK') = \operatorname{sn} u.$$

Moreover, sn is odd: $\operatorname{sn}(-u) = -\operatorname{sn} u$. It follows that sn is odd around $z = 2K$ and around $z = iK'$. The identity (11.2.4) implies that sn is even around $z = K$. It follows from this, in turn, that sn is even around $z = K + iK'$. Note that a function f is even (resp. odd) around a point $z = a$ if and only if $f(z + 2a) = f(-z)$ (resp. $f(z + 2a) = -f(-z)$).

In summary:

$$\operatorname{sn} u = \operatorname{sn}(u + 4K) = \operatorname{sn}(u + 2iK')$$

$$= \operatorname{sn}(2K - u) = \operatorname{sn}(2K + 2iK' - u)$$

$$= -\operatorname{sn}(-u) = -\operatorname{sn}(4K - u) = -\operatorname{sn}(2iK' - u). \quad (11.3.2)$$

Because of the periodicity, it is enough to compute values of sn u for u in the *period rectangle*

$$\Pi = \{u \mid 0 \le \operatorname{Re} u < 4K, \ 0 \le \operatorname{Im} u < 2K'\}. \tag{11.3.3}$$

The various calculations of values of F give

$$\operatorname{sn}(0) = \operatorname{sn}(2K) = 0; \tag{11.3.4}$$

$$\operatorname{sn}(K) = -\operatorname{sn}(3K) = 1;$$

$$\operatorname{sn}(iK') = -\operatorname{sn}(2K + iK') = \infty;$$

$$\operatorname{sn}(K + iK') = -\operatorname{sn}(3K + iK') = k^{-1}.$$

It follows from (11.2.6) and (11.2.7) that as $t \to +\infty$,

$$iK' - F(it) \sim i \int_t^\infty \frac{ds}{ks^2} = \frac{i}{kt},$$

or, setting $\varepsilon = -1/kt$,

$$\operatorname{sn}(iK' + i\varepsilon) \sim \frac{1}{ik\varepsilon}.$$

Therefore sn has a simple pole at $u = iK'$ with residue $1/k$. Consequently, it also has a simple pole at $u = 2K + iK'$ with residue $-1/k$.

Differentiating (11.3.1) gives

$$\operatorname{sn}' u = \sqrt{1 - \operatorname{sn}^2 u}\,\sqrt{1 - k^2 \operatorname{sn}^2 u}. \tag{11.3.5}$$

This leads naturally to the introduction of two related functions

$$\operatorname{cn} u = \sqrt{1 - \operatorname{sn}^2 u}, \tag{11.3.6}$$

$$\operatorname{dn} u = \sqrt{1 - k^2 \operatorname{sn}^2 u}.$$

The three functions sn, cn, and dn are the *Jacobi elliptic functions*.

The only zeros of $1 - \operatorname{sn}^2$ (resp. $1 - k^2 \operatorname{sn}^2$) in the period rectangle (11.3.3) are at K and $3K$ (resp. $K + iK'$ and $3K + iK'$). These are double zeros, so we may choose branches of the square roots that are holomorphic near these points. We choose the branches with value 1 at $u = 0$. The resulting functions are, like sn itself, meromorphic in the complex plane.

Since sn is even or odd around each of the points 0, K, $2K$, iK', and $K + iK'$, it follows that the functions cn and dn are each even or odd around each of these points. Since neither function vanishes at $u = 0$ or $u = 2K$, they are even around 0 and $2K$. Similarly, cn is even around $K + iK'$ and dn is even around K. Since cn has a simple zero at $u = K$ and a simple pole at

$u = iK'$, it is odd around these points. Similarly, dn is odd around iK' and $K + iK'$. It follows that cn has periods $4K$ and $2K + 2iK'$, while dn has periods $2K$ and $4iK'$. Therefore cn is even around $u = 2K$ and dn is even around $u = 2iK'$. (See the exercises.)

Combining these observations with the computations (11.3.4), we obtain the following:

$$\operatorname{cn} u = \operatorname{cn}(u + 4K) = \operatorname{cn}(u + 2K + 2iK')$$

$$= \operatorname{cn}(-u) = -\operatorname{cn}(2K - u) = -\operatorname{cn}(2iK' - u); \qquad (11.3.7)$$

$$\operatorname{dn} u = \operatorname{dn}(u + 2K) = \operatorname{dn}(u + 4iK') = \operatorname{dn}(2K - u)$$

$$= \operatorname{dn}(-u) = -\operatorname{dn}(2iK' - u) = -\operatorname{dn}(2K + 2iK' - u).$$

We have established most of the values in the table

	0	K	$2K$	$3K$	iK'	$K + iK'$	$2K + iK'$	$3K + iK'$
sn	0	1	0	-1	∞	k^{-1}	∞	$-k^{-1}$
cn	1	0	-1	0	∞	$-ik'k^{-1}$	∞	$ik'k^{-1}$
dn	1	k'	1	k'	∞	0	∞	0

It follows from (11.3.5) and (11.3.6) that

$$\operatorname{sn}'(u) = \operatorname{cn} u \, \operatorname{dn} u; \qquad (11.3.8)$$

$$\operatorname{cn}'(u) = -\operatorname{sn} u \, \operatorname{dn} u;$$

$$\operatorname{dn}'(u) = -k^2 \operatorname{sn} u \, \operatorname{cn} u.$$

Since

$$\operatorname{sn} u = \sqrt{1 - \operatorname{cn}^2 u} = \frac{1}{k}\sqrt{1 - \operatorname{dn}^2 u}$$

and

$$\operatorname{dn} u = \sqrt{1 - k^2 + k^2 \operatorname{cn}^2 u}; \quad \operatorname{cn} u = \frac{1}{k}\sqrt{k^2 - 1 + \operatorname{dn}^2 u},$$

it follows that cn and dn can be defined implicitly by the integrals

$$u = \int_{\operatorname{cn} u}^{1} \frac{d\zeta}{\sqrt{(1 - \zeta^2)(1 - k^2 + k^2\zeta^2)}}; \qquad (11.3.9)$$

$$u = \int_{\operatorname{dn} u}^{1} \frac{d\zeta}{\sqrt{(1 - \zeta^2)(\zeta^2 - k'^2)}}.$$

From the point of view of the Jacobi elliptic functions, the Landen transformation, the Gauss transformation, Jacobi's imaginary transformation, and the addition formula take the following forms.

The identity (11.2.8) is equivalent to

$$\text{sn}\left([1+k']u, k_1\right) = (1+k')\frac{\text{sn}\,(u,k)\,\text{cn}\,(u,k)}{\text{dn}\,(u,k)}; \quad k_1 = \frac{1-k'}{1+k'}. \quad (11.3.10)$$

It follows that

$$\text{cn}\left([1+k']u, k_1\right) = (1+k')\frac{\text{dn}^2(u,k) - k'}{k^2\,\text{dn}\,(u,k)}; \quad (11.3.11)$$

$$\text{dn}\left([1+k']u, k_1\right) = (1-k')\frac{\text{dn}^2(u,k) + k'}{k^2\,\text{dn}\,(u,k)}.$$

The identity (11.2.9) is equivalent to

$$\text{sn}\left([1+k]u, k_1\right) = \frac{(1+k)\,\text{sn}\,(u,k)}{1+k\,\text{sn}^2(u,k)}; \quad k_1 = \frac{2\sqrt{k}}{1+k}. \quad (11.3.12)$$

It follows that

$$\text{cn}\left([1+k]u, k_1\right) = \frac{\text{cn}\,(u,k)\,\text{dn}\,(u,k)}{1+k\,\text{sn}^2(u,k)}; \quad (11.3.13)$$

$$\text{dn}\left([1+k]u, k_1\right) = \frac{1 - k\,\text{sn}^2(u,k)}{1+k\,\text{sn}^2(u,k)}.$$

The identity (11.2.12) is equivalent to

$$\text{sn}\,(iu, k) = i\,\frac{\text{sn}\,(u, k')}{\text{cn}\,(u, k')}. \quad (11.3.14)$$

It follows that

$$\text{cn}\,(iu, k) = \frac{1}{\text{cn}\,(u, k')}; \quad (11.3.15)$$

$$\text{dn}\,(iu, k) = \frac{\text{dn}\,(u, k')}{\text{cn}\,(u.k')}.$$

The addition formula (11.2.13) is equivalent to

$$\text{sn}\,(u + v) = \frac{\text{sn}\,u\,\text{cn}\,v\,\text{dn}\,v + \text{sn}\,v\,\text{cn}\,u\,\text{dn}\,u}{1 - k^2\text{sn}^2u\,\text{sn}^2v}. \quad (11.3.16)$$

It follows that

$$\operatorname{cn}(u+v) = \frac{\operatorname{cn} u \operatorname{cn} v - \operatorname{sn} u \operatorname{sn} v \operatorname{dn} u \operatorname{dn} v}{1 - k^2 \operatorname{sn}^2 u \operatorname{sn}^2 v}; \qquad (11.3.17)$$

$$\operatorname{dn}(u+v) = \frac{\operatorname{dn} u \operatorname{dn} v - k^2 \operatorname{sn} u \operatorname{sn} v \operatorname{cn} u \operatorname{cn} v}{1 - k^2 \operatorname{sn}^2 u \operatorname{sn}^2 v}.$$

These formulas imply the product formulas

$$\operatorname{sn}(u+v) \operatorname{sn}(u-v) = \frac{\operatorname{sn}^2 u - \operatorname{sn}^2 v}{1 - k^2 \operatorname{sn}^2 u \operatorname{sn}^2 v}; \qquad (11.3.18)$$

$$\operatorname{cn}(u+v) \operatorname{cn}(u-v) = \frac{1 - \operatorname{sn}^2 u - \operatorname{sn}^2 v + k^2 \operatorname{sn}^2 u \operatorname{sn}^2 v}{1 - k^2 \operatorname{sn}^2 u \operatorname{sn}^2 v}.$$

A commonly used notation for reciprocals and quotients of the Jacobi elliptic functions is due to Glaisher:

$$\operatorname{ns} = \frac{1}{\operatorname{sn}}, \quad \operatorname{nc} = \frac{1}{\operatorname{cn}}, \quad \operatorname{nd} = \frac{1}{\operatorname{dn}};$$

$$\operatorname{sc} = \frac{\operatorname{sn}}{\operatorname{cn}}, \quad \operatorname{sd} = \frac{\operatorname{sn}}{\operatorname{dn}}, \quad \operatorname{cd} = \frac{\operatorname{cn}}{\operatorname{dn}};$$

$$\operatorname{cs} = \frac{\operatorname{cn}}{\operatorname{sn}}, \quad \operatorname{ds} = \frac{\operatorname{dn}}{\operatorname{sn}}, \quad \operatorname{dc} = \frac{\operatorname{dn}}{\operatorname{cn}}.$$

To complete this section we note that the change of variables $\zeta = \operatorname{sn} s$ converts the elliptic integrals of the first, second, and third kinds, (11.1.4), (11.1.5), (11.1.6) to

$$F(z) = \int_0^{\operatorname{sn}^{-1} z} ds = \operatorname{sn}^{-1} z; \qquad (11.3.19)$$

$$E(z) = \int_0^{\operatorname{sn}^{-1} z} \left(1 - k^2 \operatorname{sn}^2 s\right) ds = \int_0^{\operatorname{sn}^{-1} z} \operatorname{dn}^2(s) \, ds;$$

$$\Pi(a, z) = \int_0^{\operatorname{sn}^{-1} z} \frac{ds}{1 + a \operatorname{sn}^2 s}.$$

11.4 Theta functions

The Jacobi elliptic functions are examples of the general notion of an *elliptic function*: a function f that is meromorphic in the complex plane and doubly periodic:

$$f(z) = f(z + 2\omega_1) = f(z + 2\omega_2),$$

with periods $2\omega_j \neq 0$, such that ω_2/ω_1 is not real. Such a function is determined by its values in any *period parallelogram*

$$\Pi_a = \{z \mid z = a + 2s\omega_1 + 2t\omega_2,\ 0 \le s, t < 1\}.$$

If f is entire, then it is bounded on a period parallelogram, therefore bounded on the plane, and therefore constant, by Liouville's theorem. Otherwise it has at least two poles, counting multiplicity, in each Π_a. To see this, note first that by changing a slightly we may assume that there is no pole on the boundary. Periodicity implies that the integral over the boundary C_a of Π_a vanishes:

$$\int_{C_a} f(z)\,dz = 0.$$

Therefore the sum of the residues is zero, so there are at least two simple poles or one multiple pole.

A non-constant elliptic function takes each complex value the same number of times in each Π_a. To see this, suppose that f does not take the value c on the boundary and has no poles on the boundary. Again, periodicity implies that

$$\frac{1}{2\pi i} \int_{C_a} \frac{f'(z)}{f(z) - c}\,dz = 0.$$

But the integral is equal to the number of times (counting multiplicity) that f takes the value c in Π_a, minus the number of poles (counting multiplicity) of f in Π_a. By continuity this number is independent of c, and by varying a we may ensure that any given value c is not taken on the boundary. The Jacobi elliptic functions illustrate this. For example, in the period rectangle (11.3.3) sn takes the value zero twice, takes the values 1 and -1 once each but with multiplicity two, and has two simple poles ($c = \infty$).

One consequence of the Weierstrass factorization theorem is that any function meromorphic in the plane is a quotient of entire functions. Since a doubly periodic entire function is constant, a non-constant elliptic function cannot be the quotient of doubly periodic entire functions. However, it can be expressed as a quotient of entire functions that are "nearly" periodic. The basic such function is a *theta function*.

Up to a linear transformation of the independent variable, we may consider the periods of a doubly periodic function to be 1 and τ, with $\operatorname{Im} \tau > 0$. Thus the basic period parallelogram is

$$\Pi = \{z \mid z = s + t\tau,\ 0 \le s, t < 1\}, \tag{11.4.1}$$

with oriented boundary Γ. Following Jacobi, we look for an entire function Θ that has period 1 and comes close to having period τ:

$$\Theta(z+1) = \Theta(z); \quad \Theta(z+\tau) = a(z)\,\Theta(z),$$

where a is entire, nonzero, and has period 1. This amounts to requiring that a be a constant times an integer power of $e^{2i\pi z}$. If Θ has no zeros on Γ, the number of zeros in Π is

$$\frac{1}{2\pi i} \int_\Gamma \frac{\Theta'(\zeta)}{\Theta(\zeta)}\,d\zeta = \frac{1}{2\pi i} \left\{ \int_0^1 + \int_1^{1+\tau} + \int_{1+\tau}^\tau + \int_\tau^0 \right\} \frac{\Theta'(\zeta)}{\Theta(\zeta)}\,d\zeta$$

$$= \frac{1}{2\pi i} \left\{ \int_0^1 - \int_\tau^{1+\tau} + \int_1^{1+\tau} - \int_0^\tau \right\} \frac{\Theta'(\zeta)}{\Theta(\zeta)}\,d\zeta$$

$$= -\frac{1}{2\pi i} \int_0^1 \frac{a'(s)}{a(s)}\,ds.$$

Thus the simplest choice is $a(z) = c\,e^{-2\pi i z}$, which implies a single zero in each period parallelogram. With this choice we would have

$$\Theta(z+1) = \Theta(z); \quad \Theta(z+\tau) = c\,e^{-2i\pi z}\,\Theta(z). \tag{11.4.2}$$

To construct such a function, we note that for Θ to have period 1, it must have the form

$$\Theta(z) = \sum_{n=-\infty}^\infty a_n\,p(z)^{2n}, \quad p(z) = e^{i\pi z}.$$

Now $p(z+\tau) = q\,p(z)$ where $q = q(\tau) = e^{i\pi\tau}$, so the second equation in (11.4.2) implies

$$a_n\,q^{2n} = c\,a_{n+1}. \tag{11.4.3}$$

Taking $c = -1$ and $a_0 = 1$, we find that $\Theta(z)$ should be given by

$$\Theta(z) = \sum_{n=-\infty}^\infty (-1)^n\,p^{2n}\,q^{n(n-1)}, \quad p = e^{i\pi z}, \quad q = e^{i\pi\tau}. \tag{11.4.4}$$

The assumption $\mathrm{Im}\,\tau > 0$ implies that this series converges very rapidly, uniformly on bounded sets. Therefore Θ is an entire function and

$$\Theta(z+1) = \Theta(z); \quad \Theta(z+\tau) = -e^{-2\pi i z}\Theta(z). \tag{11.4.5}$$

It follows from (11.4.4) that

$$\Theta(0) = \sum_{n=-\infty}^{\infty} (-1)^n q^{n(n-1)} \tag{11.4.6}$$

and replacing n by $-m$ for $n \le 0$ shows that $\Theta(0) = 0$. By construction, Θ has no other zeros in Π. The properties (11.4.5) and $\Theta(0) = 0$ characterize Θ up to a constant: if Ψ were another such entire function, then Ψ/Θ would be a doubly periodic entire function, thus constant.

Suppose that an elliptic function f with periods 1 and τ has zeros $\{a_1, a_2, \ldots, a_k\}$ and poles $\{b_1, b_2, \ldots, b_k\}$ in Π, repeated according to multiplicity. If any lie on the boundary of Π, we may translate slightly so that they lie in the interior. The residue theorem gives

$$\frac{1}{2\pi i} \int_\Gamma \frac{z\, f'(z)}{f(z)}\, dz = (a_1 + a_2 + \cdots + a_k) - (b_1 + b_2 + \cdots + b_k).$$

Because of periodicity,

$$\frac{1}{i} \int_\Gamma \frac{z\, f'(z)}{f(z)}\, dz = \left[\frac{1}{i} \int_1^{1+\tau} \frac{f'(z)\, dz}{f(z)}\right] + \tau \left[\frac{1}{i} \int_{1+\tau}^{\tau} \frac{f'(z)\, dz}{f(z)}\right].$$

Each integral in brackets is the change in the argument of f along a segment. Periodicity implies that the change is an integer multiple of 2π. Therefore

$$(a_1 + a_2 + \cdots + a_k) - (b_1 + b_2 + \cdots + b_k) \in \Lambda, \tag{11.4.7}$$

where Λ is the period lattice

$$\Lambda = \big\{m + n\tau, \quad m, n = 0, \pm 1, \pm 2, \ldots \big\}.$$

Conversely, suppose that the disjoint sets of points $\{a_j\}$ and $\{b_j\}$ in Π satisfy condition (11.4.7). Then there is an elliptic function with precisely these zeros and poles in Π [1] and it can be represented as an exponential times a quotient of translates of Θ [140]. In fact, let

$$(a_1 + a_2 + \cdots + a_k) - (b_1 + b_2 + \cdots + b_k) = m + n\tau.$$

(Since the a_j and b_j belong to Π, this condition implies that $k > 1$.) Then

$$f(z) = e^{-2n\pi i z} \frac{\Theta(z - a_1)\Theta(z - a_2) \cdots \Theta(z - a_k)}{\Theta(z - b_1)\Theta(z - b_2) \cdots \Theta(z - b_k)} \tag{11.4.8}$$

is the desired function. It is unique up to a constant factor.

In addition to quotients like (11.4.8), it is convenient to represent elliptic functions by using the functions

$$Z(z) = \frac{\Theta'(z)}{\Theta(z)}, \quad Z'(z) = \frac{\Theta''(z)}{\Theta(z)} - Z(z)^2.$$

It follows from (11.4.5) that

$$Z(z + \tau) = Z(z) - 2\pi i,$$

so Z' and linear combinations of translates

$$c_1 Z(z - b_1) + c_2 Z(z - b_2) + \cdots + c_n Z(z - b_n), \quad c_1 + c_2 + \cdots + c_n = 0$$

have period τ and thus are elliptic functions. Since these are derivatives, they can be integrated immediately (in terms of functions of Θ and its translates and derivatives). Note that Z has a simple pole at each lattice point, while for $m \geq 1$ the derivative $Z^{(m)}$ is an elliptic function with a pole of order $m + 1$ at each lattice point.

This leads to an integration procedure for any elliptic function f. We may suppose that f has periods 1 and τ. If f has a pole of order $k \geq 2$ at $z = b$, then there is a constant c such that $f(z) - c\, Z^{(k-1)}(z - b)$ has a pole of order $< k$ at $z = b$. Thus the integration problem can be reduced to the case of functions f that have only simple poles. Suppose that the poles in Π are $b_1, \ldots, b_n$, with residues $\beta_1, \ldots, \beta_n$. We know that $\sum \beta_j = 0$, so there is a linear combination of translates of Z that has the same poles and residues in Π as f. It follows that the difference of f and this linear combination of translates of Z is constant.

The function Θ has exactly one zero in the period parallelogram Π, at $z = 0$. This fact and the properties (11.4.5) imply that the zeros of Θ are precisely the points of the lattice Λ. It follows that

$$\Theta\left(z + \frac{1}{2}\tau\right) = 0$$

if and only if $z = m + \left(n - \frac{1}{2}\right)\tau$, for some integers m and n, or equivalently $p^2 q^{2n-1} = 1$ for some integer n. The product

$$\prod_{n=1}^{\infty} \left(1 - p^2 q^{2n-1}\right)\left(1 - p^{-2} q^{2n-1}\right), \quad p = p(z),$$

converges for all z and has the same zeros as $\Theta\left(z + \frac{1}{2}\tau\right)$, so

$$\Theta\left(z + \frac{1}{2}\tau\right) = c(z, \tau) \prod_{n=1}^{\infty} \left(1 - p^2 q^{2n-1}\right)\left(1 - p^{-2} q^{2n-1}\right), \quad (11.4.9)$$

where $c(z, \tau)$ is an entire function of z. It can be shown that $c(z, \tau) = G(\tau)$ is independent of z and can be evaluated as a product involving powers of $q = e^{i\pi\tau}$ (see the exercises). The result is one version of Jacobi's *triple product formula*

$$\Theta\left(z + \frac{1}{2}\tau\right) = \prod_{n=1}^{\infty} \left(1 - q^{2n}\right) \left(1 - p^2 q^{2n-1}\right) \left(1 - p^{-2} q^{2n-1}\right).$$

$$(11.4.10)$$

This implies another version

$$\Theta(z) = \prod_{n=1}^{\infty} \left(1 - q^{2n}\right) \left(1 - p^2 q^{2n-2}\right) \left(1 - p^{-2} q^{2n}\right). \qquad (11.4.11)$$

11.5 Jacobi theta functions and integration

According to the results of the preceding section, the Jacobi elliptic functions can be expressed as quotients of translates of Θ, after a linear change of variables. As we shall see, it is convenient for this purpose to introduce the *Jacobi theta functions*. These are normalized versions of Θ and of translations of Θ by half-periods:

$$\theta_1(z) = i \frac{q^{\frac{1}{4}}}{p} \Theta(z) = i \sum_{n=-\infty}^{\infty} (-1)^n p^{2n-1} q^{\left(n-\frac{1}{2}\right)^2}; \qquad (11.5.1)$$

$$\theta_2(z) = \frac{q^{\frac{1}{4}}}{p} \Theta\left(z + \frac{1}{2}\right) = \sum_{n=-\infty}^{\infty} p^{2n-1} q^{\left(n-\frac{1}{2}\right)^2};$$

$$\theta_3(z) = \Theta\left(z + \frac{1}{2} + \frac{1}{2}\tau\right) = \sum_{n=-\infty}^{\infty} p^{2n} q^{n^2};$$

$$\theta_4(z) = \Theta\left(z + \frac{1}{2}\tau\right) = \sum_{n=-\infty}^{\infty} (-1)^n p^{2n} q^{n^2}.$$

(There are various other notations and normalizations; see Whittaker and Watson [315].) Note that because of the factor p^{-1}, θ_1 and θ_2 are periodic with period 2, not 1. Also, θ_1 is an odd function of z, while θ_2, θ_3, and θ_4 are even functions of z.

The triple product formula (11.4.10) implies corresponding formulas for the θ_j:

$$\theta_1(z) = 2q^{\frac{1}{4}} \sin(\pi z) \prod_{n=1}^{\infty} \left(1 - q^{2n}\right)\left(1 - p^2 q^{2n}\right)\left(1 - p^{-2} q^{2n}\right);$$

$$\theta_2(z) = 2q^{\frac{1}{4}} \cos(\pi z) \prod_{n=1}^{\infty} \left(1 - q^{2n}\right)\left(1 + p^2 q^{2n}\right)\left(1 + p^{-2} q^{2n}\right);$$

$$\theta_3(z) = \prod_{n=1}^{\infty} \left(1 - q^{2n}\right)\left(1 + p^2 q^{2n-1}\right)\left(1 + p^{-2} q^{2n-1}\right);$$

$$\theta_4(z) = \prod_{n=1}^{\infty} \left(1 - q^{2n}\right)\left(1 - p^2 q^{2n-1}\right)\left(1 - p^{-2} q^{2n-1}\right).$$

The identities (11.4.4) and (11.5.1) lead to the following table of values:

	0	$\frac{1}{2}$	$\frac{1}{2}\tau$	$\frac{1}{2} + \frac{1}{2}\tau$
θ_1	0	$\sum q^{(n-\frac{1}{2})^2}$	$iq^{-\frac{1}{4}} \sum (-1)^n q^{n^2}$	$q^{-\frac{1}{4}} \sum q^{n^2}$
θ_2	$\sum q^{(n-\frac{1}{2})^2}$	0	$q^{-1/4} \sum q^{n^2}$	$-iq^{-\frac{1}{4}} \sum (-1)^n q^{n^2}$
θ_3	$\sum q^{n^2}$	$\sum (-1)^n q^{n^2}$	$q^{-\frac{1}{4}} \sum q^{(n+\frac{1}{2})^2}$	0
θ_4	$\sum (-1)^n q^{n^2}$	$\sum q^{n^2}$	0	$q^{-\frac{1}{4}} \sum q^{(n-\frac{1}{2})^2}$

Consider now the Jacobi elliptic functions with modulus k, and let $\tau = iK'/K$. The function

$$\frac{\theta_1(u/2K)}{\theta_4(u/2K)}$$

is meromorphic as a function of u with simple zeros at $u = 0$ and $u = 2K$ and simple poles at $u = iK'$ and $u = 2K + iK'$, and has periods $4K$ and $2iK'$. It follows that it is a multiple of $\operatorname{sn} u$, and conversely:

$$\operatorname{sn} u = C \frac{\theta_1(u/2K)}{\theta_4(u/2K)}.$$

Each side of this equation may be evaluated at $u = K$, and this determines the constant C. Similar arguments apply to cn and dn. The results are:

$$\operatorname{sn} u = \frac{1}{\sqrt{k}} \frac{\theta_1(u/2K)}{\theta_4(u/2K)};$$ (11.5.2)

$$\operatorname{cn} u = \sqrt{\frac{k'}{k}} \frac{\theta_2(u/2K)}{\theta_4(u/2K)};$$

$$\operatorname{dn} u = \sqrt{k'} \frac{\theta_3(u/2K)}{\theta_4(u/2K)}.$$

Jacobi obtained a large number of formulas relating products of translations of theta functions, having a form like

$$\Theta(z + w)\,\Theta(z - w) = c_1 \Theta(z + a_1)\,\Theta(z + a_2)\,\Theta(w + a_3)\,\Theta(w + a_4)$$

$$+ c_2 \Theta(w + a_1)\,\Theta(w + a_2)\,\Theta(z + a_3)\,\Theta(z + a_4),$$

where the a_j belong to the set $\{0, \frac{1}{2}, \frac{1}{2}\tau, \frac{1}{2} + \frac{1}{2}\tau\}$. The constants are chosen so that the quotient of the right side by the left side is a doubly periodic function of z, for any given w, and the zeros of the denominator are cancelled by the zeros of the numerator. Extensive lists are given in Whittaker and Watson [315] and Rainville [236].

A deeper result is Jacobi's remarkable identity

$$\theta_1' = \pi\, \theta_2 \theta_3 \theta_4.$$ (11.5.3)

(The factor π is due to the normalization we have chosen here.) See [12, 315].

The normalizations of the Jacobi theta functions have the consequence that each one has the form

$$\theta(z, \tau) = \sum_{n=-\infty}^{\infty} a_n\, e^{2(n+c)i\pi z}\, e^{(n+c)^2 i\pi \tau},$$

for some value of c, and therefore is a solution of the partial differential equation

$$\theta_{zz}(z, \tau) = 4\pi i \theta_\tau(z, \tau).$$

If we take as variables $x = z$ and $t = -i\tau/4\pi$, this is the heat equation

$$\psi_t = \psi_{xx}.$$ (11.5.4)

The theta functions are periodic in x. Periodic solutions of the heat equation can be obtained in two different ways, a fact that provides an approach to obtaining Jacobi's imaginary transformation (11.2.12) or (11.3.14), (11.3.15) in terms of theta functions.

The fundamental solution for the heat equation on the line, i.e. the solution ψ of (11.5.4) that has the property

$$\lim_{t\to 0+} \int_{-\infty}^{\infty} \psi(x - y, t) f(y)\, dy = f(x) \tag{11.5.5}$$

for every bounded, continuous function f is

$$\psi(x, t) = \frac{e^{-x^2/4t}}{\sqrt{4\pi t}}; \tag{11.5.6}$$

see the exercises. One way to obtain the fundamental solution for the *periodic* problem is to periodize ψ, which gives

$$\sum_{n=-\infty}^{\infty} \frac{e^{-(x+n)^2/4t}}{\sqrt{4\pi t}} = \frac{e^{-x^2/4t}}{\sqrt{4\pi t}} \sum_{n=-\infty}^{\infty} e^{-nx/2t}\, e^{-n^2/4t}. \tag{11.5.7}$$

A second way to find the periodic fundamental solution is to expand in Fourier series (or to separate variables), which leads to

$$\sum_{n=-\infty}^{\infty} e^{2ni\pi x} e^{-4n^2\pi^2 t}; \tag{11.5.8}$$

see the exercises.

Let

$$z = x, \quad \tau = 4i\pi t, \quad z_1 = \frac{ix}{4\pi t} = -\frac{z}{\tau},$$

$$\tau_1 = \frac{i}{4\pi t} = -\frac{1}{\tau}, \quad q = e^{i\pi\tau}, \quad q_1 = e^{i\pi\tau_1}.$$

Then the equality of the two expressions (11.5.7) and (11.5.8) for the periodic fundamental solution takes the form

$$\sum_{n=-\infty}^{\infty} p(z)^{2n} q^{n^2} = \frac{e^{-i\pi z^2/\tau}}{\sqrt{-i\tau}} \sum_{n=-\infty}^{\infty} p(z_1)^{2n} q_1^{n^2},$$

or

$$\theta_3(z|\tau) = \frac{e^{-i\pi z^2/\tau}}{\sqrt{-i\tau}}\, \theta_3\left(-\frac{z}{\tau}\middle| -\frac{1}{\tau}\right), \tag{11.5.9}$$

where the notation makes explicit the dependence on the parameter τ.

Finally, let us return to the integration problem of Section 11.1. We have seen that the integrals to be evaluated are

$$F(z) = \operatorname{sn}^{-1} z;$$

$$E(z) = \int_0^{\operatorname{sn}^{-1}z} \operatorname{dn}^2(\zeta)\,d\zeta;$$

$$\Pi(a, z) = \int_0^{\operatorname{sn}^{-1}z} \frac{d\zeta}{1 + a\operatorname{sn}^2\zeta} = \operatorname{sn}^{-1}z - a\int_0^{\operatorname{sn}^{-1}z} \frac{\operatorname{sn}^2\zeta\,d\zeta}{1 + a\operatorname{sn}^2\zeta}.$$

It follows from previous results that the residue of dn at $u = iK'$ is $-i$. Also, dn is odd around iK', so dn^2 is even around iK' and it follows that

$$\operatorname{dn}^2(iK' + u) = -\frac{1}{u^2} + O(1).$$

The adapted theta function $\theta(u)$ is defined by

$$\theta(u) = \theta_4(u/2K) = \Theta\big((u + iK')/2K\big).$$

Jacobi's Z function is

$$Z(u) = \frac{\theta'(u)}{\theta(u)}.$$

The residue at iK' is 1, so

$$Z'(iK' + u) = -\frac{1}{u^2} + O(1).$$

Since Z' is elliptic and its only poles are at the lattice points, it follows that $\operatorname{dn}^2 - Z'$ is a constant, which is customarily written E/K. Thus

$$E(z) = \frac{\theta'\big(\operatorname{sn}^{-1}z\big)}{\theta\big(\operatorname{sn}^{-1}z\big)} + \frac{E}{K}\operatorname{sn}^{-1}z.$$

Since $\operatorname{sn}^{-1}1 = K$ and $\theta'(K) = 0$, it follows that the constant E is

$$E = E(1) = \int_0^1 \sqrt{\frac{1 - k^2\zeta^2}{1 - \zeta^2}}\,d\zeta,$$

the complete elliptic integral of the second kind.

For the elliptic integral of the third kind we use the addition formula

$$Z(u) + Z(v) - Z(u + v) = k^2\operatorname{sn}u\operatorname{sn}v\operatorname{sn}(u + v). \tag{11.5.10}$$

This is proved in the usual way, by establishing that the two sides are elliptic functions of u with the same poles and residues, and that they agree at $u = 0$.

A consequence of this and (11.3.16) is that

$$Z(u - v) - Z(u + v) + 2Z(v)$$

$$= k^2 \text{sn}\, u \, \text{sn}\, v \, [\text{sn}\,(u + v) + \text{sn}\,(u - v)]$$

$$= 2k^2 \frac{\text{sn}\, v \, \text{cn}\, v \, \text{dn}\, v \, \text{sn}^2 u}{1 - k^2 \text{sn}^2 v \, \text{sn}^2 u}. \tag{11.5.11}$$

Finding $\Pi(a, z)$ reduces to finding

$$\int \frac{\text{sn}^2 u}{1 + a \, \text{sn}^2 u} \, du.$$

If $a = 0, -1, \infty$, or $-k^2$ the integral can be reduced to integrals of the first and second kinds. Otherwise we choose b such that $a = -k^2 \text{sn}^2 b$. Up to multiplication by a constant, we are left with

$$\int_0^{\text{sn}^{-1} z} 2k^2 \frac{\text{sn}\, b \, \text{cn}\, b \, \text{dn}\, b \, \text{sn}^2 u}{1 - k^2 \text{sn}^2 b \, \text{sn}^2 u} \, du$$

$$= \int_0^{\text{sn}^{-1} z} [Z(u - b) - Z(u + b) + 2Z(b)] \, du$$

$$= \log \left[\frac{\theta(w - b)}{\theta(w + b)} \right] + 2Z(b)\, w, \quad w = \text{sn}^{-1} z.$$

11.6 Weierstrass elliptic functions

We return to the general notion of an elliptic function: a meromorphic function f with periods $2\omega_1, 2\omega_2$, where we assume that $\text{Im}\,(\omega_2/\omega_1) > 0$. As noted earlier, unless f is constant it has at least two poles, counting multiplicity, in the parallelogram

$$\Pi = \Pi(\omega_1, \omega_2) = \{u \mid u = 2s\,\omega_1 + 2t\,\omega_2,\ 0 \le s, t < 1\}.$$

Thus in some sense the simplest such function would have a double pole at each point of the period lattice

$$\Lambda = \Lambda(\omega_1, \omega_2) = \{2n_1\omega_1 + 2n_2\omega_2 \mid n_1, n_2 = 0, \pm 1, \pm 2, \ldots\},$$

and no other poles. A consequence is that for any complex c, $f(u) = c$ would have exactly two solutions u, counting multiplicity, in Π.

We show below that there is such a function, the *Weierstrass $\wp$ function*
$\wp(u) = \wp(u, \Lambda)$, which is even and has the property

$$\wp(u) = \wp(u, \omega_1, \omega_2) = \frac{1}{u^2} + O(u^2) \quad \text{as} \quad u \to 0. \tag{11.6.1}$$

This condition determines $\wp$ uniquely, since the difference of two such functions would be a bounded entire function that vanishes at the origin. The function $\wp$ satisfies a differential equation analogous to the equation (11.3.5) satisfied by Jacobi's function sn. The property (11.6.1) implies that $(\wp')^2 - 4\wp^3$ is an even elliptic function with at most a double pole at $u = 0$, so there are constants g_2 and g_3 such that

$$(\wp')^2 = 4\wp^3 - g_2\wp - g_3. \tag{11.6.2}$$

Let $\omega_3 = -\omega_1 - \omega_2$. Since $\wp$ is even with period $2\omega_j$, $j = 1, 2, 3$, it is even around ω_j, and therefore $\wp'(\omega_j) = 0$, and it follows from this and (11.6.2) that $e_j = \wp(\omega_j)$ is a root of the cubic

$$Q(t) = 4t^3 - g_2 t - g_3. \tag{11.6.3}$$

The function $\wp(u) - e_j$ has a double root at ω_j, $j = 1, 2$ and at $u = -\omega_3 = \omega_1 + \omega_2$, $j = 3$. Since each of these points is in Π, it follows that the e_j are distinct. Therefore, they are simple roots of Q. It follows that

$$e_1 + e_2 + e_3 = 0;$$

$$4(e_2 e_3 + e_3 e_1 + e_1 e_2) = -g_2;$$

$$4e_1 e_2 e_3 = g_3.$$

Any elliptic function f with periods $2\omega_1, 2\omega_2$ can be expressed as a rational function of $\wp$ and the derivative $\wp'$. To see this, suppose first that f is even. If the origin is a pole of f, we may subtract a linear combination of powers of $\wp$ so that the resulting function g is regular at the origin. The zeros and poles of g in Π can be taken to be $\pm a_1, \ldots, \pm a_n$ and $\pm b_1, \ldots, \pm b_n$ respectively, repeated according to multiplicity. The product

$$\prod_{j=1}^{n} \frac{\wp(u) - \wp(a_j)}{\wp(u) - \wp(b_j)}$$

has the same zeros and poles as g, so g is a constant multiple. Thus f is a rational function of $\wp$. If f is odd, then $f = g\wp'$ where $g = f/\wp'$ is even.

A second representation of elliptic functions can be obtained by using the *Weierstrass zeta function* ζ, which is characterized by

$$\zeta'(u) = -\wp(u), \quad \zeta(-u) = -\zeta(u). \tag{11.6.4}$$

Since $\wp$ is periodic with period $2\omega_j$, the integral

$$-\int_u^{u+2\omega_j} \wp(s)\, ds = 2\eta_j$$

is a constant. It follows that

$$\zeta(u + 2\omega_j) = \zeta(u) + 2\eta_j = \zeta(u) + 2\zeta(\omega_j). \tag{11.6.5}$$

Setting $u = -\omega_j$ shows that $\eta_j = \zeta(\omega_j)$.

Suppose now that f is an elliptic function with periods $2\omega_j$ and distinct poles a_k in Π. Let c_k be the residue at a_k; then $\sum c_k = 0$. The function

$$g(u) = f(u) - \sum c_k\, \zeta(u - a_k)$$

has no simple poles, and

$$g(u + 2\omega_j) = g(u) - 2\eta_j \sum c_k = g(u).$$

Therefore g has only multiple poles and is, up to an additive constant, a linear combination of derivatives of translates of ζ. Thus

$$f(u) = C + \sum c_k\, \zeta(u - a_k) + \sum_{\nu > 0} c_{\nu k}\, \zeta^{(\nu)}(u - a_k).$$

This reduces the problem of integrating f to the problem of integrating ζ. It is convenient at this point to introduce the *Weierstrass sigma function* $\sigma(u)$, which is characterized by the conditions

$$\frac{\sigma'}{\sigma} = \zeta; \quad \lim_{u \to 0} \frac{\sigma(u)}{u} = 1.$$

Then an integral of ζ is $\log \sigma$.

Since $\zeta(s) - 1/s$ is regular at the origin we may define

$$\sigma(u) = u \exp \int_0^u \left\{ \zeta(s) - \frac{1}{s} \right\} ds.$$

The integrand is odd, so the integral is even and σ is odd. It is not difficult to show that σ is entire, with a simple zero at each point of Λ. Equation (11.6.5) implies that the derivative of

$$\log \frac{\sigma(u + 2\omega_j)}{\sigma(u)} = \int_u^{u+2\omega_j} \zeta(s)\, ds$$

is $2\eta_j$, so

$$\log \frac{\sigma(u+2\omega_j)}{\sigma(u)} = 2\eta_j u + c_j, \quad c_j \text{ constant.}$$

Since σ is odd, taking $u = -\omega_j$ shows that $c_j = \log(-1) + 2\eta_j\omega_j$, so

$$\sigma(u+2\omega_j) = -e^{2\eta_j(u+\omega_j)}\sigma(u). \tag{11.6.6}$$

Thus σ is analogous to the theta function Θ.

The function σ allows a third representation of an elliptic function f, this time as a quotient of entire functions. Suppose that the zeros and poles of f in Π are $a_1, \ldots, a_n$ and $b_1, \ldots, b_n$. As noted at the beginning of Section 11.4, $w = \sum(a_j - b_j)$ belongs to Λ. Therefore we may replace a_1 by $a_1 - w$ and assume that $\sum(a_j - b_j) = 0$. The function

$$g(u) = \prod_{j=1}^{n} \frac{\sigma(u-a_j)}{\sigma(u-b_j)}$$

has the same zeros and poles as f and is doubly periodic by (11.6.6). Therefore f is a constant multiple of g.

The function $\wp$ has an addition formula. Given u and v in Π such that $\wp(u) \neq \wp(v)$, determine constants such that

$$\wp'(u) - B = A\wp(u), \quad \wp'(v) - B = A\wp(v).$$

Then

$$A = \frac{\wp'(u) - \wp'(v)}{\wp(u) - \wp(v)}.$$

The function $\wp' - B - A\wp$ has a unique pole, of order 3, in Π, so it has three zeros in Π. The sum of the zeros is an element of Λ (see the argument leading to (11.4.7)). By construction, u and v are zeros, so $-(u+v)$ is also a zero. Therefore $\wp(u)$, $\wp(v)$, and $\wp(u+v) = \wp(-u-v)$ are three roots of the cubic $Q(t) - (At + B)^2$. For most values of u and v these roots are distinct, so their sum is $A^2/4$:

$$\wp(u+v) = \frac{1}{4}\left[\frac{\wp'(u) - \wp'(v)}{\wp(u) - \wp(v)}\right]^2 - \wp(u) - \wp(v). \tag{11.6.7}$$

Up to this point we have been *assuming* that there is a function $\wp$ with the property (11.6.1). To construct $\wp$ we begin with the (formal) derivative

$$\wp'(u) = \sum_{p\in\Lambda} \frac{2}{(p-u)^3} = \sum_{m,n=-\infty}^{\infty} \frac{2}{(2m\omega_1 + 2n\omega_2 - u)^3}. \tag{11.6.8}$$

The series converges uniformly near any point not in Λ and defines a function that is odd, meromorphic, has a triple pole at each point of Λ, and has periods $2\omega_1$ and $2\omega_2$. Near $u = 0$,

$$\wp'(u) = -\frac{2}{u^3} + O(1).$$

Therefore we may define a function $\wp$ by

$$\wp(u) = \frac{1}{u^2} + \int_0^u \left[\wp'(s) + \frac{2}{s^3} \right] ds$$

$$= \frac{1}{u^2} + \sum_{p \in \Lambda,\, p \neq 0} \left[\frac{1}{(u-p)^2} - \frac{1}{p^2} \right]. \qquad (11.6.9)$$

This is an even meromorphic function that satisfies (11.6.1) and has a double pole at each point of Λ. The function $\wp(u + 2\omega_j)$ has the same derivative, so

$$\wp(u + 2\omega_j) - \wp(u) = c_j, \quad \text{constant}.$$

Since $\wp$ is even, setting $u = -\omega_j$ shows that the constant is zero. Thus $\wp$ has periods $2\omega_j$ and is the desired elliptic function.

Similarly we may define

$$\zeta(u) = \frac{1}{u} + \int_0^u \left[\frac{1}{s^2} - \wp(s) \right] ds = \frac{1}{u} + \sum_{p \in \Lambda,\, p \neq 0} \left[\frac{1}{u-p} + \frac{1}{p} + \frac{u}{p^2} \right]$$

and

$$\log \sigma(u) = \log u + \int_0^u \left[\zeta(s) - \frac{1}{s} \right] ds$$

$$= \log u + \sum_{p \in \Lambda,\, p \neq 0} \left[\log\left(1 - \frac{u}{p} \right) + \frac{u}{p} + \frac{u^2}{2p^2} \right],$$

so

$$\sigma(u) = u \prod_{p \in \Lambda,\, p \neq 0} \left(1 - \frac{u}{p} \right) \exp\left(\frac{u}{p} + \frac{u^2}{2p^2} \right).$$

11.7 Exercises

11.1 Show that letting $z = \sin\theta$ converts the integral of (11.1.1) to the integral of a rational function of $\sin\theta$ and $\cos\theta$. Show that setting $u = \tan\left(\frac{1}{2}\theta\right)$ converts this integral to the integral of a rational function of u.

11.2 Consider an integral $\int r(z, \sqrt{Q(z)})\,dz$ where Q is a polynomial of degree 3 (resp. 4) with distinct roots.

    (a) Show that there is a linear fractional transformation (Möbius transformation), i.e. a transformation $z = \varphi(w) = (aw + b)/(cw + d)$ that converts the integral to one of the same type, $\int r_1(w, \sqrt{Q_1(w)})\,dw$, with Q_1 of degree 4 (resp. 3). Hint: one can take $Q_1(w) = (cw + d)^4 Q(\varphi(w))$, and $\varphi(\infty)$ may or may not be a root of Q.

    (b) Show that if Q has degree 4, there is a linear fractional transformation that converts the integral to one of the same type but with the polynomial $(1 - \zeta^2)(1 - k^2\zeta^2)$. Hint: map two roots to ± 1; this leaves one free parameter.

11.3 Compute the constants in the representation (11.1.3) of K_2 in terms of integrals of K_1, J_0, J_1.

11.4 Verify that the indicated change of variables leads to (11.2.8).

11.5 Verify that the indicated change of variables leads to (11.2.9).

11.6 Verify (11.2.10) and (11.2.11).

11.7 Verify (11.2.17): let $t = \sin\theta$ and integrate the series expansion of the resulting integrands term by term, using Exercise 2.2 of Chapter 1.

11.8 Show that as $\varepsilon \to 0+$, $F(k, 1 + \varepsilon) \sim K - i\sqrt{2\varepsilon}/k'$. Deduce that $1 - \mathrm{sn}^2$ has double zeros at K and $3K$.

11.9 Show that as $\varepsilon \to 0$, $F(k, 1/k + \varepsilon) \sim K + iK' - \sqrt{2k\varepsilon}/k'$. Deduce that $1 - k^2\mathrm{sn}^2$ has double zeros at $K + iK'$ and $3K + iK'$.

11.10 Suppose that a function $f(z)$ is even (resp. odd) around $z = a$ and even (resp. odd) around $z = b$. Show that it has period $2(b - a)$.

11.11 Suppose that a function $f(z)$ is odd around $z = a$ and even around $z = b$. Show that it has period $4(b - a)$.

11.12 Use (11.3.10) to obtain (11.3.11).

11.13 Prove that $\mathrm{sn}(u, k_1)$ in (11.3.10) has periods $2K(1 + k')$, $2iK'(1 + k')$.

11.14 Use (11.3.12) to obtain (11.3.13).

11.15 What are the periods of $\mathrm{sn}(u, k_1)$ in (11.3.12)?

11.16 What are the limiting values as $k \to 0$ and as $k \to 1$ of the functions sn, cn, and dn? (See (11.2.11) and (11.2.10).) What do the formulas (11.3.10), (11.3.11) reduce to in these limits? What about (11.3.12), (11.3.13)?

11.17 Use (11.3.14) to obtain (11.3.15).

11.18 Use (11.3.16) to obtain (11.3.17).

11.19 Show that the Gauss transformation (11.3.12), (11.3.13) is the composition of the Jacobi transformation, the Landen transformation, and the Jacobi transformation.

11.20 Show that the Landen transformation (11.3.10), (11.3.11) is the composition of the Jacobi transformation, the Gauss transformation, and the Jacobi transformation.

11.21 Prove (11.3.18).

11.22 Use (11.3.16) and (11.3.17) to verify

$$\operatorname{sn}(u+K) = \frac{\operatorname{cn} u}{\operatorname{dn} u};$$

$$\operatorname{cn}(u+K) = -\frac{k' \operatorname{sn} u}{\operatorname{dn} u};$$

$$\operatorname{dn}(u+K) = \frac{k'}{\operatorname{dn} u}$$

and find corresponding formulas for translation by iK' and by $K + iK'$.

11.23 In each of the following integrals, a substitution like $t = \operatorname{sn}^2 u$ converts the integrand to a rational function of t and $\sqrt{Q(t)}$, Q quadratic, so that the integral can be found in terms of elementary functions of t. Verify the results:

$$\int \operatorname{sn} u \, du = -\frac{1}{k} \cosh^{-1}\left(\frac{\operatorname{dn} u}{k'}\right) + C$$

$$= \frac{1}{k} \log(\operatorname{dn} u - k \operatorname{cn} u) + C;$$

$$\int \operatorname{cn} u \, du = \frac{1}{k} \cos^{-1}(\operatorname{dn} u) + C;$$

$$\int \operatorname{dn} u \, du = \frac{1}{k} \sin^{-1}(\operatorname{sn} u) + C;$$

$$\int \frac{du}{\operatorname{sn} u} = \log\left(\frac{\operatorname{dn} u - \operatorname{cn} u}{\operatorname{sn} u}\right) + C;$$

$$\int \frac{du}{\operatorname{cn} u} = \frac{1}{k'} \log\left(\frac{k' \operatorname{sn} u + \operatorname{dn} u}{\operatorname{cn} u}\right) + C;$$

$$\int \frac{du}{\operatorname{dn} u} = \frac{1}{k'} \cos^{-1}\left(\frac{\operatorname{cn} u}{\operatorname{dn} u}\right) + C;$$

$$\int \frac{\operatorname{cn} u}{\operatorname{sn} u} du = \log\left(\frac{1 - \operatorname{dn} u}{\operatorname{sn} u}\right) + C;$$

$$\int \frac{\operatorname{dn} u}{\operatorname{sn} u} du = \log\left(\frac{1 - \operatorname{cn} u}{\operatorname{sn} u}\right) + C;$$

$$\int \frac{\operatorname{sn} u}{\operatorname{cn} u} du = \frac{1}{k'} \log\left(\frac{\operatorname{dn} u + k'}{\operatorname{cn} u}\right) + C;$$

$$\int \frac{\operatorname{dn} u}{\operatorname{cn} u} du = \log\left(\frac{1 + \operatorname{sn} u}{\operatorname{cn} u}\right) + C;$$

$$\int \frac{\operatorname{sn} u}{\operatorname{dn} u} du = -\frac{1}{kk'} \sin^{-1}\left(\frac{k \operatorname{cn} u}{\operatorname{dn} u}\right) + C;$$

$$\int \frac{\operatorname{cn} u}{\operatorname{dn} u} du = \frac{1}{k} \log\left(\frac{1 + k\operatorname{sn} u}{\operatorname{dn} u}\right) + C.$$

11.24 In each of the following integrals, a substitution like $v = (\operatorname{sn}^{-1}t)^2$ converts the integrand to a rational function of v and $\sqrt{Q(v)}$, Q quadratic. Verify

$$\int \operatorname{sn}^{-1}t \, dt = t \operatorname{sn}^{-1}t + \frac{1}{k} \cosh^{-1}\left(\frac{\sqrt{1 - k^2t^2}}{k'}\right) + C$$

$$= t \operatorname{sn}^{-1}t + \frac{1}{k} \log\left(\sqrt{1 - k^2t^2} + k\sqrt{1 - t^2}\right) + C;$$

$$\int \operatorname{cn}^{-1}t \, dt = t \operatorname{cn}^{-1}t - \frac{1}{k} \cos^{-1}\left(\sqrt{k'^2 + k^2t^2}\right) + C;$$

$$\int \operatorname{dn}^{-1}t \, dt = t \operatorname{dn}^{-1}t - \sin^{-1}\left(\frac{\sqrt{1 - t^2}}{k}\right) + C.$$

11.25 Deduce (11.4.3) from (11.4.2).
11.26 Deduce (11.4.4) from (11.4.3).
11.27 Prove that (11.4.8) has period τ.
11.28 Show that the product in (11.4.9) has period 1 as a function of z. Show that the effect of changing z to $z + \tau$ in the product is to multiply the

product by $-p^{-2}q^{-1}$. Deduce that the function $c(z, \tau) = G(\tau)$ depends only on τ.

11.29 Show that the function G in Exercise 11.28 has limit 1 as $\operatorname{Im} \tau \to +\infty$.

11.30 Show that

$$\Theta\left(\frac{1}{4} + \frac{1}{2}\tau\right) = G(\tau) \prod_{n=1}^{\infty} \left(1 + q^{4n-2}\right)$$

$$= \sum_{n=-\infty}^{\infty} (-1)^n i^n q^{n^2} = \sum_{m=-\infty}^{\infty} (-1)^m q^{(2m)^2}.$$

11.31 Use Exercise 11.30 to show that

$$\frac{G(\tau)}{G(4\tau)} = \prod_{n=1}^{\infty} \frac{\left(1 - q^{8n-4}\right)^2}{1 + q^{4n-2}} = \prod_{n=1}^{\infty} \left(1 - q^{8n-4}\right)\left(1 - q^{4n-2}\right).$$

11.32 Show that for $|w| < 1$,

$$\prod_{n=1}^{\infty} \left(1 - w^{2n-1}\right) = \frac{\prod_{n=1}^{\infty} \left(1 - w^n\right)}{\prod_{n=1}^{\infty} \left(1 - w^{2n}\right)}.$$

11.33 Use Exercises 11.31 and 11.32 to show that

$$\frac{G(\tau)}{G(4\tau)} = \frac{\prod_{n=1}^{\infty} \left(1 - q^{2n}\right)}{\prod_{n=1}^{\infty} \left(1 - q^{8n}\right)}.$$

Iterate this identity to get

$$\frac{G(\tau)}{G(4^m\tau)} = \frac{\prod_{n=1}^{\infty} \left(1 - q^{2n}\right)}{\prod_{n=1}^{\infty} \left(1 - q^{4^m\,2n}\right)}.$$

11.34 Deduce from Exercises 11.28 and 11.33 that $G(\tau) = \prod_{n=1}^{\infty}(1 - q^{2n})$.

11.35 Find the Fourier sine expansion of θ_1 (expressing it as a combination of the functions $\sin(m\pi z)$) and the Fourier cosine expansions of $\theta_2, \theta_3, \theta_4$.

11.36 Use (11.4.10) and (11.5.1) to verify the product expansions of the θ_j.

11.37 Verify the table of values of the Jacobi theta functions.

11.38 Use the tables of values of the Jacobi elliptic functions and Jacobi theta functions to obtain (11.5.2).

11.39 Prove that the quotient

$$\frac{\Theta(2z|2\tau)}{\Theta(z|\tau)\,\Theta\left(z + \frac{1}{2}|\tau\right)}$$

is constant. Evaluate at $z = \frac{1}{2}\tau$ to show that the constant is $\prod(1 + q^{2n})/(1 - q^{2n})$.

11.40 Show that the periods τ and τ_1 associated with the Landen transformation are related by $\tau_1 = 2\tau$. It follows that the Landen transformation relates theta functions with parameter τ to theta functions with parameter 2τ.

11.41 Suppose that $\psi(x, t)$ satisfies the heat equation (11.5.4) and also has the property (11.5.5), say for every continuous function f that vanishes outside a bounded interval. Assume that these properties determine ψ uniquely.

(a) Show that

$$u(x, t) = \int_{-\infty}^{\infty} \psi(x - y, t) f(y) \, dy, \quad t > 0,$$

is a solution of the heat equation with $u(x, t) \to f(x)$ as $t \to 0$. Assume that these properties determine u uniquely.

(b) Show that for $\lambda > 0$, $u(\lambda x, \lambda^2 t)$ is a solution of the heat equation with limit $f(\lambda x)$ as $t \to 0$ and deduce that

$$\int_{-\infty}^{\infty} \psi(x - y, t) f(\lambda y) \, dy = u(\lambda x, \lambda^2 t)$$

$$= \int_{-\infty}^{\infty} \psi(\lambda x - \lambda y, \lambda^2 t) f(\lambda y) \, d(\lambda y),$$

so $\psi(\lambda x, \lambda^2 t) = \lambda^{-1}\psi(x, t)$.

(c) Deduce from (b) that $\psi(x, t)$ has the form $t^{-\frac{1}{2}} F(x^2/t)$ and use the heat equation to show that $F(s)$ satisfies the equation

$$\left\{ s\frac{d}{ds} + \frac{1}{2} \right\} (4F' + F) = 0$$

with solution $F(s) = Ae^{-s/4}$.

(d) Deduce from the preceding steps that ψ should be given by (11.5.6). (The constant A can be determined by taking $f \equiv 1$ in (11.5.5).)

11.42 Suppose that $\psi_p(x, t)$ is the fundamental solution for the periodic heat equation, i.e. that ψ_p satisfies the heat equation and that for any continuous periodic function f, the function

$$u(x, t) = \int_0^1 \psi_p(x - y, t) f(y) \, dy$$

is a solution of the heat equation with the property that $u(x, t) \to f(x)$ as $t \to 0$. It is not difficult to show that $u(\cdot, t)$ is continuous and periodic for each $t > 0$, hence has a Fourier expansion

$$u(x, t) = \sum_{n=-\infty}^{\infty} a_n(t) \, e^{2n\pi i x}, \quad a_n(t) = \int_0^1 u(x, t) \, e^{-2n\pi i x} \, dx;$$

see Appendix B.

(a) Assuming that the Fourier expansion can be differentiated term by term, find the coefficients a_n. (Use the condition at $t = 0$.)

(b) Use the result from part (a) to write $u(x, t)$ as an integral and thus show that $\psi_p(x, t)$ is given by (11.5.8).

11.43 Verify (11.5.10) and (11.5.11).

11.44 Integrate the Weierstrass zeta function ζ over the boundary of Π to show that $2\eta_1\omega_2 - 2\eta_2\omega_1 = \pi i$.

11.45 Express sn in terms of the Weierstrass $\wp$ function; the Weierstrass zeta function; the Weierstrass sigma function.

11.46 Show that

$$\frac{1}{\text{sn}\,(u, k)^2} = \wp(u, K, iK') + \frac{1 + k^2}{3}.$$

11.47 Use Exercise 11.46 to show that any elliptic function with periods $2K$ and $2iK'$ is a rational function of sn, cn, and dn.

11.48 Determine the coefficients of u^2 and u^4 in the McLaurin expansion (Taylor expansion at $u = 0$) of $\wp(u) - 1/u^2$. Use this to show that the coefficients g_2, g_3 in (11.6.2) are

$$g_2 = 60 \sum_{p \in \Lambda, p \neq 0} \frac{1}{p^4}; \quad g_3 = 140 \sum_{p \in \Lambda, p \neq 0} \frac{1}{p^6}.$$

11.8 Summary

11.8.1 Integration

If $P(z)$ is a polynomial of degree 3 or 4 with simple roots, the problem of integrating a rational function of z and $P(z)$ can be reduced to the case

$$P(z) = (1 - z^2)(1 - k^2 z^2), \quad k^2 \neq 1,$$

and then to the case of elliptic integrals of the first, second, and third kinds:

$$F(z) = F(k, z) = \int_0^z \frac{d\zeta}{\sqrt{(1 - \zeta^2)(1 - k^2\zeta^2)}};$$

$$E(z) = E(k, z) = \int_0^z \sqrt{\frac{1 - k^2\zeta^2}{1 - \zeta^2}} \, d\zeta;$$

$$\Pi(a, z) = \Pi(a, k, z) = \int_0^z \frac{d\zeta}{(1 + a\zeta^2)\sqrt{(1 - \zeta^2)(1 - k^2\zeta^2)}}.$$

11.8.2 Elliptic integrals

Assume for convenience that $0 < k < 1$. The complementary modulus k' is $\sqrt{1 - k^2}$. The elliptic function $F(z) = F(k, z)$ is multi-valued, the value depending on the path of integration in the Riemann surface of $P(z)$. Starting from a value $F(z)$, the set of all values is

$$\{F(z) + 4mK + 2niK' | \, m, n = 0, \pm 1 \, \pm 2, \dots\},$$

where $K = K(k)$ is given by the integral

$$K = \int_0^1 \frac{dz}{\sqrt{(1 - z^2)(1 - k^2z^2)}}$$

and $K' = K(k')$. Thus values of F are determined only up to addition of elements

$$\Lambda = \{4mK + 2inK' \, | \, m, n = 0, \pm 1, \pm 2, \dots\}.$$

With this convention, F is an odd function: $F(-z) = -F(z)$. Particular values:

$$F(k, 1) = K, \quad F(k, 1/k) = K + iK', \quad F(k, \infty) = iK'.$$

Landen's transformation:

$$F(k_1, z_1) = (1 + k') \, F(k, z);$$

$$k_1 = \frac{1 - k'}{1 + k'}, \quad z_1 = (1 + k')z \sqrt{\frac{1 - z^2}{1 - k^2z^2}}.$$

Gauss's transformation:

$$F(k_1, z_1) = (1 + k) \, F(k, z);$$

$$k_1 = \frac{2\sqrt{k}}{1 + k}, \quad z_1 = \frac{(1 + k)z}{1 + kz^2}.$$

Jacobi's imaginary transformation:

$$i \, F(k, y) = F\left(k', \frac{iy}{\sqrt{1 - y^2}}\right).$$

Euler's addition formula:

$$F(k, x) + F(k, y) = F\left(k, \frac{x\sqrt{P(y)} + y\sqrt{P(x)}}{1 - k^2 x^2 y^2}\right),$$

where as before $P(x) = (1 - x^2)(1 - k^2 x^2)$.

The complete elliptic integrals can be expressed as

$$K(k) = \frac{\pi}{2} \, F\left(\frac{1}{2}, \frac{1}{2}, 1; k^2\right); \quad E(k) = \frac{\pi}{2} \, F\left(-\frac{1}{2}, \frac{1}{2}, 1; k^2\right).$$

11.8.3 Jacobi elliptic functions

The inverse of the multi-valued function F is the single-valued function sn defined by

$$u = \int_0^{\operatorname{sn} u} \frac{d\zeta}{\sqrt{(1 - \zeta^2)(1 - k^2 \zeta^2)}}.$$

Related functions cn and dn are defined by

$$\operatorname{cn} u = \sqrt{1 - \operatorname{sn}^2 u}, \quad \operatorname{dn} u = \sqrt{1 - k^2 \operatorname{sn}^2 u},$$

or implicitly by the integral formulas

$$u = \int_{\operatorname{cn} u}^1 \frac{d\zeta}{\sqrt{(1 - \zeta^2)(1 - k^2 + k^2 \zeta^2)}};$$

$$u = \int_{\operatorname{dn} u}^1 \frac{d\zeta}{\sqrt{(1 - \zeta^2)(\zeta^2 - k'^2)}}.$$

The derivatives satisfy

$$\mathrm{sn}'(u) = \mathrm{cn}\,u\,\mathrm{dn}\,u;$$

$$\mathrm{cn}'(u) = -\mathrm{sn}\,u\,\mathrm{dn}\,u;$$

$$\mathrm{dn}'(u) = -k^2\mathrm{sn}\,u\,\mathrm{cn}\,u.$$

Periodicity and related properties:

$$\mathrm{sn}\,u = \mathrm{sn}\,(u + 4K) = \mathrm{sn}\,(u + 2iK');$$

$$= \mathrm{sn}\,(2K - u) = \mathrm{sn}\,(2K + 2iK' - u);$$

$$= -\mathrm{sn}\,(-u) = -\mathrm{sn}\,(4K - u) = -\mathrm{sn}\,(2iK' - u);$$

$$\mathrm{cn}\,u = \mathrm{cn}\,(u + 4K) = \mathrm{cn}\,(u + 2K + 2iK');$$

$$= \mathrm{cn}\,(-u) = -\mathrm{cn}\,(2K - u) = -\mathrm{cn}\,(2iK' - u);$$

$$\mathrm{dn}\,u = \mathrm{dn}\,(u + 2K) = \mathrm{dn}\,(u + 4iK') = \mathrm{dn}\,(2K - u);$$

$$= \mathrm{dn}\,(-u) = -\mathrm{dn}\,(2iK' - u) = -\mathrm{dn}\,(2K + 2iK' - u).$$

Particular values:

	0	K	$2K$	$3K$	iK'	$K + iK'$	$2K + iK'$	$3K + iK'$
sn	0	1	0	-1	∞	k^{-1}	∞	$-k^{-1}$
cn	1	0	-1	0	∞	$-ik'k^{-1}$	∞	$ik'k^{-1}$
dn	1	k'	1	k'	∞	0	∞	0

Landen's transformation relates moduli k and $k_1 = (1 - k')/(1 + k')$:

$$\mathrm{sn}\,\big((1 + k')u, k_1\big) = (1 + k')\,\frac{\mathrm{sn}\,(u, k)\,\mathrm{cn}\,(u, k)}{\mathrm{dn}\,(u, k)};$$

$$\mathrm{cn}\,\big((1 + k')u, k_1\big) = (1 + k')\,\frac{\mathrm{dn}^2(u, k) - k'}{k^2\,\mathrm{dn}\,(u, k)};$$

$$\mathrm{dn}\,\big((1 + k')u, k_1\big) = (1 - k')\,\frac{\mathrm{dn}^2(u, k) + k'}{k^2\,\mathrm{dn}\,(u, k)}.$$

Gauss's transformation relates moduli k and $k_1 = 2\sqrt{k}/(1+k)$:

$$\operatorname{sn}\big((1+k)u, k_1\big) = \frac{(1+k)\operatorname{sn}(u,k)}{1+k\operatorname{sn}^2(u,k)};$$

$$\operatorname{cn}\big((1+k)u, k_1\big) = \frac{\operatorname{cn}(u,k)\operatorname{dn}(u,k)}{1+k\operatorname{sn}^2(u,k)};$$

$$\operatorname{dn}\big((1+k)u, k_1\big) = \frac{1-k\operatorname{sn}^2(u,k)}{1+k\operatorname{sn}^2(u,k)}.$$

Jacobi's imaginary transformation:

$$\operatorname{sn}(iu,k) = i\,\frac{\operatorname{sn}(u,k')}{\operatorname{cn}(u,k')};$$

$$\operatorname{cn}(iu,k) = \frac{1}{\operatorname{cn}(u,k')};$$

$$\operatorname{dn}(iu,k) = \frac{\operatorname{dn}(u,k')}{\operatorname{cn}(u.k')}.$$

Addition formulas:

$$\operatorname{sn}(u+v) = \frac{\operatorname{sn}u\operatorname{cn}v\operatorname{dn}v + \operatorname{sn}v\operatorname{cn}u\operatorname{dn}u}{1-k^2\operatorname{sn}^2u\operatorname{sn}^2v};$$

$$\operatorname{cn}(u+v) = \frac{\operatorname{cn}u\operatorname{cn}v - \operatorname{sn}u\operatorname{sn}v\operatorname{dn}u\operatorname{dn}v}{1-k^2\operatorname{sn}^2u\operatorname{sn}^2v};$$

$$\operatorname{dn}(u+v) = \frac{\operatorname{dn}u\operatorname{dn}v - k^2\operatorname{sn}u\operatorname{sn}v\operatorname{cn}u\operatorname{cn}v}{1-k^2\operatorname{sn}^2u\operatorname{sn}^2v}.$$

These imply the product formulas

$$\operatorname{sn}(u+v)\operatorname{sn}(u-v) = \frac{\operatorname{sn}^2u - \operatorname{sn}^2v}{1-k^2\operatorname{sn}^2u\operatorname{sn}^2v};$$

$$\operatorname{cn}(u+v)\operatorname{cn}(u-v) = \frac{1-\operatorname{sn}^2u - \operatorname{sn}^2v + k^2\operatorname{sn}^2u\operatorname{sn}^2v}{1-k^2\operatorname{sn}^2u\operatorname{sn}^2v},$$

and also

$$\operatorname{sn}(u+K) = \frac{\operatorname{cn}u}{\operatorname{dn}u};$$

$$\operatorname{cn}(u+K) = -\frac{k'\operatorname{sn}u}{\operatorname{dn}u};$$

$$\operatorname{dn}(u+K) = \frac{k'}{\operatorname{dn}u}.$$

Glaisher's notation:

$$\text{ns} = \frac{1}{\text{sn}}, \quad \text{nc} = \frac{1}{\text{cn}}, \quad \text{nd} = \frac{1}{\text{dn}};$$

$$\text{sc} = \frac{\text{sn}}{\text{cn}}, \quad \text{sd} = \frac{\text{sn}}{\text{dn}}, \quad \text{cd} = \frac{\text{cn}}{\text{dn}};$$

$$\text{cs} = \frac{\text{cn}}{\text{sn}}, \quad \text{ds} = \frac{\text{dn}}{\text{sn}}, \quad \text{dc} = \frac{\text{dn}}{\text{cn}}.$$

The change of variables $\zeta = \text{sn}\,s$ converts the elliptic integrals of the first, second, and third kinds to

$$F(z) = \int_0^{\text{sn}^{-1}z} ds = \text{sn}^{-1}z;$$

$$E(z) = \int_0^{\text{sn}^{-1}z} (1 - k^2\text{sn}^2 s)\,ds = \int_0^{\text{sn}^{-1}z} \text{dn}^2(s)\,ds;$$

$$\Pi(a, z) = \int_0^{\text{sn}^{-1}z} \frac{ds}{1 + a\,\text{sn}^2 s}.$$

Integrals (see the exercises):

$$\int \text{sn}\,u\,du = -\frac{1}{k}\cosh^{-1}\left(\frac{\text{dn}\,u}{k'} + C\right)$$

$$= \frac{1}{k}\log(\text{dn}\,u - k\,\text{cn}\,u) + C;$$

$$\int \text{cn}\,u\,du = \frac{1}{k}\cos^{-1}(\text{dn}\,u) + C;$$

$$\int \text{dn}\,u\,du = \frac{1}{k}\sin^{-1}(\text{sn}\,u) + C;$$

$$\int \frac{du}{\text{sn}\,u} = \log\left(\frac{\text{dn}\,u - \text{cn}\,u}{\text{sn}\,u}\right) + C;$$

$$\int \frac{du}{\text{cn}\,u} = \frac{1}{k'}\log\left(\frac{k'\,\text{sn}\,u + \text{dn}\,u}{\text{cn}\,u}\right) + C;$$

$$\int \frac{du}{\mathrm{dn}\,u} = \frac{1}{k'} \cos^{-1}\left(\frac{\mathrm{cn}\,u}{\mathrm{dn}\,u}\right) + C;$$

$$\int \frac{\mathrm{cn}\,u}{\mathrm{sn}\,u}\,du = \log\left(\frac{1-\mathrm{dn}\,u}{\mathrm{sn}\,u}\right) + C;$$

$$\int \frac{\mathrm{dn}\,u}{\mathrm{sn}\,u}\,du = \log\left(\frac{1-\mathrm{cn}\,u}{\mathrm{sn}\,u}\right) + C;$$

$$\int \frac{\mathrm{sn}\,u}{\mathrm{cn}\,u}\,du = \frac{1}{k'}\log\left(\frac{\mathrm{dn}\,u+k'}{\mathrm{cn}\,u}\right) + C;$$

$$\int \frac{\mathrm{dn}\,u}{\mathrm{cn}\,u}\,du = \log\left(\frac{1+\mathrm{sn}\,u}{\mathrm{cn}\,u}\right) + C;$$

$$\int \frac{\mathrm{sn}\,u}{\mathrm{dn}\,u}\,du = -\frac{1}{kk'}\sin^{-1}\left(\frac{k\,\mathrm{cn}\,u}{\mathrm{dn}\,u}\right) + C;$$

$$\int \frac{\mathrm{cn}\,u}{\mathrm{dn}\,u}\,du = \frac{1}{k}\log\left(\frac{1+k\,\mathrm{sn}\,u}{\mathrm{dn}\,u}\right) + C;$$

$$\int \mathrm{sn}^{-1}t\,dt = t\,\mathrm{sn}^{-1}t + \frac{1}{k}\cosh^{-1}\left(\frac{\sqrt{1-k^2t^2}}{k'}\right) + C$$

$$= t\,\mathrm{sn}^{-1}t + \frac{1}{k}\log\left(\sqrt{1-k^2t^2} + k\sqrt{1-t^2}\right) + C;$$

$$\int \mathrm{cn}^{-1}t\,dt = t\,\mathrm{cn}^{-1}t - \frac{1}{k}\cos^{-1}\left(\sqrt{k'^2+k^2t^2}\right) + C;$$

$$\int \mathrm{dn}^{-1}t\,dt = t\,\mathrm{dn}^{-1}t - \sin^{-1}\left(\frac{\sqrt{1-t^2}}{k}\right) + C.$$

11.8.4 Theta functions

An elliptic function is a meromorphic function that is doubly periodic:

$$f(z) = f(z+2\omega_1) = f(z+2\omega_2), \quad \mathrm{Im}\left(\frac{\omega_2}{\omega_1}\right) > 0.$$

If not constant it has at least two poles, counting multiplicity, in each period parallelogram

$$\Pi_a = \left\{ z \mid z = a + 2s\omega_1 + 2t\omega_2, \ 0 \le s, t < 1 \right\},$$

and takes each complex value the same number of times there. Any such function is the quotient of entire functions that are "nearly" periodic. A basic such function with periods 1 and τ, $\operatorname{Im}\tau > 0$, is the theta function

$$\Theta(z) = \Theta(z|\tau) = \sum_{n=-\infty}^{\infty} (-1)^n \, p(z)^{2n} \, q^{n(n-1)},$$

$$p(z) = e^{i\pi z}, \quad q = e^{i\pi\tau},$$

which satisfies

$$\Theta(z+1) = \Theta(z); \quad \Theta(z+\tau) = -e^{-2\pi i z}\Theta(z).$$

Jacobi's triple product formula:

$$\Theta(z) = \prod_{n=1}^{\infty} \left(1 - q^{2n}\right) \left(1 - p^2 q^{2n-2}\right) \left(1 - p^{-2} q^{2n}\right).$$

If an elliptic function f has zeros $\{a_1, a_2, \ldots, a_k\}$ and poles $\{b_1, b_2, \ldots, b_k\}$ in Π, repeated according to multiplicity, then

$$(a_1 + a_2 + \cdots + a_k) - (b_1 + b_2 + \cdots + b_k) \in \Lambda,$$

where Λ is the period lattice

$$\Lambda = \left\{ m + n\tau \mid m, n = 0, \pm 1, \pm 2, \ldots \right\}.$$

Conversely, given such points with

$$(a_1 + a_2 + \cdots + a_k) - (b_1 + b_2 + \cdots + b_k) = m + n\tau,$$

there is, up to a multiplicative constant, a unique elliptic function with these zeros and poles:

$$f(z) = e^{-2n\pi i z} \frac{\Theta(z - a_1)\Theta(z - a_2) \cdots \Theta(z - a_k)}{\Theta(z - b_1)\Theta(z - b_2) \cdots \Theta(z - b_k)}.$$

Elliptic functions can also be represented using

$$Z = \frac{\Theta'}{\Theta}.$$

Any elliptic function with periods 1 and τ is a linear combination of translates of Z and its derivatives.

11.8.5 Jacobi theta functions and integration

The Jacobi theta functions are

$$\theta_1(z) = i \frac{q^{\frac{1}{4}}}{p} \Theta(z) = i \sum_{n=-\infty}^{\infty} (-1)^n p^{2n-1} q^{\left(n-\frac{1}{2}\right)^2};$$

$$\theta_2(z) = \frac{q^{\frac{1}{4}}}{p} \Theta\left(z + \frac{1}{2}\right) = \sum_{n=-\infty}^{\infty} p^{2n-1} q^{\left(n-\frac{1}{2}\right)^2};$$

$$\theta_3(z) = \Theta\left(z + \frac{1}{2} + \frac{1}{2}\tau\right) = \sum_{n=-\infty}^{\infty} p^{2n} q^{n^2};$$

$$\theta_4(z) = \Theta\left(z + \frac{1}{2}\tau\right) = \sum_{n=-\infty}^{\infty} (-1)^n p^{2n} q^{n^2}.$$

See [315] for various notations and normalizations.

Product formulas:

$$\theta_1(z) = 2q^{\frac{1}{4}} \sin(\pi z) \prod_{n=1}^{\infty} \left(1 - q^{2n}\right) \left(1 - p^2 q^{2n}\right) \left(1 - p^{-2} q^{2n}\right);$$

$$\theta_2(z) = 2q^{\frac{1}{4}} \cos(\pi z) \prod_{n=1}^{\infty} \left(1 - q^{2n}\right) \left(1 + p^2 q^{2n}\right) \left(1 + p^{-2} q^{2n}\right);$$

$$\theta_3(z) = \prod_{n=1}^{\infty} \left(1 - q^{2n}\right) \left(1 + p^2 q^{2n-1}\right) \left(1 + p^{-2} q^{2n-1}\right);$$

$$\theta_4(z) = \prod_{n=1}^{\infty} \left(1 - q^{2n}\right) \left(1 - p^2 q^{2n-1}\right) \left(1 - p^{-2} q^{2n-1}\right).$$

Some values

	0	$\frac{1}{2}$	$\frac{1}{2}\tau$	$\frac{1}{2} + \frac{1}{2}\tau$
θ_1	0	$\sum q^{(n-\frac{1}{2})^2}$	$iq^{-\frac{1}{4}} \sum (-1)^n q^{n^2}$	$q^{-\frac{1}{4}} \sum q^{n^2}$
θ_2	$\sum q^{(n-\frac{1}{2})^2}$	0	$q^{-1/4} \sum q^{n^2}$	$-iq^{-\frac{1}{4}} \sum (-1)^n q^{n^2}$
θ_3	$\sum q^{n^2}$	$\sum (-1)^n q^{n^2}$	$q^{-\frac{1}{4}} \sum q^{(n+\frac{1}{2})^2}$	0
θ_4	$\sum (-1)^n q^{n^2}$	$\sum q^{n^2}$	0	$q^{-\frac{1}{4}} \sum q^{(n-\frac{1}{2})^2}$

The Jacobi elliptic functions in terms of θ_j:

$$\operatorname{sn} u = \frac{1}{\sqrt{k}} \frac{\theta_1(u/2K)}{\theta_4(u/2K)};$$

$$\operatorname{cn} u = \sqrt{\frac{k'}{k}} \frac{\theta_2(u/2K)}{\theta_4(u/2K)};$$

$$\operatorname{dn} u = \sqrt{k'} \frac{\theta_3(u/2K)}{\theta_4(u/2K)}.$$

For identities of the form

$$\Theta(z+w)\,\Theta(z-w) = c_1\Theta(z+a_1)\,\Theta(z+a_2)\,\Theta(w+a_3)\,\Theta(w+a_4)$$

$$+ c_2\Theta(w+a_1)\,\Theta(w+a_2)\,\Theta(z+a_3)\,\Theta(z+a_4),$$

where the a_j belong to the set $\left\{0, \frac{1}{2}, \frac{1}{2}\tau, \frac{1}{2} + \frac{1}{2}\tau\right\}$, see [236, 315]. For Jacobi's identity

$$\theta_1' = \pi\,\theta_2\theta_3\theta_4,$$

see Whittaker and Watson [315], Armitage and Eberlein [12]. (The factor π is due to the normalization we have chosen here.)

Each of the Jacobi theta functions is a solution of the partial differential equation

$$\theta_{zz}(z, \tau) = 4\pi i\theta_\tau(z, \tau),$$

leading to Jacobi's imaginary transformation:

$$\theta_3(z|\tau) = \frac{e^{-i\pi z^2/\tau}}{\sqrt{-i\tau}} \theta_3\left(-\frac{z}{\tau}\Big| -\frac{1}{\tau}\right).$$

Jacobi's Z-function:

$$Z(u) = \frac{\theta'(u)}{\theta(u)}, \quad \theta(u) = \theta_4(u/2K) = \Theta\big((u + iK')/2K\big).$$

Elliptic integrals of the first and second kind:

$$F(z) = \text{sn}^{-1}z;$$

$$E(z) = \frac{\theta'\big(\text{sn}^{-1}z\big)}{\theta\big(\text{sn}^{-1}z\big)} + \frac{E}{K}\,\text{sn}^{-1}z,$$

where

$$E = E(1) = \int_0^1 \sqrt{\frac{1 - k^2\zeta^2}{1 - \zeta^2}}\, d\zeta.$$

Finding the elliptic integral of the third kind, $\Pi(a, z)$, reduces to finding

$$\int_0^w \frac{\text{sn}^2 u\, du}{1 + a\,\text{sn}^2 u} = C\left\{\log\left[\frac{\theta(w - b)}{\theta(w + b)}\right] + 2Z(b)\,w\right\},$$

where

$$w = \text{sn}^{-1}z, \quad a = -k^2\text{sn}^2 b, \quad C = \frac{1}{2k^2\text{sn}\,b\,\text{cn}\,b\,\text{dn}\,b}.$$

11.8.6 Weierstrass elliptic functions

Let

$$\Lambda = \big\{2n_1\omega_1 + 2n_2\omega_2 \mid n_1, n_2 = 0, \pm 1, \pm 2, \ldots\big\}$$

be the period lattice associated with the periods $2\omega_1$, $2\omega_2$, with $\text{Im}\,(\omega_2/\omega_1) > 0$. The Weierstrass $\wp$ function

$$\wp(u) = \frac{1}{u^2} + \sum_{p\in\Lambda,\, p\neq 0}\left[\frac{1}{(u - p)^2} - \frac{1}{p^2}\right]$$

is an even meromorphic function with periods $2\omega_1$ and $2\omega_2$. It has a double pole at each point of Λ and

$$\wp(u) = \frac{1}{u^2} + O(u^2) \quad \text{as} \quad u \to 0.$$

Differential equation:

$$(\wp')^2 = 4\wp^3 - g_2\wp - g_3,$$

where

$$g_2 = 60 \sum_{p\in\Lambda,\, p\neq 0} \frac{1}{p^4}; \quad g_3 = 140 \sum_{p\in\Lambda,\, p\neq 0} \frac{1}{p^6}.$$

Let $\omega_3 = -(\omega_1 + \omega_2)$. The values $e_j = \wp(\omega_j)$, $j = 1, 2, 3$, are distinct and are the roots of $Q(t) = 4t^3 - g_2 t - g_3$, so

$$e_1 + e_2 + e_3 = 0;$$

$$4(e_2 e_3 + e_3 e_1 + e_1 e_2) = -g_2;$$

$$4e_1 e_2 e_3 = g_3.$$

Related functions are the Weierstrass zeta function

$$\zeta(u) = \frac{1}{u} + \sum_{p\in\Lambda,\, p\neq 0} \left[\frac{1}{u-p} + \frac{1}{p} + \frac{u^2}{p} \right]$$

and the Weierstrass sigma function

$$\sigma(u) = u \prod_{p\in\Lambda,\, p\neq 0} \left(1 - \frac{u}{p}\right) \exp\left(\frac{u}{p} + \frac{u^2}{2p^2}\right),$$

characterized by

$$\zeta'(u) = -\wp(u), \quad \zeta(-u) = -\zeta(u);$$

$$\frac{\sigma'(u)}{\sigma(u)} = \zeta(u), \quad \lim_{u\to 0} \frac{\sigma(u)}{u} = 1.$$

Any elliptic function f with periods $2\omega_j$ can be written as

$$f(u) = r_1(\wp(u)) + r_2(\wp(u))\,\wp'(u),$$

where the r_j are rational functions. It can also be written in the form

$$f(u) = C + \sum c_k \zeta(u - a_k) + \sum_{v>0} c_{vk} \zeta^{(v)}(u - a_k),$$

and thus the integral can be expressed using translates of σ, ζ, and derivatives of ζ. The function f can also be written as the quotient of entire functions

$$f(u) = A \prod_{j=1}^{n} \frac{\sigma(u - a_j)}{\sigma(u - b_j)}.$$

Addition formula:

$$\wp(u+v) = \frac{1}{4}\left[\frac{\wp'(u)-\wp'(v)}{\wp(u)-\wp(v)}\right]^2 - \wp(u) - \wp(v).$$

11.9 Remarks

The history and theory of elliptic functions is treated in the survey by Mittag-Leffler, translated by Hille [206], and in the books by Akhiezer [6], Appell and Lacour [11], Armitage and Eberlein [12], Chandrasekharan [42], Lang [174], Lawden [178], Neville [214], Prasolov and Solovyev [233], Tannery and Molk [282], Temme [284], Tricomi [288], and Walker [301]. (The extensive survey of results in Abramowitz and Stegun [3] is marred by an idiosyncratic notation.)

There are very extensive lists of formulas in Tannery and Molk [282]. Akhiezer [6] discusses the transformation theory of Abel and Jacobi. Appell and Lacour [11] give a number of applications to physics and to geometry. Chandrasekharan [42] has several applications to number theory, Lawden [178] has applications to geometry, while Armitage and Eberlein [12] have applications to geometry, mechanics, and statistics. The book by Lang [174] covers a number of modern developments of importance in algebraic number theory. By now the applications to number theory include the proof of Fermat's last theorem by Wiles and Taylor; see [283, 316]. Applications to physics and engineering are treated by Oberhettinger and Magnus [221].

The theory of elliptic functions was developed in the 18th and early 19th centuries through the work of Euler, Legendre [184], Gauss [105], Abel [1], Jacobi [139, 140], and Liouville [188], among others. Abel and Jacobi revolutionized the subject in 1827 by studying the inverse functions and developing the theory in the complex plane. (Gauss's discoveries in this direction were made earlier but were only published later, posthumously.) For an assessment of the early history, see Mittag-Leffler [206] and also Dieudonné [69], Klein [155], Stilwell [273]. Liouville introduced the systematic use of complex variable methods, including his theorem on bounded entire functions.

The version developed by Weierstrass [311] later in the 19th century is simpler than the Jacobi approach via theta functions [142]. Mittag-Leffler [206], Neville [214], and Tricomi [288] use the Weierstrass functions rather than theta functions to develop the theory of Abel and Jacobi. On the other hand, the work of Abel and Jacobi generalizes to curves of arbitrary genus, i.e. to polynomial equations of arbitrary degree. A classic treatise on the subject is

Baker [19]; see also Kempf [154] and Polischuk [232]. Theta functions have become important in the study of certain completely integrable dynamical systems, e.g. in parametrizing special solutions of the periodic Korteweg-de Vries equation and other periodic problems; see Krichever's introduction to [19] and the survey article by Dubrovin [72].

Appendix A: Complex analysis

This section contains a brief review of terminology and results from complex analysis that are used in the text.

If $z = x + iy$ is a complex number, x and y real, then

$$z = x + iy = r \cos\theta + ir \sin\theta = r e^{i\theta}, \quad \bar{z} = x - iy,$$

where $r = \sqrt{x^2 + y^2}$ is the *modulus* $|z|$ of z and θ is the *argument* $\arg z$ of z. The logarithm

$$\log z = \log r + i\theta$$

is multiple-valued: defined only up to integer multiples of $2\pi i$. The power

$$z^a = \exp(a \log z)$$

is also multiple-valued, unless a is an integer.

Typically, one makes these functions single-valued by restricting the domain, usually by choosing a range for the argument. The resulting domain is the complex plane minus a ray from the origin. Examples:

$$\mathbf{C} \setminus (-\infty, 0] = \{z : -\pi < \arg z < \pi\};$$
$$\mathbf{C} \setminus [0, +\infty) = \{z : 0 < \arg z < 2\pi\}.$$

This is referred to as choosing a *branch* of the logarithm or of the power. The *principal branch* is the one with $\arg x = 0$ for $x > 0$.

A *region* is an open, non-empty subset of the plane which is *connected*: any two points in the set can be joined by a continuous curve that lies in the set. A region Ω is said to be *simply connected* if any closed curve lying in Ω can be continuously deformed, within Ω, to a point. The plane $\mathbf{C}$ and the disc $\{z : |z| < 1\}$ are simply connected. The annulus $\{z : 1 < |z| < 2\}$ and the punctured plane $\mathbf{C} \setminus \{0\}$ are not simply connected.

A function $f(z)$ is said to be *analytic* or *holomorphic* in a region Ω if the derivative

$$f'(z) = \lim_{h \to 0} \frac{f(z+h) - f(z)}{h}$$

exists for each point z in Ω. An equivalent condition is that for each point z_0 in Ω, the function can be represented at nearby points by its Taylor series: if $|z - z_0| < \varepsilon$, then

$$f(z) = \sum_{n=0}^{\infty} a_n (z - z_0)^n, \quad a_n = \frac{f^{(n)}(z_0)}{n!}.$$

Conversely, a function that is defined in a disc by a convergent power series can often be extended to a larger region as a holomorphic function. For example,

$$f(z) = \sum_{n=0}^{\infty} z^n = \frac{1}{1-z}, \quad |z| < 1,$$

extends to the complement of the point $z = 1$. This is an example of *analytic continuation*.

It can be deduced from the local power series representation, using connectedness, that if two functions f and g are holomorphic in a region Ω and coincide in some open subset of Ω (or on a sequence of points that converges to a point of Ω), then they coincide throughout Ω. This is one version of the principle of *uniqueness of analytic continuation*. This principle is used several times above, often in the following form. Suppose functions $u_j(a, x)$, $j = 1, 2, 3$, are holomorphic with respect to a parameter a in a region Ω and satisfy a linear relation

$$u_3(a, x) = A_1(a) u_1(a, x) + A_2(a) u_2(a, x)$$

with holomorphic or meromorphic coefficients. Then to determine the coefficients A_j throughout Ω, it is enough to determine A_1 on a subregion Ω_1 and A_2 on a subregion Ω_2. (In the cases encountered here, the form of a coefficient throughout Ω is clear once one knows the form on any subregion.)

A function f that is holomorphic in a punctured disc $\{0 < |z - z_0| < \varepsilon\}$ is said to have a *pole of order n* at z_0, n a positive integer, if

$$f(z) = \frac{g(z)}{(z - z_0)^n}, \quad 0 < |z - z_0| < \varepsilon,$$

where $g(z)$ is holomorphic in the disc $|z - z_0| < \varepsilon$ and $g(z_0) \neq 0$. An equivalent condition is that $(z - z_0)^n f(z)$ has a nonzero limit at $z = z_0$. The function f is said to have a *removable singularity* at $z = z_0$ if it has a limit at $z = z_0$.

In that case, taking $f(z_0)$ to be the limit, the resulting extended function is holomorphic in the disc.

A function f that is holomorphic in a region Ω except at isolated points, each of which is a pole or removable singularity, is said to be *meromorphic* in Ω. In particular, if f and g are holomorphic in Ω and g is not identically zero, then the quotient f/g is meromorphic in Ω.

A basic result of complex analysis is the *Cauchy integral theorem*: suppose that C is a closed curve that bounds a region Ω, and suppose that f is holomorphic on Ω and continuous up to the boundary C. Then the integral vanishes:

$$\int_C f(z)\,dz = 0.$$

A typical use of the Cauchy integral theorem occurs in Appendix B: the integral

$$\int_{-\infty}^{\infty} e^{-(x+iy)^2/2}\,dx \tag{A.0.1}$$

is independent of y. To see this, take values $a < b$ for y, and consider the integral of this integrand over the rectangle C_R two of whose sides are $\{x + ia \ : \ |x| \le R\}$ and $\{x + ib \ : \ |x| \le R\}$, oriented counterclockwise. By Cauchy's theorem the integral is zero. As $R \to \infty$ the integral over the vertical sides approaches zero, while the integral over the other sides approaches

$$\int_{-\infty}^{\infty} e^{-(x+ia)^2/2}\,dx - \int_{-\infty}^{\infty} e^{-(x+ib)^2/2}\,dx.$$

Therefore the integral (A.0.1) is independent of y.

One can use the Cauchy integral theorem to derive the *Cauchy integral formula*: suppose that C is oriented so that Ω lies to the left; for example, if C is a circle oriented counterclockwise and Ω is the enclosed disc. Then for any $z \in \Omega$,

$$f(z) = \frac{1}{2\pi i} \int_C \frac{f(\zeta)}{\zeta - z}\,d\zeta.$$

A consequence is *Liouville's theorem*: a bounded entire function f is constant. (An *entire function* is one that is holomorphic in the entire plane $\mathbf{C}$.) To see this, observe that the Cauchy integral formula for f can be differentiated under the integral sign. The derivative is

$$f'(z) = \frac{1}{2\pi i} \int_C \frac{f(\zeta)}{(\zeta - z)^2}\,d\zeta.$$

We may take C to be a circle of radius $R > |z|$, centered at the origin. Taking $R \to \infty$, the integrand is $O(1/R^2)$ and the length of the curve is $2\pi R$, so $|f'(z)|$ is at most $O(1/R)$. Therefore $f'(z) = 0$ for every $z \in \mathbf{C}$, so f is constant.

The Cauchy integral formula is one instance of the residue theorem. Suppose that f has a pole at z_0. Then near z_0 it has a Laurent expansion

$$f(z) = \frac{a_{-n}}{(z - z_0)^n} + \frac{a_{1-n}}{(z - z_0)^{n-1}} + \cdots + \frac{a_{-1}}{z - z_0} + a_0 + a_1(z - z_0) + \ldots$$

The *residue* of f at z_0, denoted $\mathrm{res}(f, z_0)$, is the coefficient a_{-1} of the $1/(z - z_0)$ term in the Laurent expansion:

$$\mathrm{res}(f, z_0) = a_{-1} = \frac{1}{2\pi i} \int_C f(\zeta) \, d\zeta,$$

where C is a sufficiently small circle centered at z_0.

Suppose as before that C is an oriented curve that bounds a region Ω lying to its left. Suppose that f is meromorphic in Ω and continuous up to the boundary C, and suppose that the poles of f in Ω are $z_1, \ldots, z_m$. Then the *residue theorem* says that the integral of f over C is $2\pi i$ times the sum of the residues:

$$\int_C f(z) \, dz = 2\pi i \big[\mathrm{res}(f, z_1) + \cdots + \mathrm{res}(f, z_m) \big].$$

Suppose that f has no zeros on the boundary curve C. Then the quotient $g = f'/f$ is continuous on the boundary and meromorphic inside. It is easy to see that if z_0 is a zero of f with multiplicity m, then the residue of f'/f at z_0 is m. If z_0 is a pole of order n, then the residue of f'/f at z_0 is $-n$. Therefore the number of zeros (counting multiplicity) minus the number of poles (counting multiplicity) enclosed by C is

$$\frac{1}{2\pi i} \int_C \frac{f'(z)}{f(z)} \, dz.$$

In particular, if f is holomorphic in the enclosed region, then this integral counts the number of zeros.

The first use of the residue theorem in the main text is in the proof of Theorem 2.2.3, where we evaluated

$$\int_C \frac{t^{z-1} \, dt}{1 + t}, \quad 0 < \mathrm{Re}\, z < 1. \tag{A.0.2}$$

Here C was the curve from $+\infty$ to 0 with $\arg t = 2\pi$ and returning to $+\infty$ with $\arg t = 0$.

This is taken as a limiting case of the curve C_R, where the part of C with $t \geq R > 1$ is replaced by the circle $\{|t| = R\}$, oriented counterclockwise, and the part with $0 \leq t \leq 1/R$ is replaced by the circle $\{|t| = 1/R\}$, oriented clockwise. (This is a typical example of how the Cauchy integral theorem, the Cauchy integral formula, and the residue theorem can be extended beyond our original formulation, which assumed a bounded curve and continuity of the integrand at each point of the boundary.) The residue calculus applies to each curve C_R, and the contribution of the integration over the circles goes to zero as $R \to \infty$, so the value of (A.0.2) is $2\pi i$ times the (unique) residue at $t = -1$. With our choice of branch, the residue is $\exp[i(z - 1)\pi] = -\exp(iz\pi)$. On the other hand, over the first part of C the value of t^z differs from the value on the second part by a factor $\exp(2\pi i z)$. This gives the result

$$(1 - e^{2\pi i z}) \int_0^\infty \frac{t^{z-1}\, dt}{1 + t} = -2\pi i\, e^{i\pi z}.$$

A *linear fractional transformation* or Möbius transformation is a function of the form

$$f(z) = \frac{az + b}{cz + d}, \quad ad - bc \neq 0.$$

It may be assumed that $ad - bc = 1$. The inverse and the composition of Möbius transformations are Möbius transformations, so the set of Möbius transformations is a group. Given any two ordered triples of distinct points in the Riemann sphere $\mathbf{S} = \mathbf{C} \cup \{\infty\}$, there is a unique Möbius transformation that takes one triple to the other. For example, the transformation

$$f(z) = \frac{az - az_0}{z - z_2}, \quad a = \frac{z_1 - z_2}{z_1 - z_0}$$

takes the triple (z_0, z_1, z_2) to $(0, 1, \infty)$, and its inverse is

$$g(w) = \frac{z_2(z_1 - z_0)w + (z_2 - z_1)z_0}{(z_1 - z_0)w + z_2 - z_1}.$$

The group of Möbius transformations that permute the points $\{0, 1, \infty\}$ is generated by the two transformations

$$z \to 1 - z = \frac{-z + 1}{0z + 1}, \quad z \to \frac{1}{z} = \frac{0z + 1}{z + 0}$$

and consists of these two transformations and

$$z \to z, \quad z \to \frac{z}{z - 1}, \quad z \to \frac{1}{1 - z}, \quad z \to 1 - \frac{1}{z}.$$

The *Weierstrass factorization theorem*, mentioned in Chapter 11, implies that for every sequence of points $\{z_n\}$ in the complex plane with limit ∞, there is an entire function that has zeros (repeated according to multiplicity) at these points and no others. In particular, if f is meromorphic in $\mathbf{C}$ and h is chosen so that its zeros match the poles of f, then $g = fh$ is entire. (More precisely, g has only removable singularities, so when they are removed it is an entire function.) Thus $f = g/h$ is the quotient of two entire functions.

Appendix B: Fourier analysis

This section contains a brief account of the facts from classical Fourier analysis and their consequences that are used at various points above.

Suppose that $f(x)$ is a (real or) complex-valued function that is absolutely integrable:

$$\int_{-\infty}^{\infty} |f(x)|\,dx < \infty. \tag{B.0.1}$$

The Fourier transform $\widehat{f}$ is defined by

$$\widehat{f}(\xi) = \frac{1}{\sqrt{2\pi}} \int_{-\infty}^{\infty} e^{-ix\xi} f(x)\,dx, \quad \xi \in \mathbf{R}.$$

The condition (B.0.1) implies that $\widehat{f}$ is bounded and continuous. It can also be shown that $\widehat{f}(x) \to 0$ as $|x| \to \infty$, so $\widehat{f}$ is uniformly continuous.

A particularly useful example is $f(x) = \exp(-\frac{1}{2}x^2)$, which is its own Fourier transform:

$$\frac{1}{\sqrt{2\pi}} \int_{-\infty}^{\infty} e^{-ix\xi}\, e^{-\frac{1}{2}x^2}\,dx = \frac{e^{-\frac{1}{2}\xi^2}}{\sqrt{2\pi}} \int_{-\infty}^{\infty} e^{-\frac{1}{2}(x+i\xi)^2}\,dx.$$

To see this, take $z = x + i\xi$ in the integral on the right, so that it is an integral over the line $\{\operatorname{Im} z = \xi\}$. By the Cauchy integral theorem, the path of integration can be changed to the real line, so

$$\frac{1}{\sqrt{2\pi}} \int_{-\infty}^{\infty} e^{-ix\xi}\, e^{-\frac{1}{2}x^2}\,dx = \frac{e^{-\frac{1}{2}\xi^2}}{\sqrt{2\pi}} \int_{-\infty}^{\infty} e^{-\frac{1}{2}x^2}\,dx = e^{-\frac{1}{2}\xi^2}.$$

(See Appendix A.) Extensive tables of Fourier transforms are given in [220].

If $g(\xi)$ is absolutely integrable, then the inverse Fourier transform is defined by

$$\check{g}(x) = \frac{1}{\sqrt{2\pi}} \int_{-\infty}^{\infty} e^{ix\xi} g(\xi)\, d\xi, \quad x \in \mathbf{R}.$$

The argument just given shows that $\exp\left(-\frac{1}{2}\xi^2\right)$ has inverse Fourier transform $\exp\left(-\frac{1}{2}x^2\right)$.

The terminology "inverse Fourier transform" is justified as follows. Suppose that f is a bounded, uniformly continuous, and absolutely integrable function, and suppose that its Fourier transform $\hat{f}$ itself is absolutely integrable. We show now that $(\hat{f})^{\vee} = f$:

$$f(x) = \frac{1}{2\pi} \int_{-\infty}^{\infty} e^{ix\xi} \left[\int_{-\infty}^{\infty} e^{-iy\xi} f(y)\, dy \right] d\xi. \qquad (\text{B.0.2})$$

We introduce a convergence factor $\exp\left(-\frac{1}{2}(\varepsilon\xi)^2\right)$, $\varepsilon > 0$:

$$\int_{-\infty}^{\infty} e^{ix\xi} \left[\int_{-\infty}^{\infty} e^{-iy\xi} f(y)\, dy \right] d\xi$$

$$= \lim_{\varepsilon \to 0} \int_{-\infty}^{\infty} \int_{-\infty}^{\infty} e^{i(x-y)\xi - \frac{1}{2}(\varepsilon\xi)^2} f(y)\, dy\, d\xi.$$

The convergence factor allows us to change the order of integration. By what we have just shown about the function $\exp\left(-\frac{1}{2}\xi^2\right)$,

$$\frac{1}{2\pi} \int_{-\infty}^{\infty} e^{i(x-y)\xi - \frac{1}{2}(\varepsilon\xi)^2}\, d\xi = \frac{1}{2\pi} \int_{-\infty}^{\infty} e^{i[(x-y)/\varepsilon]\zeta}\, e^{-\frac{1}{2}\zeta^2}\, \frac{d\zeta}{\varepsilon}$$

$$= \frac{1}{\varepsilon\sqrt{2\pi}} \exp\left[-\frac{(x-y)^2}{2\varepsilon^2} \right] \equiv G_\varepsilon(x-y).$$

The functions $\{G_\varepsilon\}$ are easily seen to have the properties

$$G_\varepsilon(x) > 0; \qquad (\text{B.0.3})$$

$$\int_{-\infty}^{\infty} G_\varepsilon(x)\, dx = 1;$$

$$\lim_{\varepsilon \to 0} \int_{|x|>\delta} G_\varepsilon(x)\, dx = 0, \quad \delta > 0.$$

According to the calculations above,

$$\frac{1}{2\pi} \int_{-\infty}^{\infty} e^{ix\xi} \left[\int_{-\infty}^{\infty} e^{-iy\xi} f(y)\, dy \right] d\xi - f(x)$$

$$= \lim_{\varepsilon \to 0} \int_{-\infty}^{\infty} G_\varepsilon(x - y) \left[f(y) - f(x) \right] dy. \qquad (B.0.4)$$

The assumptions on f and the conditions (B.0.3) imply that the limit of (B.0.4) is zero.

The Fourier inversion result used in Section 7.7 is the two-dimensional version of this result. The preceding proof can be adapted easily, or the result can be proved in two steps by taking the transform in one variable at a time.

Proof of Theorem 4.1.5 Suppose that w is a positive weight function on the interval (a, b) and

$$\int_a^b e^{2c|x|}\, w(x)\, dx < \infty, \qquad (B.0.5)$$

for some $c > 0$. Note that this implies that all moments are finite, so orthonormal polynomials $\{P_n\}$ necessarily exist. Given $f \in L_w^2$, let

$$f_n(x) = \sum_{k=0}^{n} (f, P_k)\, P_k(x).$$

Then for $m < n$,

$$\| f_n - f_m \|^2 = \sum_{k=m+1}^{n} |(f, P_k)|^2.$$

By (4.1.10)

$$\sum_{k=0}^{\infty} |(f, P_k)|^2 \le \| f \|^2 < \infty,$$

so the sequence $\{ f_n \}$ is a Cauchy sequence in L_w^2. Therefore it has a limit g, and we need to show that $g = f$. For any m, $(f_n, P_m) = (f, P_m)$ for $n \ge m$, and it follows from that $(g, P_m) = (f, P_m)$, all m. Thus $h = f - g$ is orthogonal to every P_m and, therefore, orthogonal to every polynomial. We want to show that $h \equiv 0$ or, equivalently, that $hw \equiv 0$.

Extend h and w to the entire real line if necessary, by taking them to vanish outside (a, b). Note that $|hw|$ is absolutely integrable, by the Cauchy–Schwarz inequality, since it is the product of square integrable functions $|h|\sqrt{w}$ and $\sqrt{w}$. By (B.0.5) the Fourier transform

$$H(\xi) = \widehat{hw}(\xi) = \frac{1}{\sqrt{2\pi}} \int_{-\infty}^{\infty} e^{-ix\xi} h(x)\, w(x)\, dx$$

has an extension to the strip $\{|\mathrm{Im}\,\xi| < c\}$. Moreover, H is holomorphic in the strip, with derivatives

$$\frac{d^n H}{dz^n}(0) = (-i)^n \int_{-\infty}^{\infty} x^n h(x)\, w(x)\, dx.$$

Since h is orthogonal to polynomials, all derivatives of H vanish at $z = 0$. Since H is holomorphic in the strip, this implies that $H \equiv 0$ and therefore the inverse Fourier transform hw is 0. $\square$

In Section 4.8 we used the Riemann–Lebesque lemma, in the following form: if f is a bounded function on a finite interval (a, b), then

$$\lim_{\lambda \to +\infty} \int_a^b \cos(\lambda x)\, f(x)\, dx = 0.$$

Suppose first that f is differentiable and that f and f' are continuous on the closed interval $[a, b]$. Integration by parts gives

$$\int_a^b \cos(\lambda x)\, f(x)\, dx = \frac{\sin(\lambda x)}{\lambda}\, f(x) \Big|_a^b - \frac{1}{\lambda} \int_a^b \sin(\lambda x)\, f'(x)\, dx$$

$$= O(\lambda^{-1}).$$

For general f, given $\varepsilon > 0$ choose a continuously differentiable function g such that

$$\int_a^b |f(x) - g(x)|\, dx < \varepsilon$$

and apply the previous argument to g to conclude that

$$\left| \int_a^b \cos(\lambda x)\, f(x)\, dx \right| < 2\varepsilon$$

for large λ.

In Section 11.4 we tacitly used the finite interval analogue of the Fourier inversion result, namely that the functions

$$e_n(x) = e^{2n\pi i x} = e_1(x)^n, \quad n = 0, \pm 1, \pm 2, \ldots,$$

which are orthonormal in the space $L^2(I)$, $I = (0, 1)$ with weight 1, are complete in this space. Define

$$G_n(x) = c_n\, 4^n \cos^{2n}(\pi x) = c_n\, 2^n \big[\cos 2\pi x + 1\big]^n = c_n(e_1 + e_{-1} + 2)^n,$$

where c_n is chosen so that $\int_0^1 G_n(x)\,dx = 1$. The G_n have period 1, are non-negative, and satisfy the analogue of (B.0.3):

$$\lim_{n\to\infty} \int_\delta^{1-\delta} G_n(x)\,dx = 0, \quad 0 < \delta < \frac{1}{2}. \tag{B.0.6}$$

It follows that for any continuous function g with period 1, the sequence

$$g_n(x) = \int_0^1 G_n(x-y)\,g(y)\,dy$$

converges uniformly to g. Now G_n is in the span of $\{e_k\}_{|k|\le n}$, so g_n is a linear combination of these functions. It follows that any function h in $L^2(I)$ that is orthogonal to each e_k is also orthogonal to each g_n. Taking limits, h is orthogonal to each continuous periodic g. The function h itself can be approximated in L^2 norm by continuous periodic functions, so $\|h\|^2 = (h, h) = 0$. As in the proof of Theorem 4.1.5, this implies that the $\{e_n\}$ are dense in $L^2(I)$. Note that in this case, since the functions $\{e_n\}$ are complex-valued, we use the complex inner product

$$(f, g) = \int_0^1 f(x)\,\overline{g(x)}\,dx.$$

The Fourier expansion of $f \in L^2(I)$ takes the form

$$f = \sum_{n=-\infty}^{\infty} a_n\,e_n, \quad a_n = (f, e_n) = \int_0^1 f(x)\,e^{-2n\pi i x}\,dx.$$

The partial sums

$$f_n(x) = \sum_{m=-n}^{n} a_m\,e^{2m\pi i x}$$

converge to f in $L^2(I)$ norm. They can also be shown to converge to f at any point at which f is differentiable, by an argument similar to the arguments used in Section 4.8.

Notation

431

References

1. Abel, N.H., Recherches sur les fonctions elliptiques, *J. Reine Angew. Math.* **2** (1827), 101–81; **3** (1828), 160–90; *Oeuvres 1*, 263–388.
2. Abel, N.H., Sur une espèce de fonctions entières nées du developpement de la fonction $(1 - v)^{-1} e^{-\frac{xv}{1-v}}$ suivant les puissances de v, *Oeuvres 2* (1881), 284.
3. Abramowitz, M. and Stegun, I., *Handbook of Mathematical Functions*, Dover, Mineola, NY, 1965.
4. Airy, G.B., On the intensity of light in the neighborhood of a caustic, *Trans. Camb. Phil. Soc.* **VI** (1838), 379–402.
5. Akhiezer, N.I., *The Classical Moment Problem and Some Related Questions in Analysis*, Hafner, New York, 1965.
6. Akhiezer, N.I., *Elements of the Theory of Elliptic Functions*, Translations of Mathematical Monographs, American Mathematical Society, Providence, RI, 1990.
7. Andrews, G.E., Askey, R., and Roy, R., *Special Functions*, Cambridge University Press, Cambridge, 1999.
8. Andrews, L.C., *Special Functions of Mathematics for Engineers*, 2nd edn, McGraw-Hill, New York, 1992.
9. Aomoto, K., Jacobi polynomials associated with Selberg integrals, *SIAM J. Math. Anal.* **18** (1987), 545–9.
10. Appell, P. and Kampé de Fériet, J., *Fonctions Hypergeometriques de Plusieurs Variables. Polynômes d'Hermite*, Gauthier-Villars, Paris, 1926.
11. Appell, P. and Lacour, E., *Principes de la Théorie des Fonctions Elliptiques et Applications*, Gauthier-Villars, Paris, 1922.
12. Armitage, J.V. and Eberlein, W.F., *Elliptic Functions*, Cambridge University Press, Cambridge, 2006.
13. Artin, E., *The Gamma Function*, Holt, Rinehart and Winston, New York, 1964.
14. Askey, R., *Orthogonal Polynomials and Special Functions*, Society for Industrial and Applied Mathematics, Philadelphia, PA, 1975.
15. Askey, R., Handbooks of special functions, *A Century of Mathematics in America, Vol. 3*, P. Duren (ed.), American Mathematical Society, Providence, RI, 1989, 369–91.

16. Ayoub, R., The lemniscate and Fagnano's contribution to elliptic integrals, *Arch. Hist. Exact Sci.* **29** (1983/84), 131–49.

17. Baik, J., Kriecherbauer, T., McLaughlin, K., and Miller, P. D., *Discrete Orthogonal Polynomials: Asymptotics and Applications*, Princeton University Press, Princeton, NJ, 2007.

18. Bailey, W. N., *Generalized Hypergeometric Series*, Cambridge University Press, Cambridge, 1935.

19. Baker, H. F., *Abelian Functions: Abel's Theory and the Allied Theory of Theta Functions*, Cambridge University Press, New York, 1995.

20. Barnes, E. W., A new development of the theory of the hypergeometric functions, *Proc. London Math. Soc.* **50** (1908), 141–77.

21. Bassett, A. B., On the potentials of the surfaces formed by the revolution of limaçons and cardioids, *Proc. Camb. Phil. Soc.* **6** (1889), 2–19.

22. Beals, R., Gaveau, B., and Greiner, P. C., Uniform hypoelliptic Green's functions, *J. Math. Pures Appl.* **77** (1998), 209–48.

23. Beals, R. and Kannai, Y., Exact solutions and branching of singularities for some equations in two variables, *J. Diff. Equations* **246** (2009), 3448–70.

24. Bernoulli, D., Solutio problematis Riccatiani propositi in Act. Lips. Suppl. Tom VIII p. 73, *Acta Erud. publ. Lipsiae* (1725), 473–5.

25. Bernoulli, D., Demonstrationes theorematum suorum de oscillationibus corporum filo flexili connexorum et catenae verticaliter suspensae, *Comm. Acad. Sci. Imp. Petr.* **7** (1734–35, publ. 1740), 162–73.

26. Bernouilli, Jac., Curvatura laminae elasticae, *Acta Erud.*, June 1694, 276–80; *Opera I*, 576–600.

27. Bernoulli, Jac., *Ars Conjectandi*, Basil, 1713.

28. Bernoulli, Joh., Methodus generalis construendi omnes aequationes differentialis primi gradus, *Acta Erud. publ. Lipsiae* (1694), 435–7.

29. Bernstein, S., Sur les polynomes orthogonaux relatifs à un segment fini, *J. de Mathématiques* **9** (1930), 127–77; **10** (1931), 219–86.

30. Bessel, F. W., Untersuchung des Theils der planetarischen Störungen aus der Bewegung der Sonne entsteht, *Berliner Abh.* 1824, 1–52.

31. Binet, J. P. M., Mémoire sur les intégrales définies eulériennes et sur leur application à la théorie des suites ainsi qu'à l'évaluation des fonctions des grandes nombres, *J. de l'École Roy. Polyt.* **16** (1838–9), 123–43.

32. Bleistein, N. and Handelsman, R. A., *Asymptotic Expansions of Integrals*, Dover, New York, 1986.

33. Bochner, S., Über Sturm-Liouvillesche Polynomsysteme, *Math. Z.* **29** (1929), 730–36.

34. Bohr, H. and Mollerup, J., *Laerebog i Matematisk Analyse*, Vol. 3, Copenhagen, 1922.

35. Brillouin, L., Remarques sur la mécanique ondulatoire, *J. Phys. Radium* **7** (1926), 353–68.

36. Brychkov, Yu. A., *Handbook of Special Functions: Derivatives, Integrals, Series and Other Formulas*, CRC Press, Boca Raton, FL, 2008.

37. Buchholz, H., *The Confluent Hypergeometric Function with Special Emphasis on its Applications*, Springer, New York, 1969.

38. Burkhardt, H., Entwicklungen nach oscillierenden Funktionen und Integration der Differentialgleichungen der mathematischen Physik, *Deutsche Math. Ver.* **10** (1901–1908), 1–1804.

39. Campbell, R., *Les Intégrales Euleriennes et leurs Applications*, Dunod, Paris, 1966.

40. Carlini, F., *Ricerche sulla Convergenza della Serie che Serva alla Soluzione del Problema di Keplero*, Milan, 1817.

41. Carlson, B. C., *Special Functions of Applied Mathematics*, Academic Press, New York, 1977.

42. Chandrasekharan, K., *Elliptic Functions*, Springer, Berlin, 1985.

43. Charlier, C. V. L., Über die Darstellung willkürlichen Funktionen, *Ark. Mat. Astr. Fysic* **2**, no. 20 (1905–6), 1–35.

44. Chebyshev, P. L., Théorie des mécanismes connus sous le nom de parallélogrammes, *Publ. Soc. Math. Warsovie* **7** (1854), 539–68; *Oeuvres 1*, 109–43.

45. Chebyshev, P. L., Sur les fractions continues, *Utzh. Zap. Imp. Akad. Nauk* **3** (1855), 636–64; *J. Math. Pures Appl.* **3** (1858), 289–323; *Oeuvres 1*, 201–30.

46. Chebyshev, P. L., Sur une nouvelle série, *Bull. Phys. Math. Acad. Imp. St. Pét.* **17** (1858), 257–61; *Oeuvres 1*, 381–4.

47. Chebyshev, P. L., Sur le développment des fonctions à une seule variable, *Bull. Phys.-Math. Acad. Imp. Sci. St. Pét.* **1** (1859), 193–200; *Oeuvres 1*, 499–508.

48. Chebyshev, P. L., Sur l'interpolation des valeurs équidistantes, *Zap. Akad. Nauk* **4** (1864); *Oeuvres 1*, 219–42.

49. Chester, C., Friedman, B., and Ursell, F., An extension of the method of steepest descents, *Proc. Cambridge Philos. Soc.* **53** (1957), 599–611.

50. Chihara, T. S., *An Introduction to Orthogonal Polynomials*, Gordon and Breach, New York, 1978.

51. Christoffel, E. B., Ueber die Gaussische Quadratur und eine Verallgemeinerung derselben, *J. Reine Angew. Math.* **55** (1858), 61–82.

52. Christoffel, E. B., Ueber die lineare Abhängigkeit von Funktionen einer einzigen Veränderlichen, *J. Reine Angew. Math.* **55** (1858), 281–99.

53. Christoffel, E. B., Sur une classe particulière de fonctions entières et de fractions continues, *Ann. Mat. Pura Appl. 2* **8** (1877), 1–10.

54. Chu, S.-C., *Ssu Yuan Yü Chien* (Precious Mirror of the Four Elements), 1303.

55. Coddington, E. and Levinson, N., *Theory of Ordinary Differential Equations*, McGraw-Hill, New York, 1955.

56. Copson, E. T., *An Introduction to the Theory of Functions of a Complex Variable*, Oxford University Press, Oxford, 1955.

57. Copson, E. T., On the Riemann–Green function, *Arch. Rat. Mech. Anal.* **1** (1958), 324–48.

58. Copson, E. T., *Asymptotic Expansions*, Cambridge University Press, Cambridge, 2004.

59. Courant, R. and Hilbert, D., *Methods of Mathematical Physics, Vol. 1*, Wiley, New York, 1961.

60. Cuyt, A., Petersen, V. B., Verdonk, B., Waadeland, H., and Jones, W. B., *Handbook of Continued Fractions for Special Functions*, Springer, New York, 2008.

61. D'Antonio, L., Euler and elliptic integrals, *Euler at 300, An Appreciation*, Mathematical Association of America, Washington, DC, 2007, 119–29.

62. Darboux, G., Mémoire sur l'approximation des fonctions de très-grandes nombres et sur une classe étendue de développements en série, *J. Math. Pures Appl.* **4** (1878), 5–57.
63. Darboux, G., *Théorie Générale des Surfaces*, Vol. 2, Book 4, Paris, 1889.
64. Davis, P. J., Leonhard Euler's Integral: a historical profile of the gamma function: in memoriam Milton Abramowitz, *Amer. Math. Monthly* **66** (1959), 849–69.
65. de Bruijn, N. G., *Asymptotic Methods in Analysis*, North-Holland, Amsterdam, 1961.
66. Debye, P., Näherungsformeln für die Zylinderfunktionen für große Werte des Arguments und unbeschränkt veränderliche Werte des Index, *Math. Annal.* **67** (1909), 535–58.
67. Deift, P. A., *Orthogonal Polynomials and Random Matrices: a Riemann–Hilbert Approach*, American Mathematical Society, Providence, RI, 1999.
68. Delache, S. and Leray, J., Calcul de la solution élémentaire de l'opérateur d'Euler–Poisson–Darboux et de l'opérateur de Tricomi–Clairaut hyperbolique d'ordre 2, *Bull. Soc. Math. France* **99** (1971), 313–36.
69. Dieudonné, J., *Abrégé d'Histoire des Mathématiques, Vols 1, 2*, Hermann, Paris, 1978.
70. Dieudonné, J., *Special Functions and Linear Representations of Lie Groups*, CBMS Series, Vol. 42, American Mathematical Society, Providence, RI, 1980.
71. Dirichlet, G. L., Sur les séries dont le terme général dépend de deux angles, et qui servent à exprimer des fonctions arbitraires entre les limites données, *J. für Math.* **17** (1837), 35–56.
72. Dubrovin, B., Theta functions and nonlinear equations, *Russian Math. Surv.* **36** (1981), 11–80.
73. Dunkl, C., A Krawtchouk polynomial addition theorem and wreath products of symmetric groups, *Indiana Univ. Math. J.*, **25** (1976), 335–58.
74. Dunster, T. M., Uniform asymptotic expansions for Charlier polynomials, *J. Approx. Theory* **112** (2001), 93–133.
75. Dutka, J., The early history of the hypergeometric function, *Arch. Hist. Exact Sci.* **31** (1984/85), 15–34.
76. Dutka, J., The early history of the factorial function, *Arch. Hist. Exact Sci.* **43** (1991/92), 225–49.
77. Dutka, J., The early history of Bessel functions, *Arch. Hist. Exact Sci.* **49** (1995/96), 105–34.
78. Dwork, B., *Generalized Hypergeometric Functions*, Oxford University Press, New York, 1990.
79. Edwards, H. M., *Riemann's Zeta Function*, Dover, Mineola, NY, 2001.
80. Erdélyi, A., Über eine Integraldarstellung der $M_{k,m}$-Funktion und ihre aymptotische Darstellung für grosse Werte von $\Re(k)$, *Math. Ann.* **113** (1937), 357–61.
81. Erdélyi, A., *Asymptotic Expansions*, Dover, New York, 1956.
82. Erdélyi, A., Magnus, W., Oberhettinger, F., and Tricomi, F. G., *Higher Transcendental Functions, Vols I, II, III*, Robert E. Krieger, Melbourne, 1981.
83. Erdélyi, A., Magnus, W., and Oberhettinger, F., *Tables of Integral Transforms, Vols I, II*, McGraw-Hill, New York, 1954.
84. Euler, L., De progressionibus transcendentibus seu quarum termini generales algebraice dari nequent, *Comm. Acad. Sci. Petr.* **5** (1738), 36–57; *Opera Omnia I*, Vol. 14, 1–24.

85. Euler, L., De productis ex infinitus factoribus ortis, *Comm. Acad. Sci. Petr.* **11** (1739), 3–31; *Opera Omnia I*, Vol. 14, 260–90.

86. Euler, L., De motu vibratorum tympanorum, *Comm. Acad. Sci. Petr.* **10** (1759), 243–60; *Opera Omnia II*, Vol. 10, 344–58.

87. Euler, L., De integratione aequationis differentialis $m \, dx / \sqrt{1 - x^4} = n \, dy / \sqrt{1 - y^4}$, *Nov. Comm. Acad. Sci. Petr.* **6** (1761), 37–57; *Opera Omnia I*, Vol. 20, 58–79.

88. Euler, L., *Institutiones Calculi Integralis, Vol. 2*, St. Petersburg 1769; *Opera Omnia I*, Vol. 12, 221–30.

89. Euler, L., Evolutio formulae integralis $\in x^{f-1}(lx)^{m/n} dx$ integratione a valore $x = 0$ ad $x = 1$ extensa, *Novi Comm. Acad. Petrop.* **16** (1771), 91–139; *Opera Omnia I*, Vol. 17, 316–57.

90. Euler, L., Demonstratio theorematis insignis per conjecturam eruti circa integrationem formulae $\int \frac{d\varphi \cos\varphi}{(1+aa-2a\cos\varphi)^{n+1}}$, *Institutiones Calculi Integralis, Vol. 4*, St. Petersburg 1794, 242–59; *Opera Omnia I*, Vol. 19, 197–216.

91. Euler, L., Methodus succincta summas serierum infintarum per formulas differentiales investigandi, *Mém. Acad. Sci. St. Pét.* **5** (1815), 45–56; *Opera Omnia I*, Vol. 16, 200–13.

92. Fabry, E., Sur les intégrales des équations différentielles à coefficients rationnels, Dissertation, Paris, 1885.

93. Fagnano del Toschi, G. C., *Produzioni Matematiche*, 1750.

94. Fejér, L., Sur une méthode de M. Darboux, *C. R. Acad. Sci. Paris* **147** (1908), 1040–42.

95. Fejér, L., Asymptotikus értékek meghatározáráról (On the determination of asymptotic values), *Math. és termész. ért.* **27** (1909), 1–33; *Ges. Abh. I*, 445–503.

96. Ferrers, N. M., *An Elementary Treatise on Spherical Harmonics and Functions Connected with Them*, MacMillan, London, 1877.

97. Fine, N. J., *Basic Hypergeometric Series and Applications*, American Mathematical Society, Providence, RI, 1988.

98. Forrester, P. J. and Warnaar, S. O., The importance of the Selberg integral, *Bull. Amer. Math. Soc.* **45** (2008), 489–534.

99. Forsyth, A. R., *A Treatise on Differential Equations*, Dover, Mineola, NY, 1996.

100. Fourier, J. B. J., *La Théorie Analytique de la Chaleur*, Paris, 1822.

101. Freud, G., *Orthogonal Polynomials*, Pergamon, New York, 1971.

102. Gasper, G. and Rahman, M., *Basic Hypergeometric Series*, Cambridge University Press, Cambridge, 2004.

103. Gauss, C. F., Disquisitiones generales circa seriem infinitam $1 + \frac{\alpha \cdot \beta}{1 \cdot \gamma} x + \frac{\alpha(\alpha+1)\beta(\beta+1)}{1 \cdot 2 \cdot \gamma(\gamma+1)} x^2 + \ldots$, *Comm. Soc. Reg. Sci. Gött. rec.* **2** (1812), 123–62; *Werke 3*, 123–62.

104. Gauss, C. F., Determinatio attractionis quam in punctum quodvis positionis datae exerceret platea si ejus massa per totam orbitam ratione temporis quo singulae partes discribuntur uniformiter esset dispertita, *Comm. Soc. Reg. Sci. Gott.* **4** (1818), 21–48; *Werke 3*, 332–57.

105. Gauss, C. F., Lemniscatische Funktionen dargestellt durch Potenz-Reihen, *Werke 3*, 404–12.

106. Gauss, C. F., Kugelfunktionen, *Werke 5*, 630–32.

107. Gautschi, W., Some elementary inequalities relating to the gamma and incomplete gamma function, *J. Math. Phys.* **38** (1959), 77–81.

108. Gautschi, W., *Orthogonal Polynomials: Computation and Approximation*, Oxford University Press, New York, 2004.

109. Gautschi, W., Leonhard Euler: his life, the man, and his works, *SIAM Review* **50** (2008), 3–33.

110. Gegenbauer, L., Über die Bessel'schen Functionen, *Wiener Sitzungsber.* **70** (1874), 6–16.

111. Gegenbauer, L., Über einige bestimmte Itegrale, *Wiener Sitzungsber.* **70** (1874), 434–43.

112. Gegenbauer, L., Über die Funktionen $C_n^\nu(x)$, *Wiener Sitzungsber.* **75** (1877), 891–6; **97** (1888), 259–316.

113. Gegenbauer, L., Das Additionstheorem der Funktionen $C_n^\nu(x)$, *Wiener Sitzungsber.* **102** (1893), 942–50.

114. Geronimus, L. Y., *Orthogonal Polynomials*, American Mathematical Society, Providence, RI, 1977.

115. Godefroy, *La Fonction Gamma*, Gauthier-Villars, Paris, 1901.

116. Gordon, W., Über den Stoss zweier Punktladungen nach der Wellenmechanik, *Z. für Phys.* **48** (1928), 180–91.

117. Goursat, E., Sur l'équation différentielle linéaire qui admet pour intégrale la série hypergéométrique, *Ann. Sci. École Norm. Sup.* **10** (1881), 3–142.

118. Gradshteyn, I. S. and Ryzhik, I. M., *Table of Integrals, Series, and Products*, A. Jeffrey and D. Zwillinger (eds), Academic Press, San Diego, 2007.

119. Green, G., *An Essay on the Application of Mathematical Analysis to the Theories of Electricity and Magnetism*, Nottingham 1828; *J. für Math.* **39** (1850), 74–89.

120. Green, G., On the motion of waves in a variable canal of small depth and width, *Trans. Camb. Phil. Soc.* **6** (1837), 457–62.

121. Hahn, W., Bericht über die Nullstellungen der Laguerreschen und Hermiteschen Polynome, *Jahresber. Deutschen Math. Verein* **44** (1934), 215–36; ibid. **45** (1935), 211.

122. Hahn, W., Über Orthogonalpolynomen die q-Differenzgleichungen genügen, *Math. Nachr.* **2** (1949), 4–34.

123. Hankel, H., Die Cylinderfunktionen erster und zweiter Art, *Math. Ann.* **1** (1869), 467–501.

124. Hansen, P. A., Ermittelung der absoluten Störungen in Ellipsen von beliebiger Excentricität und Neigung, I, *Schriften der Sternwarte Seeberg*, Gotha, 1843.

125. Heine, H. E., *Handbuch der Kugelfunktionen*, Reimer, Berlin, 1878, 1881.

126. Hermite, C., Sur une nouveau développement en série de fonctions, *C. R. Acad. Sci. Paris* **58** (1864), 93–100.

127. Higgins, J. R., *Completeness and Basis Properties of Sets of Special Functions*, Cambridge University Press, Cambridge, 1977.

128. Hilbert, D., Über die Diskriminanten der in endlichen abbrechenden hypergeometrischen Reihe, *J. Reine Angew. Math.* **103**, 337–45.

129. Hille, E., *Ordinary Differential Equations in the Complex Domain*, Dover, Mineola, NY, 1976.

130. Hille, E., Shohat, J., and Walsh, J. L., *A Bibliography on Orthogonal Polynomials*, Bulletin National Research Council 103, Washington, DC, 1940.

131. Hobson, E. W., On a type of spherical harmonics of unrestricted degree, order, and argument, *Phil. Trans.* **187** (1896), 443–531.

132. Hobson, E. W., *The Theory of Spherical and Ellipsoidal Harmonics*, Chelsea, New York, 1965.

133. Hochstadt, H., *The Functions of Mathematical Physics*, Wiley, New York, 1971.

134. Hoëné-Wronski, J., *Réfutation de la Théorie des Fonctions Analytiques de Lagrange*, Paris, 1812.

135. Ince, E. L., *Ordinary Differential Equations*, Dover, New York, 1956.

136. Ismail, M. E. H., *Classical and Quantum Orthogonal Polynomials in One Variable*, Cambridge University Press, Cambridge, 2005.

137. Ivić, A., *The Riemann Zeta-Function, Theory and Applications*, Dover, Mineola, NY, 2003.

138. Iwasaki, K., Kimura, T., Shimomura, S., and Yoshida, M., *From Gauss to Painlevé: a modern theory of special functions*, Vieweg, Braunschweig, 1991.

139. Jacobi, C. G. J., Extraits de deux letters de M. Jacobi de l'Université de Königsberg à M. Schumacher, *Schumacher Astron. Nachr.* **6** (1827); *Werke 1*, 31–6.

140. Jacobi, C. G. J., *Fundamenta Nova Theoriae Functionum Ellipticarum*, Königsberg 1829; Werke 1, 49–239.

141. Jacobi, C. G. J., Versuch einer Berechnung der grossen Ungleichheit des Saturns nach einer strengen Entwicklung, *Astr. Nachr.* **28** (1849), col. 65–80, 81–94; *Werke 7*, 145–74.

142. Jacobi, C. G. J., Theorie der elliptischen Functionen aus den eigenschaften der Thetareihen abgeleitet, *Werke 1*, 497–538.

143. Jacobi, C. G. J., Untersuchungen über die Differentialgleichung der hypergeometrischen Reihe, *J. Reine Angew. Math.* **56** (1859), 149–65; *Werke 6*, 184–202.

144. Jahnke, E. and Emde, F., *Tables of Higher Functions*, McGraw-Hill, New York, 1960.

145. Jeffreys, H., On certain approximate solutions of linear differential equations of the second-order, *Proc. London Math. Soc.* **23** (1924), 428–36.

146. Jeffreys, H., Asymptotic solutions of linear differential equations, *Philos. Mag.* **33** (1942), 451–6.

147. Jin, X.-S. and Wong, R., Uniform asymptotic expansions for Meixner polynomials, *Constr. Approx.* **14** (1998), 113–50.

148. Jin, X.-S. and Wong, R., Asymptotic formulas for the Meixner polynomials, *J. Approx. Theory* **14** (1999), 281–300.

149. Johnson, D. E. and Johnson, J. R., *Mathematical Methods in Engineering and Physics. Special Functions and Boundary-Value Problems*, Ronald Press, New York, 1965.

150. Kamke, E., *Differentialgleichungen: Lösungsmethoden und Lösungen Band 1: Gewöhnliche Differentialgleichungen*, Chelsea, New York, 1959.

151. Kanemitsu, S. and Tsukada, H., *Vistas of Special Functions*, World Scientific, Hackensack, NJ, 2007.

152. Kelvin, Lord (W. Thomson), On the wave produced by a single impulse in water of any depth or in a dispersive medium, *Philos. Mag.* **23** (1887), 252–5.

153. Kelvin, Lord (W. Thomson), Presidential address to the Institute of Electrical Engineers, 1889, *Math. and Phys. Papers III*, 484–515.

154. Kempf, G. R., *Complex Abelian Varieties and Theta Functions*, Springer, Berlin, 1991.

155. Klein, F., *Vorlesungen über die Entwicklung der Mathematik im 19. Jahrhundert, I*, Springer, Berlin, 1927.

156. Klein, F., *Vorlesungen Über die Hypergeometrische Funktion*, Springer, Berlin, 1981.

157. Knopp, K., *Funktionentheorie II*, 5th edn, de Gruyter, Berlin, 1941.

158. Koekoek, R. and Swarttouw, R. F., *The Askey-scheme of Hypergeometric Orthogonal Polynomials and its q-analogue*, Tech. Univ. Delft, Faculty of Technical Mathematics and Informatics Report 98-17, Delft, 1998.

159. Koepf, W., *Hypergeometric Summation. An algorithmic approach to summation and special function identities*, Vieweg, Braunschweig, 1998.

160. Koornwinder, T., Jacobi polynomials. III. An analytical proof of the addition formula, *SIAM J. Math. Anal.* **6** (1975), 533–40.

161. Koornwinder, T., Yet another proof of the addition formula for Jacobi polynomials, *J. Math. Anal. Appl.* **61** (1977), 136–41.

162. Korenev, B. G., *Bessel functions and their applications*, Taylor & Francis, London, 2002.

163. Koshlyakov, N. S., On Sonine's polynomials, *Messenger Math.*, **55** (1926), 152–60.

164. Krall, A. M., *Hilbert Space, Boundary Value Problems and Orthogonal Polynomials*, Birkhäuser, Basel, 2002.

165. Kramers, H. A., Wellenmechanik und halbzahlige Quantisierung, *Z. für Phys.* **39** (1926), 828–40.

166. Krawtchouk, M., Sur une généralisation des polynomes d'Hermite, *C. R. Acad. Sci.* **189** (1929), 620–22.

167. Khrushchev, S., *Orthogonal Polynomials and Continued Fractions from Euler's Point of View*, Cambridge University Press, Cambridge, 2008.

168. Kummer, E. E., Über die hypergeometrische Reihe, *J. Reine Angew. Math.* **15** (1836), 127–72.

169. Laforgia, A., Further inequalities for the gamma function, *Math. Comp.* **42** (1984), 597–600.

170. Lagrange, J. L., Des oscillations d'un fil fixé par une de ses extrémités et chargés d'un nombre quelconque de poids, in the memoir Solution de différents problèmes de calcul intégral, *Misc. Taur.* **3** (1762–1765), *Oeuvres 1*, 471–668.

171. Lagrange, J. L., Sur le problème de Kepler, *Hist. Acad. Sci. Berlin* **25** (1769), 204–33; *Oeuvres 3*, 113–38.

172. Laguerre, E. N., Sur l'intégrale $\int_x^\infty x^{-1} e^{-x}\, dx$, *Bull. Soc. Math. France* **7** (1879), 72–81.

173. Landen, J., An investigation of a general theorem for finding the length of any arc of any conic hyperbola, by means of two elliptic arcs, with some other new and useful theorems deduced therefrom, *Phil. Trans. Royal Soc.* **65** (1775), 283–9.

174. Lang, S., *Elliptic Functions*, Springer, New York, 1987.

175. Laplace, P. S., Théorie d'attraction des sphéroïdes et de la figure des planètes, *Mém. Acad. Royal. Sci. Paris* (1782, publ. 1785), 113–96.

176. Laplace, P. S., *Mémoire sur les Intégrales Définies, et leurs Applications aux Probabilités*, Mém. de l'Acad. Sci. 11, Paris 1810–1811; *Oeuvres 12*, Paris 1898, 357–412.

177. Laplace, P. S., *Théorie Analytique des Probabilités*, 2nd edn, *Oeuvres 7*.
178. Lawden, D. F., *Elliptic Functions and Applications*, Springer, New York, 1989.
179. Lebedev, N. N., *Special Functions and Their Applications*, Dover, New York, 1972.
180. Legendre, A. M., Recherches sur les figures des planètes, *Mém. sav. étr. Acad. Roy. Sci. Paris* (1784, publ. 1787), 370–89.
181. Legendre, A. M., Recherches sur l'attraction des sphéroides homogènes, *Mém. Math. Phys. Acad. Sci.* **10** (1785), 411–34.
182. Legendre, A. M., Recherches sur diverses sortes d'intégrales définies, *Mém. de l'Inst. France* **10** (1809), 416–509.
183. Legendre, A. M., *Exercises de Calcul Intégrale*, Paris, 1811.
184. Legendre, A. M., *Traité des Fonctions Elliptiques et des Intégrales Eulériennes*, Huzard-Courcier, Paris, 1825–8.
185. Levin, A. L. and Lubinsky, D. S., *Orthogonal Polynomials for Exponential Weights*, Springer, New York, 2001.
186. Liouville, J., Mémoire sur la théorie analytique de la chaleur, *Math. Ann.* **21** (1830–1831), 133–81.
187. Liouville, J., Second mémoire sur de développement des fonctions ou parties de fonctions en séries dont les divers termes sont assujeties à satisfaire à une même équation de seconde ordre, contenant un paramètre variable, *J. de Math.* **2** (1837), 16–35.
188. Liouville, J., Lectures published by C. W. Borchardt, *J. für Math. Reine Angew.* **88** (1880), 277–310.
189. Lommel, E. C. J. von, *Studien über die Bessel'schen Funktionen*, Leipzig, 1868.
190. Luke, Y. L., *The Special Functions and Their Approximations*, 2 vols, Academic Press, New York, 1969.
191. Lützen, J., The solution of partial differential equations by separation of variables: a historical survey, *Studies in the History of Mathematics*, Mathematical Association of America, Washington, DC, 1987.
192. Macdonald, H. M., Note on Bessel functions, *Proc. London Math. Soc.* **22** (1898), 110–15.
193. Macdonald, I. G., *Symmetric Functions and Orthogonal Polynomials*, American Mathematical Society, Providence, RI, 1998.
194. MacRobert, T. M., *Spherical Harmonics. An Elementary Treatise on Harmonic Functions with Applications*, Pergamon, Oxford, 1967.
195. Magnus, W. and Oberhettinger, F., *Formulas and Theorems for the Special Functions of Mathematical Physics*, Chelsea, New York, 1949.
196. Magnus, W., Oberhettinger, F., and Soni, R. P., *Formulas and Theorems for the Special Functions of Mathematical Physics*, Springer, New York, 1966.
197. Mathai, A. M. and Saxena, R. K., *Generalized Hypergeometric Functions with Applications in Statistics and Physical Sciences*, Lect. Notes Math. 348, Springer, Berlin, 1973.
198. *Mathematics of the 19th Century. Function Theory According to Chebyshev, Ordinary Differential Equations, Calculus of Variations, Theory of Finite Differences*, Kolmogorov, A. N. and Yushkevich, A. P. (eds), Birkhäuser, Basel, 1998.
199. Matsuda, M., *Lectures on Algebraic Solutions of Hypergeometric Differential Equations*, Kinokuniya, Tokyo, 1985.
200. McBride, E. B., *Obtaining Generating Functions*, Springer, New York, 1971.

201. Mehler, F. G., Notiz über die Dirichlet'schen Integralausdrucke für die Kugelfuntion $P_n(\cos\theta)$ und über einige analoge Integralform für die Cylinderfunktion $J(z)$, *Math. Ann.* **5** (1872), 141–4.

202. Mehta, M. L., *Random Matrices*, Elsevier, Amsterdam, 2004.

203. Meixner, J., Orthogonale Polynomsysteme mit einer besonderen Gestalt der erzeugenden Funktion, *J. London Math. Soc.* **9** (1934), 6–13.

204. Miller, Jr, W., *Lie Theory and Special Functions*, Academic Press, New York, 1968.

205. Miller, Jr, W., *Symmetry and Separation of Variables*, Addison-Wesley, Reading, MA, 1977.

206. Mittag-Leffler, G., An introduction to the theory of elliptic functions, *Ann. Math.* **24** (1923), 271–351.

207. Motohashi, Y., *Spectral Theory of the Riemann Zeta-Function*, Cambridge University Press, Cambridge, 1997.

208. Mott, N. F., The solution of the wave equation for the scattering of a particle by a Coulombian centre of force, *Proc. Royal Soc. A* **118** (1928), 542–9.

209. Müller, C., *Analysis of Spherical Symmetries in Euclidean Spaces*, Springer, New York, 1998.

210. Murphy, R., *Treatise on Electricity*, Deighton, Cambridge, 1833.

211. Murphy, R., On the inverse method of definite integrals with physical applications, *Camb. Phil. Soc. Trans.* **4** (1833), 353–408; **5** (1835), 315–93.

212. Neumann, C. G., *Theorie der Bessel'schen Funktionen. Ein Analogon zur Theorie der Kugelfunctionen*, Leipzig, 1867.

213. Nevai, P. G., *Orthogonal Polynomials*, Memoirs American Mathematical Society, Providence, RI, 1979.

214. Neville, E. H., *Jacobian Elliptic Functions*, Oxford University Press, Oxford, 1951.

215. Newman, F. W., On $\Gamma(a)$ especially when a is negative, *Camb. Dublin Math. J.* **3** (1848), 59–60.

216. Nielsen, N., *Handbuch der Theorie der Zylinderfunktionen*, Leipzig, 1904.

217. Nielsen, N., *Handbuch der Theorie de Gammafunktion*, Chelsea, New York, 1965.

218. Nikiforov, A. F., Suslov, S. K., and Uvarov, V. B., *Classical Orthogonal Polynomials of a Discrete Variable*, Springer, Berlin, 1991.

219. Nikiforov, A. F. and Uvarov, V. B., *Special Functions of Mathematical Physics, a Unified Introduction with Applications*, Birkhäuser, Basel, 1988.

220. Oberhettinger, F., *Tables of Fourier Transforms and Fourier Transforms of Distributions*, Springer, Berlin, 1990.

221. Oberhettinger, F. and Magnus, W., *Anwendung der Elliptische Funktionen in Physik und Technik*, Springer, Berlin, 1949.

222. Olver, F. W. J., *Asymptotics and Special Functions*, Peters, Wellesley, 1997.

223. Olver, F. W. J., Lozier, D. W., Clark, C. W., and Boisvert, R. F., *Digital Library of Mathematical Functions*, National Institute of Standards and Technology, Gaithersburg, 2007.

224. Olver, F. W. J., Lozier, D. W., Clark, C. W., and Boisvert, R. F., *Handbook of Mathematical Functions*, Cambridge University Press, Cambridge, 2010.

225. Painlevé, P., Gewöhnliche Differentialgleichungem: Existenz der Lösungen, *Enzykl. Math. Wiss.* **2** no. 2/3 (1900), 189–229.

226. Patterson, S. J., *An Introduction to the Theory of the Riemann Zeta-Function*, Cambridge University Press, Cambridge, 1988.

227. Petiau, G., *La Théorie des Fonctions de Bessel Exposée en vue de ses Applications à la Physique Mathématique*, Centre National de la Recherche Scientifique, Paris, 1955.

228. Petkovšek, M., Wilf, H. S., and Zeilberger, D., $A = B$, A. K. Peters, Wellesley, MA, 1996.

229. Pfaff, J. F., *Nova disquisitio de integratione aequationes differentio-differentialis*, *Disquisitiones Analyticae*, Helmstadt, 1797.

230. Picard, É., Sur l'application des méthodes d'approximations successives à l'étude de certaines équations différentielles, *Bull. Sci. Math. 2* **12** (1888), 148–56.

231. Poisson, S. D., Mémoire sur la distribution de la chaleur dans les corps solides, *J. École Roy. Polyt.* **19** (1823), 1–162, 249–403.

232. Polishchuk, A., *Abelian Varieties, Theta Functions and the Fourier Transform*, Cambridge University Press, Cambridge, 2003.

233. Prasolov, V. and Solovyev, Yu., *Elliptic Functions and Elliptic Integrals*, American Mathematical Society, Providence, RI, 1997.

234. Prosser, R. T., On the Kummer solutions of the hypergeometric equation, *Amer. Math. Monthly* **101** (1994), 535–43.

235. Qiu, W.-Y. and Wong, R., Asymptotic expansion of the Krawtchouk polynomials and their zeros, *Comput. Methods Funct. Theory* **4** (2004), 189–226.

236. Rainville, E. D., *Special Functions*, Chelsea, Bronx, NY, 1971.

237. Remmert, R., Wielandt's theorem about the Γ-function, *Amer. Math. Monthly* **103** (1996), 214–20.

238. Riccati, Count J. F., Animadversiones in aequationes differentiales secundi gradus, *Acta Erud. Lipsiae publ. Suppl.* **8** (1724), 66–73.

239. Riemann, B., Beiträge zur Theorie der durch Gauss'she Reihe $F(\alpha, \beta, \gamma, x)$ darstellbaren Funktionen, *Kön. Ges. Wiss. Gött.* **7** (1857), 1–24; *Werke*, 67–90.

240. Robin, L., *Fonctions Sphériques de Legendre et Fonctions Sphéroïdales, Vols I, II, III*, Gauthier-Villars, Paris, 1957–9.

241. Rodrigues, O., Mémoire sur l'attraction des sphéroides, *Corr. École Roy. Polyt.* **3** (1816), 361–85.

242. Routh., E., On some properties of certain solutions of a differential equation of the second-order, *Proc. London Math. Soc.* **16** (1885), 245–61.

243. Russell, A., The effective resistance and inductance of a concentric main, and methods of computing the Ber and Bei and allied functions, *Phil. Mag.* **17** (1909), 524–52.

244. Sachdev, P. L., *A Compendium on Nonlinear Ordinary Differential Equations*, Wiley, New York, 1997.

245. Saff, E. B. and Totik, V., *Logarithmic Potentials with External Fields*, Springer, New York, 1997.

246. Sansone, G., *Orthogonal Functions*, Dover, New York, 1991.

247. Šapiro, R. L., Special functions related to representations of the group SU(n) of class I with respect to SU($n - 1$) $n \geq 3$ (Russian), *Izv. Vysš. Učebn. Zaved. Mat.* **4** (1968), 97–107.

248. Sasvari, Z., An elementary proof of Binet's Formula for the Gamma function, *Amer. Math. Monthly* **106** (1999), 156–8.

249. Schläfli, L., Sull'uso delle linee lungo le quali il valore assoluto di una funzione è constante, *Annali Mat.* **2** (1875), 1–20.
250. Schläfli, L., *Ueber die zwei Heine'schen Kugelfunktionen*, Bern, 1881.
251. Schlömilch. O., *Analytische Studien*, Leipzig, 1848.
252. Schlömilch, O., Ueber die Bessel'schen Funktion, *Z. Math. Phys.* **2** (1857), 137–65.
253. Schmidt, H., Über Existenz und Darstellung impliziter Funktionen bei singulären Anfangswerten, *Math. Z.* **43** (1937/38), 533–56.
254. Schwarz, H. A., Über diejenigen Fälle, in welchen die Gaussische hypergeometrische Reihe eine algebraische Function ihres vierten Elementes darstellt, *J. Reine Angew. Math.* **75** (1873), 292–335.
255. Seaborn, J. B., *Hypergeometric Functions and Their Applications*, Springer, New York, 1991.
256. Seaton, M. J., Coulomb functions for attractive and repulsive potentials and for positive and negative energies, *Comput. Phys. Comm.* **146** (2002), 225–49.
257. Selberg, A., Bemerkninger om et multipelt integral, *Norsk. Mat. Tidsskr.* **24** (1944), 159–71.
258. Sharapudinov, I. I., Asymptotic properties and weighted estimates for orthogonal Chebyshev–Hahn polynomials, *Mat. Sb.* **182** (1991), 408–20; *Math. USSR-Sb.* **72** (1992), 387–401.
259. Simon, B., *Orthogonal Polynomials on the Unit Circle, Parts 1, 2*, Amer. Math. Soc., Providence, RI, 2005.
260. Slater, L. J., *Confluent Hypergeometric Functions*, Cambridge University Press, New York, 1960.
261. Slater, L. J., *Generalized Hypergeometric Functions*, Cambridge University Press, Cambridge, 1966.
262. Slavyanov, S. Yu. and Lay, W., *Special Functions, a Unified Theory Based on Singularities*, Oxford University Press, Oxford, 2000.
263. Sneddon, I. N., *Special Functions of Mathematical Physics and Chemistry*, 3rd edn, Longman, London, 1980.
264. Sommerfeld, A. J. W., Mathematische Theorie der Diffraction, *Math. Ann.* **47** (1896), 317–74.
265. Sonine, N. J., Recherches sur les fonctions cylindriques et le développement des fonctions continues en séries, *Math. Ann.* **16** (1880), 1–80.
266. Stahl, H. and Totik, V., *General Orthogonal Polynomials*, Cambridge University Press, Cambridge, 1992.
267. Stanton, D., Orthogonal polynomials and Chevalley groups, *Special Functions: Group Theoretic Aspects and Applications*, R. Askey, T. Koornwinder, and W. Schempp (eds), Reidel, New York, 1984, pp. 87–128.
268. Steklov, V. A., Sur les expressions asymptotiques de certaines fonctions, définies par les équations différentielles linéaires de seconde ordre, et leurs applications au problème du développement d'une fonction arbitraire en séries procédant suivant les-dites fonctions, *Comm. Soc. Math. Kharkhow* (2) **10** (1907), 197–200.
269. Sternberg, W. J. and Smith, T. L., *The Theory of Potential and Spherical Harmonics*, University of Toronto Press, Toronto, 1944.
270. Stieltjes, T. J., Sur quelques théorèmes d' algèbre, *C. R. Acad. Sci. Paris* **100** (1885), 439–40; *Oeuvres 1*, 440–41.

271. Stieltjes, T. J., Sur les polynômes de Jacobi, *C. R. Acad. Sci. Paris* **100** (1885), 620–22; *Oeuvres 1*, 442–4.

272. Stieltjes, T. J., Sur les polynômes de Legendre, *Ann. Fac. Sci. Toulouse* **4** (1890), 10pp; *Oeuvres 2*, 236–52.

273. Stilwell, J., *Mathematics and its History*, Springer, New York, 2002.

274. Stirling, J., *Methodus Differentialis*, London, 1730.

275. Stokes, G. G., On the numerical calculation of a class of definite integrals and infinite series, *Camb. Philos. Trans.* **9** (1850), 166–87.

276. Sturm, J. C. F., Mémoire sur les équations différentielles linéaires du second ordre, *J. Math. Pures Appl.* **1** (1836), 106–86.

277. Sturm, J. C. F., Mémoire sur une classe d'équations à différences partielles, *J. Math. Pures Appl.* **1** (1836), 373–444.

278. Szegő, G., On an inequality of P. Turán regarding Legendre polynomials, *Bull. Amer. Math. Soc.* **54** (1949), 401–5.

279. Szegő, G., *Orthogonal Polynomials*, 4th edn, American Mathematical Society, Providence, RI, 1975.

280. Szegő, G., An outline of the history of orthogonal polynomials, *Orthogonal Expansions and their Continuous Analogues*, Southern Illinois University Press, Carbondale, IL, 1968, pp. 3–11.

281. Talman, J., D., *Special Functions: a Group Theoretic Approach*, Benjamin, New York, 1968.

282. Tannery, J. and Molk, J., *Éléments de la Théorie des Fonctions Elliptiques*, Vols 1–4, Gauthier-Villars, Paris, 1893–1902; Chelsea, New York, 1972.

283. Taylor, R. and Wiles, A., Ring-theoretic properties of certain Hecke algebras, *Ann. Math.* **141** (1995), 553–72.

284. Temme, N. M., *Special Functions, an Introduction to the Classical Functions of Mathematical Physics*, Wiley, New York, 1996.

285. Titchmarsh, E. C., *The Theory of the Riemann Zeta-Function*, 2nd edn, Oxford University Press, New York, 1986.

286. Tricomi, F., *Serie Ortogonale di Funzioni*, S. I. E. Istituto Editoriale Gheroni, Torino 1948.

287. Tricomi, F., *Funzioni Ipergeometrichi Confluenti*, Ed. Cremonese, Rome, 1954.

288. Tricomi, F., *Funzioni Ellittiche*, Zanichelli, Bologna, 1951.

289. Truesdell, C., *An Essay Toward a Unified Theory of Special Functions Based upon the Functional Equation* $(\partial/\partial z)F(z, \alpha) = F(z, \alpha + 1)$, Annals of Mathematics Studies, no. 18, Princeton University Press, Princeton, NJ, 1948.

290. Turán, P., On the zeros of the polynomials of Legendre, *Čas. Pěst. Mat. Fys.* **75** (1950), 113–22.

291. Uspensky, J. V., On the development of arbitrary functions in series of orthogonal polynomials, *Ann. Math.* **28** (1927), 563–619.

292. Van Assche, W., *Asymptotics for Orthogonal Polynomials*, Lect. Notes Math. no. 1265, Springer, Berlin, 1987.

293. van der Corput, J. G., *Asymptotic Expansions. I, II, III*, Department of Mathematics, University of California, Berkeley, CA, 1954–5.

294. van der Laan, C. G. and Temme, N. M., *Calculation of Special Functions: the Gamma Function, the Exponential Integrals and Error-like Functions*, Centrum voor Wiskunde en Informatica, Amsterdam, 1984.

295. Vandermonde, A., Mémoire sur des irrationelles de différens ordres avec une application au cercle, *Mém. Acad. Royal Sci. Paris* (1772), 489–98.

296. Varadarajan, V. S., Linear meromorphic differential equations: a modern point of view, *Bull. Amer. Math. Soc.* **33** (1996), 1–42.

297. Varchenko, A., *Multidimensional Hypergeometric Functions and Representation Theory of Lie Algebras and Quantum Groups*, World Scientific Publishing, River Edge, NJ. 1995.

298. Varchenko, A., *Special Functions, KZ Type Equations, and Representation Theory*, American Mathematics Society, Providence, RI, 2003.

299. Vilenkin, N. Ja., *Special Functions and the Theory of Group Representations*, American Mathematical Society, Providence, RI, 1968.

300. Vilenkin, N. Ja. and Klimyk, A. U., *Representation of Lie Groups and Special Functions*, 3 vols, Kluwer, Dordrecht, 1991–3.

301. Walker, P. L., *Elliptic Functions*, Wiley, Chichester, 1996.

302. Wallis, J., *Arithmetica Infinitorum*, Oxford, 1656.

303. Wang, Z. X. and Guo, D. R., *Special Functions*, World Scientific, Teaneck, NJ, 1989.

304. Wasow, W., *Asymptotic Expansions for Ordinary Differential Equations*, Dover, New York, 1987.

305. Watson, G. N., Bessel functions and Kapteyn series, *Proc. London Mat. Soc.* **16** (1917), 150–74.

306. Watson, G. N., *A Treatise on the Theory of Bessel Functions*, Cambridge University Press, Cambridge, 1995.

307. Wawrzyńczyk, A., *Group Representations and Special Functions*, Reidel, Dordrecht, 1984.

308. Weber, H., Über die Integration der partiellen Differentialgleichung: $\frac{\partial^2 u}{\partial x^2} + \frac{\partial^2 u}{\partial y^2} + k^2 u^2 = 0$, *Math. Ann.* **1** (1869), 1–36.

309. Weber, H., Über eine Darstellung willkürlicher Funktionen durch Bessel'sche Funktionen, *Math. Ann.* **6** (1873), 146–61.

310. Weber, M. and Erdélyi, A., On the finite difference analogue of Rodrigues' formula, *Amer. Math. Monthly* **59** (1952), 163–8.

311. Weierstrass, K. L., *Formeln und Lehrsätze zum Gebrauch der Elliptische Funktionen*, Kaestner, Göttingen, 1883–5.

312. Wentzel, G., Eine Verallgemeinerung der Quantenbedingungen für die Zwecke der Wellenmechanik, *Z. für Phys.* **38** (1926), 518–29.

313. Whitehead, C. S., On the functions ber x, bei x, ker x, and kei x, *Quarterly J.* **42** (1909), 316–42.

314. Whittaker, E. T., An expression of certain known functions as generalized hypergeometric functions, *Bull. Amer. Math. Soc.* **10** (1903), 125–34.

315. Whittaker E. T. and Watson, G. N., *A Course of Modern Analysis*, Cambridge University Press, Cambridge, 1969.

316. Wiles, A., Modular elliptic curves and Fermat's last theorem, *Ann. Math.* **141** (1995), 443–551.

317. Wilf, H. S. and Zeilberger, D., An algebraic proof theory for geometric (ordinary and "q") multisum/integral identities, *Invent. Math.* **108** (1992), 575–633.

318. Wong, R., *Asymptotic Approximations of Integrals*, Academic Press, Boston 1989; SIAM, Philadelphia, 2001.
319. Wong, R., Orthogonal polynomials and their asymptotic behavior, *Special Functions (Hong Kong, 1999)*, World Scientific, River Edge, NJ, 2000, pp. 139–55.
320. Yost, F. L., Wheeler, J. A., and Breit, G., Coulomb wave functions in repulsive fields, *Phys. Rev.* **49** (1936), 174–89.
321. Zhang, S. and Jin, J., *Computation of Special Functions*, Wiley, New York, 1996.
322. Zwillinger, D., *Handbook of Differential Equations*, Wiley, New York, 1998.

Author index

Index